Ecology

Ecology

PN MICHAEL

CBS Publishers & Distributors Pvt Ltd

New Delhi • Bengaluru • Chennai • Kochi • Kolkata • Mumbai
Bhubaneswar • Hyderabad • Jharkhand • Nagpur • Patna • Pune • Uttarakhand

Ecology

ISBN: 978-81-239-2651-3

Copyright © Publisher

First Edition: 2016
Reprint: 2018

Published by Satish Kumar Jain and produced by Varun Jain for

CBS Publishers & Distributors Pvt Ltd

4819/XI Prahlad Street, 24 Ansari Road, Daryaganj, New Delhi 110 002, India.
Ph: 23289259, 23266861, 23266867   Website: www.cbspd.com
Fax: 011-23243014   e-mail: delhi@cbspd.com; cbspubs@airtelmail.in.
Corporate Office: 204 FIE, Industrial Area, Patparganj, Delhi 110 092
Ph: 4934 4934   Fax: 4934 4935   e-mail: publishing@cbspd.com; publicity@cbspd.com

Branches

- **Bengaluru:** Seema House 2975, 17th Cross, K.R. Road,
  Banasankari 2nd Stage, Bengaluru 560 070, Karnataka
  Ph: +91-80-26771678/79   Fax: +91-80-26771680   e-mail: bangalore@cbspd.com
- **Chennai:** 7, Subbaraya Street, Shenoy Nagar, Chennai 600 030, Tamil Nadu
  Ph: +91-44-26680620, 26681266   Fax: +91-44-42032115   e-mail: chennai@cbspd.com
- **Kochi:** Ashana House, No. 39/1904, AM Thomas Road, Valanjambalam,
  Ernakulam 682 016, Kochi, Kerala
  Ph: +91-484-4059061-65   Fax: +91-484-4059065   e-mail: kochi@cbspd.com
- **Kolkata:** 6/B, Ground Floor, Rameswar Shaw Road, Kolkata-700 014, West Bengal
  Ph: +91-33-22891126, 22891127, 22891128   e-mail: kolkata@cbspd.com
- **Mumbai:** 83-C, Dr E Moses Road, Worli, Mumbai-400018, Maharashtra
  Ph: +91-22-24902340/41   Fax: +91-22-24902342   e-mail: mumbai@cbspd.com

Representatives

- **Bhubaneswar** 0-9911037372
- **Patna** 0-9334159340
- **Hyderabad** 0-9885175004
- **Pune** 0-9623451994
- **Jharkhand** 0-9811541605
- **Uttarakhand** 0-9716462459
- **Nagpur** 0-9021734563

Printed at India Binding House, Noida, UP

Preface

Ecology is the scientific study of the relations that living organisms have with respect to each other and their natural environment. Variables of interest to ecologists include the composition, distribution, amount (biomass), number, and changing states of organisms within and among ecosystems. Ecosystems are hierarchical systems that are organised into a graded series of regularly interacting and semi-independent parts (e.g. species) that aggregate into higher orders of complex integrated wholes (e.g. communities).

Ecosystems create biophysical feedback mechanisms between living (biotic) and nonliving (abiotic) components of the planet that regulates and sustains systems as large as continental climates and global biogeochemical cycles. Ecosystems provide goods and services that sustain human societies and general well-being. Ecosystems are sustained by biodiversity within them. Biodiversity is the full-scale of life and its processes, including genes, species and ecosystems forming lineages that integrate into a complex and regenerative spatial arrangement of types, forms, and interactions.

An understanding of how biodiversity affects ecological function is an important focus area in ecological studies. Ecosystems sustain every life-supporting function on the planet, including climate regulation, water filtration, soil formation (pedogenesis), food, fibres, medicines, erosion control, and many other natural features of scientific, historical or spiritual value.

This reference textbook on ecology is divided into eight sections and contains 31 chapters. Each chapter covers an important aspects of ecology. Section I discusses basic concepts of ecology. Chapter 1 is devoted to ecology and environment. Ecology is the scientific study of relationships between organisms and their environment. Chapter 2 focuses on productivity ecology which refers to the rate of generation of biomass in ecosystem. Chapter 3 deals with ecological concepts of species. Chapter 4 deals with green chemistry and environmentally friendly technologies. Green chemistry is a responsible way of using science and engineering that strive to improve the public image of chemistry, not as a goal in itself but as a consequence of its achievements. During the twentieth century, chemists were able to master synthetic chemistry and could virtually make all possible chemicals found in nature. Chapter 5 concentrates on industrial ecology which is the study of material and energy flows through industrial system.

Section II is devoted to climatic and topographic factors. Chapter 6 acquaints the readers with ecology of light and temperature. Light and temperature have variety of effects on plant growth. Without the sun, plants would not be able to survive, and such light is pivotal to a number of crucial processes that a plant undergoes on regular basis in order to sustain life. Chapter 7 deals with ecology of wind which is regarded as an important ecological factor in forests owning to the dramatic damage hurricanes can wreak. Chapter 8 acquaints the readers with fire ecology which is concerned with the processes linking the natural incidence of fire in an ecosystem and the ecological effects of the fire.

Section III concentrates on community ecology. Chapter 9 focuses on community structure and classification. A community is the set of all populations that inhabit a certain area. Chapter 10 is devoted to community dynamics. Community dynamics is the process of change and development within communities of living organisms. It is sometimes also referred to as biological-succession. The two extremes of biological-succession are a true desert, where there are no species (although most deserts have a few species present in very low numbers), right through to a rain forest where there is a huge diversity of species. Chapter 11 discusses population dynamics. A population is all the organisms that both belong to the same species and live in the same geographical area.

Section IV concentrates on plant ecology. Chapter 12 deals with nature and structure of plant community which is a recognisable assemblage of plant species which interact with each other as well as with the elements of their environment and is distinct from other assemblages. Chapter 13 acquaints the readers with plant succession. Chapter 14 focuses on biotic factors of ecology. Biotic components are the living things that shape an ecosystem. Chapter 15 is devoted to halophytes and other type of plants that grow where it is affected by salinity in the root area by salt spray. Chapter 16 deals with phytogeography which is the branch of biogeography that is concerned with the geographic distribution of plant species.

Section V discusses animal ecology. Chapter 17 and 18 concentrate on animal interrelationships and animal adaptations respectively. Chapter 19 is devoted to zoogeography which is the study of patterns of past, present and future distribution of animals in nature and the processes that regulate these distributions. Chapter 20 focuses on animal communities which highlight the relationship between living and nonliving things in a particular environment. Chapter 21 concentrates on marine ecology which deals with the interdependence of all organisms living in the ocean, in shallow coastal waters, and on the seashore.

Section VI deals with soil and water ecology. Chapter 22 is devoted to water ecology. Chapter 23 concentrates on soil ecology which is the study of the interactions among soil organisms, and between biotic and abiotic aspects of soil environment. Chapter 24 focuses on soil and water conservation.

Section VII focuses on biodiversity and ecological perspective. Chapter 25 focuses on biodiversity which is the degree of variation of life forms within a given ecosystem, biome, or an entire planet. Chapter 26 deals with threats to global biodiversity. Biodiversity is threatened by the sum of all human activities.

Section VIII focuses on environmental pollution and its control. Chapter 27 deals with conservation of natural resources. Conservation is an ethic of resource use, allocation, and protection. Conservation of nature means nothing but protecting and preserving of natural resources for further generation. Chapter 28 discusses air pollution and its control. Air pollution is the presence of substances in air in sufficient concentration and for sufficient time, so as to be, or threaten to be injurious to human, plant or animal life. Chapter 29 concentrates on water pollution and its control. Water pollution affects plants and organisms living in these bodies of water, and in almost all cases the effect is damaging not only to individual species and populations, but also to the natural biological communities. Chapter 30 is devoted to thermal and radioactive pollution and chapter 31 discusses wastes from the anthrosphere.

Glossary and index have been provided at the end for quick reference. Diagrams, figures and tables supplement the text. All the topics have been covered in a cogent and lucid style to help the reader grasp the information quickly and easily. It may not be wrong to hold that the present reference textbook on *Ecology* is a complete treatise on environment and ecology. It is essential reading for all students and teachers of environment engineering and life sciences. In addition, researchers in environmental and allied fields will also find it highly useful and informative.

The reference textbook also caters to the requirement of the syllabus prescribed by various Indian universities for undergraduate student pursuing engineering, life sciences, environment and allied courses. It has been prepared with meticulous care, aiming at making the book error-free. Constructive suggestions are always welcome from users of this book.

PN Michael

Contents at a Glance

SECTION V

SECTION VI

SECTION VII

SECTION VIII

Contents

SECTION II

8. Ecology of Fire

SECTION III

SECTION IV

SECTION V

Animal Ecology   *287–358*

17. Animal Interrelationships   289–300

SECTION VI

SECTION VII

SECTION VIII

SECTION I

Basic Concepts of Ecology

Chapter 1

Ecology and Environment

INTRODUCTION

The origin of ecology as a subject of study can be traced back to the mid-nineteenth century. Though, there is uncertainty about the exact year in which the term 'ecology' was developed, but the years 1866 to 1896 are considered to be specially relevant. However, the existence of ecology is as old as man himself. Ernst Haeckel, a renowned German zoologist, is known as father of the word 'ecology'. He redefined ecology as a science that treats reciprocal relations of an organism and the external world. According to the concise Oxford Dictionary, ecology is 'that branch of biology dealing with organisms relation to one another and to their surroundings'. Undoubtedly, it was developed as a branch of biology but in the modern system ecology has developed as an integrated discipline covering more scope than biology and has a close relation with other social sciences like economics, etc. Thus, ecology is the scientific study of the relationships between organisms and their environment.

BASIC CONCEPTS

In order to study ecology with regard to environment, the following components play an important role in helping us to appreciate the scope of ecology and environmental destruction. The basics of ecology has also been indicated in Fig. 1.1.

Population

The large and growing population in India is a major cause of environmental destruction. The population pressure has outstripped the country's capacity to meet its requirements. The biotic potential and carrying capacity play a major role in determining the size of natural population. The major question before us is, can India's land support such a huge and ever increasing population? Is it the number of people or animals that an area of land can support on a sustainable basis? Man being the dominant species in the ecosystem, with a highly developed brain, has created new environmental situations. In order to fulfil his social and economic needs, changes in natural ecosystem are inevitable. Within an ecosystem, a species exist as a population whose growth or decline is affected by the capacity of the ecosystem to provide requisite necessities of life.

Community

In ecology the concept of community means the sum of population of different species within a given area. A country may have large area but lack of habitation because of the extensive occurrence of forests, mountains or deserts. These is less protection of traditional communities in areas of significant biodiversity.

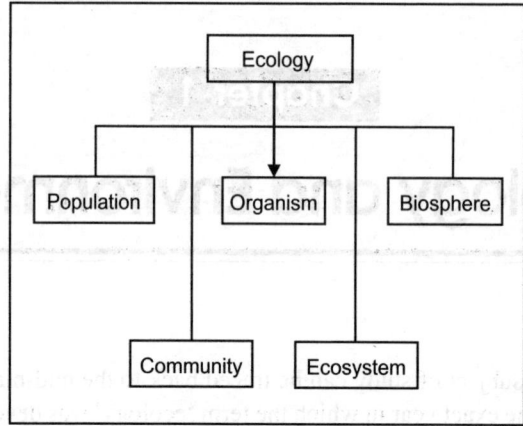

Fig. 1.1. Basics of ecology.

Organism

Organism is an important concept for study of ecology. In fact, ecology is the study of an organism. It includes individuals, plants and animals. Millions of species of plants and animals, including human, form the part of an organism.

Ecosystem

The ecosystem is the basic functional unit in ecology and includes both organisms and abiotic environment, each influencing properties of each other and both necessary for maintenance of life. Ecosystem is the sum of the communities and the non-living environment in an area.

Biosphere

Biosphere is the part of earth's surface capable of supplying life. It includes:
1. Living matter which includes various organisms.
2. The organic and inorganic mineral products created by living matters like coal, soil, etc.
3. Mineral materials formed by the association of organisms and non-living nature.

The biosphere consists of interacting ecological functional units called ecosystem. It may be defined as the sum of all ecosystems.

ECOSYSTEM

The term 'ecosystem' includes 'not only the organism-complex, but the whole complex of physical factors forming what we call the environment'. The ecosystem is the basic concept of ecology. In simple words, ecosystem is the sum of communities and the non-living environment in an area.

An ecosystem is a system formed by the interactions of variety of individual organisms with each other and with their physical environment. Ecosystem are nearly self-contained so that the exchange of nutrients within the system is much greater than the exchange with other systems.

Characteristics of Ecosystem

The main characteristics of an ecosystem based on structural and functional classification represented in Fig. 1.2 are.

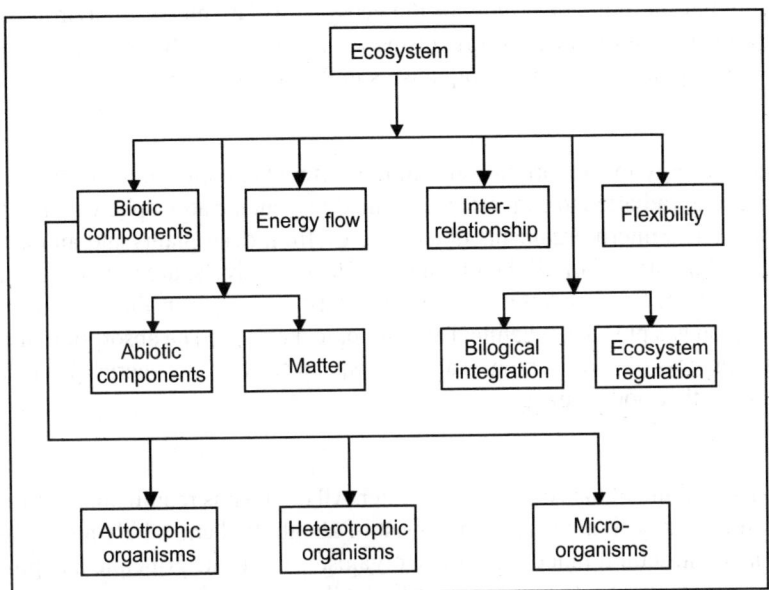

Fig. 1.2. Characteristics of ecosystem.

Biotic components

Ecosystem has both biotic and abiotic components. Biotic components include producers, consumers and reducers. The biotic sub-components may be plants, animals and micro-organisms. The brief description of the sub-components are:

Autotrophic organisms

Autotrophic means self-nourishing. In the natural system of solar radiation, autotrophic organisms are green plants. In the oceans, these exist as microscopic free floating phytoplankton. All the autotrophs manufacture their own food through the process of photosynthesis and are called producers.

Heterotrophic organisms

Heterotrophic means nourishing on others. These organisms include mainly animals, which either eat plants or other animals. Heterotrophs are also known as consumers.

Micro-organisms

Micro-organisms includes bacteria and fungi that break down the complex organic compounds of dead plants and animals, absorbing some of the decomposed products and release mineral nutrients which are put to use by producers.

This also consists of organic compounds which may become the source of food to other organisms. Micro-organisms are also called reducers.

Abiotic components

The physical environment like land, air and water is known as the abiotic component. It comprises of water, sunlight, temperature, soil, rocks, air, minerals, etc. The physical environment not only influences the biotic structure but the functional units of the ecosystem. The physical environment controls the structure through restriction on range of organisms in an ecosystem. It also represents the degree of interaction between population of various organisms in the environment.

Energy flow

All organisms need energy to support and sustain their life. One main source of energy is sun. Green plants directly use solar radiation to convert carbon dioxide from the atmosphere and water into glucose which is plant food. The micro-organisms obtain energy from dead plants and animals. Energy is an essential ingredient for survival of all the organisms. The exclusive source of energy of all ecological system is solar radiation. Solar radiation reaches the earth's surface in the form of electromagnetic waves. Solar radiation has physical, chemical and biological effects. The absorption of sun's energy by plants through the process of photosynthesis and subsequent flow of this energy from plants to animals are subject to laws of thermodynamics.

Matter

Another functional attribute of the ecosystem is matter. All organisms require matter to sustain life. The biological molecules of any living organism consist of about 40 chemical elements. These chemicals play a vital role in the structural functioning of an organism. The basic principle that governs the use of matter by a living organism and its circulation in the ecological system are:
1. The law of tolerance.
2. The law of minimum.
3. The law of conservation of matter.

According to the law of tolerance, there is an optimal level of concentration of an element at which the metabolic processes occur at a maximum speed in all organisms. Hence, there is a maximum and minimum level of concentration which an organism can tolerate. The law of minimum states that a minimum quantity of each essential element is necessary for the growth of an element. The law of conservation of matter states that in an ecological system, there is a permanent recycling of matter alternating between organic and inorganic forms brought by the living organisms that is plants, animals and micro-organism.

Interrelationship

All the components of an ecosystem, either living or non-living, are interrelated. A change in one component will have an effect on all other components.

Biological integration

The biosphere consists of interacting ecological functional units called ecosystems. Thus ecosystems result on account of high level of biological integration. There is a complex relationship between structure and function of an ecosystem.

Flexibility

Ecosystems are flexible and not static. The structural as well as ecological units change frequently and influence the living and non-living environment.

Ecosystem regulation

Another functional characteristic of an ecosystem is ecoregulation. The ecosystem can be regulated by interactions within the system and by interactions with other systems. The interactions within the system can take place in the following manners:

1. Competition between the individuals of one population.
2. Inter-specific competition.
3. Inter-specific co-operation.
4. Interaction of living components with abiotic environment.

All ecosystems are open systems and interact with each other. The output from one ecosystem becomes the input for another ecosystem. Thus, an ecosystem is a sum of the input environment, system itself and the environment.

ECOSYSTEM AND THEIR FUNCTIONING

An ecosystem is a set of interacting interdependent living (organic or biotic) and non-living (inorganic or abiotic) components. It was originally coined to convey the idea of a group of organisms and the place or habitat (i.e. home), they occupy and the way the two are linked together to form a functional unit. Thus, an ecosystem is the functional unit of ecology and represents the highest level of ecological integration which is energy based. A pond, a lake, a tract of forest, a coral reef, etc. provide convenient units for the study of an ecosystem.

It is a dynamic system which includes both abiotic physico-chemical environment and its particular biotic assemblage of plants, animals and microbes influencing the properties of each other and both are necessary for the maintenance of life. Ecosystems are the subject matter of ecology and an understanding of their structure and function is the concern of an ecologist.

Components of an Ecosystem

From the functional point of view, each ecosystem has two components.

Autotrophic component

It includes mainly the green plants that fix the solar energy and manufacture food from simple inorganic substances.

Heterotrophic component

It includes animals, bacteria, fungi, etc. that utilise the stored food of an autotroph, rearrange it and finally decompose the complex material into simple inorganic compounds. From the structural point of view, each ecosystem has four components:

1. Abiotic components.
2. The producers.
3. The consumers.
4. The decomposers and transformers.

However, Odum has divided the total components of an ecosystem into following six phases:

1. Inorganic components: Carbon, nitrogen, carbon dioxide, water, etc.
2. Organic compounds: Protein, carbohydrates, lipids, etc.
3. Climatic regime: Temperature, light and other physical factors
4. Producers: Autotrophic organisms, green plants, etc.

5. Macro consumers or phagotrops (i.e. mainly animals)
6. Microconsumers: Chiefly bacteria and fungi

Principles

Biological activity involves energy utilisation, energy that comes ultimately from the sun and which is transformed from the radiant to the chemical form photosynthesis and through from the chemical to mechanical and heat forms through cellular metabolism. These conversions and sequential dependencies are at once both elementary in conception and fundamental to the energetics of organisms and ecosystems. The various components of an ecosystem can well be described by taking the example of a pond ecosystem.

Abiotic components

These may comprise of inorganic materials like oxygen, carbon dioxide, nitrogen, calcium, phosphorus, water, etc. They are taken up by the plants and with the help of sunlight are converted into food material. These are again returned to the cycle by the action of micro-organisms on the dead bodies of plants and animals.

Producers

These comprise of green plants and algae and diatoms, etc. These can take energy from non-living environment and make it available to all living organisms. These convert the light energy of the sun into potential chemical energy in the form of organic substances, such as carbohydrates, using simpler inorganic substances such as carbon, oxygen and hydrogen. This process is known as photosynthesis, carbon assimilation, or primary biological productivity and can be briefly summarised in the following equation:

$$\underset{\text{Low chemical energy}}{6CO_2 + H_2O} + \text{Light energy} \longrightarrow \underset{\text{High chemical energy}}{C_6H_{12}O_6 \text{ (Sugar)} + 6O_2}$$

Consumers

These are the living organisms (mainly animals) which obtain their energy from sources other than themselves, directly or indirectly. Plants are eaten by herbivores, which are first order consumers. These herbivores, in turn, are taken by carnivores which form second order consumers.

Decomposers

Bacteria and moulds are called as decomposers or micro-consumers. They imbide dead organic matter. These simplify the organic constituents of each dead body. Their activity makes chemical substances available for other living beings (producers). Certain bacteria also act upon decomposed products and convert them into organic and inorganic substances.

Trophic Levels, Food Chains and Food Webs

Trophic levels

The interrelationship between plants and animals, and between animals and animals in the sphere of energy production and consumption results in a definite pattern of several stages of eating and being eaten, on the basis of which a trophic structure is built.

Thus:

$$\text{Plant} \longrightarrow \text{Herbivore} \longrightarrow \text{Carnivore}$$

Here plants are producers, herbivores are primary consumers and the carnivores are secondary consumers. There may exist tertiary consumers and so on. The plants are said to be at the first trophic level, herbivores represent the second trophic level, primary carnivores constitute the third trophic level and so on.

Food chains

All living organisms which occupy the same trophic level regardless of their taxonomic position belong to that trophic level. Thus, a grasshopper, grazing cattle and grain eating birds are all primary consumers (second trophic levels) because they directly derive their energy from plants. When only one linear relationship among producers and consumers in an unit trophic level is considered it is called food chain. The species which occupy each trophic level in the food chain form a link called the food link.

Foodweb

The final level of each food chain is occupied by top consumer or top carnivore, for it is not consumed by any other organism except decomposers. A top carnivore or a top consumer is usually linked with several food chains at a time. Thus, we see that food chains are not an isolated sequence but are interconnected with one another. Such interlocking pattern (several food chains) is called a food web. The concept of food chain is useful for analysis of a community. A typical example of terrestrial food web is given in Fig. 1.3.

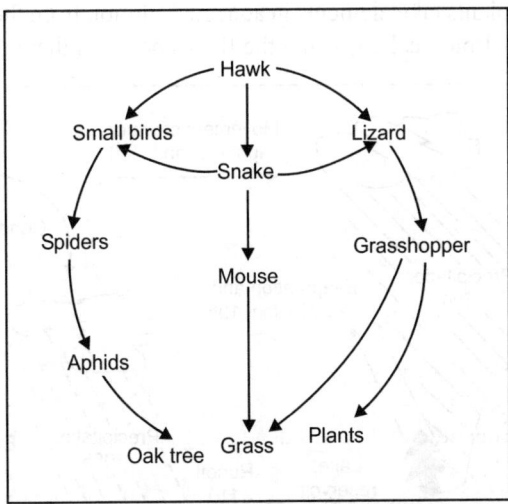

Fig. 1.3. Terrestrial food web.

NUTRIENT CYCLES IN ECOSYSTEMS

Nutrient cycles closely parallel the routes of energy flow within the biotic components of ecosystems, but an important distinction between the two processes is the relationship with the abiotic component. While energy flow is profligate in the sense of being driven by an endless solar power supply, nutrient cycling is conservative, with chemical elements being drawn from finite pools and being largely retained within ecosystems. This is also true, as for any system to continue there must always be a movement of materials in a cyclic manner. The mineral elements taken up from the environment (soil as well as atmosphere) by the green plants—the producers—are again returned to the environment. In this taking and returning processes of minerals, there are involved a number of organisms as well as some physicochemical phenomena that together make an orderly operating cycle. Thus, the movement (imports and exports) of minerals is accomplished by the operation of different chemical cycles that keep on passing the materials back and forth between organisms and their environment.

To stress the biological, geological (in rocks, soils and sediments) and chemical nature of the processes, they are sometimes called biogeochemical cycles. There are macronutrients as C, H and O, which have cycles with an atmospheric store, and some like P and K which are obtained from the soils. There are micronutrients as Cu, Fe, Co, which have soil-based or edaphic cycles. Nitrogen, a macronutrient, has inorganic pools in both the atmosphere and soil. A variety of biogeochemical processes control the links between available and unavailable forms of edaphic nutrients. The qualitative and quantitative nature of nutrient cycles in natural ecosystems is very diverse.

ATMOSPHERIC CYCLES

Hydrological Cycle

Strictly speaking, this is not an elemental cycle because it follows the course of a compound, water. Nevertheless, the movement of water within and between ecosystems is fundamental to an understanding of nutrient cycles for several reasons: (i) it is the source of hydrogen for photosynthesis in plants, (ii) plants use a large amount of water to maintain their hydrostatic skeletons and to move chemicals about their bodies, and (iii) plants take elements in aqueous solution from the soil. Figure 1.4 shows the components of the global hydrological cycle and the fluxes between them.

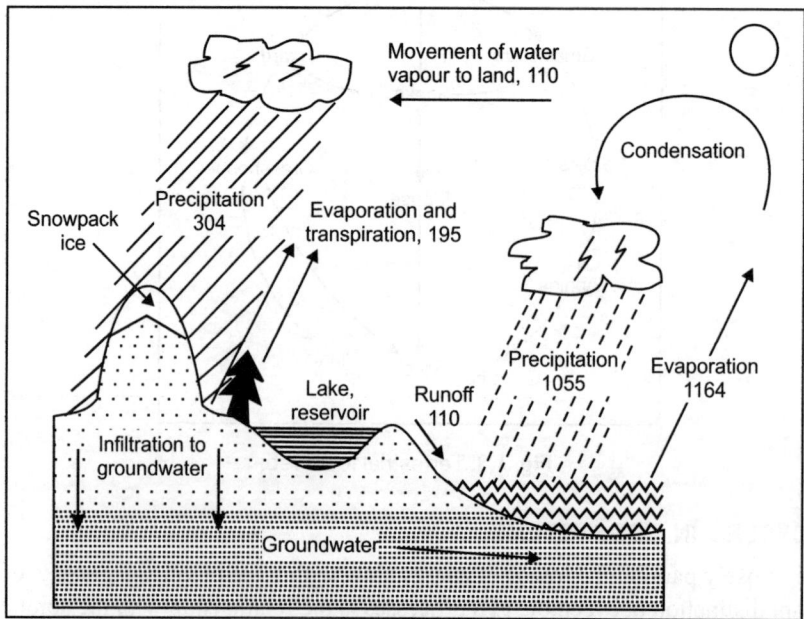

Fig. 1.4. Global hydrological cycle.

The major store is in the oceans and the major flows are evaporation from them and precipitation upon them. However, there is a net flow of water vapour, driven by winds from the oceans to land, where it falls as rain, hail or snow. The balance of the cycle, is maintained by water flowing from the land as surface run off or movement of groundwater into rivers and back to the oceans.

A major part of water is locked up in the earth's crust, and is only released in small quantities during volcanic eruptions. Similarly, the large store on polar ice caps has little effect on the hydrological cycle in the short-term due to negligible evaporation. The hydrological cycle is driven by the evaporative power of solar radiation and requires 15 per cent of the total radiation reaching the outer atmosphere. The proportion compares with the 0.2 per cent used in gross primary production worldwide.

Carbon Cycle

The simplest of nutrient cycles is the carbon cycle where general components are well recognised. Carbon is returned to the atmosphere as fast as it is removed. Carbon moves in the gaseous phase (CO_2) from the atmosphere to producers. Here, it enters the food chain and is respired out at each trophic level and finally released during decomposition, thus returning to the atmosphere. More than the amount of CO_2 getting locked up in the biomass, is now being thrown into the atmosphere by the processes of combustion of fossil fuels, burning of wood and volcanic activities. The vast ocean helps greatly in maintaining the level of CO_2 in the atmosphere by absorbing and releasing the gas when its levels are high and low, respectively. The atmospheric concentration of carbon dioxide is 0.03–0.04 per cent while oceanic water can retain up to fifty times this level, and hence is capable of regulating carbon dioxide in the atmosphere.

Carbon dioxide from the atmosphere is converted into organic compounds in plants through photosynthesis. Organic carbon from plants may go into animals. There it goes through various stages of digestion and assimilation. It may re-enter the atmosphere from plants or animals as CO_2 by oxidation or decomposition of dead organic matter. In animals, carbon may get tied up in hard parts such as shells and hence remain in the form of inorganic carbonates for long. Marine deposits of animal carbonates and inorganic precipitation of carbonates in water may result in limestone. Limestone can then return carbonates very slowly to the living carbon cycle through erosion and dissolution. Carbon may also get locked into organic deposits of coal and petroleum, where it may remain for million of years unless released by combustion (Fig. 1.5). Carbon is returned to the atmosphere in the following ways.

Biological processes

Only a small part returns through the respiratory activity of producers and consumers. Larger part is returned by the decomposers through processing of waste material and decomposition of dead remains of different trophic levels.

Non-biological processes

The main non-biological processes are combustion of fossil fuels, burning of wood and accidental fires in forests and buildings. CO_2 also comes out along with the gases escaping from the sites of diggings of fossil fuel, coal, organic gases, etc. as the geological component. Eruption of vulcano throws out lava and gases also come out in large quantities. Carbon dioxide is a major constituent of volcanic activity.

The interplay between atmospheric and aquatic CO_2 is of great significance. The interchange between the two phases, viz. atmospheric CO_2 and dissolved CO_2 in ocean occurs through diffusion. The direction of diffusion is dependent upon relative concentration. Passage into aquatic phase also takes place through precipitation. A litre of rain water contains about 0.3 cc of CO_2.

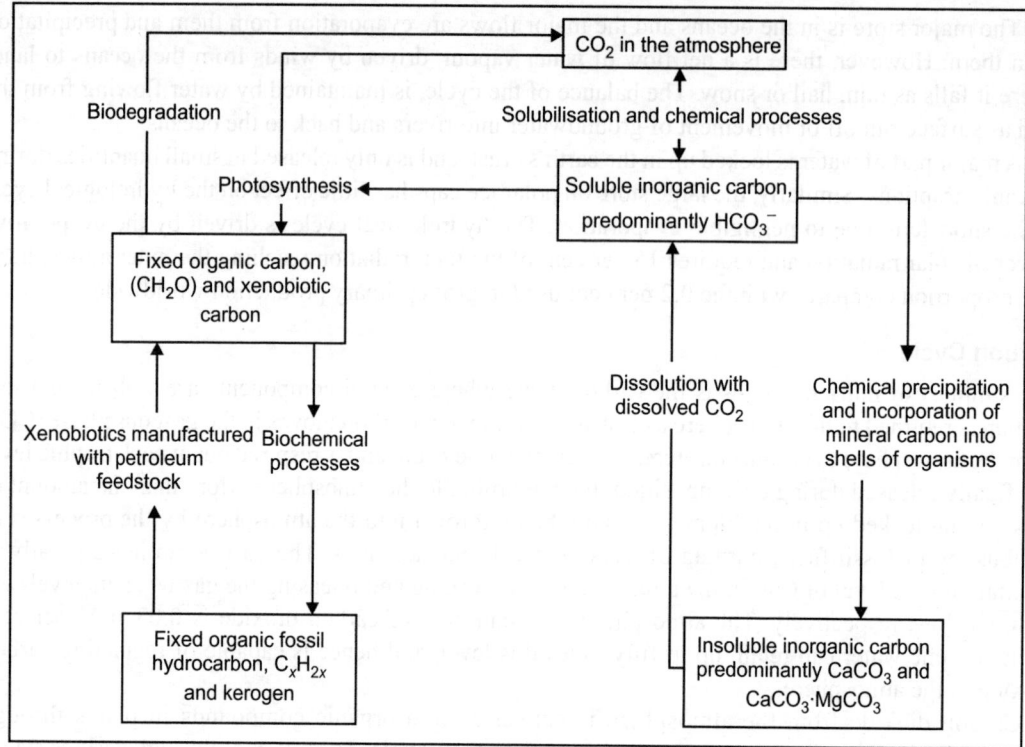

Fig. 1.5. Carbon cycle.

The dissolved CO_2 combines with water in the soil or in aquatic ecosystem to form carbonic acid (H_2CO_3) in a reversible reaction. Carbonic acid dissociates into hydrogen (H^+) and bicarbonate ions (HCO_3^-). The latter ions, in turn, dissociate in another reversible reaction into hydrogen and carbonate ions. The various reactions can be expressed as follows:

$$H_2O + CO_2 \rightleftharpoons H_2CO_3$$
$$H_2CO_3 \rightleftharpoons H^+ + HCO_3^-$$
$$HCO_3^- \rightleftharpoons H^+ + CO_3^-$$

Since all these reactions are reversible, the direction of the reaction is dependent on the concentration of critical components, i.e. the concentration of CO_2 in the atmosphere. The equilibrium is not so simple but is much more complicated, e.g. the amount of carbon present as bicarbonate is dependent on the pH of water. At higher pH (alkaline), more carbon is present as carbonate while at lower pH (acidic), more carbon is present in dissolved phase.

There are a number of avenues by which carbon is utilised and a much larger number by which it is restored to the atmosphere. Collectively, these various pathways constitute self-regulating feedback mechanisms resulting in a relatively haemostatic system. Additions and deletions are readily equilibrated at the biosphere level. To what degree the system will withstand or adapt to long-term disturbances of existing equilibrium is, of course, uncertain. There are indications that the past few decades of increased use and incomplete combustion of fossil fuels have resulted in a detectable increase in atmospheric carbon. The prediction is of 25 per cent increase by the end of year 2005. The long term effects of such an increase are of concern to humanity.

Oxygen Cycle

Oxygen is a major component of all living organisms. Hence, its adequate supply is vital for sustenance of life in the biosphere. Our atmospheric air contains about 21 per cent oxygen. It is also present, as dissolved in water. Oxygen is given out as a by-product of photosynthesis (Fig. 1.6).

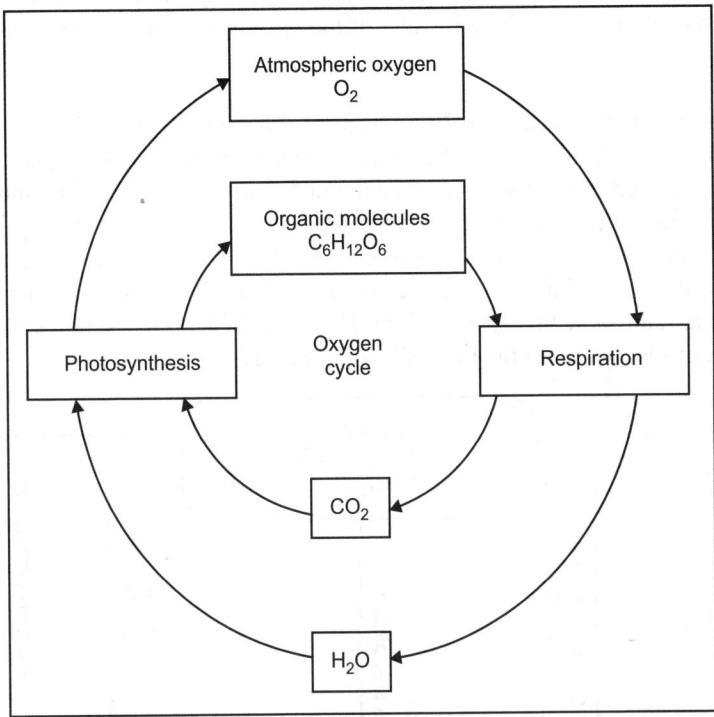

Fig. 1.6. Oxygen cycle.

Oxygen is needed by most plants, animals and all human beings for aerobic respiration or enzymatic oxidation of organic food which sustains growth and general metabolism. Thus, it is absorbed from the environment during aerobic respiration but released by plants during photosynthesis thereby setting up the oxygen cycle. There is also continuous exchange of O_2 between the atmosphere and all water surfaces on the earth. The total amount of O_2 in the biosphere is relatively constant so that the oxygen cycle is stable. The oxygen cycle is based on the exchange of O_2 among the environmental segments—atmosphere, hydrosphere, lithosphere and biosphere.

Oxygen contributes largely to the processes on the earth's surface. It participates in combustion reactions, e.g. burning of fossil fuels (C_2CH_4).

$$C + O_2 \rightarrow CO_2$$
$$CH_4 \text{ (in natural gas)} + 2O_2 \rightarrow CO_2 + 2H_2O$$

Oxygen is consumed by some oxidative weathering processes of minerals.

$$4FeO + O_2 \rightarrow 2Fe_2O_3$$

In the primitive stage of the earth, soluble iron consume bulk of O_2 giving large deposits of Fe_2O_3:

$$4Fe_2 + O_2 + 4H_2O \rightarrow 2Fe_2O_3 + 8H^+$$

Green plants return O_2 to the atmosphere through the process of photosynthesis:

$$CO_2 + H_2O + hv \rightarrow [CH_2O] + O_2$$

This process was responsible for building the original oxygen stock in the atmosphere and continues to maintain the oxygen balance in the atmosphere. It must be noted that through combustion of fossil fuels and of reducing gases (CO) from volcanoes consume large quantities of O_2, it has little impact on the total oxygen stock in the atmosphere because of the operation of the oxygen cycle.

Nitrogen Cycle

Nitrogen in its gaseous form constitutes 79 per cent of the atmosphere. However, it cannot be used directly by most forms of life. It must first be 'fixed' before it can be utilised by plants and animals. By fixation, nitrogen is converted into its chemical compounds, largely nitrates (NO_2) and ammonia (NH_3). The fixation of nitrogen takes place through both physio-chemical and biological means, although the latter is by far the much bigger contributor. The biological fixation is limited to a few, but abundant, organisms like the free living bacteria *Azetobacter* and *Clostridium*, nodule bacteria on leguminous plants like *Rhizobium* and some blue-green *Algae*. These are the keys to the movement of the nitrogen from the atmospheric reservoir into the cycle shown in Fig. 1.7.

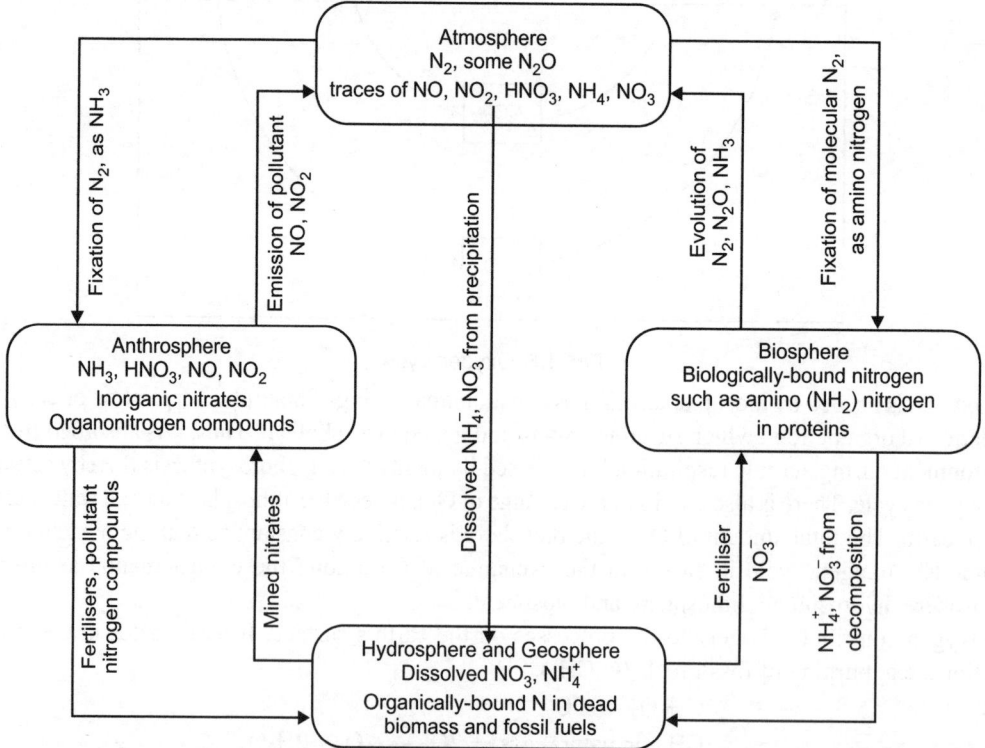

Fig. 1.7. Nitrogen cycle.

The nitrates are assimilated to form amino acids, urea and other organic residues in the producer, consumer and decomposer cycles. The amino acids and urea are then converted to ammonia through a process called 'ammonification'. To complete the cycle, denitrifying bacteria convert the ammonia into nitrites, then into nitrates and then back into gaseous nitrogen. In this way, under normal circumstances, the total amount of nitrogen fixed equals the total amount returned to the atmosphere as gas.

Man has interfered with this natural cycle by industrially fixing nitrogen. This includes production of nitrogen fertilisers and oxidation of nitrogen during fossil fuel combustion. Most of the excess nitrogen is carried off into rivers and lakes, and ultimately reaches the ocean. This increased runoff has greatly increased the productivity in many aquatic environments and has contributed to the process of eutrophication.

Phosphate Cycle

Phosphates are necessary for the growth and maintenance of animal bones and teeth while organophosphates are essential for cell division, involving the production of nuclear DNA and RNA. Plants and animals derive their nutrition via energy metabolic pathways with chemical utilising ATP (adenosine triphosphate).

Phosphate minerals are located in rocks and soil, where phosphates exist in soluble and insoluble forms. Terrestrial plants absorb inorganic phosphate salts from the soil and convert these into organic phosphate. Animals obtain their phosphate by eating plants. Plants and animals after their death and decay return phosphates to the soil, which are finally converted to humus by the action of soil micro-organisms. Bulk of the phosphate in the soil is fixed or absorbed on to soil particles, but part of it is lost by leaching out (runoff) into water courses.

In fresh water, phytoplankton (floating algae) quickly absorbs soluble inorganic phosphates and convert, them into organophosphates. *Algae* are the sources of food for zooplankton, which in turn are eaten by other aquatic animals. All these forms of life after their death and decay settle to the bottom of water. In due course the organic waste decomposes by the action of micro-organisms, releasing phosphates into the water body for recycling again (Fig. 1.8).

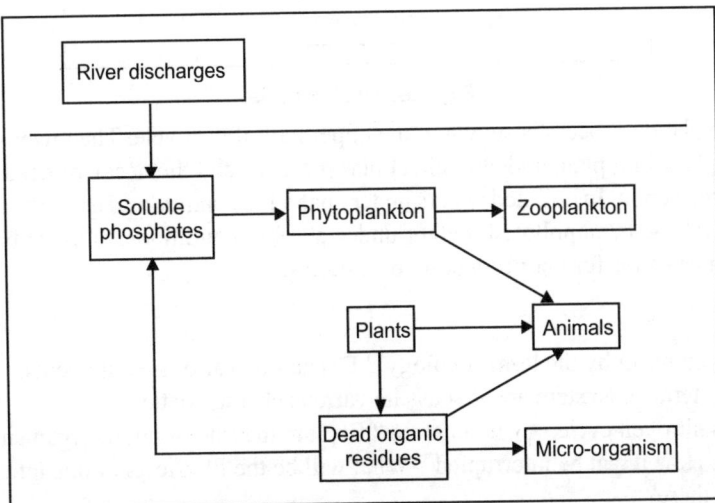

Fig. 1.8. Phosphate cycle.

The natural phosphate cycle, as discussed above, is badly affected by pollution, mainly from agricultural run-off containing superphosphate or triple superphosphate and also from domestic sewage containing phosphates derived from excreta and detergents. Phosphate pollution of rivers and lakes is the cause of algal bloom (eutrophication), which reduces the dissolved oxygen in water and disrupts natural food chains.

Sulphur Cycle

Plants and animals depend on continuous supply of sulphur and its compounds for synthesis of some amino acids and proteins. Some sulphur bacteria serve as the media for exchanges of sulphur within ecosystems. The sulphur cycle illustrates the circulation of sulphur and its compounds in the environment (Fig. 1.9).

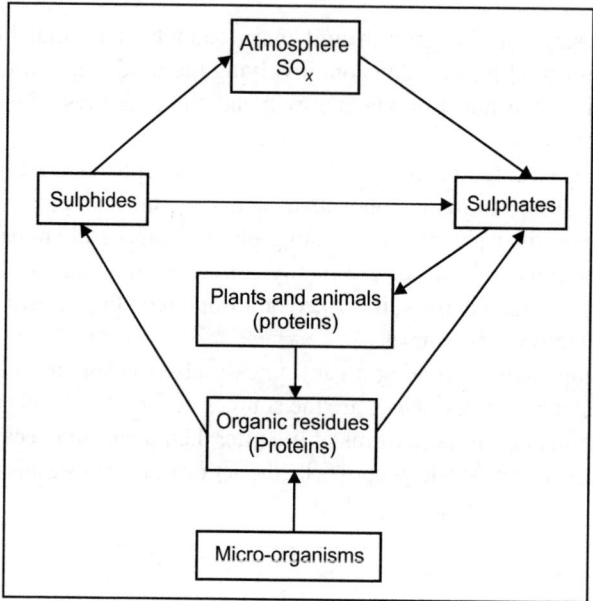

Fig. 1.9. Sulphur cycle.

The sulphur oxidation process is shown in the upper half of the cycle. The lower section shows the conversion of sulphate into plant and animal cellular proteins, and the decay of dead plant and animal material by bacterial action. In polluted waters under anaerobic conditions, H_2S is produced by bacteria, giving deposits of FeS. In unpolluted waters under aerobic conditions, sulphur bacteria transform sulphides into sulphates for further production of proteins.

Exercise

1. What do you mean by the term 'Ecology'? Discuss its various components.
2. Define the term ecosystem and discuss its various characteristics.
3. How does nitrogen cycle go on in nature ? Explain the role of micro-organisms in maintaining this cycle. How it can be interrupted ? What will be the ill-effects of this interruption and how to control these?
4. Describe the carbon cycle in nature. What are the ill-effects of its interruption? Suggest some important steps to control it.
5. Write short notes on the following:
   (a) Nutrient cycles in ecosystems.

(b) Hydrological cycle.

(c) Oxygen cycle.

(d) Sulphur cycle.

Thus, the study of ecology helps us to understand how each and every human action affects our environment and thus, in the long-run, affects ourselves. Ecology and environmental science are often viewed as synonymous, which means environmentalists are frequently considered as ecologists. Ecosystems are the subject matter of ecology and understanding of their structure and function is the concern of an ecologist. Within an ecosystem there are dynamic interrelationships between the living forms and their physical environments. These relationships are manifested as natural cycles, which provide a continuous circulation of essential constituents necessary for life. The natural cycles and ecosystems operate in a balanced manner which stabilises the entire biosphere and sustains the life process on earth.

Hydrological cycle is a continuous natural process which helps in exchange of water between the atmosphere, the land, the sea, living plants and animals. The biosphere carbon cycle is primarily concern with the atmospheric gas, carbon dioxide, its incorporation into organic matter by photosynthesis, and its subsequent release by the respiration of all the biota.

Oxygen cycle is based on the exchange of oxygen among the environmental segments—atmosphere, hydrosphere, lithosphere and biosphere. It plays a key role in atmospheric chemistry, geochemical transformation and life processes.

Nitrogen cycle unlike the carbon cycle is a very complex gaseous type cycle. It is not entirely edaphic but has an atmospheric component linked to the soil by nitrogen fixation and denitrification.

Phosphorus cycle is an example of sedimentary type of cycle having its main reservoir not in the air but in rocks and other natural deposits. Sulphur cycle illustrates the circulation of sulphur and its compounds in the environment.

ECOLOGICAL STABILITY

Ecological stability can refer to types of stability in a continuum ranging from resilience (returning quickly to a previous state) to constancy to persistence. The precise definition depends on the ecosystem in question, the variable or variables of interest, and the overall context. In the context of conservation ecology, stable populations are often defined as ones that do not go extinct. Researchers applying mathematical models from system dynamics usually use Lyapunov stability.

Types of Ecological Stability

Local stability indicates that a system is stable over small short-lived disturbances, while global stability indicates a system highly resistant to change in species composition and/or food web dynamics.

Constancy and persistence

Observational studies of ecosystems use constancy to describe living systems that can remain unchanged.

Resistance and inertia (persistence)

Resistance and inertia deal with a system's inherent response to some perturbation. A perturbation is any externally imposed change in conditions, usually happening in a short time period. Resistance is a measure of how little the variable of interest changes in response to external pressures. Inertia (or persistence) implies that the living system is able to resist external fluctuations. In the context of changing ecosystems in post-glacial North America, EC Pielou remarked at the outset of her overview,

> 'It obviously takes considerable time for mature vegetation to become established on newly exposed ice scoured rocks or glacial till...it also takes considerable time for whole ecosystems to change, with their numerous interdependent plant species, the habitats these create, and the animals that live in the habitats. Therefore, climatically caused fluctuations in ecological communities are a damped, smoothed-out version of the climatic fluctuations that cause them.'

Resilience, elasticity and amplitude

Resilience is the tendency of a system to return to a previous state after a perturbation. Elasticity and amplitude are measures of resilience. Elasticity is the speed with which a system returns. Amplitude is a measure of how far a system can be moved from the previous state and still return. Ecology borrows the idea of neighbourhood stability and a domain of attraction from dynamical systems theory.

Chapter 2

Productivity Ecology

INTRODUCTION

In ecology, productivity or production refers to the rate of generation of biomass in an ecosystem. It is usually expressed in units of mass per unit surface (or volume) per unit time, for instance grams per square metre per day. The mass unit may relate to dry matter or to the mass of carbon generated. Productivity of autotrophs such as plants is called primary productivity, while that of heterotrophs such as animals is called secondary productivity. Primary production is the synthesis of new organic material from inorganic molecules such as H_2O and CO_2. It is dominated by the process of photosynthesis which uses sunlight to synthesise organic molecules such as sugars, although chemosynthesis represents a small fraction of primary production. Organisms responsible for primary production include land plants, marine algae and some bacteria (including cyanobacteria).

Secondary production is the generation of biomass of heterotrophic (consumer) organisms in a system. This is driven by the transfer of organic material between trophic levels, and represents the quantity of new tissue created through the use of assimilated food. Secondary production is sometimes defined to only include consumption of primary producers by herbivorous consumers (with tertiary production referring to carnivorous consumers), but is more commonly defined to include all biomass generation by heterotrophs. Organisms responsible for secondary production include animals, protists, fungi and many bacteria. Secondary production can be estimated through a number of different methods including increment summation, removal summation, the instantaneous growth method and the Allen curve method. The choice between these methods will depend on the assumptions of each and the ecosystem under study. For instance, whether cohorts should be distinguished, whether linear mortality can be assumed and whether population growth is exponential.

THEORETICAL PRODUCTION ECOLOGY

Theoretical production ecology tries to quantitatively study the growth of crops. The plant is treated as a kind of biological factory, which processes light, carbon dioxide, water and nutrients into harvestable parts. Main parameters kept into consideration are temperature, sunlight, standing crop biomass, plant production distribution, nutrient and water supply.

Modelling

Modelling is essential in theoretical production ecology. Unit of modelling usually is the crop, the assembly of plants per standard surface unit. Analysis results for an individual plant are generalised to the standard surface, e.g. the leaf area index is the generalised surface of all crop leaves per surface unit.

Processes

The usual system of describing plant production divides the plant production process into at least five separate processes, which are influenced by several external parameters.

Two cycles of biochemical reactions constitute the basis of plant production, the light reaction and the dark reaction:

1. In the light reaction, sunlight photons are absorbed by chloroplasts which split water into an electron, proton and oxygen radical which is recombined with another radical and released as molecular oxygen. The recombination of the electron with the proton yields the energy carriers NADH and ATP. The rate of this reaction often depends on sunlight intensity, leaf area index, leaf angle and amount of chloroplasts per leaf surface unit. The maximum theoretical gross production rate under optimum growth conditions is approximately 250 kg per hectare per day.

2. The dark reaction or Calvin cycle ties atmospheric carbon dioxide and uses NADH and ATP to convert it into sucrose. The available NADH and ATP, as well as temperature and carbon dioxide levels determine the rate of this reaction. Together those two reactions are termed photosynthesis. The rate of photosynthesis is determined by the interaction of a number of factors including temperature, light intensity and carbon dioxide.

3. The produced carbohydrates are transported to other plant parts, such as storage organs and converted into secondary products, such as amino acids, lipids, cellulose and other chemicals needed by the plant or used for respiration. Lipids, sugars, cellulose and starch can be produced without extra elements. The conversion of carbohydrates into amino acids and nucleic acids requires nitrogen, phosphorus and sulphur. Chlorophyll production requires magnesium, while several enzymes and coenzymes require trace elements. This means, nutrient supply influences this part of the production chain. Water supply is essential for transport, hence limits this too.

4. The production centers, i.e. the leaves, are sources, the storage organs, growth tips or other destinations for the photosynthetic production are sinks. The lack of sinks can be a limiting factor for production too, as happens, e.g. in apple orchards where insects or night frost have destroyed the blossoms and the produced assimilates cannot be converted into apples. Biennial and perennial plants employ the stored starch and fats in their storage organs to produce new leafs and shoots the next year.

5. The amount of crop biomass and the relative distribution of biomass over leafs, stems, roots and storage organs determines the respiration rate. The amount of biomass in leafs determines the leaf area index, which is important in calculating the gross photosynthetic production.

6. Extensions to this basic model can include insect and pest damage, intercropping, climatical changes, etc.

Parameters

Important parameters in theoretical production models thus are:

Climate

1. Temperature: The temperature determines the speed of respiration and the dark reaction. A high temperature combined with a low intensity of sunlight means a high loss by respiration. A low temperature combined with a high intensity of sunlight means that NADH and ATP heap up but cannot be converted into glucose because the dark reaction cannot process them swiftly enough.

2. Light: Light, also called photosynthetic active radiation (PAR) is the energy source for green plant growth. PAR powers the light reaction, which provides ATP and NADPH for the conversion of carbon dioxide and water into carbohydrates and molecular oxygen. When temperature,

moisture, carbon dioxide and nutrient levels are optimal, light intensity determines maximum production level.
3. Carbon dioxide levels: Atmospheric carbon dioxide is the sole carbon source for plants. About half of all proteins in green leaves have the sole purpose of capturing carbon dioxide. Although CO_2 levels are constant under natural circumstances, CO_2 fertilisation is common in greenhouses and is known to increase yields by on average 24 per cent. C_4 plants like maize and sorghum can achieve a higher yield at high solar radiation intensities, because they prevent the leaking of captured carbon dioxide due of the spatial separation of carbon dioxide capture and carbon dioxide use in the dark reaction. This means that their photorespiration is almost zero. This advantage is sometimes offset by a higher rate of maintenance respiration. In most models for natural crops, carbon dioxide levels are assumed to be constant.

Crop
1. Standing crop biomass: Unlimited growth is an exponential process, which means that the amount of biomass determines the production. Because an increased biomass implies higher respiration per surface unit and a limited increase in intercepted light, crop growth is a sigmoid function of crop biomass.
2. Plant production distribution: Usually only a fraction of the total plant biomass consists of useful products, e.g. the seeds in pulses and cereals, the tubers in potato and cassava, the leafs in sisal and spinach, etc. The yield of usable plant portions will increase when the plant allocates more nutrients to this parts, e.g. the high-yielding varieties of wheat and rice allocate 40 per cent of their biomass into wheat and rice grains, while the traditional varieties achieve only 20 per cent, thus doubling the effective yield. Different plant organs have a different respiration rate, e.g. a young leaf has a much higher respiration rate than roots, storage tissues or stems do. There is a distinction between 'growth respiration' and 'maintenance respiration'. Sinks, such as developing fruits, need to be present. They are usually represented by a discrete switch, which is turned on after a certain condition, e.g. critical daylength has been met.

Care
1. Water supply: Because plants use passive transport to transfer water and nutrients from their roots to the leafs, water supply is essential to growth, even so that water efficiency rates are known for different crops, e.g. 5000 for sugar cane, meaning that each kilogram of produced sugar requires up to 5000 litres of water.
2. Nutrient supply: Nutrient supply has a two-fold effect on plant growth. A limitation in nutrient supply will limit biomass production as per Liebig's Law of the Minimum. With some crops, several nutrients influence the distribution of plant products in the plants. A nitrogen gift is known to stimulate leaf growth and therefore can work adversely on the yield of crops which are accumulating photosynthesis products in storage organs, such as ripening cereals or fruit-bearing fruit trees.

Phases in Crop Growth

Theoretical production ecology assumes that the growth of common agricultural crops, such as cereals and tubers, usually consists of four (or five) phases:
1. Germination: Agronomical research has indicated a temperature dependence of germination time (GT, in days).

Each crop has a unique critical temperature (CT, dimension temperature) and temperature sum (dimensions temperature times time), which are related as follows:

$$GT = \frac{TS}{\sum_{k=1}^{N}(T - T_{crit})}$$

When a crop has a temperature sum of, e.g. 150°C and a critical temperature of 10°C, it will germinate in 15 days when temperature is 20°C, but in 10 days when temperature is 25°C. When the temperature sum exceeds the threshold value, the germination process is complete.

2. Initial spread: In this phase, the crop does not cover the field yet. The growth of the crop is linearly dependent on leaf area index, which in its turn is linearly dependent on crop biomass. As a result, crop growth in this phase is exponential.

3. Total coverage of field: In this phase, growth is assumed to be linearly dependent on incident light and respiration rate, as nearly 100 per cent of all incident light is intercepted. Typically, LAI is above two to three in this phase. This phase of vegetative growth ends when the plant gets a certain environmental or internal signal and starts generative growth (as in cereals and pulses) or the storage phase (as in tubers).

4. Allocation to storage organs: In this phase, up to 100 per cent of all production is directed to the storage organs. Generally, the leafs are still intact and as a result, gross primary production stays the same. Prolonging this phase, e.g. by careful fertilisation, water and pest management results directly in a higher harvest.

5. Ripening: In this phase, leafs and other production structures slowly die off. Their carbohydrates and proteins are transported to the storage organs. As a result, the LAI and, hence, the primary production decreases.

Existing Plant Production Models

Plant production models exist in varying levels of scope (cell, physiological, individual plant, crop, geographical region, global) and of generality; the model can be crop-specific or be more generally applicable. In this section the emphasis will be on crop-level based models as the crop is the main area of interest from an agronomical point of view.

PRIMARY PRODUCTION

Primary production is the production of organic compounds from atmospheric or aquatic carbon dioxide, principally through the process of photosynthesis, with chemosynthesis being much less important. Almost all life on earth is directly or indirectly reliant on primary production. The organisms responsible for primary production are known as primary producers or autotrophs, and form the base of the food chain. In terrestrial ecoregions, these are mainly plants, while in aquatic ecoregions algae are primarily responsible. Primary production is distinguished as either net or gross, the former accounting for losses to processes such as cellular respiration, the latter not. Primary production is the production of chemical energy in organic compounds by living organisms. The main source of this energy is sunlight but a minute fraction of primary production is driven by lithotrophic organisms using the chemical energy of inorganic molecules. Regardless of its source, this energy is used to synthesise complex organic molecules from simpler inorganic compounds such as carbon dioxide (CO_2) and water (H_2O).

The following two equations are simplified representations of photosynthesis (top) and (one form of) chemosynthesis (bottom):

$$CO_2 + H_2O + \text{light} \rightarrow CH_2O + O_2$$

$$CO_2 + O_2 + 4\,H_2S \rightarrow CH_2O + 4\,S + 3\,H_2O$$

In both cases, the end point is reduced carbohydrate (CH_2O), typically molecules such as glucose or other sugars. These relatively simple molecules may be then used to further synthesise more complicated molecules, including proteins, complex carbohydrates, lipids, and nucleic acids or be respired to perform work. Consumption of primary producers by heterotrophic organisms, such as animals, then transfers these organic molecules (and the energy stored within them) up the food web, fueling all of the earth's living systems.

GPP and NPP

Gross primary production (GPP) is the rate at which an ecosystem's producers capture and store a given amount of chemical energy as biomass in a given length of time. Some fraction of this fixed energy is used by primary producers for cellular respiration and maintenance of existing tissues (i.e. growth respiration' and 'maintenance respiration'). The remaining fixed energy (i.e. mass of photosynthate) is referred to as net primary production (NPP).

$$NPP = GPP - \text{respiration [by plants]}$$

Net primary production is the rate at which all the plants in an ecosystem produce net useful chemical energy; it is equal to the difference between the rate at which the plants in an ecosystem produce useful chemical energy (GPP) and the rate at which they use some of that energy during respiration. Some net primary production goes toward growth and reproduction of primary producers, while some is consumed by herbivores. Both gross and net primary production are in units of mass/area/time. In terrestrial ecosystems, mass of carbon per unit area per year (g $C/m^2/yr$) is most often used as the unit of measurement.

Terrestrial Production

On the land, almost all primary production is now performed by vascular plants, with a small fraction coming from algae and non-vascular plants such as mosses and liverworts. Before the evolution of vascular plants, non-vascular plants likely played a more significant role. Primary production on land is a function of many factors, but principally local hydrology and temperature (the latter covaries to an extent with light, the source of energy for photosynthesis). While plants cover much of the earth's surface, they are strongly curtailed wherever temperatures are too extreme or where necessary plant resources (principally water and light) are limiting, such as deserts or polar regions.

Water is 'consumed' in plants by the processes of photosynthesis and transpiration. The latter process (which is responsible for about 90 per cent of water use) is driven by the evaporation of water from the leaves of plants. Transpiration allows plants to transport water and mineral nutrients from the soil to growth regions, and also cools the plant. Diffusion of water out of a leaf, the force that drives transpiration, is regulated by structures known as stomata. These also regulate the diffusion of carbon dioxide from the atmosphere into the leaf, such that decreasing water loss (by partially closing stomata) also decreases carbon dioxide gain. Certain plants use alternative forms of photosynthesis, called Crassulacean acid metabolism (CAM) and C4. These employ physiological and anatomical adaptations to increase water-use efficiency and allow increased primary production to take place under conditions that would normally limit carbon fixation by C3 plants (the majority of plant species).

Oceanic Production

In a reversal of the pattern on land, in the oceans, almost all primary production is performed by algae, with a small fraction contributed by vascular plants and other groups. Algae encompass a diverse range of organisms, ranging from single floating cells to attached seaweeds. They include photoautotrophs from a variety of groups. Eubacteria are important photosynthetisers in both oceanic and terrestrial ecosystems, and while some archaea are phototrophic, none are known to utilise oxygen-evolving photosynthesis. A number of eukaryotes are significant contributors to primary production in the ocean, including green algae, brown algae and red algae, and a diverse group of unicellular groups. Vascular plants are also represented in the ocean by groups such as the seagrasses.

Unlike terrestrial ecosystems, the majority of primary production in the ocean is performed by free-living microscopic organisms called phytoplankton. Larger autotrophs, such as the seagrasses and macroalgae (seaweeds) are generally confined to the littoral zone and adjacent shallow waters, where they can attach to the underlying substrate but still be within the photic zone. There are exceptions, such as *Sargassum*, but the vast majority of free-floating production takes place within microscopic organisms.

The factors limiting primary production in the ocean are also very different from those on land. The availability of water, obviously, is not an issue (though its salinity can be). Similarly, temperature, while affecting metabolic rates, ranges less widely in the ocean than on land because the heat capacity of seawater buffers temperature changes, and the formation of sea ice insulates it at lower temperatures. However, the availability of light, the source of energy for photosynthesis, and mineral nutrients, the building blocks for new growth, play crucial roles in regulating primary production in the ocean.

Light

The sunlit zone of the ocean is called the photic zone (or euphotic zone). This is a relatively thin layer (10–100 m) near the ocean's surface where there is sufficient light for photosynthesis to occur. For practical purposes, the thickness of the photic zone is typically defined by the depth at which light reaches 1 per cent of its surface value. Light is attenuated down the water column by its absorption or scattering by the water itself, and by dissolved or particulate material within it (including phytoplankton).

Net photosynthesis in the water column is determined by the interaction between the photic zone and the mixed layer. Turbulent mixing by wind energy at the ocean's surface homogenises the water column vertically until the turbulence dissipates (creating the aforementioned mixed layer). The deeper the mixed layer, the lower the average amount of light intercepted by phytoplankton within it. The mixed layer can vary from being shallower than the photic zone, to being much deeper than the photic zone. When it is much deeper than the photic zone, this results in phytoplankton spending too much time in the dark for net growth to occur. The maximum depth of the mixed layer in which net growth can occur is called the critical depth. As long as there are adequate nutrients available, net primary production occurs whenever the mixed layer is shallower than the critical depth.

Both the magnitude of wind mixing and the availability of light at the ocean's surface are affected across a range of space- and time-scales. The most characteristic of these is the seasonal cycle (caused by the consequences of the earth's axial tilt), although wind magnitudes additionally have strong spatial components. Consequently, primary production in temperate regions such as the North Atlantic is highly seasonal, varying with both incident light at the water's surface (reduced in winter) and the degree of mixing (increased in winter). In tropical regions, such as the gyres in the middle of the major basins, light may only vary slightly across the year, and mixing may only occur episodically, such as during large storms or hurricanes.

Nutrients

Mixing also plays an important role in the limitation of primary production by nutrients. Inorganic nutrients, such as nitrate, phosphate and silicic acid are necessary for phytoplankton to synthesise their cells and cellular machinery. Because of gravitational sinking of particulate material (such as plankton, dead or fecal material), nutrients are constantly lost from the photic zone, and are only replenished by mixing or upwelling of deeper water. This is exacerbated where summertime solar heating and reduced winds increases vertical stratification and leads to a strong thermocline, since this makes it more difficult for wind mixing to entrain deeper water. Consequently, between mixing events, primary production (and the resulting processes that leads to sinking particulate material) constantly acts to consume nutrients in the mixed layer, and in many regions this leads to nutrient exhaustion and decreased mixed layer production in the summer (even in the presence of abundant light). However, as long as the photic zone is deep enough, primary production may continue below the mixed layer where light-limited growth rates mean that nutrients are often more abundant.

Iron

Another factor relatively recently discovered to play a significant role in oceanic primary production is the micronutrient iron. This is used as a cofactor in enzymes involved in processes such as nitrate reduction and nitrogen fixation. A major source of iron to the oceans is dust from the earth's deserts, picked up and delivered by the wind as aeolian dust.

In regions of the ocean that are distant from deserts or that are not reached by dust-carrying winds (for example, the Southern and North Pacific oceans), the lack of iron can severely limit the amount of primary production that can occur. These areas are sometimes known as HNLC (High-Nutrient, Low-Chlorophyll) regions, because the scarcity of iron both limits phytoplankton growth and leaves a surplus of other nutrients. Some scientists have suggested introducing iron to these areas as a means of increasing primary productivity and sequestering carbon dioxide from the atmosphere.

Measurement

The methods for measurement of primary production vary depending on whether gross vs. net production is the desired measure, and whether terrestrial or aquatic systems are the focus. Gross production is almost always harder to measure than net, because of respiration, which is a continuous and ongoing process that consumes some of the products of primary production (i.e. sugars) before they can be accurately measured. Also, terrestrial ecosystems are generally more difficult because a substantial proportion of total productivity is shunted to below-ground organs and tissues, where it is logistically difficult to measure. Shallow water aquatic systems can also face this problem.

Scale also greatly affects measurement techniques. The rate of carbon assimilation in plant tissues, organs, whole plants or plankton samples can be quantified by biochemically-based techniques, but these techniques are decidedly inappropriate for large scale terrestrial field situations. There, net primary production is almost always the desired variable, and estimation techniques involve various methods of estimating dry-weight biomass changes over time. Biomass estimates are often converted to an energy measure, such as kilocalories, by an empirically determined conversion factor.

Terrestrial

In terrestrial ecosystems, researchers generally measure net primary production. Although its definition is straightforward, field measurements used to estimate productivity vary according to investigator and biome. Field estimates rarely account for below ground productivity, herbivory, decomposition, turnover, litterfall, volatile organic compounds, root exudates, and allocation to symbiotic micro-organisms. Biomass based NPP estimates result in underestimation of NPP due to incomplete accounting of these components. However, many field measurements correlate well to NPP. There are a number of comprehensive reviews of the field methods used to estimate NPP. Estimates of ecosystem respiration, the total carbon dioxide produced by the ecosystem, can also be made with gas flux measurements.

The major unaccounted for pool is below ground productivity, especially production and turnover of roots. Below ground components of NPP are difficult to measure. BNPP is often estimated based on a ratio of ANPP:BNPP rather than direct measurements.

Grasslands

Most frequently, peak standing biomass is assumed to measure NPP. In systems with persistent standing litter, live biomass is commonly reported. Measures of peak biomass are more reliable in if the system is predominantly annuals. However, perennial measurements can be reliable if there was a synchronous phenology driven by a strong seasonal climate. These methods may underestimate ANPP in grasslands by as much as 2 (temperate) to 4 (tropical) fold. Repeated measures of standing live and dead biomass provide more accurate estimates of all grasslands, particularly those with large turnover, rapid decomposition, and interspecific variation in timing of peak biomass. Wetland productivity (marshes and fens) is similarly measured. In Europe, annual mowing makes the annual biomass increment of wetlands evident.

Forests

Methods used to measure forest productivity are more diverse than those of grasslands. Biomass increment based on stand specific allometry plus litterfall is considered a suitable although incomplete accounting of above-ground net primary production (ANPP). Field measurements used as a proxy for ANPP include annual litterfall, diameter or basal area increment (DBH or BAI), and volume increment.

Aquatic

In aquatic systems, primary production is typically measured using one of four main techniques:
1. Variations in oxygen concentration within a sealed bottle (developed by Gaarder and Gran in 1927).
2. Incorporation of inorganic carbon-14 (^{14}C in the form of sodium bicarbonate) into organic matter.
3. Stable isotopes of oxygen (^{16}O, ^{18}O and ^{17}O).
4. Fluorescence kinetics (technique still a research topic).

The technique developed by Gaarder and Gran uses variations in the concentration of oxygen under different experimental conditions to infer gross primary production. Typically, three identical transparent vessels are filled with sample water and stoppered. The first is analysed immediately and used to determine the initial oxygen concentration; usually this is done by performing a Winkler titration. The other two vessels are incubated, one each in under light and darkened. After a fixed period of time, the experiment ends, and the oxygen concentration in both vessels is measured. As photosynthesis has not taken place in the dark vessel, it provides a measure of ecosystem respiration. The light vessel permits both photosynthesis and respiration, so provides a measure of net photosynthesis (i.e. oxygen production via photosynthesis subtract oxygen consumption by respiration). Gross primary production is then obtained by adding oxygen consumption in the dark vessel to net oxygen production in the light vessel.

The technique of using ^{14}C incorporation (added as labelled Na_2CO_3) to infer primary production is most commonly used today because it is sensitive, and can be used in all ocean environments. As ^{14}C is radioactive (via beta decay), it is relatively straightforward to measure its incorporation in organic material using devices such as scintillation counters. Depending upon the incubation time chosen, net or gross primary production can be estimated. Gross primary production is best estimated using relatively short incubation times (1 hour or less), since the loss of incorporated ^{14}C (by respiration and organic material excretion/exudation) will be more limited. Net primary production is the fraction of gross production remaining after these loss processes have consumed some of the fixed carbon.

Loss processes can range between 10–60 per cent of incorporated ^{14}C according to the incubation period, ambient environmental conditions (especially temperature) and the experimental species used. Aside from those caused by the physiology of the experimental subject itself, potential losses due to the activity of consumers also need to be considered. This is particularly true in experiments making use of natural assemblages of microscopic autotrophs, where it is not possible to isolate them from their consumers.

Global

As primary production in the biosphere is an important part of the carbon cycle, estimating it at the global scale is important in earth system science. However, quantifying primary production at this scale is difficult because of the range of habitats on earth, and because of the impact of weather events (availability of sunlight, water) on its variability.

Using satellite-derived estimates of the normalised difference vegetation index (NDVI) for terrestrial habitats and sea-surface chlorophyll for the oceans, it is estimated that the total (photoautotrophic) primary production for the earth was 104.9 Gt C yr^{-1}. Of this, 56.4 Gt C yr^{-1} (53.8 per cent), was the product of terrestrial organisms, while the remaining 48.5 Gt C yr^{-1}, was accounted for by oceanic production. In areal terms, it was estimated that land production was approximately 426 g C m^{-2} yr^{-1} (excluding areas with permanent ice cover), while that for the oceans was 140 g C m^{-2} yr^{-1}. Another significant difference between the land and the oceans lies in their standing stocks — while accounting for almost half of total production, oceanic autotrophs only account for about 0.2 per cent of the total biomass.

Human impact and appropriation

Extensive human land use results in various levels of impact on actual NPP (NPP_{act}). In some regions, such as the Nile valley, irrigation has resulted in a considerable increase in primary production. However, these regions are exceptions to the rule, and in general there is a NPP reduction due to land changes (ΔNPP_{LC}) of 9.6 per cent across global land-mass. In addition to this, end consumption by people raises the total human appropriation of net primary production (HANPP) to 23.8 per cent of potential vegetation (NPP_0). It is estimated that, in 2000, 34 per cent of the earth's ice-free land area (12 per cent cropland; 22 per cent pasture) was devoted to human agriculture. This disproportionate amount reduces the energy available to other species, having a marked impact on biodiversity, flows of carbon, water and energy, and ecosystem services, and scientists have questioned how large this fraction can be before these services begin to break down.

Measurement of Primary Production

Organic matter in soils (by wet combustion colorimetric method)

Soil organic matter is the organic fraction of soil which includes plant, animal and microbial residues and soil humus. An estimate of soil organic matter content is often used as an indicator of soil fertility.

Organic matter in the soil is oxidised by treatment with a mixture of potassium dichromate and sulphuric acid. The amount of dichromate remaining after reaction with soil organic matter is estimated by colorimetry after removal of the soil by filtration. The dichromate reduced during the reaction with the soil is assumed to be equivalent to a proportion (~75 per cent) of the organic carbon present in the sample. The method also assumes that organic carbon is the only substance present that reduces dichromate.

This method is applicable to mineral soils containing up to 0.1 per cent to 7 per cent organic matter. This method can be used to estimate organic matter content in the range 7–10 per cent, but variability will be up to 25 per cent of the value. The method should not be used for organic matter contents >10 per cent.

Harvest method

This method is employed by harvesting vegetation at periodic intervals and weighing the material. The plants are usually clipped at ground level but recent studies have shown that roots and rhizomes, root-stocks should also be removed if possible. After drying to constant weight, the harvest can be expressed in terms of the biomass as mass per unit area per unit time, e.g. as gms/sq.m/yr. The major drawbacks of this method include: (i) fails to account for material consumed by the herbivores, and (ii) fails to account for the energy utilised by the plant is its own metabolism, growth and development.

Leaf area index

Leaf area index (LAI) is the ratio of total upper leaf surface of vegetation divided by the surface area of the land on which the vegetation grows. LAI is a dimensionless value, typically ranging from 0 for bare ground to 6 for a dense forest.

LAI in silviculture

Foresty scientists define leaf area index as the one-sided green leaf area per unit ground surface area in broadleaf canopies. In conifers, three different definitions have been used:
1. Total needle surface area per unit ground area.
2. Half of the total needle surface area per unit ground area.
3. Projected needle area per unit ground area.

Interpretation and application of LAI

LAI is used to predict photosynthetic primary production and as a reference tool for crop growth. As such, LAI plays an essential role in theoretical production ecology. An inverse exponential relation between LAI and light interception, which is linearly proportional to the primary production rate, has been established:

$$P = P_{max} (1 - e^{-c \cdot LAI})$$

where, P_{max} designates the maximum primary production and c designates a crop-specific growth coefficient. This inverse exponential function is called the primary production function.

Determining LAI

LAI is determined directly by taking a statistically significant sample of foliage from a plant canopy, measuring the leaf area per sample plot and dividing it by the plot land surface area. Indirect methods measure canopy geometry or light extinction and relate it to LAI.

Direct methods: Direct methods require stripping and measuring the foliage of plant canopy samples. LAI for clip plots or individual plants is measured by hand or by using an LAI meter. Traditional LAI meters require each plant leaf to be stripped and fed through the entrance of the machine, which can be likened to a kind of crude image scanner.

Indirect methods: Indirect methods of estimating LAI *in situ* can be divided in two categories: (i) indirect contact LAI measurements such as plumb lines and inclined point quadrats, and (ii) indirect non-contact measurements. Due to the subjectivity and labour involved with the first method, indirect non-contact measurements are typically preferred. Non-contact LAI tools, such as hemispherical photography, the LAI-2000 from LI-COR Biosciences and the LP-80 LAI ceptometer from Decagon Devices, measure LAI in a non-destructive way. Hemispherical photography methods estimate LAI and other canopy structure attributes from analysing upward-looking fisheye photographs taken beneath the plant canopy. The LAI-2000 calculates LAI and other canopy structure attributes from solar radiation measurements made with a wide-angle optical sensor. Measurements made above and below the canopy are used to determine canopy light interception at five angles, from which LAI is computed using a model of radiative transfer in vegetative canopies. The LP-80 calculates LAI by means of measuring the difference between light levels above the canopy and at ground level, and factoring in the leaf angle distribution, solar zenith angle, and plant extinction coefficient. Such indirect methods, where LAI is calculated based upon observations of other variables (canopy geometry, light interception, leaf length and width, etc.) are generally faster, amenable to automation, and thereby allow for a larger number of spatial samples to be obtained. For reasons of convenience when compared to the direct (destructive) methods, these tools are becoming more and more important.

Disadvantages of methods

The disadvantage of the direct method is that it is destructive, time consuming and expensive, especially if the study area is very large.

The disadvantage of the indirect method is that in some cases it can underestimate the value of LAI in very dense canopies, as it does not account for leaves that lay on each other, and essentially act as one leaf according to the theoretical LAI models.

CHLOROPHYLL ESTIMATION

Chlorophyll is a green pigment found in almost all plants, algae, and cyanobacteria. Its name is derived from the Greek words. Chlorophyll is an extremely important biomolecule, critical in photosynthesis, which allows plants to obtain energy from light. Chlorophyll absorbs light most strongly in the blue portion of the electromagnetic spectrum, followed by the red portion. However, it is a poor absorber of green and near-green portions of the spectrum; hence the green colour of chlorophyll-containing tissues.

Measuring Chlorophyll

The chlorophyll content of leaves can be non-destructively measured using hand-held, battery-powered meters. Chlorophyll content meters measure the optical absorption of a leaf to estimate it's chlorophyll content. Chlorophyll molecules absorb in the blue and red bands, but not the green and infra-red bands. Chlorophyll content meters measure the amount of absorption at the red band to estimate the amount of chlorophyll present in the leaf. To compensate for varying leaf thickness, Chlorophyll Meters also measure absorption at the infrared band which is not significantly affected by chlorophyll. For instance, the CCM200 plus Chlorophyll Meter measures the transmittance at 653 nm (in the red band) and transmittance at 931 nm (in the infrared band). The percentage of transmittance at 931 nm, relative to the percentage of transmittance at 653 nm, estimates the relative chlorophyll content of the leaf.

The measurements made by these devices are simple, quick and relatively inexpensive. They now, typically, have large data storage capacity, averaging and graphical displays.

ECOSYSTEM PRODUCTIVITY

Primary and Secondary Productivity

Every ecosystem has a level of productivity, which helps discover the potential of an ecosystem for food production.

Primary productivity: Plant productivity.

Secondary productivity: Animal productivity.

Gross primary productivity: The measure of all photosynthesis that occurs in an ecosystem.

Net productivity: Energy left after losses as a result of respiration, growth, excreta.

Net primary productivity (NPP): Amount of energy made available by plants to animals, only at the herbivore level, and is expressed as kg/m²/yr.

This means that once the rate of primary productivity in an ecosystem is established it is possible to compare ecosystems, as a figure for the potential of each ecosystem for food production can be found. The NPP of an ecosystem depends on the levels of heat, moisture, nutrients available, competition, amount of sunlight, age and health of plants. In broad terms NPP increases towards the equator and decreases away from it, towards the poles.

Examples of plant primary productivity and biomass for various ecosystems are given in Table 2.1.

Table 2.1. Plant primary productivity and biomass for various ecosystems.

Ecosystem	Area (million km²)	Mean NPP g/m²/yr	World net primary productivity (billion per year)	mean biomass tons (kg/m²)
Tropical rain forest	17.0	2200	37.4	45
Savannah	15.0	900	13.5	4
Tundra and Alpine	8.0	140	1.1	0.6
Temperate grassland	9.0	600	5.4	1.6

Reasons for decrease in energy at each level within an ecosystem: Energy is lost at each stage in an ecosystem (at each transfer). Figure 2.1 shows how energy is lost within an ecosystem.

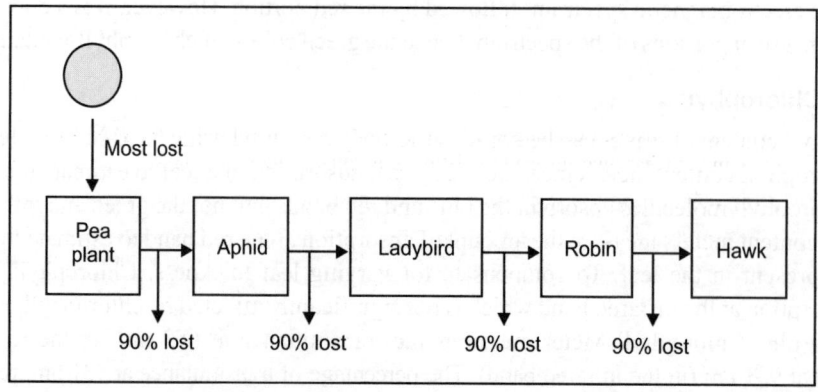

Fig. 2.1. Energy is lost within an ecosystem.

High Productivity Ecosystems

Temperate deciduous forest

It could be argued that the temperate deciduous forest is not a true example of a high productivity ecosystem, but in the UK it is one of the most productive. Table 2.2 highlights its main characteristics.

Table 2.2. Highlights main characteristics of temperature deciduous forest.

Productivity	High NPP, 1.2 kg/m^2/year, result of high summer temps and large amounts of daylight. Large amount of biomass as a result of woody material.
Vegetation	Varies with soil type acidic = birch and rowan trees, alkaline = box and maple, elm common on clay, willow on gleyed soils. Oak often dominant due to tolerance of wider pH range. More ground vegetation beneath oak trees, as small leaves allow more light to ground.
Climate	Mild/wet up to 1500 mm per year, more in winter often from depressions. More precipitation than evapotranspiration. Temps above freezing in winter. Average summer temperature = 15–20.
Soils	Fertile brown earths, with a mildly acidic mull humus. Wide range of flora and fauna in litter layer, soil mixing encouraged by earthworms. Blurred soil horizons due to worms.
Nutrient cycle	Fast rates of leaching balanced by fast rates of weathering. Many nutrients in soil as a result of slow winter growth and low density of vegetation.
Animals	Adaptations occur in the winter because of low temperatures. Many animals either migrate or hibernate.
Human interference	Few natural areas are left, and many areas have been cleared for agriculture or recreation.

Low Productivity Ecosystems

Temperate coniferous (boreal) forest

Although found in the UK the boreal forest is not as common as the deciduous, and is more common on Canada and Eastern/Central Europe as shown on the Table 2.3.

Table 2.3. Temperate coniferous (boreal) forest.

Productivity	Low, NPP = 0.8 kg/m^2/year. High biomass from the woody material.
Vegetation	Trees 20–30 m high, mainly pine, larch and spruce. Evergreen to allow photosynthesis all year. Needle leaves reduce evapotranspiration, conical shape removes snow easily.
Climate	Either cool temperate or cold continental. Small amounts of rainfall (below 500 mm per year). Summer frosts, much winter snowfall, precipitation above evapotranspiration. Reduced growing season, but 16–20 hrs of sunlight in summer increases photosynthesis.
Soils	Normally podsols. Leaching is a result of snowmelt. Humus is acidic (pH 4.5–5.5). Iron pans may form. Few earthworms lead to distinct horizons. A thick litter layer.
Nutrient cycle	Controlled by the low temperatures which limit rates of weathering in transfer, resulting in many nutrients in the litter.
Animals	Sparse as little available food. Human interference In the UK many are planted and used for forestry.

Chapter 3

Ecological Concepts of Species

INTRODUCTION

A species is often defined as a group of organisms capable of interbreeding and producing fertile offspring. While in many cases this definition is adequate, more precise or differing measures are often used, such as similarity of DNA, morphology or ecological niche. Presence of specific locally adapted traits may further subdivide species into subspecies.

Species that are believed to have the same ancestors are grouped together, and this group is called a genus. A species can only belong to one genus that it was grouped into. The belief is best checked by a similarity of their DNA, but for practical reasons, other similar properties are used.

BIOLOGICAL CLOCK

In biological clock the mechanism, presumed to exist within many animals and plants, that produces regular periodic changes in behaviour or physiology. Biological clocks underlie many of the biorhythms seen in organisms (e.g. hibernation in animals). They continue to run even when conditions are kept artificially constant, but eventually drift out of step with the natural environment without the specific signals that normally keep them synchronised. Studies in the fruit fly *Drosophila* have revealed the molecular basis of the biological clock, and similar mechanisms are thought to occur in other animals, including mammals. It involves various proteins, some of which serve as transcription factors for their own genes, particularly PER (encoded by the per gene) and TIM (encoded by the tim gene). These form part of a negative feedback loop in which the concentration of the proteins cyclically rises and falls. The timing of each cycle is determined by the time required for transcription, export of messenger RNA to the cytoplasm, translation, and, crucially, the formation of PER–TIM dimers—the only form in which these two proteins can enter the nucleus. Also, some of the proteins, including TIM, are sensitive to light and are degraded during the day. Hence, the biological clock is entrained to the day–night cycle.

Circadian Rhythm and Biological Clock

Diversity among plants is wide and varied. The genetic make up that is responsible for structural and functional variations is always dynamic. Apart from interactions with its immediate cellular environment, it also interacts with the extracellular environment. The three way interaction between the genetic material, cytoplasmic factor and external environmental factor sustains the life vibrant and dynamic. The day, on which the life originated on this planet, it is subjected to the vagaries of nature. In this struggle against nature, organisms have learnt to adopt by changing their functions and forms thus living beings progressed, during which process they accumulated more and more of informational materials and information material itself has undergone dramatic changes, yet it is still retaining the basic information genetic code. Living beings during the course of millions of years of their and structural existence on this planet have developed physiological systems which really respond to the changes in the environment with equal vigour and dynamicity. Such regulated physiological processes which control growth and development exhibits a distinct pattern in their life cycle.

The behavioural pattern of them is well adapted to environmental factors like day period, dark period, temperature, water availability, nutrition, etc. Like many animals, plants also show certain behavioural patterns, which again depends upon the changes in the environment. Flowering, opening and closing of stomata, sleeping movements of leaves, opening and closing of petals, mitotic cycle are some of the examples of behavioural patterns exhibited by plants.

As man has adopted his sleeping habits to day and nights periods, plants also exhibit diurnal rhythm in closing and opening of stomata, folding and unfolding of leaflets, and flowers, etc. For example in *Kalanchoe blossefeldiana* the floral petals open during day period and close at nights.

The leaflets of *Phaseolus multiflorus* spread out horizontally during day and fold upwards at nights showing sleeping movements. Such periodic movements are adapted to 24 hours cycle of day and night. Such daily behavioural pattern of plants is called circadian rhythm (circa-about; dian-day).

The daily periodicity has a profound influence on the physiological properties of the plant, which manifest in their morphological changes. Under such condition of the physiological processes, if there is a change or disturbance in the daily periodicity, plants still continue to behave in the same pattern, which means the circadian rhythm persists for sometime. For example, if the seedlings of phaseolus multiflorus are grown under a photoperiodic cycle of 12 hours a day and 12 hours a night for a number of days, the leaves exhibit rhythmic circadian movements. If such well adopted plants are transferred and maintained under continuous dark conditions, the opening and closing of leaves continue for a few cycles. Later the rhythmic behaviour peters off and ultimately vanishes as if it is exhausted.

A similar behavioural pattern is found in the case of stomata in many plants and floral parts of Kalanchoe. In phyllocactus, flowers open at night and close in the day times. The above said behavioural pattern is attributed to the presence of an inbuilt time measuring devices and such a device is called Biological clock.

The material basis for such a biological clock is known to by phytochrome. The chromophore protein complex by absorbing red light transforms into PfR form which slowly undergoes decay to PR form in dark. The accumulation of PR and PfR forms in cells is known to perform many physiological activities within the cell(s). Probably such pigments may act as allosteric modulators. Changes in the concentration of PR or PfR form of pigments can affect the permeability properties of cell membranes and thus they can bring about turgour movements. If a plant that is already adapted to a day length, if changed to different photoperiodic conditions, the existing active pigments still continue to operate for some time till they are completely exhausted or rendered inactive. That is why when a plant which is exhibiting rhythmic behaviour in particular periodic cycles is subjected to continuous dark conditions, the rhythmic movements persist and continues for few more cycles and then vanishes with time (Fig. 3.1).

The operation of biological clock in photoperiodic plants has been explained by Bunning. Accordingly, plants exhibit two phases of growth, i.e. photophilic phase (light loving) and scotophilic phase (dark loving) (Fig. 3.2). But the physiological process that operates in such phases is not clearly explained. However, Hyde proposed a theoretical model to explain the operation of biological clock.

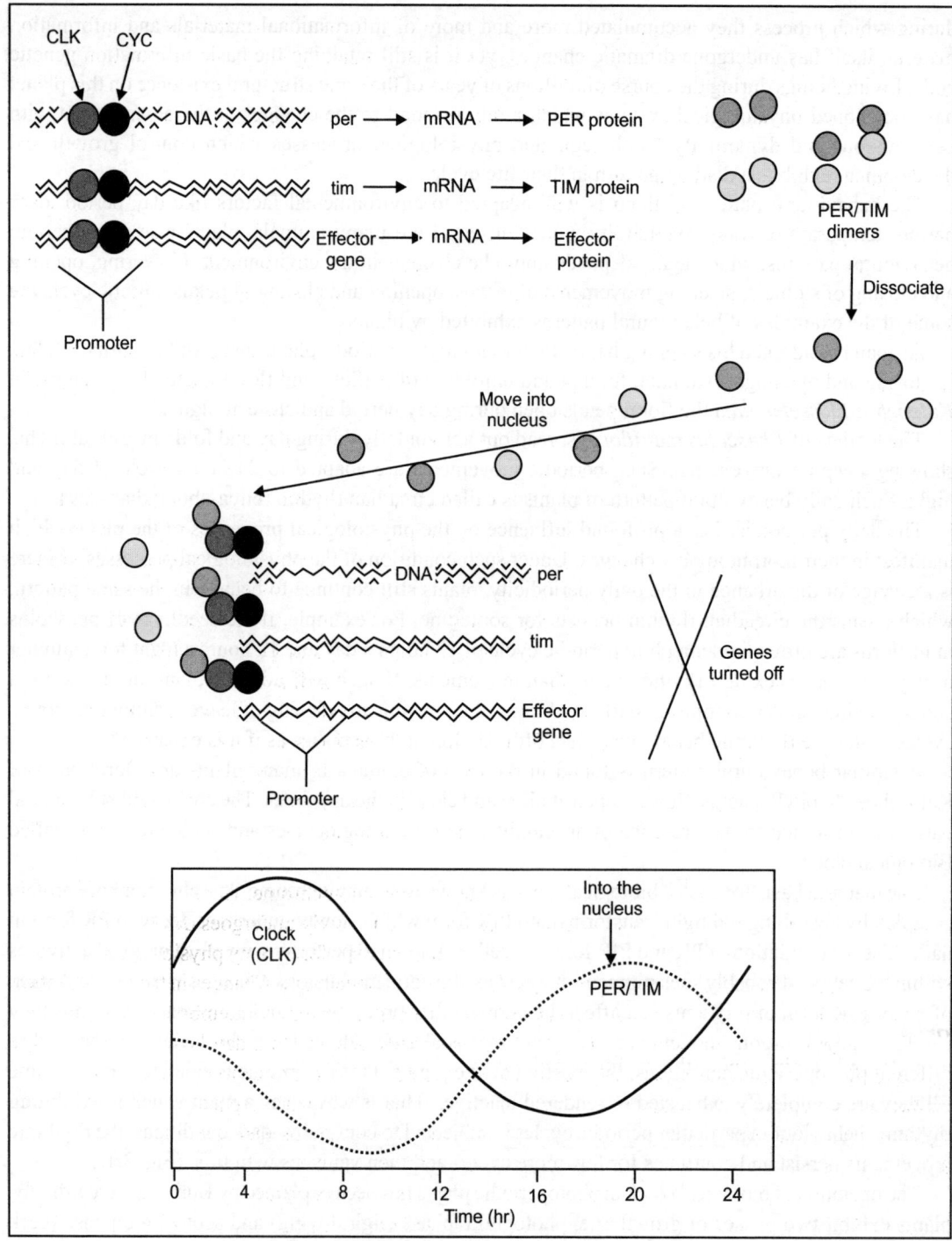

Fig. 3.1. Circadian rhythm and biological clock.

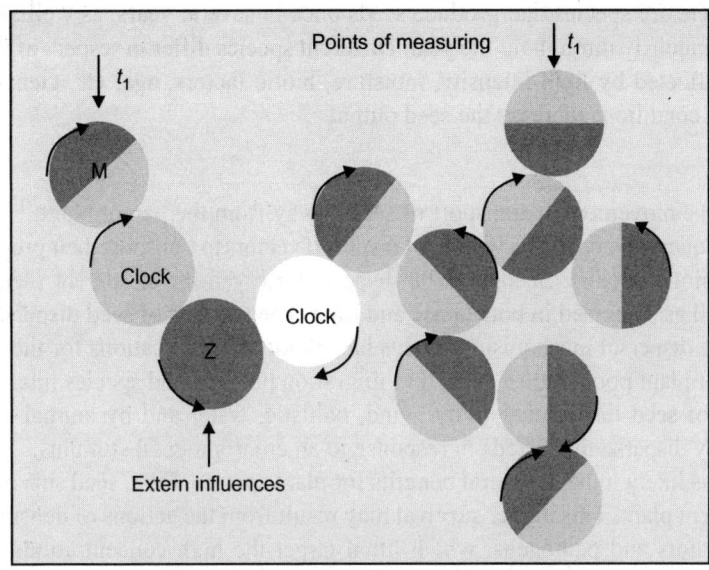

Fig. 3.2. Operation of biological clock.

This is actually based on the relative concentrations of active and inactive phytochrome accumulated within the plant structures during photophilic period and scotophilic period. Ultimately, it is the quantity of excited phytochrome that determines the rhythmic behavioural patterns and this is further believed to operate through allosteric modulations by active and inactive forms of phytochromes and other protein factors.

Name, morphology, geographic distribution, etc.

Botanical and local names of the species, its geographical distribution and history, morphological variations, if any, fossil evidences, centre of its origin and migration routes, etc. are all dealt with.

Natural distribution

This deals with the habitat of the species, soil characteristics, climatic relationships, light relations, inter as well as intraspecific competitions, and various modifications as a result of changing environmental conditions. Some species are found distributed over a limited natural area, whereas others have a wider range of natural distribution. It depends on the ecological amplitude of the species.

Regeneration

This depends mainly upon the average seed output, viability of seeds, seed dormancy, reproductive capacity, seed dispersal, seedling growth, vegetative propagation, vegetative growth, and reproductive growth.

Seed output

Plants generally produce more seeds than the habitat can sustain, as many of them are wasted, destructed or consumed in various ways. The number of seeds produced in each flush is the seed output. Annuals generally produce seeds once in their life time, whereas perennial shrubs and trees do so usually once in a year although there are species that produce seeds once in several years, as well as those producing several times continuously throughout the year. Different species differ in respect of their average seed output, which is affected by light intensity, moisture, biotic factors, age, etc. Generally higher light intensities and dry conditions increase the seed output.

Seed dispersal

Seed dispersal is the movement or transport of seeds away from the parent plant. Plants have limited mobility and consequently rely upon a variety of dispersal vectors to transport their propagules, including both abiotic and biotic vectors. Seeds can be dispersed away from the parent plant individually or collectively, as well as dispersed in both space and time. The patterns of seed dispersal are determined in large part by the dispersal mechanism and this has important implications for the demographic and genetic structure of plant populations, as well as migration patterns and species interactions. There are five main modes of seed dispersal: gravity, wind, ballistic, water and by animals. Some plants are serotinous and only disperse their seeds in response to an environmental stimulus.

Seed dispersal is likely to have several benefits for plant species. First, seed survival is often higher away from the parent plant. This higher survival may result from the actions of density-dependent seed and seedling predators and pathogens, which often target the high concentrations of seeds beneath adults. Competition with adult plants may also be lower when seeds are transported away from their parent. Seed dispersal also allows plants to reach specific habitats that are favourable for survival, a hypothesis known as directed dispersal. For example, *Ocotea endresiana* (Lauraceae) is a tree species from Latin America which is dispersed by several species of birds, including the three-wattled bellbird. Male bellbirds perch on dead trees in order to attract mates, and often defecate seeds beneath these perches where the seeds have a high chance of survival because of high light conditions and escape from fungal pathogens. In the case of fleshy-fruited plants, seed-dispersal in animal guts (endozoochory) often enhances the amount, the speed, and the asynchrony of germination, which can have important plant benefits. Seeds dispersed by ants (myrmecochory) are not only dispersed to short distances but are also buried underground by the ants. These seeds can thus avoid adverse environmental effects such as fire or drought, reach nutrient-rich microsites and survive longer than other seeds. These features are peculiar to myrmecochory, which may thus provide additional benefits not present in other dispersal modes. Finally, at another scale, seed dispersal may allow plants to colonise vacant habitats and even new geographic regions.

Seed viability

The viability of the seed accession is a measure of how many seeds are alive and could develop into plants which will reproduce themselves, given the appropriate conditions. It is important to know that the seeds that are stored in a genebank will grow to produce plants. Therefore they must have a high viability at the start and during storage. The viability of seeds at the start of storage will also determine, within the environmental conditions, the storage life of the accession.

Viability will need to be determined at the start of storage and at regular intervals during storage to predict the correct time for regeneration of the accession. The viability test takes from a few days to weeks or even months to give an accurate result. If possible the results should be available before the seeds are packaged and placed in the genebank so that poor quality seeds can be identified and regenerated before storage. Where the viability cannot be determined before storage, the seeds should be placed into long-term storage to ensure their safety whilst awaiting the results of the test.

Seed dormancy

Seed dormancy is a condition of plant seeds that prevents germinating under optimal environmental conditions. Living, non-dormant seeds germinate when soil temperatures and moisture conditions are suited for cellular processes and division; dormant seeds do not. One important function of most seeds is delayed germination, which allows time for dispersal and prevents germination of all the seeds at same time. The staggering of germination safeguards some seeds and seedlings from suffering damage or death from short periods of bad weather or from transient herbivores; it also allows some seeds to germinate when competition from other plants for light and water might be less intense. Another form of delayed seed germination is seed quiescence, which is different than true seed dormancy and occurs when a seed fails to germinate because the external environmental conditions are too dry or warm or cold for germination. Many species of plants have seeds that delay germination for many months or years, and some seeds can remain in the soil seed bank for more than 50 years before germination.

True dormancy or innate dormancy is caused by conditions within the seed that prevent germination under normally ideal conditions. Often seed dormancy is divided into two major categories based on what part of the seed produces dormancy: exogenous and endogenous. There are three types of dormancy based on their mode of action: physical, physiological and morphological.

Physical dormancy occurs when seeds are impermeable to water or the exchange of gases. Physiological dormancy prevents embryo growth and seed germination until chemical changes occur. These chemicals include inhibitors that often retard embryo growth to the point where it is not strong enough to break through the seed coat or other tissues.

Morphological dormancy. Embryo underdeveloped or undifferentiated. Some seeds have fully differentiated embryos that need to grow more before seed germination, or the embryos are not differentiated into different tissues at the time of fruit ripening.

Seed germination

A seed certainly looks dead. It does not seem to move, to grow, nor do anything. In fact, even with biochemical tests for the metabolic processes we associate with life (respiration, etc.) the rate of these processes is so slow that it would be difficult to determine whether there really was anything alive in a seed. Indeed if a seed is not allowed to germinate (sprout) within some certain length of time, the embryo inside will die. Each species of seed has a certain length of viability. Some maple species have seeds that need to sprout within two weeks of being dispersed or they die.

Common vegetable garden seeds generally lack any kind of dormancy. The seeds are ready to sprout. All they need is some moisture to get their biochemistry activated, and temperature warm enough to allow the chemistry of life to proceed. Seeds taken from the wild, however, are frequently endowed with deeper forms of dormancy.

Seeds with truly dormant embryos

There are several mechanisms that permit seeds to be truly dormant.

Thick seed coat. Many kinds of seeds have very thick seed coats. These obviously keep water out of the seed, so the embryo cannot get the water needed to activate its metabolism and start growing. The lotus seeds are an example of this. An outstanding example from the northern temperate zone is the Kentucky coffee tree (*Gymnocladus dioica*).

Thin seed coat. A thin seed coat is so thin that it is no barrier to water. Some other kind of dormancy mechanism is needed. Knowing that light can penetrate thin layers of plant tissue (leaves for example) should give you the idea that light might be a signal. That plants can absorb light and respond biochemically is a fact you know from your study of photosynthesis. All we need is a pigment molecule that can absorb light and cause a change in the behaviour of the embryo.

Insufficient development: If a seed's embryo is not completely developed, some additional maturation may be needed before the seed can sprout. This happens in seeds with little-to-no storage material invested in the seed. Examples include orchid seeds. They are the size of dust and have almost nothing but a very immature embryo on-board. Such a seed needs an association with fungi in the soil or other environments to feed the developing embryo until the embryo is mature enough to actually penetrate the seed coat. These seeds are also likely to have a very brief viability.

Inhibitors present: Many plant species invest chemicals in the developing seeds, and these chemicals inhibit the development of the embryos. They keep the embryos dormant. Obviously the seed must have some way to eliminate these chemicals before they can sprout.

Phenolic compounds: Plants that live in deserts have a different problem. There is no cold, moist, winter to allow vernalisation of abscisic acid. These plants instead use more potent toxins, phenolic compounds, to keep their seeds dormant until the proper season for germination. Phenolic compounds are freely water-soluble, the plant is living in a desert. Deserts typically have very long dry seasons and a short wet season accompanied by flash floods and so on. How do you think the phenolic compounds are lost? How would the mechanism ensure that seeds do not sprout in the dry season, but only after the seed could be sure it is in the wet season? The word 'leaching' might give you a hint?

Vegetative growth

The seedling growth stage lasts for about two to three weeks after seeds have germinated. Once a strong root system is established and foliage growth increases rapidly, seedlings enter the vegetative growth stage. When chlorophyll production is full speed ahead, a vegetative plant will produce as much green, leafy foliage as it is genetically possible to manufacture as long as light, CO_2, nutrients, and water are not limited. Properly maintained, marijuana will grow from one-half to two inches per day. A plant stunted now could take weeks to resume normal growth. A strong, unrestricted root system is essential to supply much needed water and nutrients.

Unrestricted vegetative growth is the key to a healthy harvest. A plant's nutrient and water intake changes during vegetative growth. Transpiration is carried on at a more rapid rate, requiring more water. High levels of nitrogen are needed; potassium, phosphorus, calcium, magnesium, sulphur, and trace elements are used at much faster rates. The larger a plant gets and the bigger the root system, the faster the soil will dry out. The key to strong vegetative growth and a heavy harvest is supplying roots and plants with the perfect environment.

Vegetative growth is maintained with 16 or more hours of light. It is believed that a point of diminishing returns was reached after 18 hours of light, but further research shows that vegetative plants grow faster under 24 hours of light. Marijuana will continue vegetative growth a year or longer (theoretically forever), as long as an 18-hour photoperiod is maintained. Cannabis is photoperiodic-reactive; flowering can be controlled with the light and dark cycle. This allows indoor horticulturists to control vegetative and flowering growth. Once a plant's sex is determined, it can become a mother, clone, or breeding male, and can be harvested or even rejuvenated. *Note*: Plants show early male or female 'pre-flowers' about the fourth week of vegetative growth. Cloning, transplanting, pruning, and bending are all initiated when plants are in the vegetative growth stage.

LIEBIG'S LAW (OR THE LAW OF THE MINIMUM)

An important concept to understand when dealing with plants (like algae) is Liebig's Law or the Law of the Minimum. This concept was formulated by German chemist Justus von Liebig, often called the 'father of the fertiliser industry'.

Liebig's Law of the Minimum, states that plant growth will continue as long as all required factors are present (e.g. light, water, nitrogen, phosphorus, potassium, etc.). When one of those factors is depleted, growth stops. Increasing the amount of the 'limiting' component will allow growth to continue until that component (or another) is depleted. The nutrient most typically 'limiting' algae growth in lakes is phosphorus. If phosphorus concentrations can be controlled, then algae can be controlled usually. Sometimes, other nutrients or conditions can limit algae.

Liebig's Law has been extended to biological populations (and is commonly used in ecosystem models). For example, the growth of an organism such as a plant may be dependent on a number of different factors, such as sunlight or mineral nutrients (e.g. nitrate or phosphate). The availability of these may vary, such that at any given time one is more limiting than the others. Liebig's Law states that growth only occurs at the rate permitted by the most limiting.

For instance, in the equation below, the growth of population O is a function of the minimum of three Michaelis–Menten terms representing limitation by factors I, N and P.

$$\frac{dO}{dt} = O\left(\min\left(\frac{\mu_I I}{k_I + I}, \frac{\mu_N N}{k_N + N}, \frac{\mu_P P}{k_P + P} \right) - m \right)$$

The use of the equation is limited to a situation where there are steady state conditions, and factor interactions are tightly controlled.

Other Applications

More recently Liebig's Law is starting to find an application in natural resource management where it surmises that growth in markets dependent upon natural resource inputs is restricted by the most limited scarcest input. As the natural capital upon which growth depends is limited in supply due to the finite nature of the planet, Liebig's Law encourages scientists and natural resource managers to calculate the scarcity of essential resources in order to allow for a multigenerational approach to resource consumption.

Neo-liberal economic theory has sought to confute the issue of resource scarcity by the universal application of the law of substitutability and technological innovation. The substitutability 'law', which has a powerful influence on the discourse of ideas despite the lack of an empirical evidence, states that as one resource is exhausted—and prices rise due to a lack of surplus—new markets based on alternative resources appear at certain prices in order to satisfy demand. Technological innovation implies that humans are able to use technology to fill the gaps in situations where resources are imperfectly substitutable.

Economic theories based on ceteris paribus deal only with a small selection of variables and are a weak planning tool when applied to the complex web of interconnectedness of the global market responsible for the primary consumption of natural resources. While the theory of the law of substitutability may seem to reinforce the postulation that economic growth is invulnerable to resource limits it fails to consider the underlying implication of Liebig's Law. If the resource which is limited in supply is also essential to the establishment of substitute markets, then substitution cannot occur.

Biotechnology

One example of technological innovation is in plant genetics whereby the biological characteristics of species can be changed by employing genetic modification to alter biological dependence on the most limiting resource. Biotechnological innovations are thus able to extend the limits for growth in species by an increment until a new limiting factor is established, which can then be challenged through technological innovation. Theoretically there is no limit to the number of possible increments towards an unknown productivity limit. This would be either the point where the increment to be advanced is so small it cannot be justified economically or where technology meets an invulnerable natural barrier. It may be worth adding that biotechnology itself is totally dependent on external sources of natural capital.

Natural barriers to growth

The only perfect limiting factor is that which enables life to exist, energy.

SHELFORD'S LAW OF TOLERANCE

Shelford's Law of Tolerance: A law, proposed by VE Shelford, that states that the presence and success of an organism depend upon the extent to which a complex of conditions are satisfied. The absence or failure of an organism can be controlled by the qualitative or quantitative deficiency or excess or any one of several factors which may approach the limits of tolerance for that organism. Shelford's Law of Tolerance states that the distribution of a species will be limited by its range of tolerance for local environmental factors (Fig. 3.3). Though the principle behind both Liebig's and Shelford's Laws is important in ecology, the assumption that a single factor is always limiting is potentially misleading. In nature, the various environmental factors interact in so many ways that it is often impossible to describe any one factor as the liming one. When one condition is not optimal—though tolerable—for a species, the limits of tolerance for other factors may be reduced. Moreover, unless the Law of Tolerance is extended to such biotic limiting factors as predation and competition, it has only restricted applicability.

Seedling

A seedling is a young plant sporophyte developing out of a plant embryo from a seed. Seedling development starts with germination of the seed. A typical young seedling consists of three main parts: the radicle (embryonic root), the hypocotyl (embryonic shoot), and the cotyledons (seed leaves). The two classes of flowering plants are distinguished by their numbers of seed leaves: Monocotyledons (monocots) have one blade-shaped cotyledon, whereas dicotyledons (dicots) possess two round cotyledons. Gymnosperms are more varied.

For example, pine seedlings have up to eight cotyledons. The seedlings of some flowering plants have no cotyledons at all. These are said to be acotyledons.

During germination, the young plant emerges from its protective seed coat with its radicle first, followed by the cotyledons. The radicle orients towards gravity, while the hypocotyl orients away from gravity and elongates through cell expansion to push the cotyledons out of the ground.

Seedling growth and maturation

Once the seedling starts to photosynthesise, it is no longer dependent on the seed's energy reserves. The apical meristems start growing and give rise to the root and shoot. The first 'true' leaves expand and can often be distinguished from the round cotyledons through their species-dependent distinct shapes. While the plant is growing and developing additional leaves, the cotyledons eventually senesce and fall off.

Seedling growth is also affected by mechanical stimulation, such as by wind or other forms of physical contact, through a process called thigmomorphogenesis.

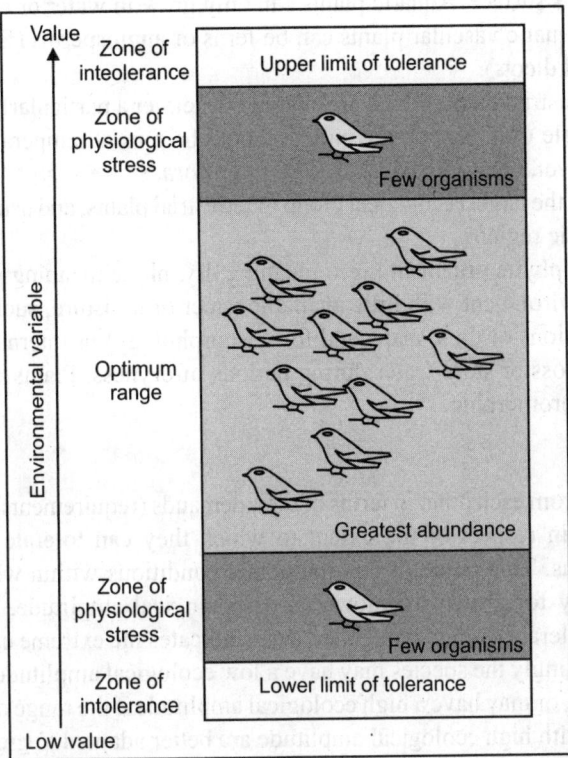

Fig. 3.3. A diagrammatic illustration of the Law of Tolerance. The species in question is most abundant in areas where the environmental variable is within the optimum range for that species. The species is rare in areas where it experiences physiological stress because the environmental variable has either too high or too low a value. The species does not occur at all in areas beyond its upper and lower limits of tolerance.

ECOLOGICAL SPECIES CONCEPT

The ecological species concept is a concept of species in which a species is a set of organisms adapted to a particular set of resources, called a niche, in the environment. According to this concept, populations form the discrete phenetic clusters that we recognise as species because the ecological and evolutionary processes controlling how resources are divided up tend to produce those clusters. Ecological research, particularly with closely related species living in the same area, has abundantly demonstrated that the differences between species in form and behaviour are often related to differences in the ecological resources the species exploit. The ecological species concept should be contrasted with the biological, recognition and cladistic species concepts.

Autecological Level (Genecology)

The foregoing discussion on various ecological factors and the species growing in such an environmental complex makes it clear that plants in one way or the other keep themselves adjusted (adapted in their existing environmental conditions.

Aquatic plants are plants that have adapted to living in aquatic environments. They are also referred to as hydrophytes or aquatic macrophytes. These plants require special adaptations for living submerged in water or at the water's surface. Aquatic plants can only grow in water or in soil that is permanently saturated with water. Aquatic vascular plants can be ferns or angiosperms (from a variety of families, including monocots and dicots).

Mesophytes are terrestrial plants which are adapted to neither a particularly dry nor particularly wet environment. An example of a mesophytic habitat would be a rural temperate meadow, which might contain Goldenrod, Clover, Oxeye Daisy, and *Rosa multiflora*.

Mesophytes make up the largest ecological group of terrestrial plants, and usually grow under moderate to hot and humid climatic regions.

A **xerophyte** or xerophytic organism (xero meaning dry, phyte meaning plant) is a plant which is able to survive in an environment with little available water or moisture, such as a desert. Xerophytic plants may have adaptations of their shape and form (morphology) or internal functions (physiology) that reduce their water loss or store water during periods of dryness. Plants with such morphological adaptations are called xeromorphic.

Ecological amplitude

Different species differ from each other in terms of their demands (requirements) from their environment, and consequently also in respect of the extent to which they can tolerate the fluctuation in their environmental conditions. This range of environmental conditions within which a species shows its characteristic potentially for growth is known as its ecological amplitude. Ecological amplitude is sometimes also called tolerance range though the latter indicates the extreme conditions within which a species survives. Accordingly the species may have a low ecological amplitude, if its range of demands and tolerance are narrow, or may have a high ecological amplitude if the range of demands and tolerance are wider. The species with high ecological amplitude are better adapted to greater fluctuations in their environments. These species are found in nature into several morphological forms in various habitat conditions.

Limiting factors

Whatever limits the growth in size of an individual or in numbers of a population is known as a limiting factor to that individual or population. The ecological principle of limiting factors is stated by EP Odum as follows: 'The presence and success of an organism or a group of organisms depends upon a number of complex conditions. Any condition which approaches or exceeds the limits of tolerance is said to be a limiting condition or a limiting factor'.

Limiting factors are of two types, viz. physical and biological. Physical factors that limit population growth would include factors of climate and weather, the absence of water or presence of excess water, the availability of essential soil minerals, and so on. Biological factors involve competition, predation, parasitism, disease, and other interactions between or within a species that are limiting to growth or increases.

Ecads: These are variable produced in an essentially homogenous genetic stock under changed environmental conditions. These variations are simply environmentally induced and are temporary or reversible.

Ecotypes: These are locally adapted populations of species adjusted to local environmental conditions. These variations are heritable. Turesson described ecotypes as 'genetic varieties within a species'.

Ecospecies

Ecospecies are species consisting of different subspecies, or breeds, of an organism which despite being adapted to slightly different environments and/or having distinctly different appearances and behaviours, can still successfully interbreed. Pertaining to a species which has a diversified over time, where each subspecies involved is still capable of reproducing fertile offspring with one another.

Tolerance range, acclimation

Habitat represents a particular set of environmental factors in any physical space where any species live in nature. Different factors of a habitat are investigated. We may be able to determine the whole range over which the species is able to live, for one of many factors. We then will know the range of tolerance of that species for that factor. A goldfish, for example, might be able to live in waters in which the temperature ranges from 2° to 34°C but it will die of heat or cold at temperatures above or below this range. Tolerance ranges differ from one species to another. Even for the same species, range of tolerance for any factor may vary from season to season. For example, the range of temperature tolerance of a fish or insect or pine tree may vary from season to season, depending on the temperature at which it has recently been living. In midwinter, a fish may be able to live in the range from 0° to 24°C but in summer the same individual may have a tolerance range extending up to 33°C but down only to 15°C. Such adjustments in the ecological response to a changed environment are known as acclimation. Acclimation is indeed of much help in allowing organisms to survive permanently in changing environments. Without this ability many organisms would either die or be forced to migrate to other areas during unfavourable seasons. Of course, some organisms as birds do show these two adaptations.

ECOSYSTEM-LEVEL PROCESSES AND THE CONSEQUENCES OF BIOLOGICAL INVASIONS

Numerous studies demonstrate that biological invasions by exotic species can alter the population dynamics and community structure of native ecosystems. However, considerably less information is available for the ecosystem-level consequences of biological invasions. In this section, we define ecosystem-level changes as those that alter the fluxes of water and energy or the cycling and loss of material. Commonly studied characteristics such as primary productivity, decomposition, secondary production, mineral nutrient availability, and hydrological balances would be included in this definition, as would the type and frequency of disturbance. Changes in these properties can alter the conditions of life for all of the organisms in an ecosystem. Our emphasis on ecosystem-level characteristics of invasions is based in part on the need to evaluate the potential impacts of invasions on ecosystems from the point of view of managing or mitigating those effects. However, we believe that this emphasis also addresses fundamental concerns in ecosystem-level ecology. For example, there is a considerable debate on the functional significance of individual species in ecosystems. We believe that a demonstration of widespread ecosystem-level consequences of biological invasions would constitute an explicit demonstration that species make a difference on the ecosystem level, and would further suggest ways in which to integrate the often disparate approaches of population biology and ecosystem-level ecology.

Limited research on ecosystem-level consequences of biological invasions has been carried out, and some of the generalisations that can be drawn from that work are summarised by Vitousek. Exotic animals, especially mammals, clearly alter ecosystems in many areas.

Determining the ecosystem-level consequences of invasions by plants is more difficult, although some clear examples of major effects are well documented. In part, this distinction between the ecosystem-level effects of invading plants and animals probably reflects a real difference—most exotic animals have a large effect on native ecosystems than do most exotic plants.

One reason for this difference may be that invading plants most often occupy disturbed habitats, especially in sites altered by humans. It can be difficult to separate the ecosystem-level effects of biological invasions from those of the disturbance that created the invaded habitats. The consequences of biological invasions in disturbed sites and secondary succession represent at once the most difficult and the most important area of research.

It is here that the effects of invasions are most easily confounded with those of the massive, prolonged or novel disturbances which often form the invaded habitat, here that economic effects on humans are most important, and here that changes in ecosystem-level characteristics are most rapid even in the absence of biological invasions. In this section, we will briefly review the ecosystem-level consequences of biological invasions into intact native ecosystems and into primary succession.

Invasions of Intact Ecosystems

The generalisation that biological invasions are primarily successful in disturbed habitats is well supported as it applies to plants, but exotic animals frequently invade intact native ecosystems and cause significant changes in ecosystem-level properties of the areas invaded. Pigs probably provide the best example.

Many other examples of the effects of exotic animals have been documented. Goats remove vegetative cover and cause increased soil erosion on many oceanic islands. Even exotic insects are likely to be important. The ant *Pheidole megacephala* totally changed the lowland invertebrate fauna (including many of the pollinators and organisms on the decomposer food chain) of Hawaii about 100 years ago and the Argentinian ant (*Iridomyrmex humilis*) is now doing so at higher elevations.

Invasions by exotic plants into intact native ecosystems are less common, but they do occur. For example, Eurasian phreatophytes of the genus *Tamarix* invade both natural water courses and reservoir margins in the arid southwestern United States. Their transpiration is much more rapid than that of native communities, and they can convert marshes which support surface water during part of the year into wholly dry areas. In one case, removal of *Tamarix* led to regeneration of a marsh. The floating aquatic plant *Salvinia molesta* can change parts of tropical river systems into thick masses of live and dead plant material. Exotic grasses in seasonal montane forests or shrublands can increase fire frequency or intensity, thereby altering ecosystems in a way that favours increased dominance by the grasses. Finally, exotic trees can invade certain native shrublands and woodlands, thereby altering rates of productivity and nutrient circulation. Exotic plants can also interact with exotic animals in ways that facilitate invasions of disturbed habitats. Invasions of seemingly intact mature ecosystems by exotic plants often take place in close association with the activities of exotic animals. For example, guavas (*Psidium guajava* and *Psidium cattleianum*) are common and serious invaders of oceanic islands worldwide. Their dispersal and success are closely tied to the movements and soil-disturbing activities of cattle and pigs. Similarly, feral pigs consume and disperse the exotic vine *Passiflora mollisima* in Hawaii, and then deposit the seeds within mounds of organic fertiliser in seedbeds cleared by the pigs rooting activity. *Passiflora mollisima* then apparently affects patterns of mineral cycling in montane Hawaiian rainforests.

Overall, detailed studies of the population biology and plant—animal interactions of that subset of invaders able to colonise intact native ecosystems would be most useful. To the extent that these species are able to alter ecosystem-level characteristics, they are likely to have impacts disproportionate to their numbers.

Invasions of Primary Succession

Primary succession involves the development of ecosystems on areas that are free of the influence of previous biotic communities. Primary succession can be initiated by glacial recession, eolian activity, river meanders or vocanic activity; some human activities (such as strip mining) reproduce it quite closely.

The effects of biological invasions upon primary succession are relatively little-studied, in large part because primary succession does not occur over large areas at present. It is slower and often much simpler (fewer species in the early stages, more predictable patterns of soil-plant interaction) than secondary succession, however, so it may lend itself better to studies of the effects of biological invasions. For example, nitrogen availability is generally low early in primary succession. The exotic symbiotic nitrogen fixer *Myrica faya* is now invading young volcanic substrates in Hawaii and substantially altering nitrogen availability there; it is likely to alter the course of soil development substantially. Australian species of Acacia may have a similar effect in South Africa.

Casuarina equisetiifolia invades beaches in south Florida and in many oceanic islands. Its effects on nitrogen availability have been not studied, but it significantly alters the form of shorelines and patterns of beach erosion.

Invasions of Secondary Succession

Exotic species are prominent in disturbed sites throughout the world, on continents as well as islands and in the tropics as well as the temperate zone. The association between disturbed sites and invaders may be due to greater invasiveness on the part of successional species or it may simply reflect the association between humans and disturbance which gives early successional species greater opportunities to invade. In either case, it is often difficult or impossible to determine how invaders have altered secondary succession because it is difficult to know what secondary succession was like before humans altered disturbance regimes and brought in exotic species.

Nevertheless, there are several documented cases in which biological invaders have altered ecosystem processes during secondary succession. The phenology of exotic *Andropogon* in some areas of Hawaii does not match the seasonal distribution of rainfall, and in consequence boggy conditions develop in invaded sites. The ice-plant *Mesembryanthemum crystallinum* invades degraded pastures in California and Australia; once established it redistributes salt from throughout the rooting zone onto the soil surface, thereby interfering with the growth of other species and eventually increasing soil erosion. In fact, invaders may alter ecosystem processes in secondary succession wherever they can obtain resources the natives cannot or wherever they differ substantially from the natives in resource use efficiency— but it is difficult to document the effects of many invasions for the reasons outlined above.

One type of disturbance which lends itself well to studies of the effects of invasion on secondary succession is shifting cultivation. It is practiced over large areas of the tropics, and in many of these areas it has been the major form of disturbance for millenia. The consequences of biological invasions in these relatively well-defined successional systems are therefore more understandable than is true in systems with novel disturbance regimes.

Ecological niche

Ecological niche is a term for the position of a species within an ecosystem, describing both the range of conditions necessary for persistence of the species, and its ecological role in the ecosystem. Ecological niche subsumes all of the interactions between a species and the biotic and abiotic environment, and thus represents a very basic and fundamental ecological concept. The tentative definition presented above indicates that the concept of niche has two sides which are not so tightly related: one concerns the effects environment has on a species, the other the effects a species has on the environment. In most of ecological thinking, however, both meanings are implicitly or explicitly mixed. The reason is that ecology is about interactions between organisms, and if persistence of a species is determined by the presence of other species (food sources, competitors, predators, etc.), all species are naturally both affected by environment, and at the same time affect the environment for other species.

If we want to treat both of these aspects of ecological niche within one framework, we can define it more formally as the part of ecological space (defined by all combinations of biotic and abiotic environmental conditions) where the species population can persist and thus utilise resources and impact on its environment. It is useful, however, to distinguish three main approaches to the niche. The first approach emphasises environmental conditions necessary for a species presence and maintenance of its population, the second approach stresses the functional role of species within ecosystems, and the third one a dynamic position of species within a local community, shaped by species' biotic and abiotic requirements and by coexistence with other species.

Concepts of niche

Niche as the description of a species' habitat requirements

The first formulations of the concept of ecological niche were close to the general meaning of the term: the ecological niche was defined by a place a species can take in nature, determined by its abiotic requirements, food preferences, microhabitat characteristics (for example a foliage layer), diurnal and seasonal specialisation, or predation avoidance. This concept is associated mostly with Joseph Grinnell, who first introduced the term. He was especially interested in factors determining where we can find a given species and how niches, generated by the environment, are filled. The knowledge of a species niche determined by its habitat requirements is essential for understanding and even predicting its geographic distribution; this concept of the niche is thus more relevant in biogeography and macroecology than in community or ecosystem ecology.

Niche as ecological function of the species

In this concept of niche, each species has a particular role in an ecosystem and its dynamics, and one such role can be fulfilled by different species in different places. The observation of distant species adapted to equivalent ecological roles (the resemblance between jerboa and kangaroo rat, between many eutherian and marsupial species or the Galapagos finches diversifying to highly specialised roles including those normally taken by woodpeckers) was clearly influential to Charles Elton, who emphasised the functional roles of species. According to Elton, there is the niche of burrowing detritivores, the niche of animals specialising in cleaning ticks or other parasites, or the pollination niche. Elton's niche can apply to several species, for example 'the niche filled by birds of prey which eat small mammals'. This functional niche therefore refers to a species position in food webs and trophic chains, and the concept is thus especially relevant for ecosystem ecology.

Niche as a species position in a community—formalisation of ecological niche concept

The emphasis on the diversity of ecological communities and interspecific competition within them in the second half of 20th century has lead to the formalisation of niche concept, and an emphasis on the properties of the niches which enable species coexistence within a habitat. George Evelyn Hutchinson postulated that niche is a 'hypervolume' in multidimensional ecological space, determined by a species requirements to reproduce and survive. Each dimension in the niche space represents an environmental variable potentially or actually important for a species persistence. These variables are both abiotic and biotic, and can be represented by simple physical quantities as temperature, light intensity or humidity, but also more sophisticated quantities such as soil texture, ruggedness of the terrain, vegetation complexity or various measures of resource characteristics. This could be viewed simply as a formalisation of original Grinnellian niche, i.e. the exact descriptions of a species habitat requirements. However, in the Hutchinsonian view ecological niches are dynamic, as the presence of one species constrains the presence of another species by interspecific competition, modifying the position of species' niches within the multidimensional space. This concept therefore combines the ecological requirements of the species with its functional role in the local community.

Ecological equivalent species

Ecologically equivalent species should exhibit great similarities in species attributes (e.g. life-form, size, life-span, fruit type, dispersal agent, etc.), in site and climatic requirements as well as in successional status. Generally, only detailed investigations can differentiate ecologically equivalent species, thus inferences can only be drawn from a limited number of examples as the autecology of invasive and native plants is usually lacking.

A good example of ecologically equivalent species is provided by the two European timber trees *Acer pseudoplatanus* L. and *Fraxinus excelsior* L. both native to Europe, with the former being invasive in many parts of the world. Their autecology is markedly similar but they differ mainly in terms of duration in seed dormancy, tolerance to flooding, susceptibility to grass competition and variation in latitudinal and altitudinal distributions.

Ecological equivalence in the tropics

A review of the literature of well-documented invasive woody species in the tropics shows that in only one case is the existence of an ecologically equivalent species a possibility. In tropical Africa the introduced *Cecropia peltata* is difficult to differentiate from the native *Musanga cecropioides* from a morphological and autecological point of view. However, the former species has been reported as displacing the latter. In all other cases the existence of ecologically equivalent species in the invaded region is wanting.

A number of introduced species appear to possess attributes not found in the local flora, such as mycorrhizal associations and life-history characteristics or have life-forms not encountered at all. These unique characteristics must provide the invasive species with some competitive advantage over native species, although this has yet to be demonstrated.

Chapter 4

Green Chemistry and Environmental Friendly Technologies

INTRODUCTION

Green chemistry is the universally accepted term to describe the movement towards more environmentally acceptable chemical processes and products. It encompasses education, research, and commercial application across the entire supply chain for chemicals. Green chemistry can be achieved by applying environmentally friendly technologies—some old and some new. While Green chemistry is widely accepted as an essential development in the way that we practice chemistry, and is vital to sustainable development, its application is fragmented and represents only a small fraction of actual chemistry. It is also important to realise that Green chemistry is not something that is only taken seriously in the developed countries. Again, however, it is important to realise that there were many more good examples of Green chemistry at work long before this—for example, commercial, no-solvent processes were operating in Germany and renewable catalysts were being used in processes in the UK but they did not get the same publicity as those in the United States.

The developing countries that are rapidly constructing new chemical manufacturing facilities have an excellent opportunity to apply the catchphrase of Green chemistry 'Benign by Design' from the ground upwards. It is much easier to build a new, environmentally compatible plant from scratch than to have to deconstruct before reconstructing, as is the case in the developed world.

Green chemistry is the utilisation of a set of principles that reduces or eliminates the use or generation of hazardous substances in the design, manufacture and application of chemical products. The goal of Green chemistry is to design synthetic methods that reduce or eliminate the use of toxic substances, wastes, solvents and other auxiliaries; it is a tool to minimise the negative impacts of the chemicals and processes involved in the production of the chemicals.

Green chemistry is a responsible way of using science and engineering that strive to improve the public image of chemistry, not as a goal in itself but as a consequence of its achievements. During the twentieth century, chemists were able to master synthetic chemistry and could virtually make all possible chemicals found in nature. While this has been a dazzling accomplishment, the information and education on toxicity and ecotoxicity of the materials and molecules has been almost completely neglected. As a result, chemistry has long been practiced without limiting its harmful consequences upon the environment and society. The net result of this gross negligence is that today issues related to sustainability have become quite grim. Against this scenario, the approach of green chemistry comes as a major respite.

The constant and at times even bitter fight between industry and environmental organisations need not be contentious anymore because green chemistry enables the environmental movements and the industrialists to work cooperatively. Green chemistry gives equal respect to the cause of environment protection and industrial success.

Various pesticides, insecticides and colours are drained out from agricultural lands, chemical/ pharmaceutical and other industries, and tanneries, etc. which ultimately contribute to all three: water, air and soil pollution. These and many more such examples are enough reasons for incorporation of green chemistry methodologies, techniques and practices.

Green chemistry not only includes the shift of the use of harmful chemicals but also directs to set the usage of systems where the use of chemicals could be minimised with maximum atom economy—a concept introduced by Green chemistry. Thus, this science involves the design and redesign of chemical synthesis and processes to prevent pollution, rendering the products cost effective as well as environment-friendly.

EMERGENCE OF GREEN CHEMISTRY

In this chapter we shall start by exploring the drivers behind the movement towards Green and Sustainable Chemistry. These can all be considered to be 'costs of waste' that effectively penalise current industries and society as a whole. It is important that, while reading this, we see Green chemistry in the bigger picture of sustainable development as we seek to somehow satisfy society's needs without compromising the survival of future generations.

Green chemistry, is very appropriately known as sustainable chemistry. The reason for its rapid adoption around the world is the realisation of it being a pathway to ensure economic and environmental prosperity. Green chemistry starts at the molecular level and ultimately delivers more environmentally benign products and processes.

PRINCIPLES OF GREEN CHEMISTRY

The twelve principles of green chemistry are:

1. Prevent waste: Design chemical syntheses to prevent waste, leaving no waste to treat or clean up.
2. Design safer chemicals and products: Design chemical products to be fully effective, yet have little or no toxicity.
3. Design less hazardous chemical syntheses: Design syntheses to use and generate substances with little or no toxicity to humans and the environment.
4. Use renewable feedstocks: Use raw materials and feedstocks that are renewable rather than depleting. Renewable feedstocks are made from agricultural products or the wastes of other processes. Depleting feedstocks are made from fossil fuels (petroleum, natural gas, or coal) or are mined.
5. Use catalysts, not stoichiometric reagents: Minimise waste by using catalytic reactions. Catalysts are used in small amounts and can carry out a single reaction many times. They are preferable to stoichiometric reagents, which are used in excess and work only once.
6. Avoid chemical derivatives: Avoid using blocking or protecting groups or any temporary modifications if possible. Derivatives use additional reagents and generate waste.
7. Maximise atom economy: Design syntheses so that the final product contains the maximum proportion of the starting materials. There should be few, if any, wasted atoms.
8. Use safer solvents and reaction conditions: Avoid using solvents, separation agents or other auxiliary chemicals. If these chemicals are necessary, use innocuous chemicals.

9. Increase energy efficiency: Run chemical reactions at ambient temperature and pressure whenever possible.

10. Design chemicals and products to degrade after use: Design chemical products to break down to innocuous substances after use so they do not accumulate in the environment.

11. Analyse in real time to prevent pollution: Include in-process real-time monitoring and control during syntheses to minimise or eliminate the formation of by-products.

12. Minimise the potential for accidents: Design chemicals and their forms (solid, liquid, or gas) to minimise the potential for chemical accidents including explosions, fires, and releases to the environment.

According to the Michigan Green Chemistry Roundtable, the scope and boundaries of a Green Chemistry Program should:

1. Consider all the stages of the life cycle of a chemical.

2. Focus on 'hazard reduction' as the primary impact category of interest in each life cycle stage with a focus on the design stage. Other life-cycle impacts of innovation should also be considered.

3. Reduce hazards to human and ecosystem health.

OBJECTIVES FOR GREEN CHEMISTRY: THE COSTS OF WASTE

The basic building blocks in multi-step synthesis are mostly aromatic compounds with incorporation of such groups as halo, $-NH_2$, $-OH$, $-NO_2$, $-SO_3H$, alkyl, acryl, aryl, etc. which change the reactivity of the cyclic compound and sometimes the colour and odour. Multi-step synthesis is typically done via routes which do not take into account the atom economy or the amount and nature of co-products or by-products. Catalysis, if any, is limited to homogeneous catalysis involving highly corrosive substances posing disposal problems. Use of solvents is also very common. Low boiling, halogen containing hazardous inflammable solvents are used.

Some of the important precursors used in these industries are: aniline, acetanilide, benzyl chloride, benzoic acid, monochlorobenzene, chloronitrobenzenes, cresols, cresylic acid, cumene, cyclohexane, cyclohexanone, dichlorobenzenes, dihydroxybenzenes, dimethyl sulphate, ethylbenzene, bisphenol-A, alpha-methylstyrene, nitrobenzene, nonyl-phenol, xylenes, phenol, phthalic anhydride, pyridine, picolines, salicylic acid, etc.

Most of these are hazardous to handle and wasted during separation or washings. Irrespective of the best solutions to reduce waste at source, it may not be totally avoided and in such cases, it should provide an impetus to convert the waste into an asset. Plant engineers should look for integrated facility to convert liabilities into assets. The environmental challenges faced by these industries during the 21st century will be:

1. Development of theoretically zero waste (or minimum waste) processes.

2. Minimisation of hazardous products and greenhouse gases.

3. Replacement of corrosive and nonreusable catalysts.

4. Evolution of sustainable systems; zero energy input processes.

5. Development of single-pot or cascade engineered processes.

6. Reduction in number of steps.

7. Minimum use of solvents, environmentally benign solvents or solvent-free synthesis.

8. Design of electrically engineered systems (catalysts, different forms of energies, ultrasound; microwave radiation).

9. Replacement of petroleum feedstock by renewable resources.

Hundreds of tons of hazardous waste are released to the air, water, and land by industry every hour of every day. The chemical industry is the biggest source of such waste. Ten years ago less than 1 per cent of commercial substances in use were classified as hazardous, but it is now clear that a much higher proportion of chemicals presents a danger to human health or to the environment. The relatively small number of chemicals formally identified as being hazardous was due to very limited testing regulations, which effectively allowed a large number of chemicals to be used in everyday products without much knowledge of their toxicity and environmental impact. New legislation will dramatically change that situation.

In Europe, Registration, Evaluation, Assessment of Chemicals (REACH) will come into force in the first decade of the twenty-first century. REACH will considerably extend the number of chemicals covered by regulations, notably those that have been on market since 1981 (previously exempt), will place the responsibility for chemicals testing with industry, and will require testing whether the chemical is manufactured in Europe or imported for use there. Apart from the direct costs to industry of testing, REACH is likely to result in some chemical substances becoming restricted, prohibitively expensive, or unavailable. This will have dramatic effects on the supply chain for many consumer goods that rely on multiple chemical inputs.

Increased knowledge about chemicals, and the classification of an increasing number of chemical substances as being in some way 'hazardous', will have health and safety implications, again making the use of those substances more costly and difficult. Furthermore, it will undoubtedly cause local authorities and governments to restrict and increase the costs of disposal of waste containing those substances (or indeed waste simply coming from processes involving such substances). Thus, legislation will increasingly force industry and the users of chemicals to change—both through substitution of hazardous substances in their processes or products and through the reduction in the volume and hazards of their waste. The costs of waste to a chemical manufacturing company are high and diverse (Fig. 4.1) and, for the foreseeable future, they will get worse. These costs and other pressures are now evident throughout the supply chain for a chemical product—from the increasing costs of raw materials, as petroleum becomes more scarce and carbon taxes penalise their use, to a growing awareness amongst end-users of the risks that chemicals are often associated with, and the need to disassociate themselves from any chemical in their supply chain that is recognised as being hazardous (e.g. phthalates, endocrine disrupters, polybrominated compounds, heavy metals, etc. Fig. 4.2).

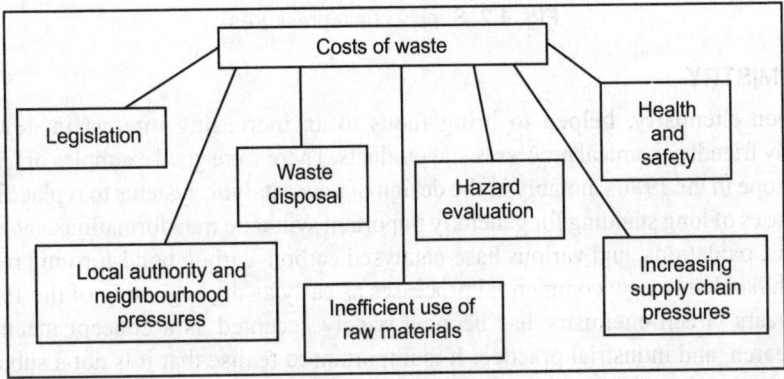

Fig. 4.1. The cost of waste.

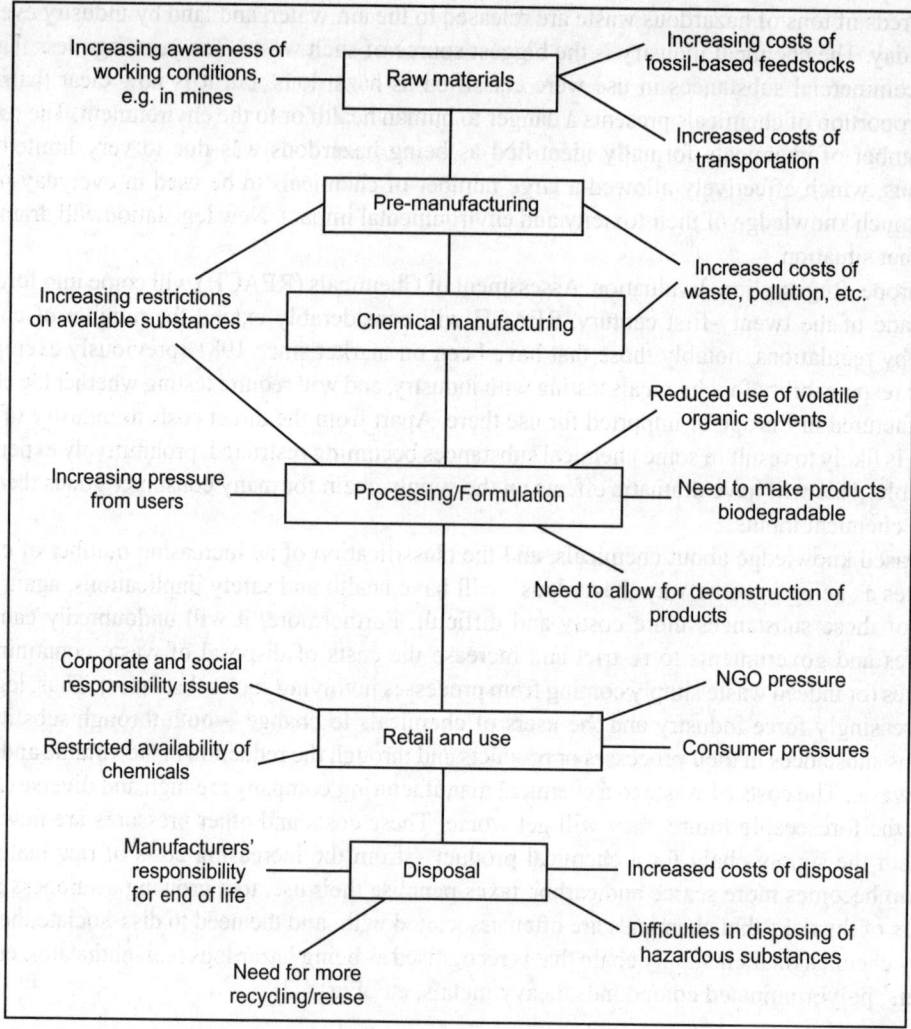

Fig. 4.2. Supply chain pressures.

GREEN CHEMISTRY

The term Green chemistry, helped to bring focus to an increasing interest in developing more environmentally friendly chemical processes and products. There were good examples of Green chemistry research in Europe in the 1980s, notably in the design of new catalytic systems to replace hazardous and wasteful processes of long standing for generally important synthetic transformations, including Friedel–Crafts reactions, oxidations, and various base-catalysed carbon–carbon bond-forming reactions. Some of this research had led to new commercial processes as early as the beginning of the 1990s.

In recent years Green chemistry has become widely accepted as a concept meant to influence education, research, and industrial practice. It is important to realise that it is not a subject area in the way that organic chemistry is. Rather, Green chemistry is meant to influence the way that we practice chemistry—be it in teaching children, researching a route to an interesting molecule, carrying out an analytical procedure, manufacturing a chemical or chemical formulation, or designing a product. Green chemistry has been promoted worldwide by an increasing but still small number of dedicated individuals and through the activities of some key organisations. These include the Green Chemistry Network (GCN).

At about the same time as the establishment of the GCN, the Royal Society of Chemistry (RSC) launched the journal 'Green chemistry'. The intention for this journal was always to keep its readers aware of major events, initiatives, and educational and industrial activities, as well as leading research from around the world.

Green chemistry can be considered as a series of reductions (Fig. 4.3). These reductions lead to the goal of triple bottom-line benefits of economic, environmental, and social improvements. Costs are saved by reducing waste (which is becoming increasingly expensive to dispose of, especially when hazardous) and energy use (likely to represent a larger proportion of process costs in the future) as well as making processes more efficient by reducing materials consumption. These reductions also lead to environmental benefit in terms of both feedstock consumption and end-of-life disposal. Furthermore, an increasing use of renewable resources will render the manufacturing industry more sustainable. The reduction in hazardous incidents and the handling of dangerous substances provides additional social benefit—not only to plant operators but also to local communities and through to the users of chemical-related products.

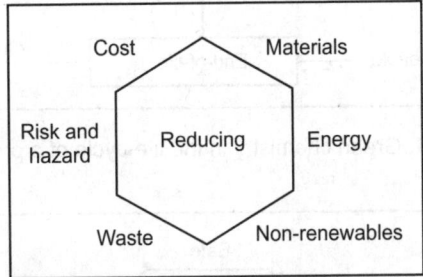

Fig. 4.3. 'Reducing': The heart of Green chemistry.

It is particularly important to seek to apply Green chemistry throughout the life-cycle of a chemical product (Fig. 4.4).

Scientists and technologists need to routinely consider life-cycles when planning new synthetic routes, when changing feedstocks or process components, and, fundamentally, when designing new products. Many of the chemical products in common use today were not constructed for end-of-life nor were full supply-chain issues of resource and energy consumption and waste production necessarily considered. The Green chemistry approach of 'benign by design' should, when applied at the design stage, help assure the sustainability of new products across their full life-cycle and minimise the number of mistakes we make. Much of the research effort relevant to Green chemistry has focused on chemical manufacturing processes. Here we can think of Green chemistry as directing us towards the 'ideal synthesis' (Fig. 4.5).

Yield is the universally accepted metric in chemistry research for measuring the efficiency of a chemical synthesis. It provides a simple and understandable way of measuring the success of a synthetic route and of comparing it to others. Green chemistry teaches us that yield is not enough. It fails to allow for reagents that have been consumed, solvents and catalysts that will not be fully recovered, and, most importantly, the often laborious and invariably resource- and energy-consuming separation stages such as water quenches, solvent separations, distillations, and recrystallisations. Green chemistry metrics are now available and commonly are based on 'atom efficiency' whereby we seek to maximise the number of atoms introduced into a process into the final product.

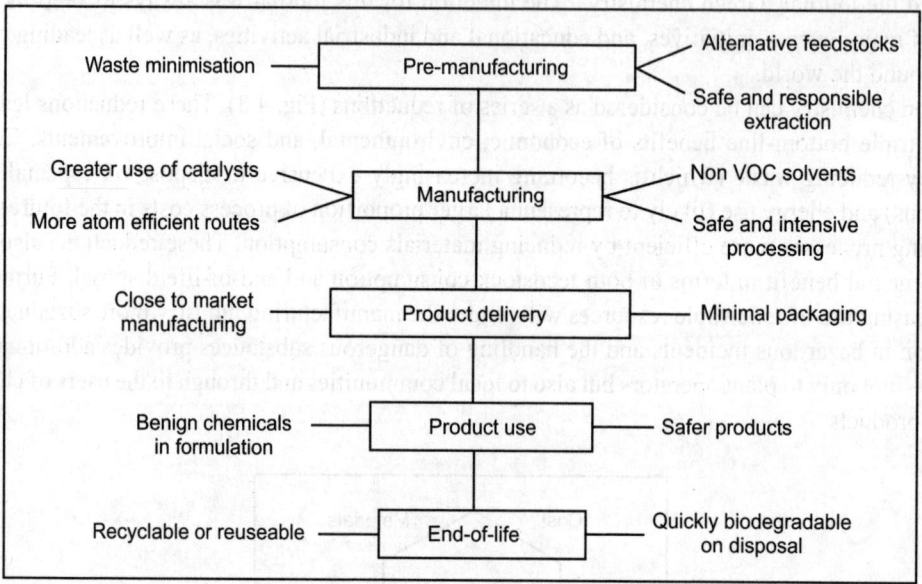

Fig. 4.4. Green chemistry in the life-cycle of a product.

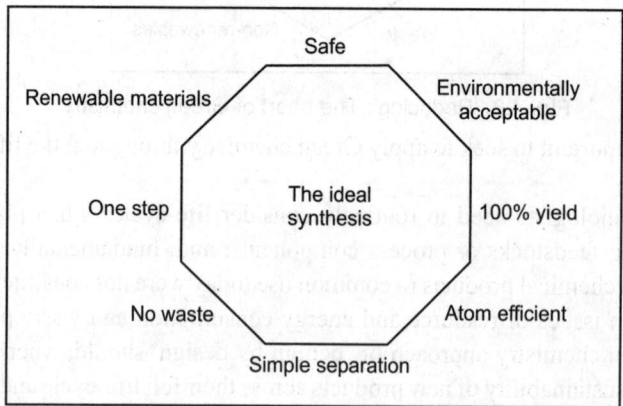

Fig. 4.5. Features of the 'ideal synthesis'.

These are discussed in more detail later in this chapter. As indicated, simple separation with minimal input and additional outputs is an important target. An ideal reaction from a separation standpoint would be one where the substrates are soluble in the reaction solvent but the product is insoluble. The process would, of course, be further improved if no solvent was involved at all. Some of the worst examples of atom inefficiency and relative quantities of waste are to be found in the pharmaceutical industry. The so-called E factor (total waste/product by weight) is a simple but quite comprehensive measure of process efficiency and commonly shows values of 100+ in drug manufacture. This can be largely attributed to the complex, multistep nature of these processes. Typically, each step in the process is carried out separately with workup, isolation, and purification all adding to the inputs and amount of waste produced. Simplicity in chemical processes is vital to good Green chemistry. Steps can be 'telescoped' together for example, reducing the number of discrete stages in the process.

To achieve greener chemical processes we will need to make increasing use of technologies, some old and some new, which are becoming proven as clean technologies.

ENVIRONMENTAL FRIENDLY TECHNOLOGIES

There is a pool of technologies that are becoming the most widely studied or used in seeking to achieve the goals of Green chemistry. The major 'clean technologies' are summarised in Fig. 4.6. They range from well-established and proven technologies through to new and largely unproven technologies.

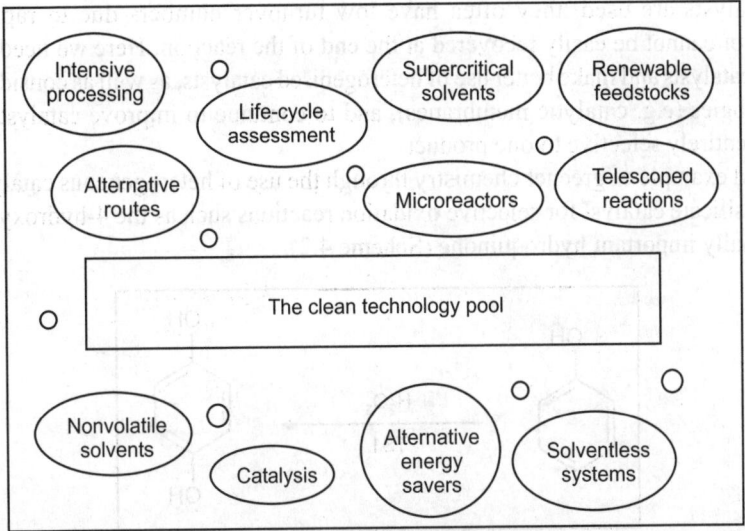

Fig. 4.6. The major clean technologies.

Catalysis is truly a well-established technology, well proven at the largest volume end of the chemicals industry. In petroleum refineries, catalysts are absolutely fundamental to the success of many processes and have been repeatedly improved over more than 50 years. Acid catalysts, for example, have been used in alkylations, isomerisations and other reactions for many years and have progressively improved from traditional soluble or liquid systems, through solid acids such as clay, to structurally precise zeolite materials, which not only give excellent selectivity in reactions but are also highly robust, with modern catalysts having lifetimes of up to 2 years. In contrast, the lower volume but higher value end of chemical manufacturing—specialities and pharmaceutical intermediates—still relies on hazardous and difficult routes to separate soluble acid catalysts such as H_2SO_4 and $AlCl_3$ and is only now beginning to apply modern solid acids. Cross-sector technology transfer can greatly accelerate the greening of many highly wasteful chemical processes. A good, if sadly rare, example of this is the use of a zeolite to catalyse the Friedel–Crafts reaction of anisole with acetic anhydride (Scheme 4.1).

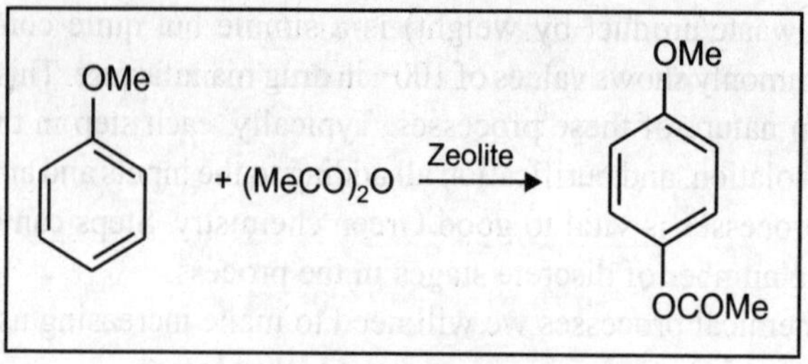

Scheme 4.1.

In comparison to the traditional route using AlCl$_3$, the zeolite-based method is more selective. However, anisole is highly activated and the method is not applicable to most substrates—zeolites tend to be considerably less reactive than conventional catalysts such as AlCl$_3$.

Many speciality chemical processes continue to operate using traditional and problematic stoichiometric reagents (e.g. in oxidations), which we should aim to replace with catalytic systems. Even when catalysts are used, they often have low turnover numbers due to rapid poisoning or decomposition, or cannot be easily recovered at the end of the reaction. Here we need to develop new longer-lifetime catalysts and make better use of heterogenised catalysts, as well as considering alternative catalyst technologies (e.g. catalytic membranes), and to continue to improve catalyst design so as to make reactions entirely selective to one product.

Another good example of greener chemistry through the use of heterogeneous catalysis is the use of TS1, a titanium silicate catalyst for selective oxidation reactions such as the 4-hydroxylation of phenol to the commercially important hydroquinone (Scheme 4.2).

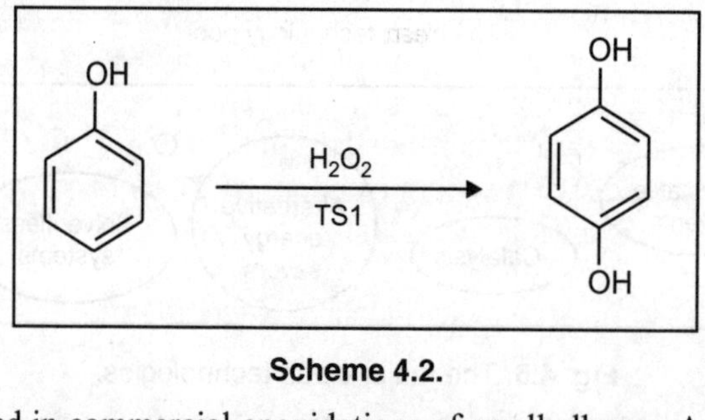

Scheme 4.2.

TS1 has also been used in commercial epoxidations of small alkenes. A major limitation with this catalyst is its small pore size, typical of many zeolite materials. This makes it unsuitable for larger substrates and products. Again like many zeolites, it is also less active than some homogeneous metal catalysts and this prevents it from being used in what would be a highly desirable example of a green chemistry process—the direct hydroxylation of benzene to phenol. The commercial routes to this continue to be based on atom-inefficient and wasteful processes such as decomposition of cumene hydroperoxide, or via sulphonation (Scheme 4.3).

Of course, the direct reaction of oxygen with benzene to give phenol would be 100 per cent atom efficient and based on the most sustainable oxidant—truly an ideal synthesis if we can only devise a good enough catalyst to make it viable.

The increased use of catalysis in the manufacture of low volume, high value chemicals will surely extend to biotechnology and, in particular, the use of enzymes. Enzymes provide highly selective routes to chemical products, often under mild conditions and usually in environmentally benign aqueous media. Drawbacks to their more widespread introduction include slow reactions, low space–time yields and, perhaps most importantly, a lack of familiarity with and even suspicions of the technology from many chemical compounds.

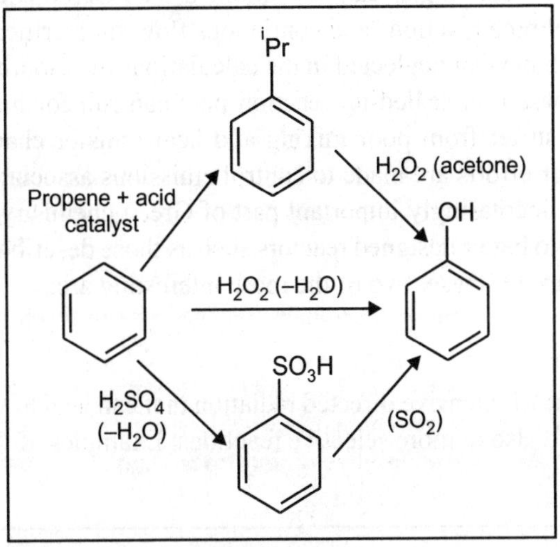

Scheme 4.3.

The replacement of hazardous volatile organic compounds (VOCs) as solvents is one of the most important targets for countless process companies including those operating in chemical manufacturing, cleaning, and formulation. Some VOCs such as carbon tetrachloride and benzene have been widely prohibited and replaced but other problematic solvents, notably dichloromethane (DCM), continue in widespread use. While in many cases other, less harmful, VOCs are used to remove the immediate problems (e.g. ozone depletion) due to such compounds as DCM, more fundamental technology changes have included the use of non-organic compounds such as supercritical carbon dioxide or water, the use of nonvolatile solvents such as ionic liquids (molten salts), and the total avoidance of solvent (e.g. through using a surface-wetting catalyst in a reaction, or simply relying on interfacial reaction occurring between solids).

Carbon dioxide has also been successfully introduced into some dry-cleaning processes and various consumer formulations now no longer contain a VOC solvent.

Green chemistry needs to be combined with more environmentally friendly technologies if step-change improvements are to be made in chemical manufacturing processes. Synthetic chemists have traditionally not been adventurous in their choice of reactors—the familiar round-bottomed flask with a magnetic stirrer remains the automatic choice for most, even when the chemistry they plan to use is innovative e.g. the use of a nonvolatile ionic liquid solvent or a heterogeneous catalyst as an alternative to a soluble reagent. However, an increasing number of research articles describing green chemical reactions are based on alternative reactors including.

1. Continuous flow reactors (a technology that dominates the petrochemical industry but is little utilised in speciality chemical manufacturing).
2. Microchannel reactors whereby reaction volumes are kept small and scale is highly flexible thus reducing hazards and risk.

3. Intensive processing systems such as spinning disc reactors which combine the benefits of low reaction volumes with excellent heat transfer and mixing characteristics.

4. Membrane reactors that can maintain separation of aqueous and nonaqueous phases, hence simplifying the normally waste-intensive separation stages of a process.

These alternative reactor technologies can be combined with Green chemistry methods including, for example, catalytic membrane reactions and continuous flow supercritical fluid reactions.

Energy has often been somewhat neglected in the calculations of resource utilisation for a chemical process. Batch processes based on scaled-up reaction pots can run for many hours or even days to maximise yield and often suffer from poor mixing and heat transfer characteristics. As the cost of energy increases and greater efforts are made to control emissions associated with generating energy, energy use will become an increasingly important part of Green chemistry metrics calculations. This will open the door not only to better designed reactors such as those described earlier but also to the use of alternative energy sources. Of these, two of the more interesting are:

1. Ultrasonic reactors.
2. Microwave reactors.

Both are based on the use of intensive directed radiation that can lead to very short reaction times or increased product yields and also to more selective reactions. Examples of the use of these reactors are shown in Scheme 4.4.

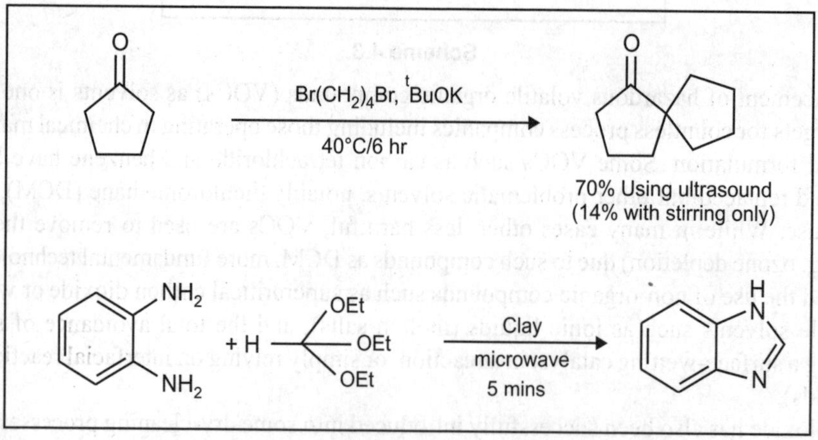

Scheme 4.4.

A life-cycle approach to the environmental performance and sustainability of chemical products demands a proper consideration of pre-manufacturing and specifically the choice of feedstocks. Today's chemical industry is largely based on petroleum-derived starting materials, a consequence of the rapid growth in the new petroleum-based energy industry in the early twentieth century. This industry was based on an apparently inexhaustible supply of cheap oil, which we could afford to use on a once-only basis for burning to produce energy. Petrochemicals was a relatively small (around 10 per cent) part of the business, generating a disproportionately high income and helping to keep energy costs down, which in turn maintained ultra-high demand for the raw material even when extraction became more difficult and transportation more controversial. The parallel and mutually supportive growth in petro-energy and petrochemicals from the petro-refineries of the Middle East, Americas, Africa, and elsewhere is surely past its peak. It now seems likely that as we try to tackle the inevitable decline of oil as an energy source, so shall we attempt to seek alternatives for the manufacture of at least some of the many chemicals we use today. While forecasts seem to change every day and political parties can selectively use bits of the overwhelming amount of conflicting data to suit their own agenda, no one will argue that these changes must occur in the twenty-first century— 'one hundred years of petroleum' is beginning to look about right. The use of sustainable, plant-based chemicals for future manufacturing can involve several approaches (Fig. 4.7).

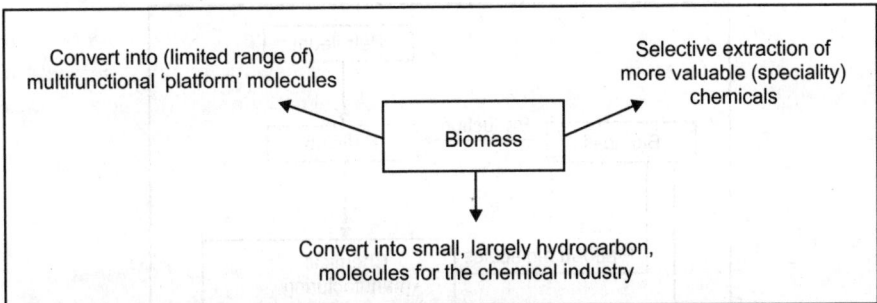

Fig. 4.7. Approaches to the use of plant-based chemicals.

Many of the earlier plans in this area were based on the bulk conversion of large quantities of biomass into the type of starting materials that the chemical industry has grown up on (CO, H_2, C_2H_4, C_6H_6, etc.). On one hand the logic behind this approach is clear—the manufacturing industries are equipped to work with such simple small molecules. On the other hand, it is perverse to consume resources and generate waste in removing functionality from albeit a soup of molecules, just so that we can then apply our chemical technology toolkit to consume more resources and generate more waste in converting the intermediate simpler molecules into ones we can use in the many industries that use chemicals. The scale of operation, and the added costs of the extra steps, will always make this technology expensive and of limited appeal except in those situations where a large volume of waste biomass is in close proximity to suitable industrial plant.

Nature manufacturers an enormous array of chemicals to perform the many functions that its creatures need to survive, grow, and propagate. A tree contains some 30,000 different molecules ranging from simple hydrocarbons to polyfunctional organics and high molecular weight polymers. Many of these molecules have immediate and sometimes very high value, for example as pharmaceutical intermediates. The selective extraction of compounds from such complex mixtures is, however, often impractical and uneconomic and may lead to a very high environmental impact product as a result of enormous inputs of energy and outputs of waste. The extraction of families of compounds with high value themselves or through Green chemistry modification is a more likely approach to take advantage of some of nature's gifts of sustainable and interesting molecular entities.

The third approach of using a large proportion of biomass to produce so-called 'platform molecules' is worth close consideration. Here, we need to learn how to make best use of a number of medium-sized, usually multifunctional, organic molecules that can be obtained relatively easily by controlled enzymatic fermentation or chemical hydrolysis. The simplest of these is (bio) ethanol; others include levulinic acid, vanillin, and lactic acid. These are chemically interesting molecules in the sense that they can be used themselves or can quite easily be converted into other useful molecules—building on rather than removing functionality.

One of these products, polylactic acid, has become the basis of one of the best recent commercial illustrations of the potential value of this approach. Cargill-Dow now manufacture polylactic acid polymer materials using a starch feedstock. The materials are finding widespread use as versatile, sustainable, and (importantly) biodegradable alternatives to petro-plastics.

Making more direct use of the chemicals in biomass and the functionality they contain, rather than reducing them to simpler, smaller starting materials for synthesis, makes sense from a life-cycle point of view as well as economically (Fig. 4.8).

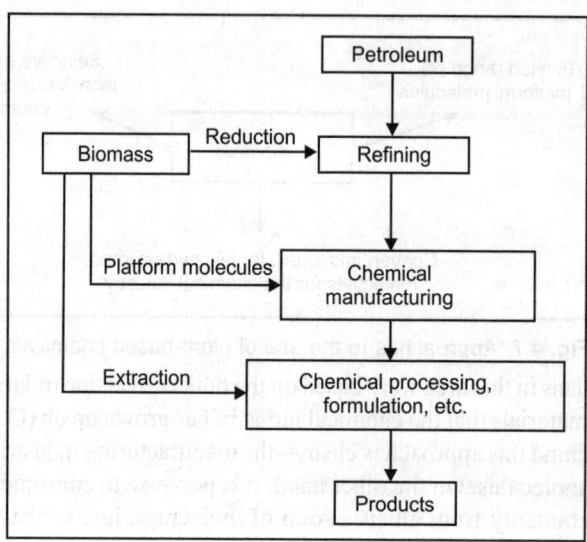

Fig. 4.8. The use of biomass chemicals in traditional chemical industry processes.

Thus, Green chemistry has been heavily focused on developing new, cleaner, chemical processes using the technologies described in this chapter. Increasing legislation will force an increasing emphasis on products but it is important that these in turn are manufactured by green chemical methods. Industry is becoming more aware of these issues and some companies can see the business edge and competitive advantage that Green chemistry can bring. However, the rate of uptake of Green chemistry into commercial application remains very small. While the reasons for this are understandably complex, and also dependent on the economic vitality of the industry, it is important that the advantages offered by Green chemistry can be quantified.

Chapter 5

Industrial Ecology and Its Importance

INTRODUCTION

Industrial ecology conceptualises industry as a man-made ecosystem that operates in a similar way to natural ecosystems, where the waste or by product of one process is used as an input into another process. Industrial ecology interacts with natural ecosystems and attempts to move from a linear to cyclical or closed loop system. Like natural ecosystems, industrial ecology is in a continual state of flux. Industrial ecology is the study of the physical, chemical, and biological interactions and interrelationships both within and between industrial and ecological systems. Additionally, some researchers feel that industrial ecology involves identifying and implementing strategies for industrial systems to more closely emulate harmonious, sustainable, ecological ecosystems Environmental problems are systemic and thus require a systems approach so that the connections between industrial practices/ human activities and environmental/ecological processes can be more readily recognised. A systems approach provides a holistic view of environmental problems, making them easier to identify and solve; it can highlight the need for and advantages of achieving sustainability. Table 5.1 depicts hierarchies of political, social, industrial, and ecological systems. Industrial ecology studies the interaction between different industrial systems as well as between industrial systems and ecological systems. The focus of study can be at different system levels.

Table 5.1. Organisational hierarchies.

Political entities	Social organisations	Industrial organisations	Industrial systems	Ecological systems
UNEP	World population	ISO	Global human material	Ecosphere
US (EPA, DOE)	Cultures	Trade associations	and energy flows	Biosphere
State of Michigan	Communities	Corporations	Sectors (e.g. transpor-	Biogeographical
(Michigan DEQ)	Product systems	Divisions	tation or health care)	region
Washtenaw County	Households	Product develop-	Corporations/institutions	Biome landscape
City of Ann Arbor	Individuals/	ment teams	Product systems	Ecosystem
Individual voter	Consumers	Individuals	Life-cycle stages/unit steps	Organism

One goal of industrial ecology is to change the linear nature of our industrial system, where raw materials are used and products, by-products, and wastes are produced, to a cyclical system where the wastes are reused as energy or raw materials for another product or process.

Fundamental to industrial ecology is identifying and tracing flows of energy and materials through various systems. This concept, sometimes referred to as industrial metabolism, can be utilised to follow material and energy flows, transformations, and dissipation in the industrial system as well as into natural systems. The mass balancing of these flows and transformations can help to identify their negative impacts on natural ecosystems. By quantifying resource inputs and the generation of residuals and their fate, industry and other stakeholders can attempt to minimise the environmental burdens and optimise the resource efficiency of material and energy use within the industrial system.

Industrial ecology is an emerging field. There is much discussion and debate over its definition as well as its practicality. Questions remain concerning how it overlaps with and differs from other more established fields of study. It is still uncertain whether industrial ecology warrants being considered its own field or should be incorporated into other disciplines. This mirrors the challenge in teaching it.

Industrial ecology is rooted in systems analysis and is a higher level systems approach to framing the interaction between industrial systems and natural systems. This systems approach methodology can be traced to the work of Jay Forrester at MIT in the early 1960s and 70s; he was one of the first to look at the world as a series of interwoven systems. Donella and Dennis Meadows and others furthered this work in their seminal book 'Limits to Growth'. Using systems analysis, they simulated the trends of environmental degradation in the world, highlighting the unsustainable course of the then-current industrial system.

In 1989, Robert Ayres developed the concept of industrial metabolism: the use of materials and energy by industry and the way these materials flow through industrial systems and are transformed and then dissipated as wastes. By tracing material and energy flows and performing mass balances, one could identify inefficient products and processes that result in industrial waste and pollution, as well as determine steps to reduce them. Robert Frosch and Nicholas Gallopoulos, in their important article 'Strategies for Manufacturing', developed the concept of industrial ecosystems, which led to the term industrial ecology. Their ideal industrial ecosystem would function as 'an analogue' of its biological counterparts. This metaphor between industrial and natural ecosystems is fundamental to industrial ecology. In an industrial ecosystem, the waste produced by one company would be used as resources by another. No waste would leave the industrial system or negatively impact natural systems.

DEFINING INDUSTRIAL ECOLOGY

There is still no single definition of industrial ecology that is generally accepted. However, most definitions comprise similar attributes with different emphases. These attributes include the following:

1. A systems view of the interactions between industrial and ecological systems.
2. The study of material and energy flows and transformations.
3. A multidisciplinary approach.
4. An orientation toward the future.
5. A change from linear (open) processes to cyclical (closed) processes, so the waste from one industry is used as an input for another.
6. An effort to reduce the industrial systems environmental impacts on ecological systems.
7. An emphasis on harmoniously integrating industrial activity into ecological systems.
8. The idea of making industrial systems emulate more efficient and sustainable natural systems.
9. The identification and comparison of industrial and natural systems hierarchies, which indicate areas of potential study and action (Table 5.1).

There is substantial activity directed at the product level using such tools as life-cycle assessment and life-cycle design and utilising strategies such as pollution prevention. Activities at other levels include tracing the flow of heavy metals through the ecosphere.

A cross-section of definitions of industrial ecology is provided in Appendix A. Further work needs to be done in developing a unified definition. Issues to address include the following:

1. Is an industrial system a natural system? Some argue that everything is ultimately natural.
2. Is industrial ecology focusing on integrating industrial systems into natural systems or is it primarily attempting to emulate ecological systems? Or both?
3. Current definitions rely heavily on technical, engineered solutions to environmental problems. Some authors believe that changing industrial systems will also require changes in human behaviour and social patterns. What balance between behavioural changes and technological changes is appropriate?
4. Is systems analysis and material and energy accounting the core of industrial ecology?

MAIN FEATURES OF INDUSTRIAL ECOLOGY

Industrial processes, from material extraction through to product disposal, have an adverse impact upon the environment. Industrial ecology aims to reduce environmental stress caused by industry whilst encouraging innovation, resource efficiency and sustained growth. Industrial ecology acknowledges that industry will continue operate and expand however, it supports industry that is environmentally conscious and has less burden upon the planet. It views industrial sites as part of a wider ecology rather than an external, solitary entity.

Within the industrial ecology concept, industry interacts with nature and utilises the wastes and by products of other industries as inputs into its own processes. Industrial ecology ranges from purely industrial ecosystems to purely natural ecosystems with a range of hybrid industrial/natural ecosystems in between. Covering both industrial management and technology, industrial ecology encompasses other sustainability concepts and tools such as material flows analysis; environmentally sound technologies; design for disassembly; and dematerialisation. The principles of industrial ecology as defined by Tibbs are:

1. Create industrial ecosystems—close the loop; view waste as a resource; create partnerships with other industries to trade by-products which are used as inputs to other processes.
2. Balance industrial inputs and outputs to natural levels—manage the environmental-industrial interface; increase knowledge of ecosystem behaviour, recovery time and capacity; increase knowledge of how and when industry can interact with natural ecosystems and the limitations.
3. Dematerialisation of industrial output—use less virgin materials and energy by becoming more resource efficient; reuse materials or substituting more environmentally friendly materials; do more with less.
4. Improve the efficiency of industrial processes—redesign products, processes, equipment; reuse materials to conserve resources.
5. Energy use—incorporate energy supply within the industrial ecology; use alternative sources of energy that have less or no impact upon the environment.
6. Align policies with the industrial ecology concept—incorporate environment and economics into organisational, national and international policies; internalise the externalities; use economic instruments to encourage a move towards industrial ecology; use a more appropriate discount rate; use a more comprehensive index to measure a nation's wealth rather than GNP.

BENEFITS OF INDUSTRIAL ECOLOGY

The benefits of industrial ecology include: cost savings (materials purchasing, licensing fees, waste disposal fees, etc.); improved environmental protection; income generation through selling waste or by products; enhanced corporate image; improved relations with other industries and organisations and market advantages. Limitations to industrial ecology include: no market for materials; lack of support from government and industry; reluctance of industry to invest in appropriate technology; perceived legal implications and reluctance to move to another supplier.

The formation of virtual or physical eco-parks arises from clusters of industry that agree to supply or sell waste to each other, thereby moving towards the industrial ecology concept. Most eco-parks are virtual due to the high cost associated with relocating facilities. However some physical eco-parks are being designed whereby certain industries are located on the same site.

Industrial Ecology as a Field of Ecology

The term 'Industrial ecology' implies a relationship to the field(s) of ecology. A basic understanding of ecology is useful in understanding and promoting industrial ecology, which draws on many ecological concepts.

Ecology has been defined by the ecological society of America as 'the scientific discipline that is concerned with the relationships between organisms and their past, present, and future environments. These relationships include physiological responses of individuals, structure and dynamics of populations, interactions among species, organisation of biological communities, and processing of energy and matter in ecosystems'.

Further, Eugene Odum has written that 'the word ecology is derived from the Greek *oikos*, meaning 'household,' combined with the root logy, meaning 'the study of.' Thus, ecology is, literally the study of households including the plants, animals, microbes, and people that live together as interdependent beings on Spaceship Earth. As already, the environmental house within which we place our human-made structures and operate our machines provides most of our vital biological necessities; hence we can think of ecology as the study of the earth's life-support systems'.

In industrial ecology, one focus (or object) of study is the interrelationships among firms, as well as among their products and processes, at the local, regional, national, and global system levels (as already presented in Table 5.1). These layers of overlapping connections resemble the food web that characterises the interrelatedness of organisms in natural ecological systems.

Industrial ecology perhaps has the closest relationship with applied ecology and social ecology. According to the Journal of Applied Ecology 'application of ecological ideas, theories and methods to the use of biological resources in the widest sense. It is concerned with the ecological principles underlying the management, control, and development of biological resources for agriculture, forestry, aquaculture, nature conservation, wildlife and game management, leisure activities, and the ecological effects of biotechnology'.

The institute of social ecology's definition of social ecology states that 'social ecology integrates the study of human and natural ecosystems through understanding the interrelationships of culture and nature. It advances a critical, holistic world view and suggests that creative human enterprise can construct an alternative future, reharmonising people's relationship to the natural world by reharmonising their relationship with each other'.

Ecology can be broadly defined as the study of the interactions between the abiotic and the biotic components of a system. Industrial ecology is the study of the interactions between industrial and ecological systems; consequently, it addresses the environmental effects on both the abiotic and biotic components of the ecosphere. Additional work needs to be done to designate industrial ecology's place in the field of ecology. This will occur concurrently with efforts to better define the discipline and its terminology.

GOALS OF INDUSTRIAL ECOLOGY

The primary goal of industrial ecology is to promote sustainable development at the global, regional, and local levels. Sustainable development has been defined by the United Nations World Commission on Environment and Development as 'meeting the needs of the present generation without sacrificing the needs of future generations'. Key principles inherent to sustainable development include: the sustainable use of resources, preserving ecological and human health (e.g. the maintenance of the structure and function of ecosystems), and the promotion of environmental equity (both intergenerational and intersocietal).

Sustainable Use of Resources

Industrial ecology should promote the sustainable use of renewable resources and minimal use of nonrenewable ones. Industrial activity is dependent on a steady supply of resources and thus should operate as efficiently as possible. Although in the past mankind has found alternatives to diminished raw materials, it cannot be assumed that substitutes will continue to be found as supplies of certain raw materials decrease or are degraded. Besides solar energy, the supply of resources is finite. Thus, depletion of nonrenewables and degradation of renewables must be minimised in order for industrial activity to be sustainable in the long-term.

Ecology and Human Health

Human beings are only one component in a complex web of ecological interactions: their activities cannot be separated from the functioning of the entire system. Because human health is dependent on the health of the other components of the ecosystem, ecosystem structure and function should be a focus of industrial ecology. It is important that industrial activities do not cause catastrophic disruptions to ecosystems or slowly degrade their structure and function, jeopardising the planet's life support system.

Environmental Equity

A primary challenge of sustainable development is achieving intergenerational as well as intersocietal equity. Depleting natural resources and degrading ecological health in order to meet short-term objectives can endanger the ability of future generations to meet their needs. Intersocietal inequities also exist, as evidenced by the large imbalance of resource use between developing and developed countries. Developed countries currently use a disproportionate amount of resources in comparison with developing countries. Inequities also exist between social and economic groups within the USA. Several studies have shown that low income and ethnic communities in the US for instance, are often subject to much higher levels of human health risk associated with certain toxic pollutants.

KEY CONCEPTS OF INDUSTRIAL ECOLOGY

Systems Analysis

Critical to industrial ecology is the systems view of the relationship between human activities and environmental problems. As stated earlier, industrial ecology is a higher order systems approach to framing the interaction between industrial and ecological systems. There are various system levels that may be chosen as the focus of study (Table 5.1). For example, when focusing at the product system level, it is important to examine relationships to higher-level corporate or institutional systems as well as at lower levels, such as the individual product life-cycle stages. One could also look at how the product system affects various ecological systems ranging from entire ecosystems to individual organisms. A systems view enables manufacturers to develop products in a sustainable fashion. Central to the systems approach is an inherent recognition of the interrelationships between industrial and natural systems.

In using systems analysis, one must be careful to avoid the pitfall that Kenneth Boulding has described: Seeking to establish a single, self-contained 'general theory of practically everything' which will replace all the special theories of particular disciplines. Such a theory would be almost without content, for we always pay for generality by sacrificing content, and all we can say about practically everything is almost nothing.

Material and Energy Flows and Transformations

A primary concept of industrial ecology is the study of material and energy flows and their transformation into products, by-products, and wastes throughout industrial systems. The consumption of resources is inventoried along with environmental releases to air, water, land, and biota. Figures 5.1 and 5.2 are examples of such material flow diagrams.

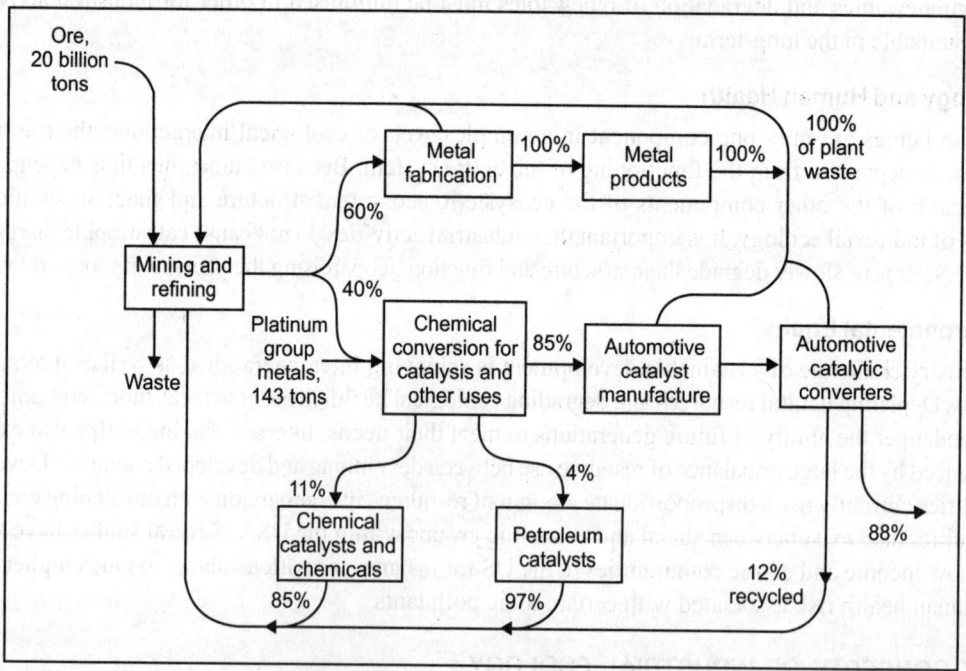

Fig. 5.1. Flow of platinum through various product systems.

One strategy of industrial ecology is to lessen the amount of waste material and waste energy that is produced and that leaves the industrial system, subsequently impacting ecological systems adversely. For instance, in Fig. 5.1, which shows the flow of platinum through various products, 88 per cent of the material in automotive catalytic converters leaves this product system as scrap. Recycling efforts could be intensified or other uses found for the scrap to decrease this waste. Efforts to utilise waste as a material input or energy source for some other entity within the industrial system can potentially improve the overall efficiency of the industrial system and reduce negative environmental impacts. The challenge of industrial ecology is to reduce the overall environmental burden of an industrial system that provides some service to society.

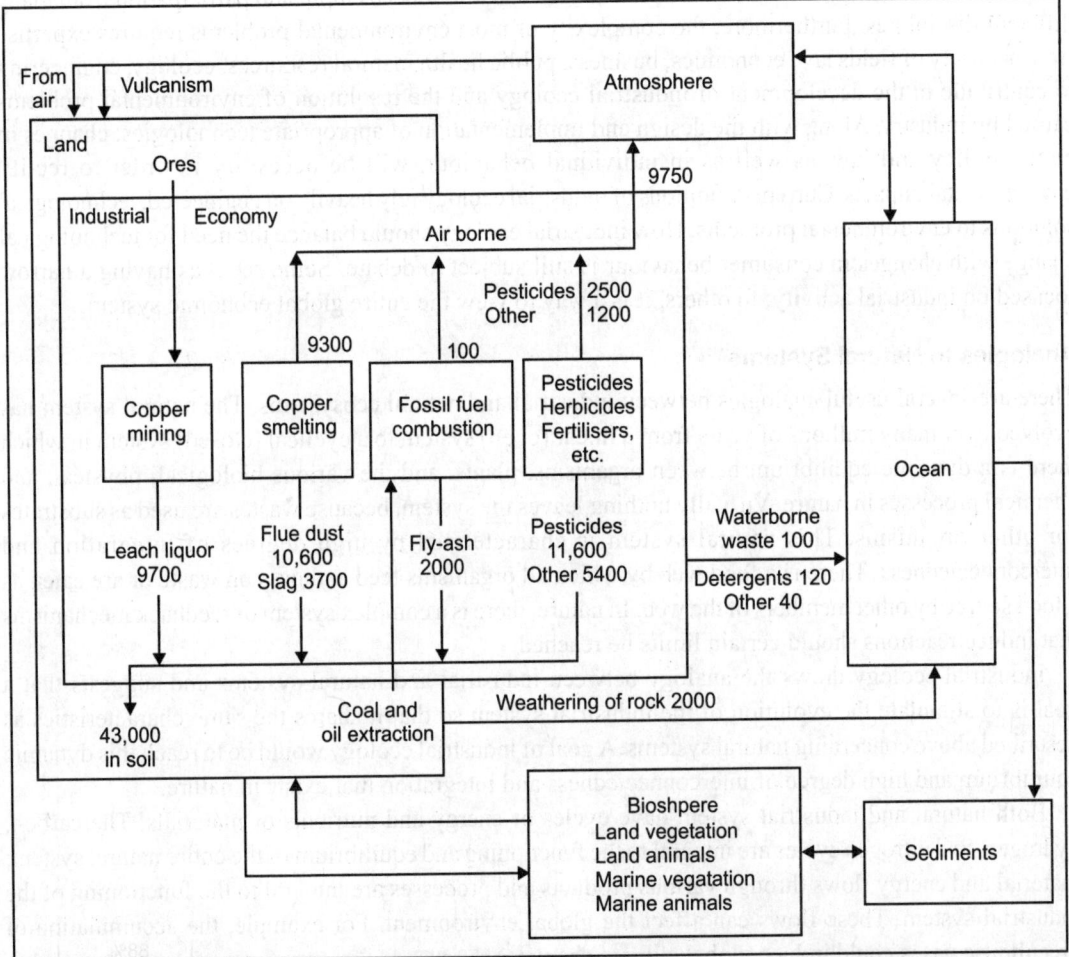

Fig. 5.2. Simplified representation of arsenic pathways in the US (metric tons).

To identify areas to target for reduction, one must understand the dissipation of materials and energy (in the form of pollutants) — how these flows intersect, interact, and affect natural systems. Distinguishing between natural material and energy flows and anthropogenic flows can be useful in identifying the scope of human-induced impacts and changes. The anthropogenic sources of some materials in natural ecosystems are much greater than natural sources.

Industrial ecology seeks to transform industrial activities into a more closed system by decreasing the dissipation or dispersal of materials from anthropogenic sources, in the form of pollutants or wastes, into natural systems. In the automobile example, it is useful to further trace what happens to these materials at the end of the products' lives in order to mitigate possible adverse environmental impacts. Some educational courses may wish to concentrate on developing skills to do mass balances and to trace the flows of certain energy or material forms in processes and products.

Multidisciplinary Approach

Since industrial ecology is based on a holistic, systems view, it needs input and participation from many different disciplines. Furthermore, the complexity of most environmental problems requires expertise from a variety of fields law, economics, business, public health, natural resources, ecology, engineering to contribute to the development of industrial ecology and the resolution of environmental problems caused by industry. Along with the design and implementation of appropriate technologies, changes in public policy and law, as well as in individual behaviour, will be necessary in order to rectify environmental impacts. Current definitions of industrial ecology rely heavily on engineered, technological solutions to environmental problems. How industrial ecology should balance the need for technological change with changes in consumer behaviour is still subject to debate. Some see it as having a narrow focused on industrial activity; to others, it is a way to view the entire global economic system.

Analogies to Natural Systems

There are several useful analogies between industrial and natural ecosystems. The natural system has evolved over many millions of years from a linear (open) system to a cyclical (closed) system in which there is a dynamic equilibrium between organisms, plants, and the various biological, physical, and chemical processes in nature. Virtually nothing leaves the system, because wastes are used as substrates for other organisms. This natural system is characterised by high degrees of integration and interconnectedness. There is a food web by which all organisms feed and pass on waste or are eaten as a food source by other members of the web. In nature, there is a complex system of feedback mechanisms that induce reactions should certain limits be reached.

Industrial ecology draws the analogy between industrial and natural systems and suggests that a goal is to stimulate the evolution of the industrial system so that it shares the same characteristics as described above concerning natural systems. A goal of industrial ecology would be to reach this dynamic equilibrium and high degree of interconnectedness and integration that exists in nature.

Both natural and industrial system have cycles of energy and nutrients or materials. The carbon, hydrogen, and nitrogen cycles are integral to the functioning and equilibrium of the entire natural system; material and energy flows through various products and processes are integral to the functioning of the industrial system. These flows can affect the global environment. For example, the accumulation of greenhouse gases could induce global climate change.

Linear (Open) versus Cyclical (Closed) Loop Systems

The evolution of the industrial system from a linear system, where resources are consumed and damaging wastes are dissipated into the environment, to a more closed system, like that of ecological systems, is a central concept to industrial ecology. Braden Allenby has described this change as the evolution from a Type I to a Type III system, as shown in Fig. 5.3.

A Type I system is depicted as a linear process in which materials and energy enter one part of the system and then leave either as products or by-products/wastes. Because wastes and by-products are not recycled or reused, this system relies on a large, constant supply of raw materials. Unless the supply of materials and energy is infinite, this system is unsustainable; further, the ability for natural systems to assimilate wastes (known as 'sinks') is also finite. In a Type II system, which characterises much of our present-day industrial system, some wastes are recycled or reused within the system while others still leave it.

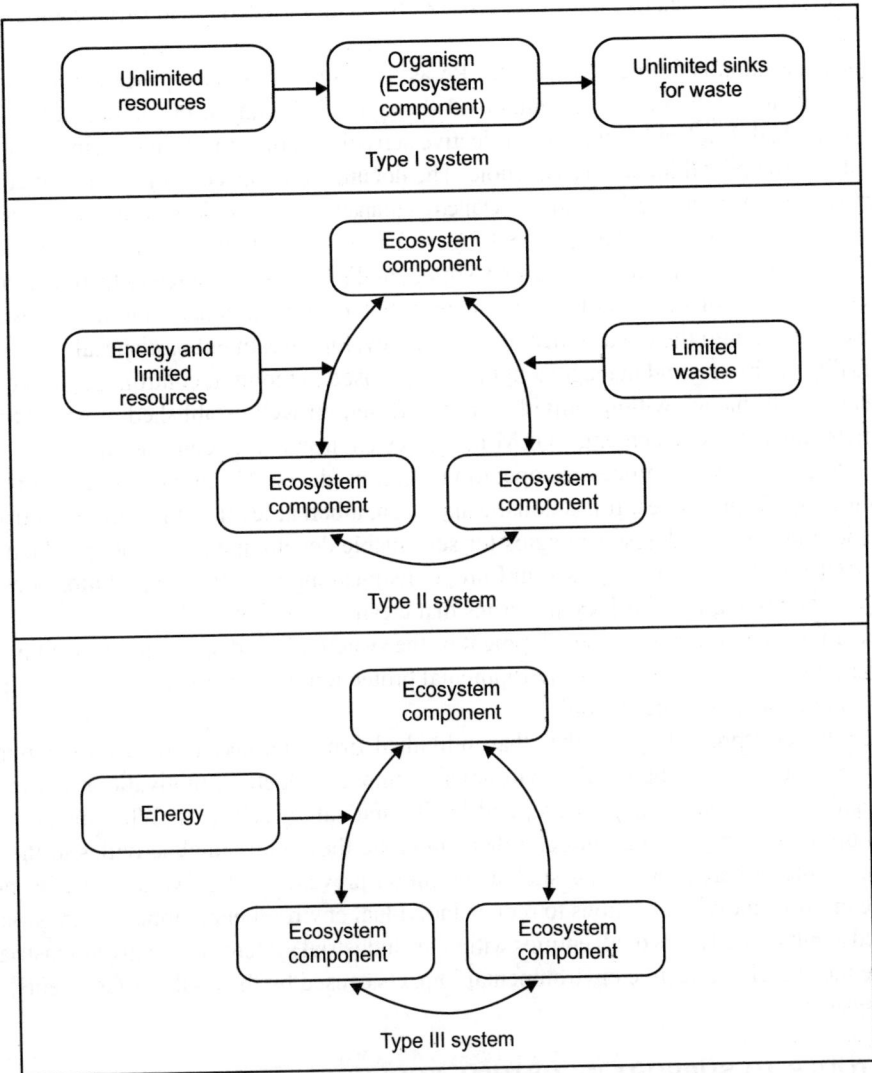

Fig. 5.3. System types.

A Type III system represents the dynamic equilibrium of ecological systems, where energy and wastes are constantly recycled and reused by other organisms and processes within the system. This is a highly integrated, closed system. In a totally closed industrial system, only solar energy would come from outside, while all by-products would be constantly reused and recycled within. A Type III system represents a sustainable state and is an ideal goal of industrial ecology.

STRATEGIES FOR ENVIRONMENTAL IMPACT REDUCTION: INDUSTRIAL ECOLOGY AS A POTENTIAL UMBRELLA FOR SUSTAINABLE DEVELOPMENT STRATEGIES

Various strategies are used by individuals, firms, and governments to reduce the environmental impacts of industry. Each activity takes place at a specific systems level. Some feel that industrial ecology could serve as an umbrella for such strategies, while others are wary of placing well-established strategies under the rubris of a new idea like industrial ecology. Strategies related to industrial ecology are briefly noted below.

Pollution prevention is defined by the US EPA as 'the use of materials, processes, or practices that reduce or eliminate the creation of pollutants at the source'. Pollution prevention refers to specific actions by individual firms, rather than the collective activities of the industrial system (or the collective reduction of environmental impacts) as a whole. The document in this compendium entitled 'Pollution Prevention Concepts and Principles' provides a detailed examination of this topic with definitions and examples.

Waste minimisation is defined by the US EPA as 'the reduction, to the extent feasible, of hazardous waste that is generated or subsequently treated, sorted, or disposed of'. Source reduction is any practice that reduces the amount of any hazardous substance, pollutant or contaminant entering any waste stream or otherwise released into the environmental prior to recycling, treatment or disposal.

Total quality environmental management (TQEM) is used to monitor, control, and improve a firm's environmental performance within individual firms. Based on well established principles from Total Quality Environmental Management, TQEM integrates environmental considerations into all aspects of a firm's decision-making, processes, operations, and products. All employees are responsible for implementing TQEM principles. It is a holistic approach, albeit at level of the individual firm.

Many additional terms address strategies for sustainable development. Cleaner production, a term coined by UNEP in 1989, is widely used in Europe. Its meaning is similar to pollution prevention. In Clean Production Strategies, Tim Jackson writes that clean production is:

'An operational approach to the development of the system of production and consumption, which incorporates a preventive approach to environmental protection. It is characterised by three principles: precaution, prevention, and integration'.

These strategies represent approaches that individual firms can take to reduce the environmental impacts of their activities. Along with environmental impact reduction, motivations can include cost savings, regulatory or consumer pressure, and health and safety concerns. What industrial ecology potentially offers is an organising umbrella that can relate these individual activities to the industrial system as a whole. Whereas strategies such as pollution prevention, TQEM, and cleaner production concentrate on firms individual actions to reduce individual environmental impacts, industrial ecology is concerned about the activities of all entities within the industrial system. The goal of industrial ecology is to reduce the overall, collective environmental impacts caused by the totality of elements within the industrial system.

SYSTEM TOOLS TO SUPPORT INDUSTRIAL ECOLOGY

Life-cycle Assessment (LCA)

Life-cycle assessment (LCA), along with 'ecobalances' and resource environmental profile analysis, is a method of evaluating the environmental consequences of a product or process 'from cradle to grave'. The society for environmental toxicology and chemistry (SETAC) defines LCA as 'a process used to evaluate the environmental burdens associated with a product, process, or activity. The US EPA has stated that an LCA 'is a tool to evaluate the environmental consequences of a product or activity holistically, 'across its entire life'. In the United States, SETAC, the US EPA and consulting firms are active in developing LCAs.

Components of an LCA

LCA methodology is still evolving. However, the three distinct components defined by SETAC and the US EPA (Fig. 5.4) are the most widely recognised:

1. Inventory analysis: Identification and quantification of energy and resource use and environmental releases to air, water, and land.
2. Impact analysis: Technical qualitative and quantitative characterisation and assessment of the consequences on the environment.
3. Improvement analysis: Evaluation and implementation of opportunities to reduce environmental burden some life-cycle assessment practitioners have defined a fourth component, the scoping and goal definition or initiation step, which serves to tailor the analysis to its intended use. Other efforts have also focused on developing streamlined tools that are not as rigorous as LCA (e.g. Canadian Standards Association).

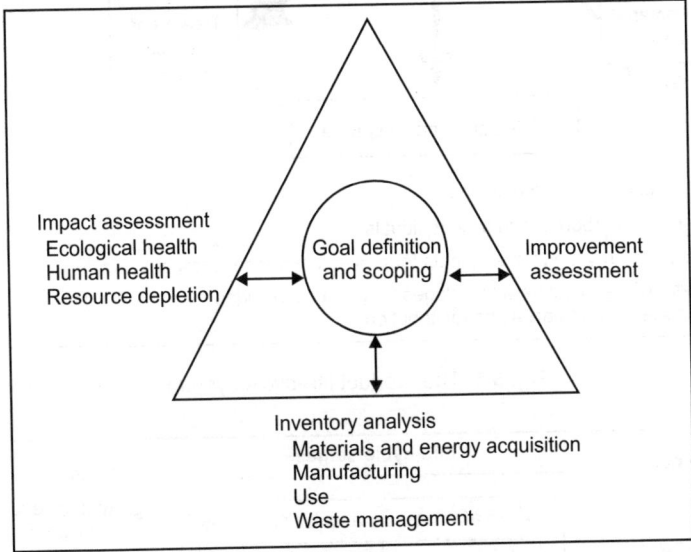

Fig. 5.4. Technical framework for life-cycle assessment.

Methodology

A life-cycle assessment focuses on the product life-cycle system as shown in Fig. 5.5. Most research efforts have been focused on the inventory stage. For an inventory analysis, a process flow diagram is constructed and material and energy inputs and outputs for the product system are identified and quantified as depicted in Fig. 5.6. A template for constructing a detailed flow diagram for each subsystem is shown in Fig. 5.7. Figure 5.8 shows the many stages involved in the life-cycle of a bar of soap, illustrating how, even for a relatively simple product, the inventory stage can quickly become complicated, especially as products increase in number of components and in complexity.

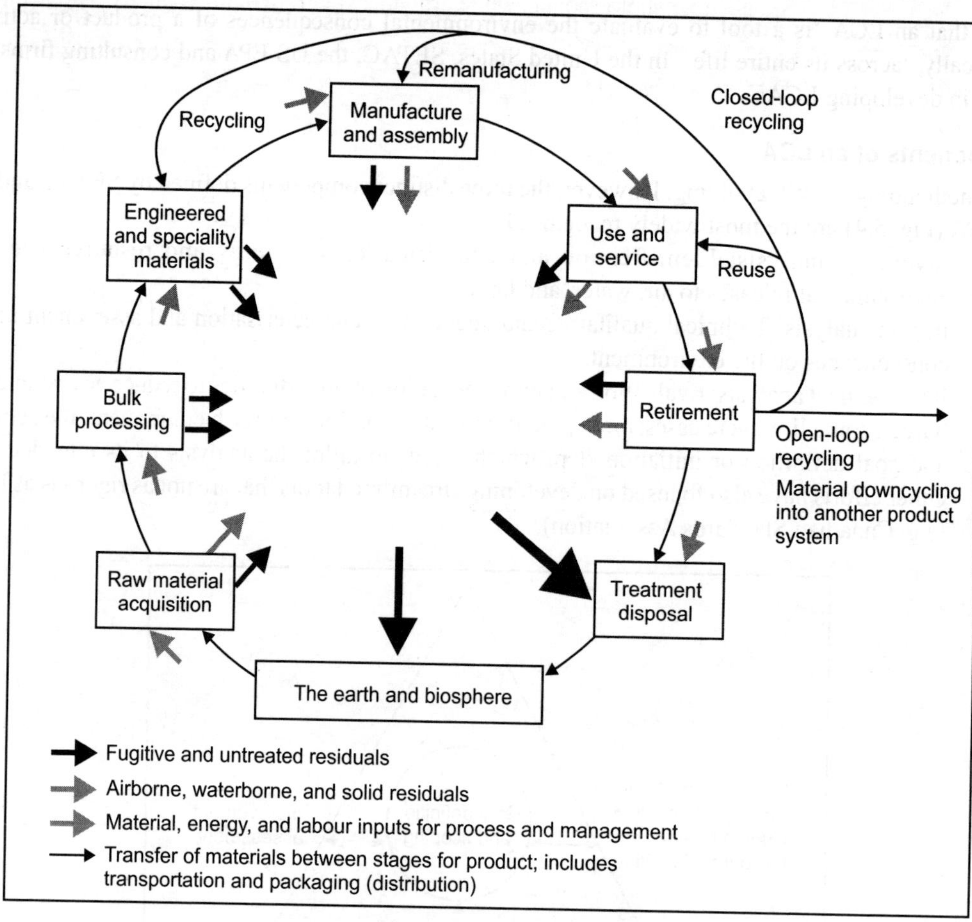

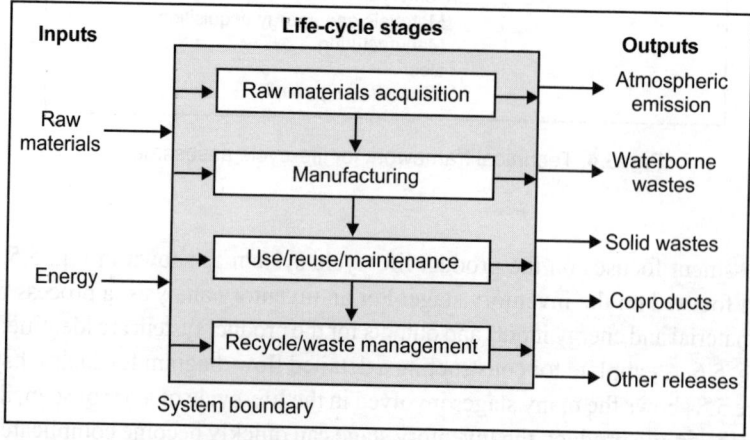

Fig. 5.5. The product life-cycle system.

Fig. 5.6. Process flow diagram.

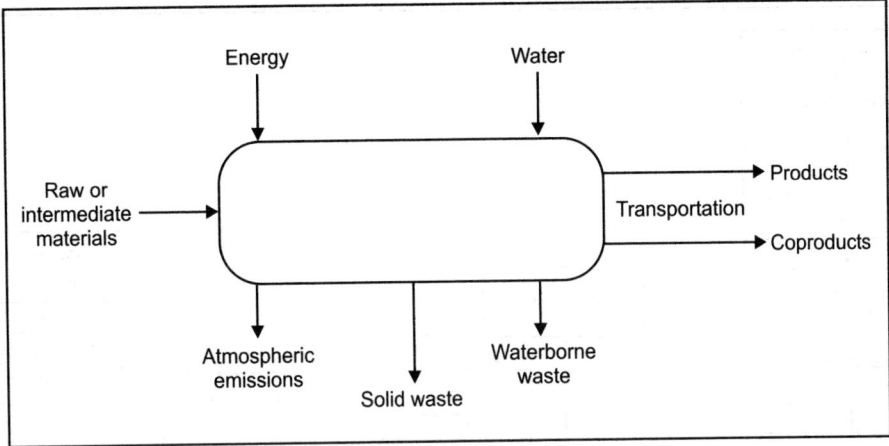

Fig. 5.7. Flow diagram template.

Once the environmental burdens haven been identified in the inventory analysis, the impacts must be characterised and assessed. The impact assessment stage seeks to determine the severity of the impacts and rank them as indicated by Fig. 5.9. As the figure shows, the impact assessment involves three stages: classification, characterisation, and valuation. In the classification stage, impacts are placed in one of four categories: resource depletion, ecological health, human health, and social welfare. Assessment endpoints must then be determined. Next, conversion models are used to quantify the environmental burden. Finally, the impacts are assigned a value and/or are ranked.

Efforts to develop methodologies for impact assessment are relatively new and remain incomplete. It is difficult to determine an endpoint. There are a range of conversion models, but many of them remain incomplete. Furthermore, different conversion models for translating inventory items into impacts are required for each impact, and these models vary widely in complexity, uncertainty, and sophistication. This stage also suffers from a lack of sufficient data, model parameters and conversion models.

The final stage of a LCA, the improvement analysis, should respond to the results of the inventory and/or impact assessment by designing strategies to reduce the identified environmental impacts. Proctor and Gamble is one company that has used life-cycle inventory studies to guide environmental improvement for several products. One of its case studies on hard surface cleaners revealed that heating water for use with the product resulted in a significant percentage of total energy use and air emissions related to cleaning. Based on this information, opportunities for reducing impacts were identified, such as designing cold-water and no-rinse formulas and educating consumers to use cold water.

Application of LCA

Life-cycle assessments can be used both internally (within an organisation) and externally (by the public and private sectors). Internally, LCAs can be used to establish a comprehensive baseline (i.e. requirements) that product design teams should meet, identify the major impacts of a product's life-cycle, and guide the improvement of new product systems toward a net reduction of resource requirements and emissions in the industrial system as a whole. Externally, LCAs can be used to compare the environmental profiles of alternative products, processes, materials, or activities and to support marketing claims. LCA can also support public policy and eco-labelling programs.

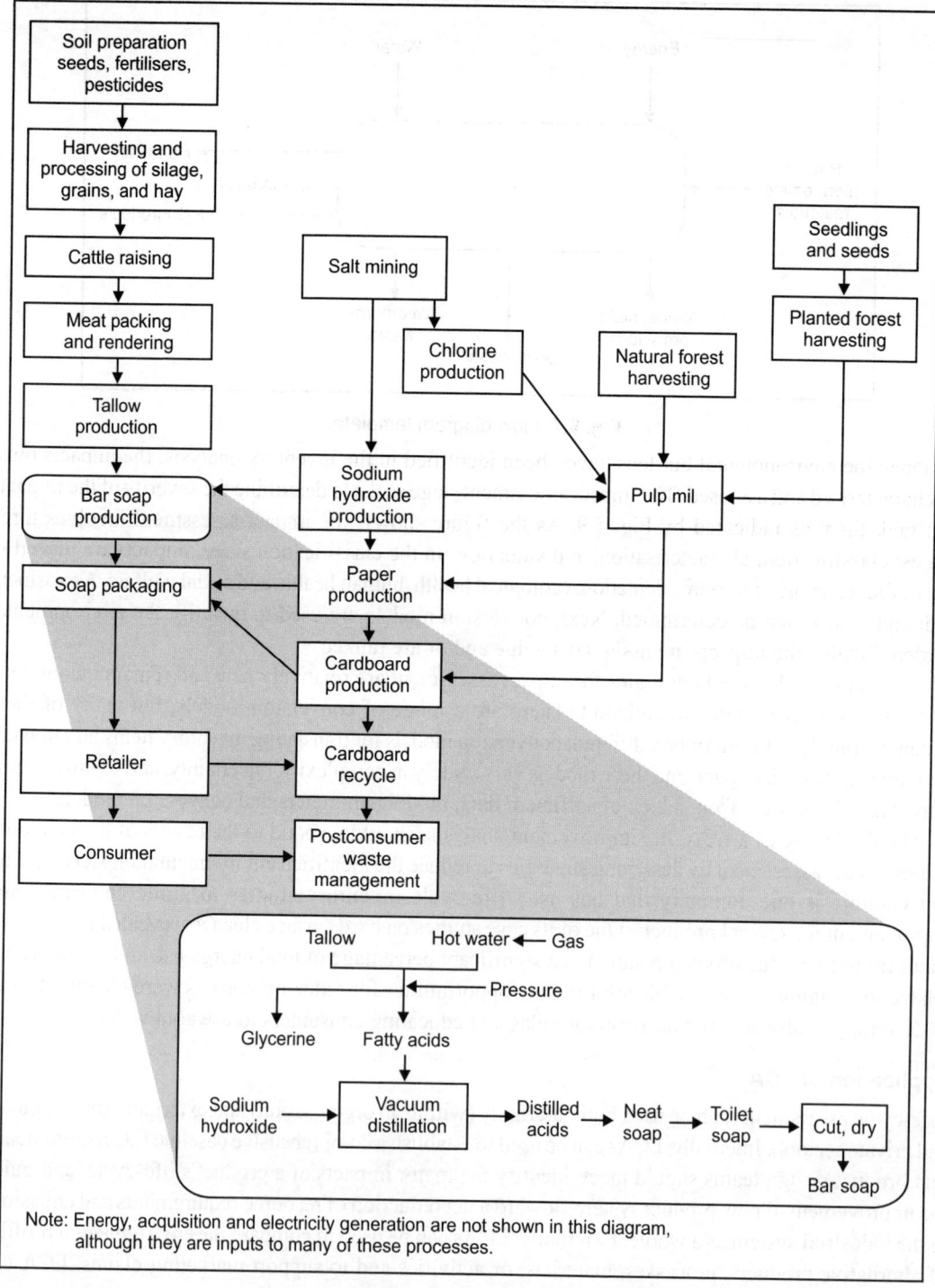

Fig. 5.8. Detailed system diagram for bar soap.

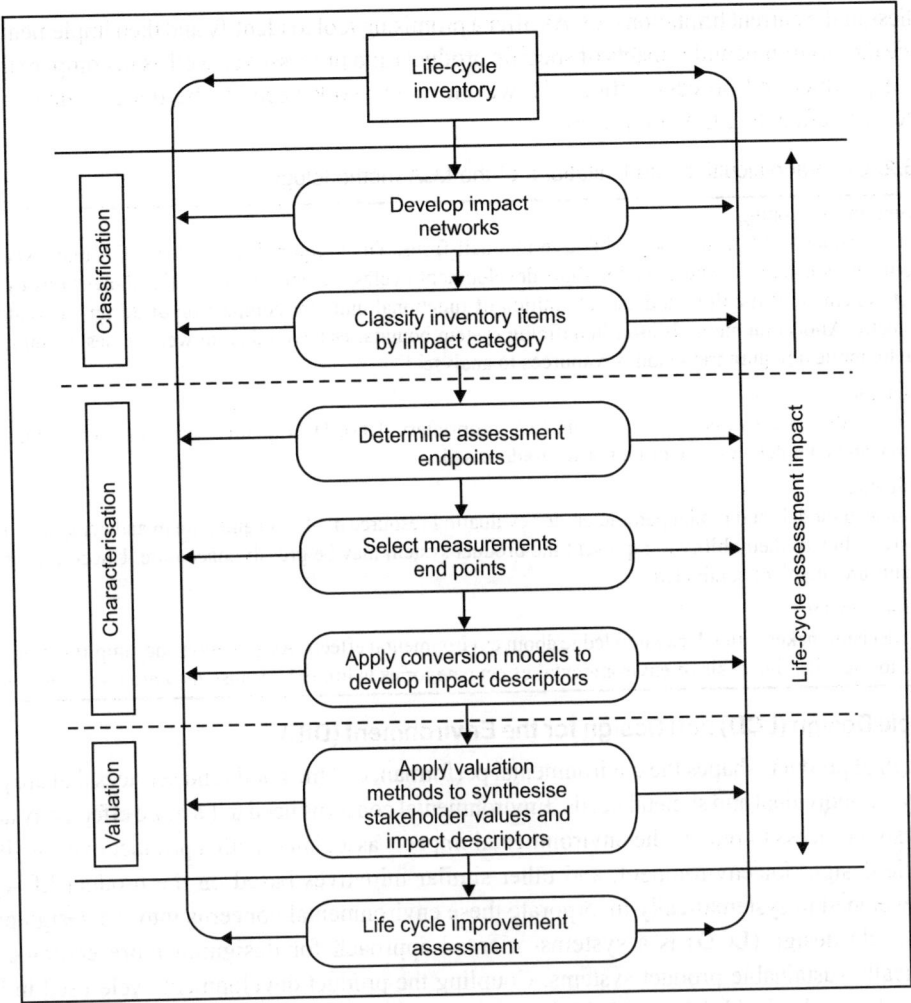

Fig. 5.9. Impact assessment conceptual framework.

Difficulties with LCA

As shown in Table 5.2, many methodological problems and difficulties inhibit use of LCAs, particularly for smaller companies. For example, the amount of data and the staff time required by LCAs can make them very expensive, and it isn't always easy to obtain all of the necessary data. Further, it is hard to properly define system boundaries and appropriately allocate inputs and outputs between product systems and stages. It is often very difficult to assess the data collected because of the complexity of certain environmental impacts. Conversion models for transforming inventory results into environmental impacts remain inadequate. In many cases there is a lack of fundamental understanding and knowledge about the actual cause of certain environmental problems and the degree of threat that they pose to ecological and human health.

In the absence of an accepted methodology, results of LCAs can differ. Order-of-magnitude differences are not uncommon. Discrepancies can be attributed to differences in assumptions and system boundaries.

Regardless of the current limitations, LCAs offer a promising tool to identify and then implement strategies to reduce the environmental impacts of specific products and processes as well as to compare the relative merits of product and process options. However, much work needs to be done to develop, utilise, evaluate, and refine the LCA framework.

Table 5.2. General difficulties and limitations of the LCA methodology.

Goal definition and scoping

Costs to conduct an LCA may be prohibitive to small firms. Time required to conduct LCA may exceed product development constraints, especially for short development cycles. Temporal and spatial dimensions of a dynamic product system are difficult to address. Definition of functional units for comparison of design alternatives can be problematic. Allocation methods used in defining system boundaries have inherent weaknesses. Complex products (e.g. automobiles) require tremendous resources to analyse.

Data collection

Data availability and access can be limiting (e.g. proprietary data). Data quality concerns such as bias, accuracy, precision, and completeness are often not well-addressed.

Data Evaluation

Sophisticated models and model parameters for evaluating resource depletion and human and ecosystem health may not be available, or their ability to represent the product system may be grossly inaccurate. Uncertainty analyses of the results are often not conducted.

Information transfer

Design decisionmakers often lack knowledge about environmental effects. Aggregation and simplification techniques may distort results. Synthesis of environmental effect categories is limited because they are incommensurable.

Life-cycle Design (LCD) and Design for the Environment (DfE)

The design of products shapes the environmental performance of the goods and services that are produced to satisfy our individual and societal needs. Environmental concerns need to be more effectively addressed in the design process to reduce the environmental impacts associated with a product over its life-cycle. Life-cycle design, for environment, and other similar initiatives based on the product life-cycle are being developed to systematically incorporate these environmental concerns into the design process.

Life-cycle design (LCD) is a systems-oriented approach for designing more ecologically and economically sustainable product systems. Coupling the product development cycle used in business with a product's physical life-cycle. LCD integrates environmental requirements into each design stage so total impacts caused by the product system can be reduced.

Design for environment (DfE) is another design strategy that can be used to design products with reduced environmental burden. DfE and LCD can be difficult to distinguish. They have similar goals but evolved from different sources. DfE evolved from the 'Design for X' approach, where X can represent manufacturability, testability, reliability, or other 'downstream' design considerations. Braden Allenby has developed a DfE framework to address the entire product life-cycle. Like LCD, DfE uses a series of matrices in an attempt to develop and then incorporate environmental requirements into the design process. DfE is based on the product life-cycle framework and focuses on integrating environmental issues into products and process design.

Life-cycle design seeks to minimise the environmental consequences of each product system component: product, process, distribution and management.

Figure 5.10 indicates the complex set of issues and decisions required in LCD. When sustainable development is the goal, the design process can be affected by both internal and external factors.

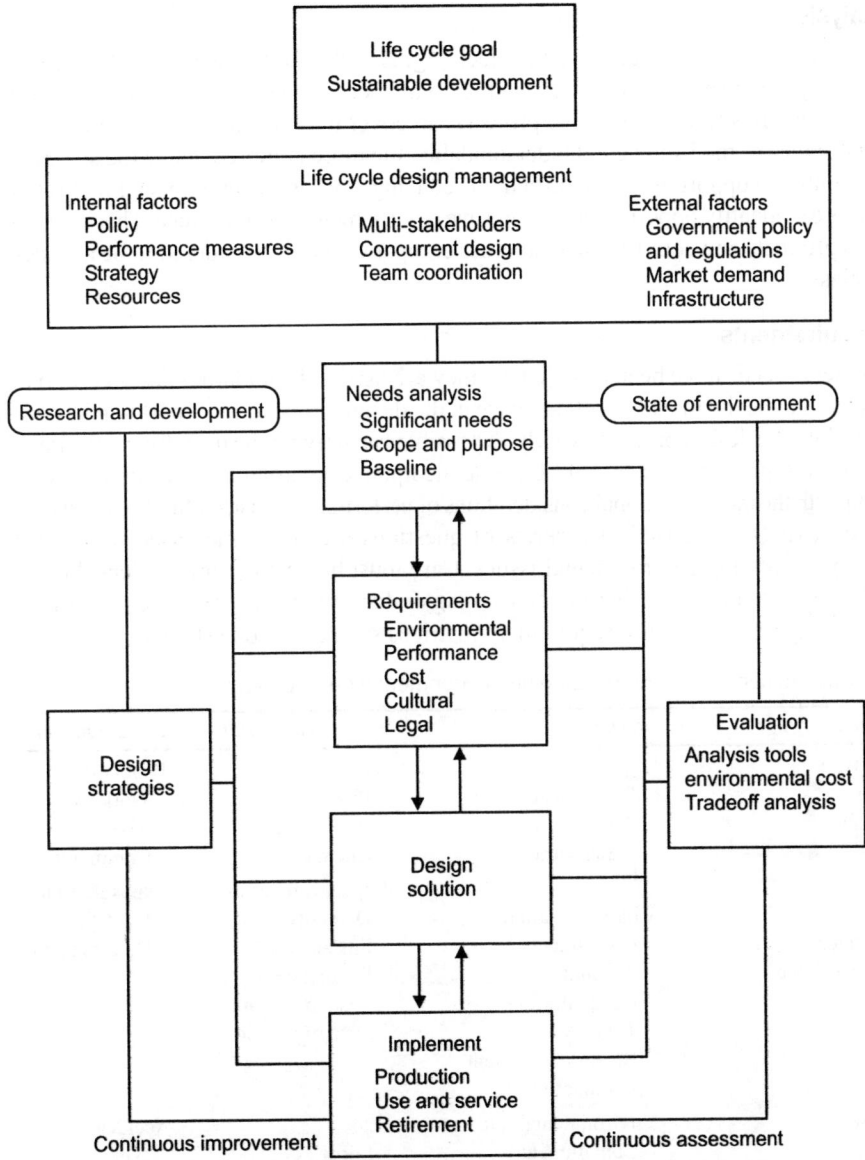

Fig. 5.10. Life-cycle design.

Internal factors include corporate policies and the companies' mission, product performance measures, and product strategies as well as the resources available to the company during the design process. For instance, a company's corporate environmental management system, if it exists at all, greatly affects the designer's ability to utilise LCD principles.

External factors such as government policies and regulations, consumer demands and preferences, the state of the economy, and competition also affect the design process, as do current scientific understanding and public perception of risks associated with the product.

Needs Analysis

As shown in the figure, a typical design project begins with a needs analysis. During this phase, the purpose and scope of the project is defined, and customer needs and market demand are clearly identified. The system boundaries (the scope of the project) can cover the full life-cycle system, a partial system, or individual stages of the life-cycle. Understandably, the more comprehensive the system of study, the greater the number of opportunities identified for reducing environmental impact. Finally, benchmarking of competitors can identify opportunities to improve environmental performance. This involves comparing a company's products and activities with another company who is considered to be a leader in the field or 'best in class'.

Design Requirements

Once the projects needs have been established, they are used in formulating design criteria. This step is often considered to be the most important phase in the design process. Incorporating key environmental requirements into the design process as early as possible can prevent the need for costly, time-consuming adjustments later. A primary objective of LCD is to incorporate environmental requirements into the design criteria along with the more traditional considerations of performance, cost, cultural, and legal requirements.

Design checklists comprised of a series of questions are sometimes used to assist designers in systematically addressing environmental issues. Care must be taken to prevent checklists, such as the one in Table 5.3, from being overly time-consuming or disruptive to the creative process. Another more comprehensive approach is to use requirement matrices such as the one shown in Fig. 5.11.

Table 5.3. Issues to consider when developing environmental requirements.

Materials and energy	Residuals	Ecological health	Human health and safety
Amount and type 　Renewable 　Nonrenewable	Type 　Solid waste 　Air emissions 　Waterborne	Stressors 　Physical 　Biological 　Chemical	Population at risk 　Workers 　Users 　Community
Character 　Virgin 　reused/recycled 　Reusable/recyclable	Characterisation 　Constituents 　Amount 　Concentration 　Toxicity 　Hazardous content 　Radioactivity	Impact categories 　Diversity 　Sustainability 　Resilience 　System structure 　System function	Exposure routes 　Inhalation, contact, ingestion 　Duration and frequency
Resource base 　Location 　Local vs. other 　Availability 　Quality 　Management 　Restoration practices	Environmental fate 　Containment 　Bioaccumulation 　Degradability 　Mobility/transport 　Ecologial impacts 　Human health impacts	Scale 　Local 　Regional 　Global	Accidents 　Type 　Frequency Toxic character 　Acute effects 　Chronic effects 　Morbidity/mortality
Impacts from extraction and use 　Material/energy use 　Residuals 　Ecosystem health 　Human health			Nuisance effects 　Noise 　Odours 　Visibility

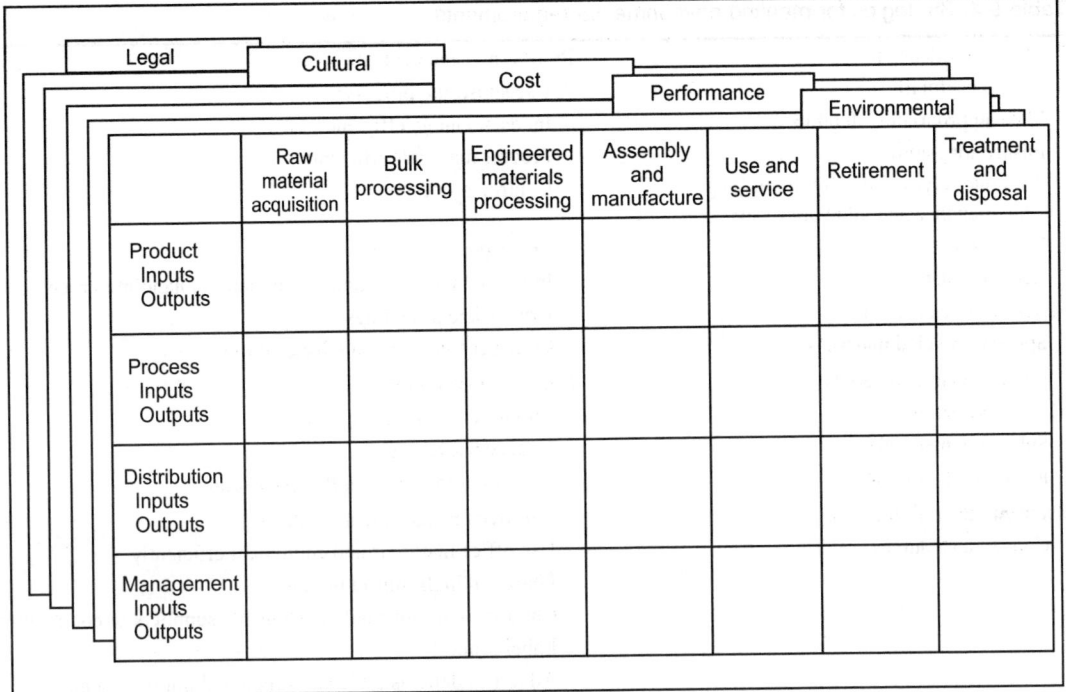

Fig. 5.11. Requirements matrices.

Matrices can be used by product development teams to study interactions between life-cycle requirements and their associated environmental impacts. There are no absolute rules for organising matrices. Development teams should choose a format that is appropriate for their project. The requirements matrices shown are strictly conceptual; in practice such matrices can be simplified to address requirements more broadly during the earliest stages of design, or each cell can be further subdivided to focus on requirements in more depth.

Government policies, along with the criteria identified in the needs analysis, also should be included. It is often useful in the long term to set environmental requirements that exceed current regulatory requirements to avoid costly design changes in the future.

Performance requirements relate to the functions needed from a product. Cost corresponds to the need to deliver the product to the marketplace at a competitive price. LCD looks at the cost to stakeholders such as manufacturers, suppliers, users and end-of-life managers. Cultural requirements include aesthetic needs such as shape, form, colour, texture, and image of the product as well as specific societal norms such as convenience or ease of use. These requirements are ranked and weighed given a chosen mode of classification.

Design Strategies

Once the criteria have been defined, the design team can then use design strategies to meet these requirements. Multiple strategies often must be synthesised in order to translate these requirements into solutions. A wide range of strategies are available for satisfying environmental requirements, including product system life extension, material life extension, material selection, and efficient distribution. A summary of these strategies are given in Table 5.4. Note that recycling is often overemphasised.

Table 5.4. Strategies for meeting environmental requirements.

Product life extension	Process management
Extend useful life	Use substitute processes
Make appropriately durable	Increase energy efficiency
Ensure adaptability	Process materials efficiently
Facilitate serviceability by simplifying	Control processes
maintenance and allowing repair	
Enable remanufacture	Improve process layout
Accommodate reuse	Improve inventory control and material handling processes
Material life extension	Plan efficient facilities
Specify recycled materials	Consider treatment and disposal too
Use recyclable materials	Efficient distribution
Material selection	Choose efficient transportation
Substitute materials	Reduce packaging
Reformulate products	Use low-impact or reusable packaging
Reduced material intensity	Improved management practices
Conserve resources	Use office materials and equipment efficiently
	Phase out high-impact products
	Choose environmentally responsible suppliers or contractors
	Label properly
	Advertise demonstrable environmental improvements

Design Evaluation

Finally, it is critical that the design is evaluated and analysed throughout the design process. Tools for design evaluation range from LCA to single-focus environmental metrics. In each case, design solutions are evaluated with respect to a full spectrum of criteria, which includes cost and performance.

DfE methods developed by Allenby use a semiquantitative matrix approach for evaluating life-cycle environmental impacts. A graphic scoring system weighs environmental effects according to available quantitative information for each life-cycle stage. In addition to an environmental matrix and toxicology/ exposure matrix, manufacturing and social/political matrices are used to address both technical and nontechnical aspects of design alternatives.

Although LCD is not yet widely practiced, it has been used by various companies and is recognised as an important approach for reducing environmental burdens. To enhance the use of LCD, appropriate government policies must be evaluated and established. In addition, environmental accounting methods must be further developed and utilised by industry (these methods are often referred to as life-cycle costing or full cost accounting (Table 5.5).

FUTURE NEEDS FOR THE DEVELOPMENT OF INDUSTRIAL ECOLOGY

Industrial ecology is an emerging framework. Thus much research and development of the field and its concepts need to be done. Future needs for the further development of industrial ecology include:

1. A clearer definition of the field and its concepts. The definition of industrial ecology, its scope and its goals need to be clarified and unified in order to be more useful. The application of systems analysis must be further refined.

2. A clearer definition of sustainable development, what constitutes sustainable development, and how it might be achieved, will help define the goals and objectives of industrial ecology. Difficult goals to address, along with the maintenance of ecological system health, are intergenerational and intersocietal equity.

3. More participation from a cross section of fields such as ecology, public health, business, natural resources and engineering should be encouraged in order to meet some of the vast research and information requirements needed to identify and implement strategies to reduce environmental burdens.

4. Increased curriculum development efforts on sustainable development in professional schools of engineering, business, public health, natural resources, and law. The role of industrial ecology in these efforts should be further explored and defined. Determining whether industrial ecology courses should be discipline specific, interdisciplinary or integrated as modules into existing courses.

5. Further research on the impacts of industrial ecosystem activities on natural ecosystems in order to identify what problems need to be resolved and how.

6. Greater recognition of the importance of the systems approach to identifying and resolving environmental problems.

7. Further development of tools such as life-cycle assessment and life-cycle design and design for the environment.

8. The improvement of governmental policies that will strengthen incentives for industry to reduce environmental burdens.

Table 5.5. Definitions of accounting and capital budgeting terms relevant to LCD.

Accounting	
Full cost accounting	A method of managerial cost accounting that allocates both direct and indirect environmental costs to a product, product line, process, service, or activity. Not everyone uses this term the same way. Some only include costs that affect the firm's bottom line; others include the full range of costs throughout the life-cycle, some of which do not have any indirect or direct effect on a firm's bottom line
Life-cycle costing	In the environmental field, this has come to mean all costs associated with a product system throughout its life-cycle, from materials acquisition to disposal. Where possible, social costs are quantified; if this is not possible, they are addressed qualitatively. Traditionally applied in military and engineering to mean estimating costs from acquisition of a system to disposal. This does not usually incorporate costs further upstream than purchase
Capital budgeting	
Total cost assessment	Long-term, comprehensive financial analysis of the full range of internal (i.e. private) costs and savings of an investment. This tool evaluates potential investments in terms of private costs, excluding social considerations. It does include contingent liability costs. Further, educational institutions must work to continue the development and the dissemination of the LCD methodology and related approaches. Key issues in environmental accounting that need to be addressed include: measurement and estimation of environmental costs, allocation procedures, and the inclusion of appropriate externalities

APPENDIX A: SELECTED DEFINITIONS OF INDUSTRIAL ECOLOGY

The idea of an industrial ecology is based upon a straightforward analogy with natural ecological systems. In nature an ecological system operates through a web of connections in which organisms live and consume each other and each other's waste. The system has evolved so that the characteristic of communities of living organisms seems to be that nothing that contains available energy or useful material will be lost. There will evolve some organism that will manage to make its living by dealing with any waste product that provides available energy or usable material. Ecologists talk of a food web: an interconnection of uses of both organisms and their wastes. In the industrial context we may think of this as being use of products and waste products. The system structure of a natural ecology and the structure of an industrial system, or an economic system, are extremely similar.

—Rober A. Forsch, 'Industrial Ecology: A Philosophical Introduction'

Somewhat teleologically, 'industrial ecology' may be defined as the means by which a state of sustainable development is approached and maintained. It consists of a systems view of human economic activity and its interrelationship with fundamental biological, chemical, and physical systems with the goal of establishing and maintaining the human species at levels that can be sustained indefinitely, given continued economic, cultural, and technological evolution.

—Braden Allenby, 'Achieving Sustainable Development Through Industrial Ecology'

Industrial ecology is a new approach to the industrial design of products and processes and the implementation of sustainable manufacturing strategies. It is a concept in which an industrial system is viewed not in isolation from its surrounding systems but in concert with them. Industrial ecology seeks to optimise the total materials cycle from virgin material to finished material to component, to product, to waste products, and to ultimate disposal. Characteristics are: (i) proactive not reactive, (ii) designed in not added on, (iii) flexible not rigid, and (iv) encompassing not insular.

—L.W. Jelinski, T.E. Graedel, R.A. Laudise, D.W. McCall, and C. Kumar, N. Patel, 'Industrial Ecology: Concepts and Approaches'

Industrial ecology can be best defined as the totality or the pattern of relationships between various industrial activities, their products, and the environment. Traditional ecological activities have focused on two time aspects of interactions between the industrial activities and the environment—the past and the present. Industrial ecology, a systems view of the environment, pertains to the future.

—C. Kumar, N. Patel, 'Industrial Ecology'

Industrial ecology is the study of how we humans can continue rearranging earth, but in such a way as to protect our own health, the health of natural ecosystems, and the health of future generations of plants and animals and humans. It encompasses manufacturing, agriculture, energy production, and transportation—nearly all of those things we do to provide food and make life easier and more pleasant than it would be without them.

—Bette Hileman, 'Industrial Ecology Route to Slow Global Change Proposed'

Industrial ecology involves designing industrial infrastructures as if they were a series of interlocking manmade ecosystems interfacing with the natural global ecosystem. Industrial ecology takes the pattern of the natural environment as a model for solving environmental problems, creating a new paradigm for the industrial system in the process. The aim of industrial ecology is to interpret and adapt an understanding of the natural system and apply it to the design of the manmade system, in order to achieve a pattern of industrialisation that is not only more efficient, but that is intrinsically adjusted to the tolerance and characteristics of the natural system. The emphasis is on forms of technology that work with natural systems, not against them.

—Hardin B.C. Tibbs, 'Industrial Ecology: An Environmental Agenda for Industry'

The heart of industrial ecology is a simple recognition that manufacturing and service systems are in fact natural systems, intimately connected to their local and regional ecosystems and the global biosphere. The ultimate goal is bringing the industrial system as close as possible to being a closed-loop system, with near complete recycling of all materials.

—Ernest Lowe, 'Industrial Ecology—An Organising Framework for Environmental Management'

Industrial ecology is the means by which humanity can deliberately and rationally approach and maintain a desirable carrying capacity, given continued economic, cultural, and technological evolution. The concept requires that an industrial system be viewed not in isolation from its surrounding systems, but in concert with them. It is a systems view in which one seeks to optimise the total materials cycle from virgin material, to finished material, to component, to product, to waste product, and to ultimate disposal. Factors to be optimised include resources, energy, and capital.

—Braden Allenby and Thomas E. Graedel, 'Industrial Ecology'

Industrial ecology provides for the first time a largescale, integrated management tool that designs industrial infrastructures 'as if they were a series of interlocking, artificial ecosystems interfacing with the natural global ecosystem.' For the first time, industry is going beyond life-cycle analysis methodology and applying the concept of an ecosystem to the whole of an industrial operation, linking the 'metabolism' of one company with that of another.

—Paul Hawken, 'The Ecology of Commerce'

SECTION II

Climatic and Topographic Factors

Chapter 6

Ecology of Light and Temperature

INTRODUCTION

Ecological factors which can affect dynamic change in a population or species in a given ecology or environment are usually divided into two groups: abiotic and biotic. Abiotic factors are geological, geographical, hydrological and climatological parameters. A biotope is an environmentally uniform region characterised by a particular set of abiotic ecological factors.

Biotic ecological factors also influence biocenose viability; these factors are considered as either intraspecific and interspecific relations. The existing interactions between the various living beings go along with a permanent mixing of mineral and organic substances, absorbed by organisms for their growth, their maintenance and their reproduction, to be finally rejected as waste. These permanent recyclings of the elements (in particular carbon, oxygen and nitrogen) as well as the water are called biogeochemical cycles. They guarantee a durable stability of the biosphere (at least when unchecked human influence and extreme weather or geological phenomena are left aside). This self-regulation, supported by negative feedback controls, ensures the perenniality of the ecosystems. It is shown by the very stable concentrations of most elements of each compartment. This is referred to as homoeostasis. The ecosystem also tends to evolve to a state of ideal balance, reached after a succession of events, the climax (for example a pond can become a peat bog).

EFFECT OF LIGHT ON PLANTS

Light has a variety of effects on plant growth. Without the sun, plants would not be able to survive, and such light is pivotal to a number of crucial processes that a plant undergoes on a regular basis in order to sustain life. There are four main processes to consider. One is well-known and also crucial to human life, and the other three are less discussed in elementary science texts, but are no less important to the life of plants. These four processes crucial to plant life are photosynthesis, photomorphogenesis, photoperiodism and phototropism.

Photosynthesis is the process whereby plants take light energy from the sun or another source, convert it into chemical based energy, and then use this energy to survive. The light energy becomes organic molecules, which make up some fundamental structures inherent to each plant. Photosynthesis also plays a part in the ever-important process of helping to sustain human life. Carbon dioxide is pulled from the environment outside of the plant and used as a part of photosynthesis to acquire energy and to run life-sustaining processes. As a by-product of all this work and of this carbon dioxide collection, oxygen is released back into the environment. This process aids human in two ways. First, it removes harmful (to humans) carbon dioxide from the outside air, and secondly, it creates new oxygen for humans to breathe. The second plant process created from light is photomorphogenesis. This is the effect of light intensity on the plant's individual growth. Some plants need a lot of light to sprout from their seeds, while others initially need little or can even find direct light detrimental to growth. Additionally, light during photomorphogenesis can even affect a plant's circadian rhythms. Lack of sufficient light in this process can often also result in plants that are lacking in normal colour due to a lack of chlorophyll being produced.

The third plant process is photoperiodism. This is a process that determines the manner in which plants flower. Plants flower differently depending upon how each day's light and dark periods affect the individual plant. Some plants need a longer period of sun during a day to flower properly, while other plants desire much less. The fourth and final plant process from light is phototropism. This determines how and in what direction a plant's individual appendages grow. Plant stems, leaves and more lean in the direction from which sunlight comes. The quality, intensity, and duration of light directly impact plant growth.

Light Quality

Light quality refers to the colour or wavelength reaching the plant's surface. A prism (or raindrops) can divide sunlight into respective colours of red, orange, yellow, green, blue, indigo and violet. Red and blue have the greatest impact on plant growth. Green light is least effective (the reflection of green light gives the green colour to plants). Blue light is primarily responsible for vegetative leaf growth. Red light, when combined with blue light, encourages flowering. Light quality is a major consideration for indoor growing (Fig. 6.1).

1. Fluorescent cool white lamps are high in the blue range, and the best choice for starting seeds indoors.
2. For flowering plants that need more red light, use broad spectrum fluorescent bulbs.
3. Incandescent lights are high in red and red-orange, but generally produce too much heat for use in supplementing plant growth.

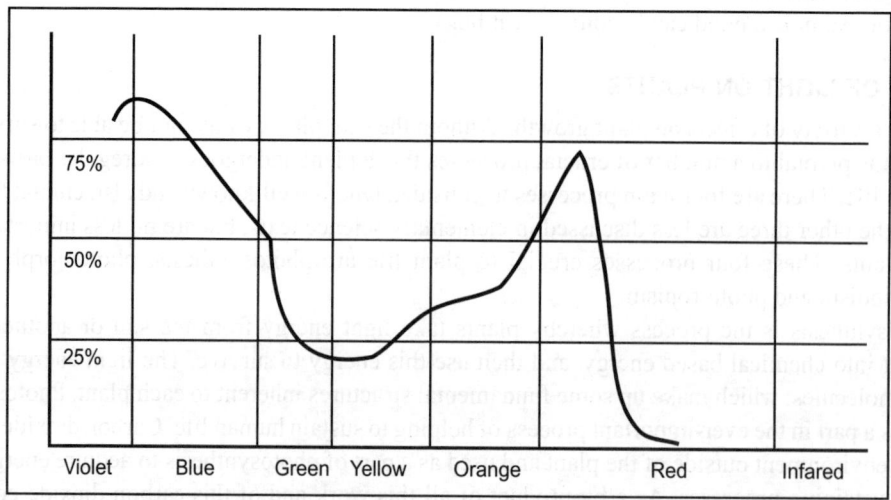

Fig. 6.1. Relative efficiency of various light colours in photosynthesis.

Light Intensity

The more sunlight a plant receives, to a degree, the higher the photosynthetic rate will be. However, leaves of plants growing in low light readily sun scorch when moved to a bright location. Over time, as the wax content on a leaf increases, it will become more sun tolerant.

As illustrated in Fig. 6.2, light levels in most homes are below that required for all but low light house plants. Except for rather bright sunny rooms, most house plants can only be grown directly in front of bright windows. Inexpensive light meters are available in many garden supply stores to help the indoor gardener evaluate light levels.

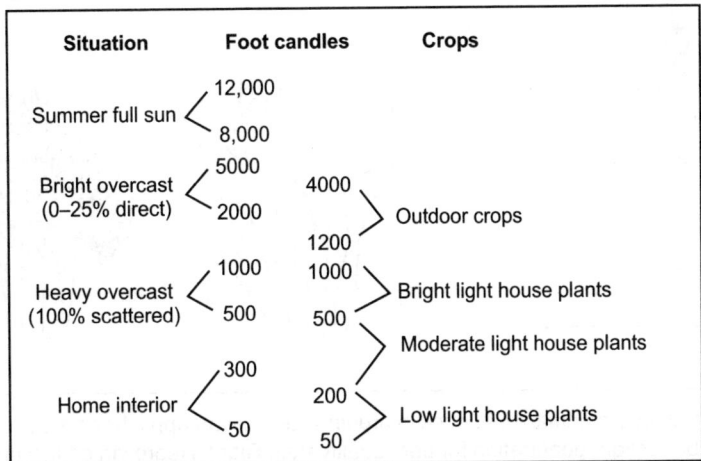

Fig. 6.2. Light intensity for various situations. Light intensity is measured in lux or foot-candles.

Landscape plants vary in their adaptation to light intensity. Many gardening texts divide plants into sun, partial sun and shade. However the experienced gardener understands the differences between these seven degrees of sun/shade:

1. Full sun—direct sun for at least 8 hours a day, including from 9 am to 4 pm.
2. Full sun with reflected heat—Where plants receive reflected heat from a building or other structure, temperatures can be extremely hot. This situation significantly limits the choice of plants for the site.
3. Morning shade with afternoon sun—This southwest and west reflected heat can be extremely hot and limiting to plant growth.
4. Morning sun with afternoon shade—This is an ideal site for many plants. The afternoon shade protects plants from extreme heat.
5. Filtered shade—Dappled shade filtered through trees can be bright shade to dark shade depending on the tree's canopy. The constantly moving shade pattern protects under-story plants from heat. In darker dappled shade, only the more shade tolerant plants will thrive.
6. Open shade—Plants may be in the situation where they have open sky above, but direct sunlight is blocked during the day by buildings, fences and other structures. Here only more shade tolerant plants will thrive.
7. Closed shade—The situation where plants are under a canopy blocking sunlight is most limiting. Only the most shade tolerant plants will survive this situation, like under a deck or covered patio.

In hot climates, temperature is often a limiting factor related to shade. Some plants, like impatiens and begonias, may require shade as an escape from heat. These plants will tolerate full sun in cooler summer climates. Light penetration is a primary influence on correct pruning. For example, prune dwarf apple trees to a Christmas tree shape. This gives better light penetration for best quality fruit. Mature fruit trees are thinned each spring for better light penetration. A hedge should be pruned with a wider base and narrow top. Otherwise the bottom thins out form the shading from above. A common mistake in pruning flowering shrubs is to shear off the top. The resulting regrowth gives a thick upper canopy that shades out the bottom foliage (Fig. 6.3).

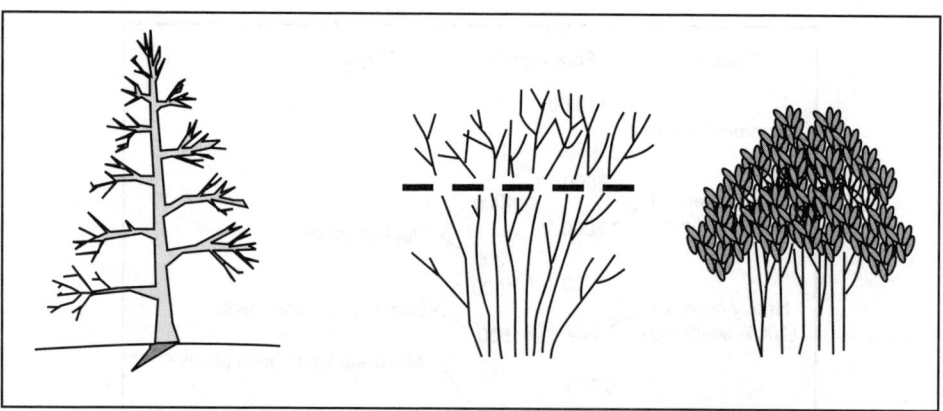

Fig. 6.3. Light penetration is a primary influence in pruning. Left: Dwarf apple trees are pruned to a Christmas tree shape to allow better light penetration for best quality fruit. Right: Regrowth on flowering shrubs that are sheared on top is a very heavy upper canopy growth. This shades out the bottom giving a woody base.

Light Duration

Light duration refers to the amount of time that a plant is exposed to sunlight. Travellers to Alaska often marvel at the giant vegetables and flowers that grow under the long days of the arctic sun even with cool temperatures. When starting transplants indoors, generally give plants 12–14 hours of light per day. Plants are generally intolerant of continuous light for 24 hours.

Photoperiod

The flowering response of many plants is controlled by the photoperiod (the length of uninterrupted darkness) (Fig. 6.4). Photoperiod response can be divided into three types.
1. Short day plants flower in response to long periods of night darkness. Examples include poinsettias, Christmas cactus, chrysanthemums, and single-crop strawberries.
2. Long day plants flower in response to short periods of night darkness. Examples include onions and spinach.
3. Day neutral plants flower without regard to the length of the night, but typically flower earlier and more profusely under long daylight regimes. Day neutral strawberries provide summer long harvesting (except during heat extremes).

TRANSPIRATION

Transpiration is a process similar to evaporation. It is a part of the water cycle, and it is the loss of water vapour from parts of plants (similar to sweating), especially in leaves but also in stems, flowers and roots. Leaf surfaces are dotted with openings which are collectively called stomata, and in most plants they are more numerous on the undersides of the foliage. The stoma are bordered by guard cells that open and close the pore. Leaf transpiration occurs through stomata, and can be thought of as a necessary 'cost' associated with the opening of the stomata to allow the diffusion of carbon dioxide gas from the air for photosynthesis. Transpiration also cools plants and enables mass flow of mineral nutrients and water from roots to shoots.

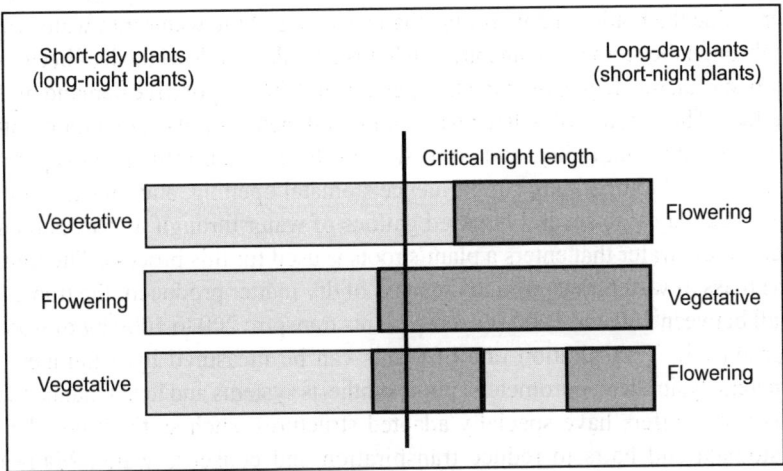

Fig. 6.4. Photoperiod and flowering, Left side: Short-day plants flower with uninterrupted long nights. Right side: Long-day plants flower with short nights or interrupted long nights.

Mass flow of liquid water from the roots to the leaves is caused by the decrease in hydrostatic (water) pressure in the upper parts of the plants due to the diffusion of water out of stomata into the atmosphere. Water is absorbed at the roots by osmosis, and any dissolved mineral nutrients travel with it through the xylem (Fig. 6.5).

Fig. 6.5. Some xerophytes will reduce the surface of their leaves during water deficiencies (left). If temperatures are cool enough and water levels are adequate the leaves expand again (right).

The rate of transpiration is directly related to the evaporation of water molecules from plant surface, especially from the surface openings, or stoma, on leaves. Stomatal transpiration accounts for most of the water loss by a plant, but some direct evaporation also takes place through the cuticle of the leaves and young stems. The amount of water given off depends somewhat upon how much water the roots of the plant have absorbed. It also depends upon such environmental conditions as light intensity, humidity, winds and temperature. A plant should not be transplanted in full sunshine because it may lose too much water and wilt before the damaged roots can supply enough water. Transpiration occurs as the sun warms the water inside the blade. The warming changes much of the water into water vapour. This gas can then escape through the stomata. Transpiration helps cool the inside of the leaf because the escaping vapour has absorbed heat, the degree of stomatal opening, and the evaporative demand of the atmosphere surrounding the leaf. The amount of water lost by a plant depends on its size, along with surrounding light intensity, temperature, humidity, and wind speed (all of which influence evaporative demand). Soil water supply and soil temperature can influence stomatal opening, and thus transpiration rate.

A fully grown tree may lose several hundred gallons of water through its leaves on a hot, dry day. About 90 per cent of the water that enters a plant's roots is used for this process. The transpiration ratio is the ratio of the mass of water transpired to the mass of dry matter produced; the transpiration ratio of crops tends to fall between 200 and 1000 (i.e. crop plants transpire 200 to 1000 kg of water for every kg of dry matter produced). Transpiration rate of plants can be measured by a number of techniques, including potometers, lysimeters, porometers, photosynthesis systems and heat balance sap flow gauges.

Desert plants and conifers have specially adapted structures, such as thick cuticles, reduced leaf areas, sunken stomata and hairs to reduce transpiration and conserve water. Many cacti conduct photosynthesis in succulent stems, rather than leaves, so the surface area of the shoot is very low. Many desert plants have a special type of photosynthesis, termed crassulacean acid metabolism or CAM photosynthesis in which the stomata are closed during the day and open at night when transpiration will be lower.

Light Absorption by Chlorophyll Induces Electron Transfer

The trapping of light energy is the key to photosynthesis. The first event is the absorption of light by a photoreceptor molecule. The principal photoreceptor in the chloroplasts of most green plants is chlorophyll *a*, a substituted tetrapyrrole. The four nitrogen atoms of the pyrroles are coordinated to a magnesium ion. Unlike a porphyrin such as heme, chlorophyll has a reduced pyrrole ring. Another distinctive feature of chlorophyll is the presence of phytol, a highly hydrophobic 20-carbon alcohol, esterified to an acid side chain.

Chlorophylls are very effective photoreceptors because they contain networks of alternating single and double bonds. Such compounds are called polyenes. They have very strong absorption bands in the visible region of the spectrum, where the solar output reaching earth also is maximal (Fig. 6.6). The peak molar absorption coefficient (ε) of chlorophyll a is higher than 10^5 M^{-1} cm^{-1}, among the highest observed for organic compounds.

What happens when light is absorbed by a molecule such as chlorophyll? The energy from the light excites an electron from its ground energy level to an excited energy level (Fig. 6.7). This high-energy electron can have several fates. For most compounds that absorb light, the electron simply returns to the ground state and the absorbed energy is converted into heat. However, if a suitable electron acceptor is nearby, the excited electron can move from the initial molecule to the acceptor (Fig. 6.8). This process results in the formation of a positive charge on the initial molecule (due to the loss of an electron) and a negative charge on the acceptor and is, hence, referred to as photoinduced charge separation. The site where the separational change occurs is called the reaction center. We shall see how the photosynthetic apparatus is arranged to make photoinduced charge separation extremely efficient. The electron, extracted from its initial site by absorption of light, can reduce other species to store the light energy in chemical forms.

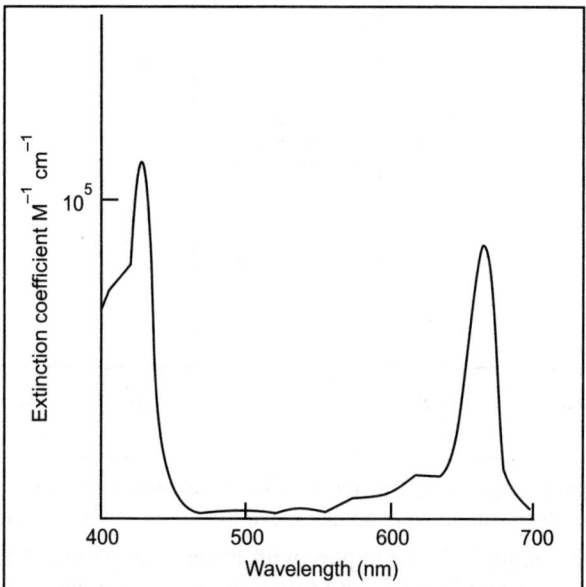

Fig. 6.6. Light absorption by chlorophyll A. Chlorophyll a absorbs visible light efficiently as judged by the extinction coefficients near 10^5 M^{-1} cm^{-1}.

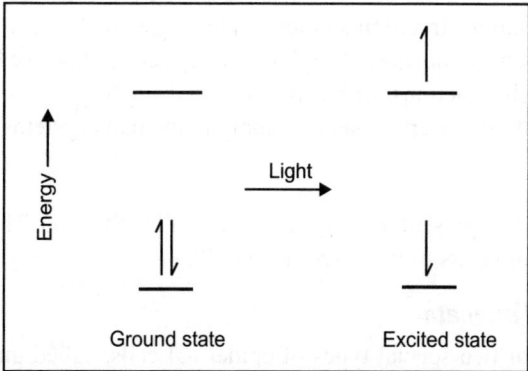

Fig. 6.7. Light absorption: The absorption of light leads to the excitation of an electron from its ground state to a higher energy level.

Photosynthetic bacteria and the photosynthetic reaction centers of green plants have a common core

Photosynthesis in green plants is mediated by two kinds of membrane-bound, light-sensitive complexes—photosystem I (PS I) and photosystem II (PS II). Photosystem I typically includes 13 polypeptide chains, more than 60 chlorophyll molecules, a quinone (vitamin K_1), and three 4Fe-4S clusters. The total molecular mass is more than 800 kd.

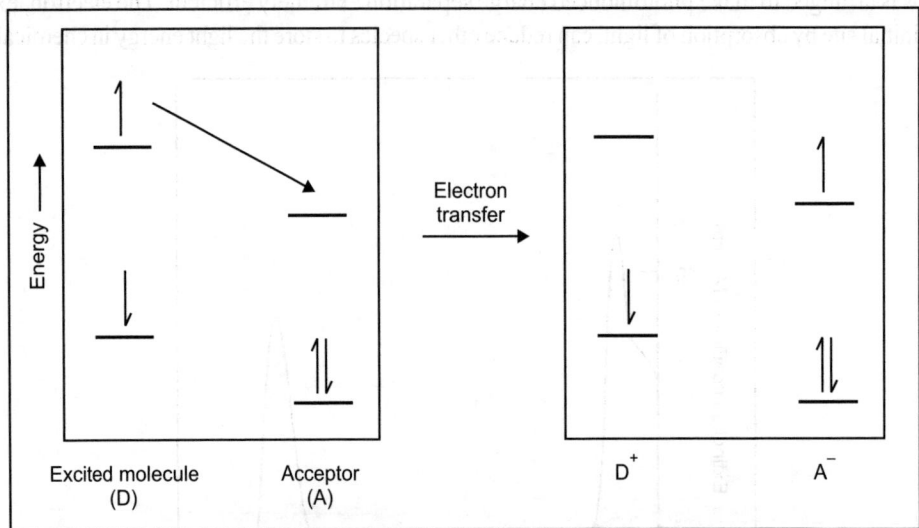

Fig. 6.8. Photoinduced charge separation. If a suitable electron acceptor is nearby, an electron that has been moved to a high energy level by light absorption can move from the excited molecule to the acceptor.

Photosystem II is only slightly less complex with at least 10 polypeptide chains, more than 30 chlorophyll molecules, a nonheme iron ion, and four manganese ions. Photosynthetic bacteria such as *Rhodopseudomonas viridis* contain a simpler, single type of photosynthetic reaction center, the structure of which was revealed at atomic resolution. The bacterial reaction center consists of four polypeptides: L (31 kd), M (36 kd), and H (28 kd) subunits and C, a *c*-type cytochrome. The results of sequence comparisons and low-resolution structural studies of photosystems I and II revealed that the bacterial reaction center is homologous to the more complex plant systems. Thus, we begin our consideration of the mechanisms of the light reactions within the bacterial photosynthetic reaction center, with the understanding that many of our observations will apply to the plant systems as well.

Stomatal Movement

The large number of tiny pores present in every leaves are called Stomata. These stomata are located on the surface of the epidermal layers of the leaves (Fig. 6.9).

Opening and closing of stomata

Each stoma is composed of two special types of epidermal cells, called guard cells. The guard cells control the opening and closing of stomata, which in turn regulate the loss of water from the leaves to the atmosphere through transpiration. The size of typical stomatal pore ranges from 3–12 micrometer in width and from 10–14 micrometer in length; the number of such tiny pores.

Stomata varies from 1000 to 60,000 per cm^2 of the leaf surface. Stomata are found on the lower (Abaxial) as well as (Adaxial) monocot leaves (grasses), the stomatal density and number is almost the same on both the surfaces. Leaves of dicot plants usually contain fewer stomata on the upper surface. Opening and closing if stomata is controlled by accumulation of solutes in the guard cells. Solutes are taken in by the guard cells from the neighbouring epidermal cells and mesophyll cells. As a result both the osmotic potential and water potential of the guard cells are lowered. This would create a water potential gradient between the guard cells and the neighbouring cells, and make the water move into the guard cells. The guard cells become turgid and swell in size, resulting in the stomatal opening. With a decline in guard cell solutes, water moves out of the guard cells, making them Flaccid. As a result, the stomata close. Thus the stomatal opening and closing is controlled by osmotic movement of water in or out of the guard cells along the water potential gradient.

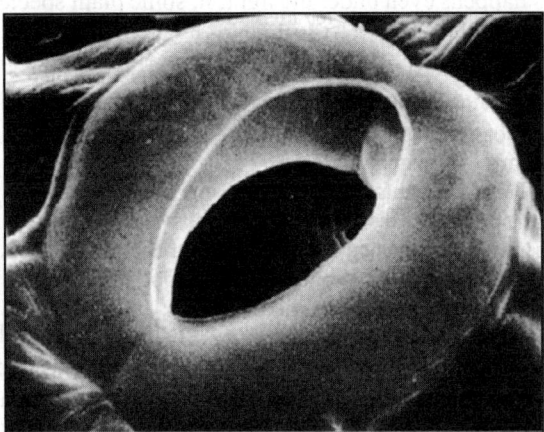

Fig. 6.9. Stomata.

The major solute, which is taken in by the guard cells from the neighbouring cells, is potassium. The rise in potassium levels causes stomatal opening and its decrease causes stoamatal closing. The uptake in the potassium controls the gradient in water potential. This in turn, triggers osmotic flow of water into the guard cells raising the turgor pressure. Role of K in the stomatal opening is now universally accepted. The extent K accumulation in the guard cells determines the size of the stomatal opening.

The accumulation of large amounts of potassium in the guard cells is electrically balanced by the uptake of negatively charged ions-chloride and malate. The high amount of malate in the guard cells of open stomata accumulates by hydrolysis of starch.

Factors affecting Stomatal movement

Many environmental factors affect stomatal movement. Most important factor includes light, temperature, water availability to plants and CO_2 concentration. Some endogenous factors like K, Cl^- and H^- ions and organic acids, also influence stomatal movement.

1. Light: Stomata open in the presence of light and close in darkness. Light intensity required to open the stomata is very low, as compared to the intensity required for photosynthesis. Even moonlight is sufficient to keep the stomata open in some plant species. In plants with Crassulacean Acid Metabolism (CAM), stomata open during the dark and close during the day. This unique kind of behaviour of stomata is a kind of adaptation to conserve moisture in CAM plants, for example pineapple, agave, etc.

2. Temperature: Stomata tend to open more with temperature and close with a decrease in temperature. In some plant species, stomata remains closed even under Stomatal opening and closing continuous light at 0°C. However if the temperature is increased, stomatal opening in

these species increases. At temperatures higher than 30°C, there is a decline in stomatal opening in some species.

3. Water availability: If the water availability to plants is less and transpiration rate is high, plants undergo water stress. Water stress also called, water deficit or moisture deficit, includes stomatal closure. This happens to conserve moisture in plants by cutting down the transpirational loss of water.

4. CO_2 concentration: With an increase in the carbon dioxide concentration inside the leaf, the stomata close. This happens even under the light. In some plant species stomata also close even if we merely breathe on their leaves. It is the internal leaf carbon dioxide concentration rather than the atmospheric carbon dioxide concentration that dictates the stomatal opening. If plants are transferred to carbon dioxide free environment, but kept in darkness, the stomata will still remain closed. This means that since the internal carbon dioxide is not utilised due to the absence of photosynthesis in the dark, it influences the stomata to remain closed. However if these plants are exposed to light, photosynthesis will utilise the carbon dioxide permitting the stomata to open.

Photomorphogenesis

Big sessile organisms like plants and fungi and small organisms whose motility can't move them that far, like bacteria and protists, have no option but to function in the environment they are found in. For photosynthetic organisms it has been adaptive for them to develop mechanisms to sense their light environment, and adjust their form and metabolism to optimise their performance under their local conditions. Since light environments change, these organisms have also developed the ability to continuously adjust their function to current conditions. Taken together these responses to light constitute the phenomenon known as photomorphogenesis.

The definition of photomorphogenesis, as applied in this module, is any change in form or function of an organism occurring in response to changes in the light environment. Photomorphogenesis is often defined as light-regulated plant development (Fig. 6.10), but there are also changes in morphology and/or cell structure and function, which occur as transient acclimatisations to a changing environment, which are also light regulated.

Particularly if this more inclusive definition is used, photomorphogenesis is a process common to organisms well beyond the plant kingdom. While there may be only a few examples of photo-morphogenesis in the animal kingdom, it is a common feature of development in fungi, protists, and bacteria, as well as plants. While this module will focus on what is known from studies of plant photomorphogenesis, there will be selected examples from other kingdoms.

Information content of light

Photomorphogenesis is an organismal response to information present in the light environment. There is information concerning: simple presence of light, light direction, light intensity, light duration, spectral quality and polarisation. Studies have demonstrated photomorphogenic effects of all of these parameters of the light environment in diverse organisms.

However, it must be acknowledged that the responses to polarisation were observed in laboratory conditions that may not have any parallel in the natural light environment. It should also be noted that the earliest responses to light exhibited by germinating seedlings are initiated by the mere presence (or absence) of light, information of the simplest sort possible.

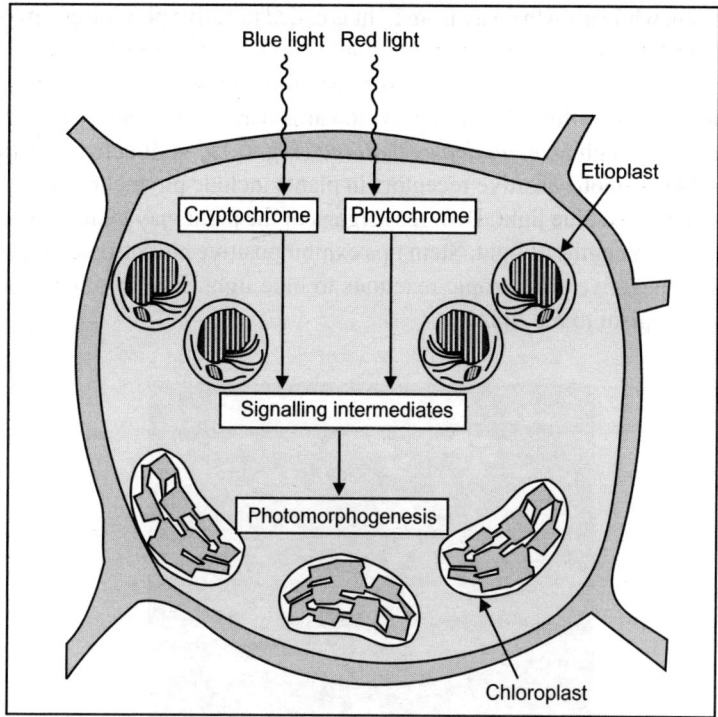

Fig. 6.10. Photomorphogenesis as a morphological and as a cellular process.

Phytochrome

Phytochrome is a photoreceptor, a pigment that plants use to detect light. It is sensitive to light in the red and far-red region of the visible spectrum. Many flowering plants use it to regulate the time of flowering based on the length of day and night (photoperiodism) and to set circadian rhythms. It also regulates other responses including the germination of seeds, elongation of seedlings, the size, shape and number of leaves, the synthesis of chlorophyll, and the straightening of the epicotyl or hypocotyl hook of dicot seedlings. It is found in the leaves of most plants.

Biochemically, phytochrome is a protein with a bilin chromophore. Phytochrome has been found in most plants including all higher plants; very similar molecules have been found in several bacteria. A fragment of a bacterial phytochrome now has a solved three-dimensional protein structure. Other plant photoreceptors include cryptochromes and phototropins, which are sensitive to light in the blue and ultraviolet regions of the spectrum.

PHOTOTROPISM

Phototropism is directional growth in which the direction of growth is determined by the direction of the light source. In other words, it is the growth and response to a light stimulus. Phototropism is most often observed in plants, but can also occur in other organisms such as fungi. The cells on the plant that are farthest from the light have a chemical called auxin that reacts when phototropism occurs. This causes the plant to have elongated cells on the farthest side from the light. Phototropism is one of the many plant tropisms or movements which respond to external stimuli. Growth towards a light source is a positive phototropism, while growth away from light is called negative phototropism (or Skototropism). Most plant shoots exhibit positive phototropism, while roots usually exhibit negative phototropism, although gravitropism may play a larger role in root behaviour and growth. Some vine shoot tips exhibit negative phototropism, which allows them to grow towards dark, solid objects and climb them.

Phototropism in plants such as *Arabidopsis thaliana* (Fig. 6.11) is directed by blue light receptors called phototropins. Other photosensitive receptors in plants include phytochromes that sense red light and cryptochromes that sense blue light. Different organs of the plant may exhibit different phototropic reactions to different wavelengths of light. Stem tips exhibit positive phototropic reactions to blue light, while root tips exhibit negative phototropic reactions to blue light. Both root tips and most stem tips exhibit positive phototropism to red light.

Fig. 6.11. The Thale Cress (*Arabidopsis thaliana*) is regulated by blue to UV light.

Phototropism is enabled by auxins. Auxins are plant hormones that have many functions. In this respect, auxins are responsible for expelling protons (by activating proton pumps) which decreases pH in the cells on the dark side of the plant. This acidification of the cell wall region activates enzymes known as expansins which break bonds in the cell wall structure, making the cell walls less rigid. In addition, the acidic environment causes disruption of hydrogen bonds in the cellulose that makes up the cell wall. The decrease in cell wall strength causes cells to swell, exerting the mechanical pressure that drives phototropic movement.

Other Light Responses

1. Etiolation is the response of a plant when light is nearly (or completely) absent.
2. Heliotropism is the diurnal motion of plant parts (flowers or leaves) in response to the direction of the sun. It is not a phototropism since it does not involve growth.
3. Photonasty involves the movement of plant parts that does not involve growth but is triggered by light. The plant movement is not determined by the direction of light so it is not a phototropism. Photonasty in prayer plant (*Maranta leuconeura*) involves the downward movement of leaves when they receive light in the morning.
4. Phototaxis is movement of an entire organism in which the direction of movement is determined by the direction of light. It occurs in some motile microbes such as Euglena and algae. It is not a phototropism because growth is not required.
5. Photo-orientation occurs within a plant cell when chloroplasts change their positions depending upon light intensity. This was discovered in 1987 by Chelsea Polevy and Kelsey Joyce when experimenting in their laboratory. When the light intensity is high, chloroplasts move to the edge of the cell to reduce photobleaching (destruction of chlorophyll). In low light, chloroplasts tend to spread out within the protoplasm to maximise their capture of light energy. Photo-orientation is also not a phototropism.

PHOTOPERIODISM

Photoperiodicity is the physiological reaction of organisms to the length of day or night. It occurs in plants and animals. Many flowering plants use a photoreceptor protein, such as phytochrome or cryptochrome, to sense seasonal changes in night length, or photoperiod, which they take as signals to flower. In a further subdivision, obligate photoperiodic plants absolutely require a long or short enough night before flowering, whereas facultative photoperiodic plants are more likely to flower under the appropriate light conditions, but will eventually flower regardless of night length.

Photoperiodic flowering plants are classified as long-day plants or short-day plants, though the regulatory mechanism is actually governed by hours of darkness, not the length of the day.

Modern biologists believe that it is the coincidence of the active forms of phytochrome or cryptochrome, created by light during the daytime, with the rhythms of the circadian clock that allows plants to measure the length of the night. Other than flowering, photoperiodism in plants includes the growth of stems or roots during certain seasons, or the loss of leaves.

Long-day Plants

A long-day plant requires fewer than a certain number of hours of darkness in each 24-hour period to induce flowering. These plants typically flower in the northern hemisphere during late spring or early summer as days are getting longer. In the northern hemisphere, the longest day of the year is on or about 21 June (solstice). After that date, days grow shorter (i.e. nights grow longer) until 21 December (solstice). This situation is reversed in the southern hemisphere (i.e. longest day is 21 December and shortest day is 21 June). In some parts of the world, however, 'winter' or 'summer' might refer to rainy versus dry seasons, respectively, rather than the coolest or warmest time of year. Some long-day obligate plants are:

1. Carnation (*Dianthus*).
2. Henbane (*Hyoscyamus*).
3. Oat (*Avena*).

4. Ryegrass (*Lolium*).
5. Clover (*Trifolium*).
6. Bellflower (*Campanula carpatica*).

Some long-day facultative plants are:
1. Pea (*Pisum sativum*).
2. Barley (*Hordeum vulgare*).
3. Lettuce (*Lactuca sativa*).
4. Wheat (*Triticum aestivum*, spring wheat cultivars).
5. Turnip (*Brassica rapa*).
6. *Arabidopsis thaliana* (model organism).

Short-day Plants

Short-day plants flower when the night is longer than a critical length. They cannot flower under long days or if a pulse of artificial light is shone on the plant for several minutes during the middle of the night; they require a consolidated period of darkness before floral development can begin. Natural nighttime light, such as moonlight or lightning, is not of sufficient brightness or duration to interrupt flowering.

In general, short-day (i.e. long-night) plants flower as days grow shorter (and nights grow longer) after 21 June in the northern hemisphere, which is during summer or fall. The length of the dark period required to induce flowering differs among species and varieties of a species. Photoperiod affects flowering when the shoot is induced to produce floral buds instead of leaves and lateral buds. Note that some species must pass through a 'juvenile' period during which they cannot be induced to flower—common cocklebur is an example of a plant species with a remarkably short period of juvenility and plants can be induced to flower when quite small. Some short-day obligate plants are:
1. Chrysanthemum.
2. Coffee.
3. Poinsettia.
4. Strawberry.
5. Tobacco, var. Maryland Mammouth.
6. Common duckweed (*Lemna minor*).
7. Cocklebur (*Xanthium*).
8. Maize—tropical cultivars only.

Some short-day facultative plants are:
1. Hemp (*Cannabis*).
2. Cotton (*Gossypium*).
3. Rice.
4. Sugar cane.

Day-neutral Plants

Day-neutral plants, such as cucumbers, roses and tomatoes, do not initiate flowering based on photoperiodism at all; they flower regardless of the night length. They may initiate flowering after attaining a certain overall developmental stage or age or in response to alternative environmental stimuli, such as vernalisation (a period of low temperature), rather than in response to photoperiod.

VERNALISATION

Vernalisation (from Latin: vernus, of the spring) is the acquisition of a plant's ability to flower or germinate in the spring by exposure to the prolonged cold of winter. After vernalisation, plants have acquired the ability to flower, but they may require additional seasonal cues or weeks of growth before they will actually flower. Many plants grown in temperate climates require vernalisation and must experience a period of low winter temperature to initiate or accelerate the flowering process. This ensures that reproductive development and seed production occurs in spring and summer, rather than in autumn. The needed cold is often expressed in chill hours. Typical vernalisation temperatures are between 5° and 10°C (40° and 50°F). For many perennial plants, such as fruit tree species, a period of cold is needed to break dormancy, prior to flowering. Many monocarpic annuals and biennials, including some ecotypes of *Arabidopsis thaliana* and winter cereals such as wheat, must go through a prolonged period of cold before flowering occurs.

ERWIN BÜNNING HYPOTHESIS

In the early-1930s, Bünning proposed that a circadian rhythm of sensitivity to light was being used to measure how much daylight was 'encroaching' into the night-time, thus measuring the photoperiod. He demonstrated that plants and insects behaved according to circadian rhythms, whether or not they were in continuous light or darkness. His crossing experiments with bean plants of different periods in 1932 demonstrated that the next generation had periods of intermediate durations, supporting the suggestion that circadian rhythms are heritable. In 1935, Erwin Bünning determined in plants the genetic origin of the 'biological clock', a term he coined. It took at least a decade of experimental work by Bunning and others for this proposal, called the Bünning Hypothesis, to get firmly accepted by chronobiologists.

Chronobiology investigates biological rhythms occurring in some prokaryotes and in all eukaryotes from fungi to humans. The most common of these rhythms, called Circadian rhythms or Biological Clocks, have a periodicity of 24 hours.

Germination

Germination is the process in which a plant or fungus emerges from a seed or spore, respectively, and begins growth. The most common example of germination is the sprouting of a seedling from a seed of an angiosperm or gymnosperm. However, the growth of a sporeling from a spore, for example the growth of hyphae from fungal spores, is also germination. In a more general sense, germination can imply anything expanding into greater being from a small existence or germ.

Seed Germination

Germination is the growth of an embryonic plant contained within a seed; it results in the formation of the seedling. The seed of a higher plant is a small package produced in a fruit or cone after the union of male and female sex cells. All fully developed seeds contain an embryo and, in most plant species some store of food reserves, wrapped in a seed coat. Some plants produce varying numbers of seeds that lack embryos, these are called empty seeds, and never germinate. Most seeds go through a period of quiescence where there is no active growth; during this time the seed can be safely transported to a new location and/or survive adverse climate conditions until circumstances are favourable for growth. Quiescent seeds are ripe seeds that do not germinate because they are subject to external environmental conditions that prevent the initiation of metabolic processes and cell growth. Under favourable conditions, the seed begins to germinate and the embryonic tissues resume growth, developing towards a seedling.

Factors affecting seed germination

Seed germination depends on both internal and external conditions. The most important external factors include temperature, water, oxygen and sometimes light or darkness. Various plants require different variables for successful seed germination, often this depends on the individual seed variety and is closely linked to the ecological conditions of a plant's natural habitat. For some seeds, their future germination response is affected by environmental conditions during seed formation; most often these responses are types of seed dormancy.

LIGHT AND ANIMALS

Artificial nighttime lighting changes animals' behaviour. Some animals shun lit areas while others take advantage of the light to forage or hunt at the expense of other species. Everyone has seen moths clustering around outdoor lights—attracted to the brightness when they should be looking for a mate. This situation looks like a feast for bats, but some species of bats shun lit areas so the natural balance of bat species can be skewed by night time lighting. Owls are supremely adapted to hunting in nearly total blackness, but lose that advantage over rodents whenever there is man-made light available.

Many large shy predators avoid lit areas altogether. These animals need dark passageways to go from one hunting field to another—a lit roadway can act as a fence, cutting their territory in half.

Lit buildings can kill birds: Most of our songbirds migrate at night. Unfortunately, warblers can be confused by artificial night time lights and often collide with lit structures. Birds are reluctant to fly out of the lighted area into the dark, and often continue to flap around in the beam of light until they drop to the ground from exhaustion. For this reason, such obstacles should not be lit overnight.

Light and hormones: Light at night affects the normal melatonin levels produced in the pineal gland. This hormone, melatonin, is responsible for controlling the day-night changes in hormones that affect sleep and the immune system. Melatonin also directs the seasonal changes in sex hormones that rule animals mating season. Change the light levels at night and one can change the course of history for a species. Light at night can allow young male Deer Mice to become sexually mature later into the Fall, when normally the shorter nights cause young mice to delay sexual maturity until the following Spring, when more food is available. Light at night increases the testosterone levels in these animals through the modulating effects of melatonin. Experiments on rats and hamsters showed that continuous night time light advanced the onset of puberty in females compared to rats that experienced normal day-night cycles. Besides a variety of rodents (which are an important food source for carnivores), this effect is also true in female cattle, sheep, goats and horses. Studies on both male and female Rhesus Monkeys indicate that significant alteration of hormones levels and the timing of sexual maturity can accomplished by modifying the natural day-night cycle length. This effect has also been noted in humans, where long nights correspond to higher melatonin levels and decreased ovarian activity and conversely, short nights (or light at night) correlate to increased ovarian activity. These effects can be passed on to offspring.

Studies on Meadow Voles show that the young of mother voles that have been exposed to light at night develop less quickly (both weight and sexual maturity) than litters of mother voles that have experienced long nights.

Pigmentation: Skin colour is indirectly affected by light through the mediation of the eyes or other receptors. The characteristic lack of pigment in cave animals is associated with darkness. Blind cave amphibians and fishes with little or no colour have been found to develop abundant pigment in the skin after exposure to normal day light.

Photokinesis: Ablility to control light. To the greater extent, one could make themselves invisible, concentrate light particles/waves into lasers and manipulate any part of the light spectrum (ultra violet, microwaves, gamma rays, etc.); making Photokinesis the 'father' of Autokinesis, Lumokinesis, Optikinesis, Magnetokinesis, etc. can be used to create illusions, this can be used to generate light shields/force fields, and project healing energies too, one could create a blast of light which impales enemies, or even create a ball or shape of light to blast.

Vision: Many aquatic animals may show some activity response to the increase or decrease of light. Studies of the diurnal vertical migration of zooplankton indicate that certain species may react to light from the surface at 800 m and possibly at 1000 m or more than half a mile down.

Orientation: The term tropism is used for orientation by growth or turgor movements exhibited by sessile forms. If the orientation is to gravity the term geotropism is issued. If it is to light, the growth movements is referred to as phototropism. On the other hand, the orientation of the locomotion of motile organisms is referred to as a taxis.

Photoperiodism in Insects and Mites

The synchronisation of growth and reproduction in insect species with the favourable seasons within their range of dispersal, and of their activity with the daily rhythm of light and darkness, are of extremely high survival value. Specific diurnal or nocturnal activity in insects is related to temperature and water requirements, feeding habits, and morphological appearance. Diapause and many forms of polymorphism in insects of the temperature zone are season-bound.

Diapause

Diapause is the delay in development in response to regularly and recurring periods of adverse environmental conditions. It is considered to be a physiological state of dormancy with very specific initiating and inhibiting conditions. Diapause is a mechanism used as a means to survive predictable, unfavourable environmental conditions, such as temperature extremes, drought or reduced food availability. Diapause is most often observed in arthropods, especially insects, and in the embryos of many of the oviparous species of fish in the order Cyprinodontiformes. (Diapause does not occur in embryos of the viviparous and ovoviviparous species of Cyprinodontiformes.)

Diapause is not only induced in an organism by specific stimuli or conditions, but once it is initiated, only certain other stimuli are capable of bringing the organism out of diapause. The latter feature is essential in distinguishing diapause as a different phenomenon from other forms of dormancy such as stratification, and hibernation.

Activity levels of diapausing stages can vary considerably among species. Diapause may occur in a completely immobile stage, such as the pupae and eggs or it may occur in very active stages that undergo extensive migrations, such as the adult Monarch butterfly, *Danaus plexippus*. In cases where the insect remains active, feeding is reduced and reproductive development is slowed or halted.

Diapause in insects is a dynamic process consisting of several distinct phases. While diapause varies considerably from one taxon of insects to another, these phases can be characterised by particular sets of metabolic processes and responsiveness of the insect to certain environmental stimuli. Diapause can occur during any stage of development in arthropods, but each species exhibits diapause in specific phases of development. Reduced oxygen consumption is typical as is reduced movement and feeding.

Seasonal Dimorphism

In temperate regions the spring and summer generations nearly always differ slightly in appearance. The Comma Polygonia *c*-album, e.g. produces a more brightly coloured form called hutchinsoni in early summer, but the progeny of this brood have darker and more sombre undersides, and a more ragged wing shape. This generation hibernates as adults, and the winter coloration provides them with a more effective camouflage when they are hiding amongst dead brown leaves at the base of bushes and trees. A more extreme example is the Map butterfly *Araschnia levana*, in which the spring generation are orange with black spots, and resemble small Fritillaries. Summer brood Map butterflies however are black with prominent white bands and resemble miniature White Admirals (Fig. 6.12).

Araschnia levana (spring generation)   Araschnia levana (summer generation)

Fig. 6.12. *Araschinia levana.*

The formation of different wing patterns in the spring and summer broods is known to be triggered by temperature and length of day during the pupal stage, but while the mechanisms are well understood, the 'purpose' and possible benefits of the dimorphism is unknown. It is likely however that such seasonal differences in appearance somehow give the species an advantage over predators. In the case of certain tropical species such as *Taygetis mermeria* from the Amazon, the advantages gained from having 'rainy season' and 'dry season' forms are more obvious. The butterflies spend long periods at rest, settled among leaf litter on the forest floor. In the dry season the leaves are desiccated and orange-brown in colour, so the butterfly has evolved an orange-brown form which simulates the appearance of dead leaves. The wet season form is much darker, with olive-brown wings that are a more effective camouflage in the tropical summer when the foliage is greener and denser, and the shadows darker.

CIRCADIAN RHYTHM

A circadian rhythm is an endogenously driven roughly 24-hour cycle in biochemical, physiological, or behavioural processes. Circadian rhythms have been widely observed, in plants, animals, fungi and cyanobacteria. The formal study of biological temporal rhythms such as daily, tidal, weekly, seasonal, and annual rhythms is called chronobiology. Although circadian rhythms are endogenous ('built-in', self-sustained), they are adjusted (entrained) to the environment by external cues called zeitgebers, the primary one of which is daylight.

Photosensitive proteins and circadian rhythms are believed to have originated in the earliest cells, with the purpose of protecting the replicating of DNA from high ultraviolet radiation during the daytime.

As a result, replication was relegated to the dark. The fungus Neurospora, which exists today, retains this clock-regulated mechanism.

Circadian rhythms allow organisms to anticipate and prepare for precise and regular environmental changes; they have great value in relation to the outside world. The rhythmicity appears to be as important in regulating and coordinating internal metabolic processes, as in coordinating with the environment. This is suggested by the maintenance (heritability) of circadian rhythms in fruit flies after several hundred generations in constant laboratory conditions, as well as in creatures in constant darkness in the wild, and by the experimental elimination of behavioural but not physiological circadian rhythms in quail.

Importance in Animals

Circadian rhythmicity is present in the sleeping and feeding patterns of animals, including human beings. There are also clear patterns of core body temperature, brain wave activity, hormone production, cell regeneration and other biological activities. In addition, photoperiodism, the physiological reaction of organisms to the length of day or night, is vital to both plants and animals, and the circadian system plays a role in the measurement and interpretation of day length.

Timely prediction of seasonal periods of weather conditions, food availability or predator activity is crucial for survival of many species. Although not the only parameter, the changing length of the photoperiod (daylength) is the most predictive environmental cue for the seasonal timing of physiology and behaviour, most notably for timing of migration, hibernation and reproduction.

Impact of light–dark cycle

The rhythm is linked to the light–dark cycle. Animals, including humans, kept in total darkness for extended periods eventually function with a free-running rhythm. Each 'day', their sleep cycle is pushed back or forward, depending on whether their endogenous period is shorter or longer than 24 hours. The environmental cues that reset the rhythms each day are called zeitgebers. It is interesting to note that totally-blind subterranean mammals (e.g. blind mole rat *Spalax* sp.) are able to maintain their endogenous clocks in the apparent absence of external stimuli. Although they lack image-forming eyes, their photoreceptors (detect light) are still functional; as well, they do surface periodically.

Free-running organisms that normally have one or two consolidated sleep episodes will still have them when in an environment shielded from external cues, but the rhythm is, of course, not entrained to the 24-hour light–dark cycle in nature.

The sleep–wake rhythm may, in these circumstances, become out of phase with other circadian or ultradian rhythms such as metabolic, hormonal, CNS electrical, or neurotransmitter rhythms. Recent research has influenced the design of spacecraft environments, as systems that mimic the light–dark cycle have been found to be highly beneficial to astronauts.

Arctic animals

Norwegian researchers at the University of Tromsø have shown that some Arctic animals (ptarmigan, reindeer) show circadian rhythms only in the parts of the year that have daily sunrises and sunsets. In one study of reindeer, animals at 70 degrees North showed circadian rhythms in the autumn, winter, and spring, but not in the summer. Reindeer at 78 degrees North showed such rhythms only autumn and spring. The researchers suspect that other Arctic animals as well may not show circadian rhythms in the constant light of summer and the constant dark of winter.

However, another study in northern Alaska found that ground squirrels and porcupines strictly maintained their circadian rhythms through 82 days and nights of sunshine. The researchers speculate that these two small mammals see that the apparent distance between the sun and the horizon is shortest once a day, and, thus, a sufficient signal to adjust by.

Butterfly migration

The navigation of the fall migration of the Eastern North American monarch butterfly (*Danaus plexippus*) to their overwintering grounds in central Mexico uses a time-compensated sun compass that depends upon a circadian clock in their antennae.

In Plants

Plant circadian rhythms tell the plant what season it is in and when to flower for the best chance of attracting insects to pollinate them and can include leaf movement, growth, germination, stomatal/gas exchange, enzyme activity, photosynthetic activity, and fragrance emission. Circadian rhythms occur as a biological rhythm with light, are endogenously generated and self sustaining, and are relatively constant over a range of ambient temperatures. Circadian rhythms feature a transcriptional feedback loop, a presence of PAS proteins, and several photoreceptors that fine-tune the clock to different light conditions. Anticipation of changes in the environment changes the physiological state that provides plants with an adaptive advantage. A better understanding of plant circadian rhythms has applications in agriculture such as helping farmers stagger crop harvests thus extending crop availability, and to secure against massive losses due to weather (Fig. 6.13).

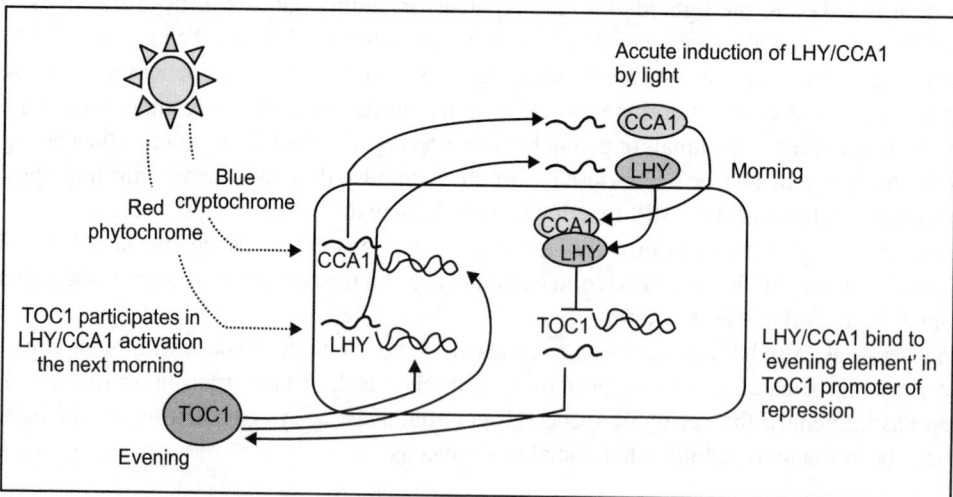

Fig. 6.13. Diagram showing a small portion of the transcriptional feedback loop in Arabidopsis. LHY and CCA1 are considered negative elements due to its repression against TOC1 in the morning while TOC1 is considered a positive element because it results in increased transcription of LHY and CCA1 during the evening because of its accumulation.

Clocks are set through signals such as light, temperature, and nutrient availability, so that the internal time matches the local time. Light is the signal and is sensed by a wide variety of photoreceptors. Red and blue light are absorbed through several phytochromes and cryptochromes.

DIURNATION

Diurnation means the habit of some animals, of sleeping, being dormant, or remaining quiescent during the day, as contrasted with their activities at night.

Diel Activities of Animals

Understanding diel patterns of habitat selection and activity is a particularly challenging ecological problem. Some species are clearly specialised for either diurnal or nocturnal foraging, but for other animals choosing whether to forage during day or night is a critical decision. During the day, food is generally easier to detect but predation risk is often higher. Habitat selection is a closely related behaviour because the value of a habitat for feeding or hiding differs between day and night. Habitat that provides a good trade-off between energy intake and mortality risk during the day might be unprofitable for nocturnal foraging, and habitat highly profitable for nocturnal foraging might be too risky during daytime.

Stream-dwelling salmonids, for example, exhibit complex and variable diel activity patterns. Summer observations have often found most, but not all, salmonids feeding during the day and hiding at night. Bremset and Cunjak observed juvenile Atlantic salmon (*Salmo salar*) generally shifting to nocturnal feeding and diurnal hiding during cold seasons, but there were many exceptions to this generality. Nocturnal feeding in winter is often attributed to reduced metabolic needs and higher predation risk at low temperature.

However, Amundsen observed nocturnal feeding in Atlantic salmon during August and September, which they attributed to seasonal changes in food availability. Nocturnal feeding has been shown to be more prevalent among larger fish in some cases but not in others. Interactions between activity and habitat selection are illustrated by observations that salmonids use different habitat for feeding vs. hiding, and that diel activity patterns vary with reach-scale habitat features. Metcalfe, after a series of controlled laboratory experiments on stream-dwelling juvenile salmon, drew this conclusion:

> Daily activity patterns are therefore suggested to be the result of a complex trade-off between growth and survival, which takes account of diel fluctuations in food availability, food capture efficiency and predation risk; individual variation in the extent of diurnal feeding in salmon may result from state-dependent differences in the benefits of rapid feeding and growth.

For convenience, we refer to the theory investigated here as SPHAST (state-based, predictive habitat- and activity-selection theory). SPHAST comprises two major assumptions:

1. Individual animals select the combination of diurnal and nocturnal habitat and activity that maximises their expected fitness over a future time horizon. (This time horizon is not the decision time step: individuals repeat their habitat and activity choice at a half-day time step, the start of each day and night, while considering its consequences over the much longer time horizon.) When individuals decide which activity to perform, they consider the consequences of both day and night activities, not just the activity selected for the current time; and they consider the best available habitat (within the area the animal is familiar with) for each activity.

2. An individual's expected fitness, for some combination of habitat and activity, is approximated by a measure (which we refer to as EF) that has two terms. First is the probability of surviving both starvation and predation risks over a future time horizon (90 d, in this study). The second term is the individual's size at the end of the time horizon, relative to two 'landmarks': the minimum size for reproduction and the size of the largest competing animal. This term represents

the importance of growth to fitness. Thus, EF depends on the individual's current physiological state and a simple projection of growth and mortality risks over the time horizon. This approximation of future fitness is especially useful for individuals not in the reproductive cycle (e.g. juveniles) because it does not explicitly represent reproduction; instead, it is assumed that eventual reproductive success is usefully approximated by the survival probability and growth rate over the upcoming 90 days.

Under these two assumptions, an individual's diel activity pattern depends on habitat availability, the individual's state, and the ways in which habitat conditions and activity affect the individual's growth and survival. The choice of activity and habitat during one diel phase (day or night) depends on expected choice of activity and habitat during the opposite phase. For example, the availability of good habitat for nocturnal feeding could result in diurnal hiding.

BIOLUMINESCENCE

Bioluminescence is the production and emission of light by a living organism. Its name is a hybrid word, originating from the Greek bios for 'living' and the Latin lumen 'light'. Bioluminescence is a naturally occurring form of chemiluminescence where energy is released by a chemical reaction in the form of light emission. Fireflies, anglerfish, and other creatures produce the chemicals luciferin (a pigment) and luciferase (an enzyme). The luciferin reacts with oxygen to create light. The luciferase acts as a catalyst to speed up the reaction, which is sometimes mediated by cofactors such as calcium ions or ATP. The chemical reaction can occur either inside or outside the cell. In bacteria, the expression of genes related to bioluminescence is controlled by an operon called the Lux operon.

Bioluminescence occurs in marine vertebrates and invertebrates, as well as micro-organisms and terrestrial animals. Symbiotic organisms carried within larger organisms are also known to bioluminesce.

Bioluminescence is a form of luminescence or 'cold light' emission; less than 20 per cent of the light generates thermal radiation. It should not be confused with fluorescence, phosphorescence or refraction of light. Ninety per cent of deep-sea marine life are estimated to produce bioluminescence in one form or another. Most marine light-emission belongs in the blue and green light spectrum, the wavelengths that can transmit through the seawater most easily. However, certain loose-jawed fishes emit red and infrared light and the genus *Tomopteris emits* yellow bioluminescence.

Non-marine bioluminescence is less widely distributed, but a larger variety in colours is seen. The two best-known forms of land bioluminescence are fireflies and glow worms. Other insects, insect larvae, annelids, arachnids and even species of fungi have been noted to possess bioluminescent abilities. Some forms of bioluminescence are brighter (or only exist) at night, following a circadian rhythm.

Adaptations for Bioluminescence

There are five main theories for bioluminescent traits:

Counterillumination camouflage

In some squid species bacterial bioluminescence is used for counterillumination so the animal matches the overhead environmental light seen from below. In these animals, photoreceptive vesicles have been found that control the contrast of this illumination to create optimal matching. Usually these light organs are separate from the tissue containing the bioluminescent bacteria. However, in one species Euprymna scolopes these bacteria make up an integral component of the animal's light organ.

Attraction

Bioluminescence is used as a lure to attract prey by several deep sea fish such as the anglerfish. A dangling appendage that extends from the head of the fish attracts small animals to within striking distance of the fish. Some fish, however, use a non-bioluminescent lure.

The attraction of mates is another proposed mechanism of bioluminescent action. This is seen actively in fireflies, which use periodic flashing in their abdomens to attract mates in the mating season. In the marine environment this has only been well documented in certain small crustaceans called ostracod. It has been suggested that pheromones may be used for long-distance communication, and bioluminescence used at close range to 'home in' on the target.

Repulsion

Certain squid and small crustaceans use bioluminescent chemical mixtures or bioluminescent bacterial slurries in the same way as many squid use ink. A cloud of luminescence is expelled, confusing or repelling a potential predator while the squid or crustacean escapes to safety. Every species of firefly has larvae that glow to repel predators.

Communication

Communication between bacteria (quorum sensing) plays a role in the regulation of luminesence in many bacterial species. Using small extracellularly secreted molecules, they are able to adapt their behaviour to only turn on genes for light production when they are at high cell densities.

Illumination

While most marine bioluminescence is green to blue, the Black Dragonfish produces a red glow. This adaptation allows the fish to see red-pigmented prey, which are normally invisible in the deep ocean environment where red light has been filtered out by the water column.

TEMPERATURE EFFECT ON PLANTS

Sometimes temperatures are used in connection with day length to manipulate the flowering of plants. Chrysanthemums will flower for a longer period of time if daylight temperatures are 50°F. The Christmas cactus forms flowers as a result of short days and low temperatures. Temperatures alone also influence flowering. Daffodils are forced to flower by putting bulbs in cold storage in October at 35° to 40°F. The cold temperature allows the bulb to mature. The bulbs are transferred to the greenhouse in midwinter where growth begins. The flowers are then ready for cutting in 3 to 4 weeks.

Thermoperiod refers to daily temperature change. Plants produce maximum growth when exposed to a day temperature that is about 10° to 15°F higher than the night temperature. This allows the plant to photosynthesise (build up) and respire (break down) during an optimum daytime temperature, and to curtail the rate of respiration during a cooler night. High temperatures cause increased respiration, sometimes above the rate of photosynthesis. This means that the products of photosynthesis are being used more rapidly than they are being produced. For growth to occur, photosynthesis must be greater than respiration. Low temperatures can result in poor growth. Photosynthesis is slowed down at low temperatures. Since photosynthesis is slowed, growth is slowed, and this results in lower yields. Not all plants grow best in the same temperature range. For example, snapdragons grow best when night time temperatures are 55°F, while the poinsettia grows best at 62°F. Florist cyclamen does well under very cool conditions, while many bedding plants grow best at a higher temperature.

Buds of many plants require exposure to a certain number of days below a critical temperature (chilling hours) before they will resume growth in the spring. Peaches are a prime example; most cultivars require 700 to 1000 hours below 45°F and above 32°F before they break their rest period and begin growth. This time period varies for different plants. The flower buds of forsythia require a relatively short rest period and will grow at the first sign of warm weather. During dormancy, buds can withstand very low temperatures, but after the rest period is satisfied, buds become more susceptible to weather conditions, and can be damaged easily by cold temperatures or frost.

Temperature Considerations

Temperature factors that figure into plant growth potentials include the following:
1. Maximum daily temperature.
2. Minimum daily temperature.
3. Difference between day and night temperatures.
4. Average daytime temperature.
5. Average night time temperature.

Microclimates

The microclimate of a garden plays a primary role in actual garden temperatures. In mountain communities, changes in elevation, air drainage, exposure, and thermal heat mass (surrounding rocks) will make some gardens significantly warmer or cooler than the temperatures recorded for the area. In mountain communities, it is important to know where the local weather station is located so gardeners can factor in the difference in their specific location to forecast temperatures more accurately. Examples of factors to consider include the following:
1. Elevation—A 300 foot rise in elevation accounts for approximately 1°F drop in temperature.
2. Drainage—At night, cool air drains to low spots. Valley floors may be more than 10°F cooler than surrounding gardens on hillsides above the valley floor. That is why fruit orchards are typically located on the benches rather than on the valley floor.
3. Exposure—Southern exposures absorb more solar radiation than northern exposures. In mountain communities, northern exposures will have shorter growing seasons. In mountain communities, gardeners often place warm season plants, like tomatoes, on the south side of buildings to capture more heat. Based on local topography, buildings, fences, plantings, and garden areas may be protected from or exposed to cold and drying winds. They may also be exposed to or protected from warm and drying winds.
4. Thermal heat mass (surrounding rocks)—In many Colorado communities, the surrounding rock formations can form heat sinks creating wonderful gardening spots for local gardeners. Nestled in among the mountains, some gardeners have growing seasons several weeks longer than neighbours only a half mile away.

In cooler locations, rock mulch may give some frost protection and increase temperatures for enhanced crop growth. In warmer locations, rock mulch can significantly increase summer temperatures and water requirements of landscape plants. In Phoenix, Arizona, the urban heat island (with all their rock mulch instead of grass and trees) has significantly raised day and night temperatures. The upward convection of heat has become so strong that summer storms are going around the city and not raining on the urban heat island.

Impact of Heat on Crop Growth

Temperature affects the growth and productivity of plants, depending on whether the plant is a warm season or cool season crop.

Photosynthesis

Within limits, rates of photosynthesis and respiration both rise with increasing temperatures. As temperatures reach the upper growing limits for the crop, the rate of food used by respiration may exceed the rate at which food is manufactured by photosynthesis. For tomatoes, growth peaks at 96°F.

Temperature influence on growth

Seeds of cool season crops germinate at 40° to 80°F. Warm season crop seeds germinate at 50° to 90°F. In the spring, cool soil temperatures are a limiting factor for plant growth. In midsummer, hot soil temperatures may prohibit seed germination.

Examples of temperature influence on flowering

1. Tomatoes.
    (a) Pollen does not develop if night temperatures are below 55°F.
    (b) Blossoms drop if daytime temperatures rise above 95°F before 10 am.
    (c) Tomatoes grown in cool climates will have softer fruit with bland flavours.
2. Spinach (a cool season, short day crop) flowers in warm weather with long days.
3. Christmas cacti and poinsettias flower in response to cool temperatures and short days.

Examples of temperature influence on crop quality

1. High temperatures increase respiration rates, reducing sugar content of produce. Fruits and vegetables grown in heat will be less sweet.
2. In heat, crop yields reduce while water demand goes up.
3. In hot weather, flower colours fade and flowers have a shorter life.

Heat zone map

A new concept in plant selection is heat zone mapping, a measurement of the typical summer heat accumulation. It will help identify geographic areas that have adequate heat accumulation to mature various crops. It should be recognised that in mountain communities, minor changes in elevation and exposure (for example, south slopes versus north slopes) make significant differences in heat accumulation. A heat zone for a community's zip code may not reflect the actual growing conditions in any specific garden.

Impact of Cold Temperatures

Hardiness zone map

Hardiness zone maps indicate the average annual minimum temperature expected for geographic areas. While this is a factor in plant selection, it is only one of many factors influencing plant hardiness.

Plant hardiness

Hardiness refers to a plant's tolerance to cold temperatures. Low temperature is only one of many factors influencing plant hardiness (ability to tolerate cold temperatures).

Key hardiness factors include the following:
1. Photoperiod.
2. Genetics (source of plant material).
3. Low temperature.
4. Recent temperature pattern.
5. Rapid temperature changes.
6. Moisture.
7. Wind exposure.
8. Sun exposure.
9. Carbohydrate reserve.

Figure 6.14 shows the influence of temperature change on winter hardiness of trees.

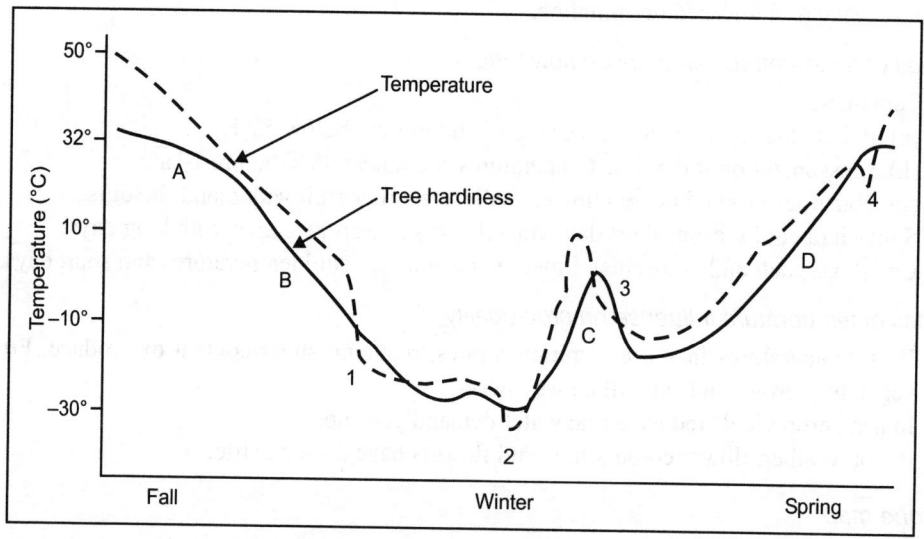

Fig. 6.14. Influence of temperature change on winter hardiness of trees—Solid line represents a tree's hardiness. Regions A-D represent various stages of hardiness through the winter season. Dotted line represents temperature. When the dotted (temperature) line drops below the solid (hardiness) line damage occurs. Points 1–4 represent damage situations. A. Increased cold hardiness induced by shorter day length of fall. B. Increased cold hardiness induced by lowering temperatures. C. Dehardening due to abnormally warm midwinter temperatures. D. Normal spring dehardening as temperatures warm. 1. Injury due to rapid drop in temperatures with inadequate fall hardening. 2. Injury at temperatures lower than hardening capability. 3. Injury due to rise and fall of midwinter temperatures. 4. Injury due to spring frosts.

Examples of Winter Injury

1. Bud kill and dieback—from spring and fall frosts.
2. Root temperature injury—Roots have limited tolerance to subfreezing temperatures. Roots receive limited protection from soil, mulch, and snow. Under extreme cold, roots may be killed by the lack of snow cover or mulch. Street trees are at high risk for root kill in extreme, long-term cold.
3. Soil heaving pushes out plants, breaking roots. Protect with snow cover or mulch.

4. Trunk injury—Drought predisposes trunks to winter injury.
   (a) Sunscald—caused by heating of bark on sunny winter days followed by a rapid temperature drop, rupturing membranes as cells freeze. Winter drought predisposes tree trunks to sunscald.
   (b) Frost shake—separation of wood along one or more growth rings, typically between phloem (inner bark) and xylem (wood), caused by sudden rise in bark temperature.
   (c) Frost crack—vertical split on tree trunk caused by rapid drop in bark temperature.

Winter injury on evergreens

1. Winter drought—water transpires from needles and can't be replaced from frozen soils. It is more severe on growing tips and on the windy side of trees.
2. Sunscald—winter sun warms needles, followed by rapid temperature drop rupturing cell membranes. It occurs typically on southwest side, side of reflected heat or with sudden shade.
3. Photo-oxidisation of chlorophyll—foliage bleaches during cold sunny days. Needles may green-up again in spring.
4. Tissue kill—tissues killed when temperatures drop below hardiness levels.

THERMAL STRATIFICATION

Biological processes within the lake water are influenced by the physical state of water which in turn is governed by the temperature. For example, at 4°C water has its maximum density. In lakes, the increase in depth with a corresponding change in the temperature leads to a more or less stable seasonal stratification called thermal stratification.

Most people know of or have heard of, 'turnover'. That's when the lake water mixes from the surface to the bottom. So what happens during the time when the lake isn't 'turning over'?

A volume of water is heaviest at 4°C (39.2°F). That is just above freezing. The same volume of water becomes lighter as it gets warmer. So in a lake, warm lake water is at the top and the colder water is at the bottom (except in winter).

As the sun continues to heat the water at the top, the difference in temperature between the top and bottom water becomes greater. Eventually there are 2 distinct layers, the epilimnion at the top and the hypolimnion at the bottom. Between these 2 layers is a third, less distinct, transition layer called the metalimnion. Because of the temperature difference (and thus density difference) between the epilimnion and hypolimnion, they don't typically mix together during the summer. It takes a major climactic event to accomplish this, though the lake will mix in the autumn as the surface water cools.

Often in the summer, the hypolimnion will become depleted of oxygen. The bacteria responsible for decomposition consume the oxygen and access to the atmosphere's oxygen is cut off by the stratification.

Then Why Does Water Freeze from the Top Down?

If water becomes more dense as it gets colder, then it should freeze from the bottom up, right? Well, water is most dense at 4°C (or 39.2°F), which is warmer than freezing. So as water continues to cool from 4°C (39.2°F), it becomes less dense and rises back to the top, leaving the slightly warmer water below. At the surface, the cooler water is exposed to freezing air temperatures and may eventually freeze. Once ice forms, the water beneath cannot be mixed by the wind. When the ice melts in the spring, the entire water column will be at approximately 4°C for a brief time. The lake will mix thoroughly (turn over) with just a bit of wind. A calm, warm day can heat the surface water and initiate the stratification process. Stratification may also occur due to changes in salt content as well as temperature. Oceans, particularly in places where freshwater enters, may be stratified by salinity. The problems with Gulf Hypoxia may be attributed partially to the inability of the dense and salty bottom water to mix with the oxygen-rich, less salty water above.

CHEMICAL STRATIFICATION

In thermally stratified lakes, nutrient-enriched hypolimnion and a nutrient-depleted epilimnion is a common feature. Vertical mixing between these layers affects geochemical and biological processes. We used chloride ion as an inert tracer to model the main factors controlling the chemical stratification and to identify lake-wide mixing processes. The stratified lake is treated as two completely mixed reservoirs separated by the thermocline.

One of the characteristics of thermally stratified eutrophic lakes is the hydrochemical differentiation of the water column into an oxic epilimnion and an anoxic hypolimnion caused by the interplay between physical forcing and a succession of microbiological process. With the onset of thermal stratification, hypoliminetic dissolved oxygen is gradually depelted, followed by nitrate reduction and the steady increase of sulphide concentrations due to microbial sulphate reduction. These reducing conditions typically favour the accumulation of ammonium and phosphate, both of which originate from organic matter decomposition and sedimentary release. Contrary to the hypolimnion, the epilimnion often becomes nutrient depleted. The result is a water column that consists of two hydrochemically diverse water layers separated by the thermocline. Due to the scarcity of nutrients in the photic zone, any mixing process between the two water layers is critical for replenishment of nutrients and for biologivity in the lake.

Temperature and Animals

One of the first signs of an infection is that an animal will have an elevated temperature. For that reason, a temperature should be taken for any animal that appears to be under the weather. It has become a common practice on many dairies to take daily temperatures for five days after calving. A bulb type thermometer or digital thermometer can be used to take the temperature. A bulb thermometer needs to be shaken down below 96°F before use. The thermometer or probe should be lubricated prior to insertion. Most people use their own saliva as a lubricant but Vaseline or other suitable lubricant can be used. A bulb thermometer should be left in the rectum for 2–3 minutes. It is a good idea to have a string attached to the hole in the top of the thermometer. After the allotted time, remove the thermometer, wipe it clean, and read the temperature. Table 6.1 gives the normal ranges for temperature, heart and respiration rates for animals at rest.

Table 6.1. Normal ranges for temperature, heart and respiration rates for animals at rest.

	Temperature	Heart rate	Respiratory rate
Newborn	101.4–104	130	56
1 month old	101.4–103.5	105	50
3 month old	101.4–103.5	99	40
6 month old	101–103.5	96	30
1 year old	99.5–103.5	80	18
Cows	99–103	80	14

Heart rates can be felt in the artery on the underside of the tail, the artery inside the hind leg, the chest of a calf or heard with a stethoscope on the chest of a cow.

Warm-blooded Animals (Homeotherms)

In humans and other mammals, temperature regulation represents the balance between heat production from metabolic sources and heat loss from evaporation (perspiration) and the processes of radiation, convection, and conduction. In a cold environment, body heat is conserved first by constriction of blood vessels near the body surface and later by waves of muscle contractions, or shivering, which serve to increase metabolism. Shivering can result in a maximum fivefold increase in metabolism. Below about 40°F (4°C) a naked person cannot sufficiently increase the metabolic rate to replace heat lost to the environment. Another heat-conserving mechanism, goose bumps, or piloerection, raises the body hairs; although not especially effective in humans, in animals it increases the thickness of the insulating fur or feather layer.

In a warm environment, heat must be dissipated to maintain body temperature. In humans, increased surface blood flow, especially to the limbs, acts to dissipate heat at the surface. At environmental temperatures above 93°F (34°C), or at lower temperatures when metabolism has been increased by work, heat must be lost through the evaporation of the water in sweat. People in active work may lose as much as 4 quarts per hour for short periods. However, when the temperature and humidity are both high, evaporation is slowed, and sweating is not effective. Most mammals do not have sweat glands but keep cool by panting (evaporation through the respiratory tract) and by increased salivation and skin and fur licking.

Temperature regulatory mechanisms act through the autonomic nervous system and are largely controlled by the hypothalamus of the brain, which responds to stimuli from nerve receptors in the skin. Continued exposure to heat or cold results in some slow acclimatisation, e.g. more active sweating in response to continued heat and an increase in subcutaneous fat deposits in response to continued cold.

Environmental extremes may result in failure to maintain normal body temperature. In both increased body temperature or hyperthermia, and decreased body temperature, or hypothermia, death may result. Controlled hypothermia is used in some types of surgery to temporarily decrease the metabolic rate. Fever, caused by a resetting of the temperature regulatory mechanism, is a response to fever-causing or pyrogenic, substances, such as bacterial endotoxins or leucocyte extracts. The upper limit of body temperature compatible with survival is about 107°F (42°C), while the lower limit varies.

In humans the inner body temperature alternates in daily activity cycles; it is usually lowest in early morning and is slightly higher at the late afternoon peak. In human females there is also a monthly temperature variation related to the ovulatory cycle. In many mammals and birds the body temperature shows more pronounced cyclic variations than in humans. For example, in hibernators the body temperature may lower to only a few degrees above the environmental temperature during the dormant periods.

Chapter 7

Ecology of Wind

INTRODUCTION

Wind has long been regarded as an important ecological factor in forests owing to the dramatic damage hurricanes can wreak. However, the long-term wind regime of a site also exerts a strong influence on the growth of trees. A relatively large amount is known about the acclimation of trees to wind but less about intra- or interspecific adaption to high winds. In fact, changes resulting from the effect of wind may have a greater effect on the ecology of forests than the more acute effects of destructive storms. Improved understanding of the mechanical effects of wind is helping foresters manage their plantations and may help us to account better for local and geographical variations in forest ecology.

FLAG TREE

A variation of the Krummholz Formation is a flag tree or banner tree. The wind kills branches on the windward side, giving the tree a characteristic flag-like appearance. Where the lower portion of the tree is protected by snow cover, only the exposed upper portion may have this appearance. Trade winds in tropical regions near the equator can also deform trees in a similar manner.

The almost constant, strong winds kill and/or deform branches on the windward side, giving the tree a characteristic flag-like appearance. Where the lower portion of the tree is protected by snow cover or rocks, only the exposed upper portion may have this appearance. This is a rather common occurrence in Red Spruce trees of the highest peaks of the central, even southern Appalachian Mountains, and is most commonly seen in the wind swept high peaks and plateaus of the Allegheny Mountains. This formation most notably occurs with high frequency in the Dolly Sods and Roaring Plains West Wilderness areas along the Allegheny Front in eastern West Virginia, typically occurring at elevations of 3800 feet (1100 m) and higher. Trade winds in tropical regions near the equator can also deform trees in a similar manner.

Leaf Damage

Leaf and needle mines are caused by insects with chewing mouthparts feeding occurs inside the leaf or needle on the mesophyll, between the lower and upper epidermis. The different types of mines are described according to their shape as linear mine, serpentine mine, blotch mine, digitate mine, leaf blister, needle mine or a combination of one or more (Fig. 7.1).

External leaf or needle damage is caused by insects with chewing mouthparts feeding externally on the leaf or needle. A leaf or needle chewed from the outside, called free feeding, is sometimes left with only the tough veins and the middle rib as skeleton (skeletonising). A small patch eaten through all layers of the leaf, creating a hole in the leaf is referred to as hole feeding. If only one surface of the leaf is affected resulting in a more or less transparent 'window', we talk about window feeding (Fig. 7.2).

Fig. 7.1. Leaf and needle mines damage.

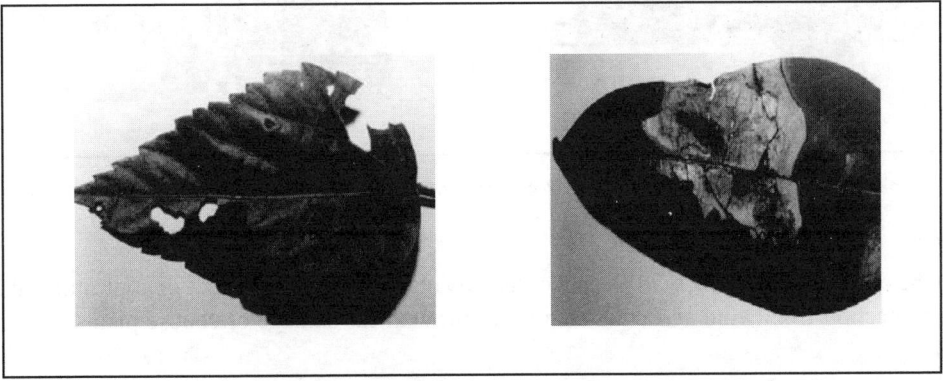

Fig. 7.2. External leaf or needle damage.

Stippling damage is caused by the piercing and sucking action of hemipterans and mites. Usually these animals inject saliva for the external digestion of the plant juices prior to their ingestion. The toxic effect of the saliva results in small, circular 'dead' spots on the leaf (Fig. 7.3).

Fig. 7.3. Stippling damage.

Shelter feeding refers to foliage that is modified into a shelter in which an insect hides and feeds. Either one or several leaves are tied together with a silk thread or web, either by one or more caterpillars (leaf/needle tying, web-enclosed foliage), or folded or rolled and tied together (leaf folding and rolling). The piercing-sucking action of insects can make the leaf crinkle and curl around the insect, thus providing shelter (crinkled leaf). Sometimes, if eggs are laid into the tissue of a leaf, the affected area starts to grow abnormally like a tumour. Those structures are called leaf or petiole galls. The immature stages of gall-forming insects develop and feed inside a gall. Galls can also be induced by saliva injected during the piercing-sucking action of Hemiptera and mites (Fig. 7.4).

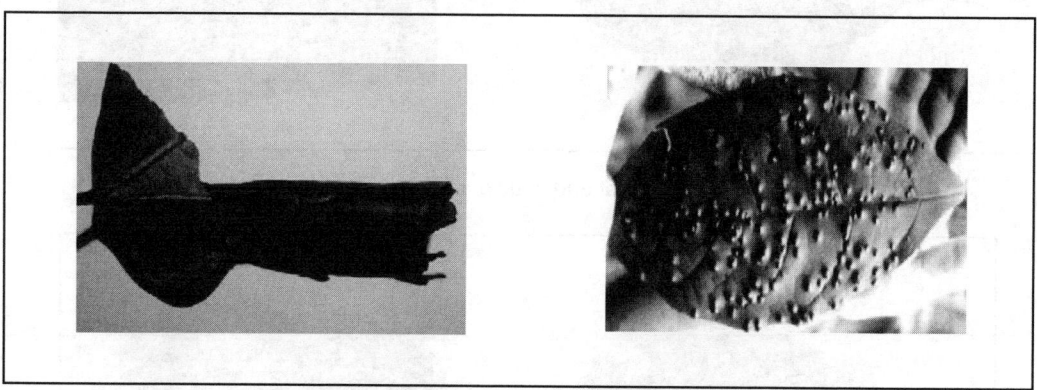

Fig. 7.4. Shelter feeding.

Abrasion

This can be caused by children, wildlife and especially above all, by the wind. For controlling abrasion provide protection from wind and other sources of mechanical damage or choose plants that are less affected.

Salt Spray

Along sea coasts salt spray is carried ashore by wind and during severe storms it can penetrate inland for many kilometres. The worst damage to plants follows storms which are not accompanied by rains, so that a film of salt is deposited on plant surfaces.

Lodging

An important problem in high winds is that the forces exerted on the plants can lead to structural failure, either the uprooting of whole plants (particularly common with trees) or else the breaking or bucking of stems. This process of plants being layed flat by wind is called lodging.

The occurrence of lodging depends on the forces exerted on the plant by wind, rain, etc. on the height from the ground at which they act, and on the strength of the stem. In a cereal, for example, the force due to the wind acts primarily on the head of the plant and induces a torque (T) or turning moment, that increases down the stem and causes bending.

Wind Erosion

Wind erosion is the movement of material by wind and occurs when the lifting power of moving air is able to exceed the force of gravity and the friction which holds an object to the surface. The movement of the sand dunes is an example of wind erosion. In arid climates, the main source of erosion is wind. The general wind circulation moves small particulates such as dust across wide oceans thousands of kilometres downwind of their point of origin, which is known as deflation. Erosion can be the result of material movement by the wind. There are two main effects. First, wind causes small particles to be lifted and therefore moved to another region. This is called deflation. Second, these suspended particles may impact on solid objects causing erosion by abrasion (ecological succession). Wind erosion generally occurs in areas with little or no vegetation, often in areas where there is insufficient rainfall to support vegetation. An example is the formation of sand dunes, on a beach or in a desert. Loess is a homogeneous, typically nonstratified, porous, friable, slightly coherent, often calcareous, fine-grained, silty, pale yellow or buff, windblown (aeolian) sediment.

It generally occurs as a widespread blanket deposit that covers areas of hundreds of square kilometres and tens of meters thick. Loess often stands in either steep or vertical faces. Loess tends to develop into highly rich soils. Under appropriate climatic conditions, areas with loess are among the most agriculturally productive in the world. Loess deposits are geologically unstable by nature, and will erode very readily. Therefore, windbreaks (such as big trees and bushes) are often planted by farmers to reduce the wind erosion of loess.

Wind Speed

Wind speed, or wind velocity, is a fundamental atmospheric rate. Wind speed affects weather forecasting, aircraft and maritime operations, construction projects, growth and metabolism rate of many plant species, and countless other implications. Wind speed is now commonly measured with an anemometer but can also be classified using the older Beaufort scale which is based on people's observation of specifically defined wind effects.

Factors affecting wind speed

Wind speed is affected by a number of factors and situations, operating on varying scales (from micro to macro scales). These include the pressure gradient, Rossby waves and jet streams, and local weather conditions. There are also links to be found between wind speed and wind direction, notably with the pressure gradient and surfaces over which the air is found.

Pressure gradient is a term to describe the difference in air pressure between two points in the atmosphere or on the surface of the earth. It is vital to wind speed, because the greater the difference in pressure, the faster the wind flows (from the high to low pressure) to balance out the variation. The pressure gradient, when combined with the Coriolis effect and friction, also influences wind direction.

Rossby waves are strong winds in the upper troposphere. These operate on a global scale and move from West to East (hence being known as Westerlies). The Rossby waves are themselves a different wind speed from what we experience in the lower troposphere.

Local weather conditions play a key role in influencing wind speed, as the formation of hurricanes, monsoons and cyclones as freak weather conditions can drastically affect the velocity of the wind.

The wind direction and speed are affected by three main factors:

1. Pressure gradient—the difference in barometric pressure between adjacent zones of high and low pressure.
2. Frictional forces—features on the earth's surface which oppose the wind; e.g.: mountains, trees, buildings, etc.

3. Coriolis effect—the earth's rotation causes winds to be deflected to the right in the Northern Hemisphere, and in the Southern Hemisphere to the left.

All three of these combined result in the spiral motion of air in both high and low pressure systems.

Windbreak

A windbreak or shelterbelt is a plantation usually made up of one or more rows of trees or shrubs planted in such a manner as to provide shelter from the wind and to protect soil from erosion. They are commonly planted around the edges of fields on farms. If designed properly, windbreaks around a home can reduce the cost of heating and cooling and save energy. Windbreaks are also planted to help keep snow from drifting onto roadways and even yards. Other benefits include providing habitat for wildlife and in some regions the trees are harvested for wood products.

A further use for a shelterbelt is to screen a farm from a main road or motorway. This improves the farm landscape by reducing the visual incursion of the motorway, mitigating noise from the traffic and providing a safe barrier between farm animals and the road.

A further use for 'windbreaks' is for a retail item used on the beach and camping to prevent wind from disturbing social enjoyment. Americans tend to use the term windbreaker whereas Europeans favour the term 'windbreak'. Normally made from cotton, nylon, canvas and recycled sails, windbreaks tend to have three or more panels, held in place with poles that slide into pockets sewn into the panel. The poles are then hammered into the ground and a windbreak is formed.

TOPOGRAPHIC FACTORS

Topography. It is the description of the physical features of a place. It describes configuration of the ground, its altitude, slope, aspect, etc. and also affects vegetation through climate, soil formation processes, soil moisture, soil nutrients, etc.

Topographic factors can be classified into:
1. Configuration of land surface.
2. Altitude.
3. Slope.
4. Aspect and exposure.

Configuration of the Land Surface

1. It influences vegetation through its effect on temperature, wind movement, etc.
2. In a hills and valley country, sunlight reaches the valley late in the morning and disappears early in the afternoon.
3. The shade of the neighbouring hills makes valley colder in winter and that of radiated heat makes the valley hot. So, diurnal and seasonal temperatures of the valley differ from the temperatures on the hills. Pool frost occurring on hills and in valleys affects the vegetation.
4. It also affects wind movements. It results in more rain in the east and less in the western Nepal. It has greater influence on humidity and temperature variance eventually affects vegetation of the site.

Altitude

1. It affects vegetation through solar radiation, temperature and rainfall.

2. The intensity of radiation goes beyond optimum limit has a dwarfing effect on shoot, the growth of root being favoured.
3. Temperature as it is higher and lower the optimum level, affects the species composition and the site quality.
4. Similarly, rainfall affected by altitude affects the temperature and moisture resulting in the change in the nature of vegetation. It has been estimated that about half the water vapour in the air lies below 2000 m while three quarters lies below 4000 m and so, high mountain range is a very effective barrier for the monsoons.

Slope

1. Slope affects runoff and drainage having a profound influence on the moisture regime of the soil. As a general rule, the steeper the slope, the greater the runoff and better the drainage.
2. Slope modifies the intensity of insolation, temperature and moisture of the surface soil.
3. Slope also affects erosion and depth of soil as greater the slope, greater the erosion. The depth of soil in the hills varies with the increasing slope.
4. Thus, slope affects vegetation of the site through affecting the runoff, insolation, temperature, moisture and depth of soil.

Aspect and Exposure

1. Aspect—the direction towards which a slope faces.
2. Exposure—the relation of a site to weather conditions, especially sun and wind.
3. Both determine the amount of insolation received by a hill slope. In Nepal, southern slope is warmer than the northern slopes and consequently temperature differs.
4. We can see different species on different aspect of a hill.
5. Similarly, different aspects receive insolation differently. The eastern slope is exposed to the sun in the earlier part of the day and so dew is seen. In the morning, soil moisture has not melted resulting in seedlings being killed. Whereas, the western aspect has desiccating effect due to noon's sun.

TREE LINE OR TIMBER LINE

The tree line is the edge of the habitat at which trees are capable of growing. Beyond the tree line, they are unable to grow because of inappropriate environmental conditions (usually cold temperatures, insufficient air pressure or lack of moisture). Some distinguish additionally a deeper timberline, where trees form a forest with a closed canopy. At the tree line, tree growth is often very stunted, with the last trees forming low, densely matted bushes. If it is caused by wind, it is known as krummholz formation, from the German for 'twisted wood'. The tree line, like many other natural lines (lake boundaries, for example), appears well-defined from a distance, but upon sufficiently close inspection, it is a gradual transition in most places. Trees grow shorter towards the inhospitable climate until they simply stop growing.

Types of Tree Lines

There are several types of tree lines defined in ecology and geology.

Alpine

The highest elevation that sustains trees; higher up, it is too cold or snow cover persists for too much of the year, to sustain trees. Usually associated with mountains, the climate above the tree line is called an alpine climate, and the terrain can be described as alpine tundra. In the northern hemisphere treelines on north-facing slopes are lower than on than south-facing slopes because increased shade means the snowpack takes longer to melt which shortens the growing season for trees. This is reversed in the southern hemisphere.

Desert

The driest places that trees can grow; drier desert areas having insufficient rainfall to sustain trees. These tend to be called the 'lower' tree line and occur below about 5000 ft (1500 m) elevation in the Desert Southwestern United States. The desert treeline tends to be higher on pole-facing slopes than equator-facing slopes, because the increased shade on a pole-facing slope keeps those slopes cooler and prevents moisture from evaporating as quickly, giving trees a longer growing season and more access to water.

Desert-alpine

In some mountainous areas, higher elevations above the condensation line or on equator-facing and leeward slopes can result in low rainfall and increased exposure to solar radiation. This dries out the soil, resulting in a localised arid environment unsuitable for trees. The slopes of Mauna Loa above 10,000 ft in Hawaii are an example of this. Many south-facing ridges of the mountains of the Western US have a lower treeline than the northern faces because of increased sun exposure and aridity.

Double

Different tree species have different tolerances to drought and cold. Mountain ranges isolated by oceans or deserts may have restricted reportoires of tree species with gaps that are above the alpine tree line for some species yet below the desert tree line for others. For example several mountain ranges in the Great Basin of North America have lower belts of Pinyon Pines and Junipers separated by intermediate brushy but treeless zones from upper belts of Limber and Bristlecone Pines.

Exposure

On coasts and isolated mountains the tree line is often much lower than in corresponding altitudes inland and in larger, more complex mountain systems, because strong winds reduce tree growth. In addition the lack of suitable soil, such as along talus slopes or exposed rock formations, prevents trees from gaining an adequate foothold and exposes them to drought and sun.

Arctic

The northernmost latitude in the Northern Hemisphere where trees can grow; farther north, it is too cold to sustain trees. Extremely cold temperatures can result in freezing of the internal sap of trees, killing them. In addition, permafrost in the soil can prevent trees from getting their roots deep enough for the necessary structural support.

Antarctic

The southernmost latitude in the Southern Hemisphere where trees can grow; further south, it is too cold to sustain trees. It is a theoretical concept that does not have any defined location. No trees grow in Antarctica or the sub-antarctic islands. This tree line would be the southernmost point in the environment at which trees can no longer grow, except there are no landmasses that have a true treeline analogous to the arctic treeline.

Other

The immediate environment is too extreme for trees to grow. This can be caused by geothermal exposure associated with hot springs or volcanoes, such as at Yellowstone, high soil acidity near bogs, high salinity associated with playas or salt lakes, or ground that is saturated with groundwater that excludes oxygen from the soil, which most tree roots need for growth. The margins of muskegs and bogs are common examples of these types of open areas. However, no such line exists for swamps, where trees, such as Bald cypress and the many mangrove species, have adapted to growing in permanently waterlogged soil. In some colder parts of the world there are tree lines around swamps, where there are no local tree species that can develop. There are also man-made pollution tree lines in weather exposed areas, where new tree lines have developed because of the increased stress of pollution. Example are around Nikel in Russia and previously in the Erzgebirge.

Typical Vegetation

Some typical Arctic and alpine tree line tree species (note the predominance of conifers).

Eurasia

1. Dahurian Larch (*Larix gmelinii*).
2. Macedonian Pine (*Pinus peuce*).
3. Swiss Pine (*Pinus cembra*).
4. Mountain Pine (*Pinus mugo*).
5. Arctic White Birch (*Betula pubescens* subsp. *tortuosa*).

North America

Subalpine fir (*Abies lasiocarpa*).
Subalpine Larch (*Larix lyallii*).
Engelmann Spruce (*Picea engelmannii*).
Whitebark Pine (*Pinus albicaulis*).
Great Basin Bristlecone Pine (*Pinus longaeva*).
Rocky Mountains Bristlecone Pine (*Pinus aristata*).
Foxtail Pine (*Pinus balfouriana*).
Limber Pine (*Pinus flexilis*).
Potosi Pinyon (*Pinus culminicola*).
Black spruce (*Picea mariana*).
Hartweg's Pine (*Pinus hartwegii*).

WIND ENERGY AND POLLUTION

Wind power consumes no fuel for continuing operation, and has no emissions directly related to electricity production. Wind power stations, however, consume resources in manufacturing and construction, as do most other power production facilities. Wind power may also have an indirect effect on pollution at other production facilities, due to the need for reserve and regulation, and may affect the efficiency profile of plants used to balance demand and supply, particularly if those facilities use fossil fuel sources. Compared to other power sources, however, wind energy's direct emissions are low, and the materials used in construction (concrete, steel, fibreglass, generation components) and transportation are straightforward.

Wind power's ability to reduce pollution and greenhouse gas emissions will depend on the amount of wind energy produced, and hence scalability.

1. Wind power is a renewable resource, which means using it will not deplete the earth's supply of fossil fuels. It also is a clean energy source, and operation does not produce carbon dioxide, sulphur dioxide, mercury, particulates, or any other type of air pollution, as do conventional fossil fuel power sources.

2. Electric power production is only part (about 39 per cent in the USA) of a country's energy use, so wind power's ability to mitigate the negative effects of energy use—as with any other clean source of electricity—is limited (except with a potential transition to electric or hydrogen vehicles). Wind power contributed less than 1 per cent of the UK's national electricity supply in 2007 and hence had negligible effects on CO_2 emissions, which continued to rise in 2002 and 2006 (Department of Trade and Industry); the growth of installed wind capacity in the UK has been impressive (installed wind capacity doubled from 2002 to 2004, and again from end-2004 to mid-2006), but from low levels. Until wind energy achieves substantially greater scale worldwide, its ability to contribute will be limited.

3. Groups such as the UN's Intergovernmental Panel on Climate Change state that the desired mitigation goals can be achieved at lower cost and to a greater degree by continued improvements in general efficiency—in building, manufacturing, and transport—than by wind power.

4. During manufacture of the wind turbine, steel, concrete, aluminium and other materials will have to be made and transported using energy-intensive processes, generally using fossil energy sources.

5. The energy return on investment (EROI) for wind energy is equal to the cumulative electricity generated divided by the cumulative primary energy required to build and maintain a turbine. The EROI for wind ranges from 5 to 35, with an average of around 18. This places wind energy in a favourable position relative to conventional power generation technologies in terms of EROI. Baseload coal-fired power generation has an EROI between 5 and 10:1. Nuclear power is probably no greater than 5:1, although there is considerable debate regarding how to calculate its EROI. The EROI for hydropower probably exceeds 10, but in most places in the world the most favourable sites have been developed.

6. Net energy gain for wind turbines has been estimated in one report to be between 17 and 39 (i.e. over its lifetime a wind turbine produces 17–39 times as much energy as is needed for its manufacture, construction, operation and decommissioning). A similar Danish study determined the payback ratio to be 80, which means that a wind turbine system pays back the energy invested within approximately 3 months. This is to be compared with payback ratios of 11 for coal power plants and 16 for nuclear power plants, though such figures do not take into account the energy content of the fuel itself, which would lead to a negative energy gain.

7. The ecological and environmental costs of wind plants are paid by those using the power produced, with no long-term effects on climate or local environment left for future generations.

Ecology

1. Because it uses energy already present in the atmosphere, and can displace fossil-fuel generated electricity (with its accompanying carbon dioxide emissions), wind power mitigates global warming. While wind turbines might impact the numbers of some bird species, conventionally

fuelled power plants could wipe out hundreds or even thousands of the world's species through climate change, acid rain, and pollution.

2. Unlike fossil fuel or nuclear power stations, which circulate or evaporate large amounts of water for cooling, wind turbines do not need water to generate electricity.

Ecological footprint

Large-scale onshore and near-shore wind energy facilities (wind farms) can be controversial due to aesthetic reasons and impact on the local environment. Large-scale offshore wind farms are not visible from land and according to a comprehensive 8-year Danish Offshore Wind study on 'Key Environmental Issues' have no discernable effect on aquatic species and no effect on migratory bird patterns or mortality rates. Modern wind farms make use of large towers with impressive blade spans, occupy large areas and may be considered unsightly at onshore and near-shore locations. They usually do not, however, interfere significantly with other uses, such as farming. The impact of onshore and near-shore wind farms on wildlife—particularly migratory birds and bats—is hotly debated, and studies with contradictory conclusions have been published. Two preliminary conclusions for onshore and near-shore wind developments seem to be supported: first, the impact on wildlife is likely low compared to other forms of human and industrial activity; second, negative impacts on certain populations of sensitive species are possible, and efforts to mitigate these effects should be considered in the planning phase. Aesthetic issues are important for onshore and near-shore locations in that the 'visible footprint' may be extremely large compared to other sources of industrial power (which may be sighted in industrially developed areas), and wind farms may be close to scenic or otherwise undeveloped areas. Offshore wind development locations remove the visual aesthetic issue by being at least 10 km from shore and in many cases much further away.

Land Use

Clearing of wooded areas is often unnecessary, as the practice of farmers leasing their land out to companies building wind farms is common. In the US, farmers may receive annual lease payments of two thousand to five thousand dollars per turbine. The land can still be used for farming and cattle grazing. Less than 1 per cent of the land would be used for foundations and access roads, the other 99 per cent could still be used for farming. Turbines can be sited on unused land in techniques such as center pivot irrigation.

The clearing of trees around onshore and near-shore tower bases may be necessary to enable installation. This is an issue for potential sites on mountain ridges, such as in the northeastern US.

Wind turbines should ideally be placed about ten times their diameter apart in the direction of prevailing winds and five times their diameter apart in the perpendicular direction for minimal losses due to wind park effects. As a result, wind turbines require roughly 0.1 square kilometres of unobstructed land per megawatt of nameplate capacity. A wind farm that produces the energy equivalent of a conventional 2 GW power plant might have turbines spread out over an area of approximately 200 square kilometres.

Areas under onshore and near-shore windfarms can be used for farming, and are protected from further development. Although there have been installations of wind turbines in urban areas (such as Toronto's exhibition place), these are generally not used. Buildings may interfere with wind, and the value of land is likely too high if it would interfere with other uses to make urban installations viable. Installations near major cities on unused land, particularly offshore for cities near large bodies of water, may be of more interest. Despite these issues, Toronto's demonstration project demonstrates that there are no major issues that would prevent such installations where practical, although non-urban locations are expected to predominate.

Offshore locations, such as that being developed on a large underwater plateau in eastern Lake Ontario by Trillium Power use no land per and avoid known shipping channels. However, that is generally not the norm for most offshore locations. Some offshore locations are uniquely located close to ample transmission and high load centres however that is not the norm for most offshore locations. Most offshore locations are at considerable distances from load centres and may face transmission and line loss challenges.

Airborne

An airborne wind turbine is a design concept for a wind turbine that is supported in the air without a tower. A tether would be used to transmit energy to the ground. These systems would have the advantage of tapping an almost constant wind and doing so without a set of slip rings or yaw mechanism, without the expense of tower construction. The main disadvantage is that kites and 'helicopters' come down when there is insufficient wind. These schemes require a very long power cable and an aircraft exclusion zone. As of 2006, no commercial airborne wind turbines are in regular operation.

Wildlife

Onshore and near-shore studies show that the number of birds killed by wind turbines is negligible compared to the number that die as a result of other human activities such as traffic, hunting, power lines and high-rise buildings and especially the environmental impacts of using non-clean power sources. For example, in the UK, where there are several hundred turbines, about one bird is killed per turbine per year; 10 million per year are killed by cars alone. In the United States, onshore and near-shore turbines kill 70,000 birds per year, compared to 57 million killed by cars and 97.5 million killed by collisions with plate glass. Another study suggests that migrating birds adapt to obstacles; those birds which don't modify their route and continue to fly through a wind farm are capable of avoiding the large offshore windmills, at least in the low-wind non-twilight conditions studied. In the UK, the Royal Society for the Protection of Birds (RSPB) concluded that 'The available evidence suggests that appropriately positioned wind farms do not pose a significant hazard for birds.' It notes that climate change poses a much more significant threat to wildlife, and therefore supports wind farms and other forms of renewable energy.

Some onshore and near-shore windmills kill birds, especially birds of prey. More recent siting generally takes into account known bird flight patterns, but some paths of bird migration, particularly for birds that fly by night, are unknown although a 2006 Danish Offshore Wind study showed that radio tagged migrating birds travelled around offshore wind farms. A Danish survey in 2007 showed that less than 1 per cent of migrating birds passing an oshore wind farm in Rønde, Denmark, got close to collision, though the site was studied only during low-wind non-twilight conditions. A survey at Altamont Pass, California, conducted by a California Energy Commission in 2007 showed that onshore turbines killed between 1766 and 4721 birds annually (881 to 1300 of which were birds of prey). Radar studies of proposed onshore and near-shore sites in the eastern US have shown that migrating songbirds fly well within the reach of large modern turbine blades. In Australia, a proposed onshore/near-shore wind farm was cancelled before production because of the possibility that a single endangered bird of prey was nesting in the area. An onshore/near-shore wind farm in Norway's Smøla islands is reported to have destroyed a colony of sea eagles, according to the British Royal Society for the Protection of Birds. The society said turbine blades killed nine of the birds in a 10 month period, including all three of the chicks that fledged that year. Norway is regarded as the most important place for white-tailed eagles. The numbers of bats killed by existing onshore and near-shore facilities has troubled even industry personnel. A study in 2007 estimated that over 2200 bats were killed by 63 onshore turbines in just six weeks at two sites in the eastern US. This study suggests some onshore and near-shore sites may be particularly hazardous to local bat populations and more research is urgently needed. Migratory bat species appear to be particularly at risk, especially during key movement periods (spring and more importantly in fall).

Lasiurines such as the hoary bat (*Lasiurus cinereus*), red bat (*Lasiurus borealis*), and the semi-migratory silver-haired bats (*Lasionycteris noctivagans*) appear to be most vulnerable at North American sites. Almost nothing is known about current populations of these species and the impact on bat numbers as a result of mortality at windpower locations. Offshore wind sites 10 km or more from shore do not interact with bat populations.

Aesthetics

Recorded experience that onshore and near-shore wind turbines are noisy and visually intrusive creates resistance to the establishment of land-based wind farms in many places. Moving the turbines far offshore (10 km or more) mitigates the problem, but offshore wind farms may be more expensive and transmission to onshore locations may present challenges in many but not all cases.

Some residents near onshore and near-shore windmills complain of 'shadow flicker', which is the alternating pattern of sun and shade caused by a rotating windmill casting a shadow over residences. Efforts are made when siting onshore and near-shore turbines to avoid this problem.

Large onshore and near-shore wind towers require aircraft warning lights, which create light pollution at night, which bothers humans and can disrupt the local ecosystem. Complaints about these lights have caused the Federal Aviation Administration (FAA) to consider allowing a less than 1:1 ratio of lights per turbine in certain areas.

Improvements in blade design and gearing have quietened modern turbines to the point where a normal conversation can be held underneath one. Newer wind farms have more widely spaced turbines due to the greater power of the individual wind turbines, and so look less cluttered.

The aesthetics of onshore and near-shore wind turbines have been compared favourably to those of pylons from conventional power stations. Offshore sites have on average a considerably higher energy yield than onshore sites, and generally cannot be seen from the shore even on the clearest of days.

Chapter 8

Ecology of Fire

INTRODUCTION

Fire ecology is concerned with the processes linking the natural incidence of fire in an ecosystem and the ecological effects of this fire. Many ecosystems, such as the North American prairie and chaparral ecosystems, and the South African savanna, have evolved with fire as a natural and necessary contributor to habitat vitality and renewal. Many plant species in naturally fire-affected environments require fire to germinate. Fire suppression can lead to the build-up of inflammable debris and the creation of less frequent but much larger and destructive wildfires.

Fire suppression, in combination with other human-caused environmental changes, has resulted in unforeseen consequences for natural ecosystems. Some uncharacteristically large wildfires in the United States have been caused as a consequence of years of fire suppression and the continuing expansion of people into fire-adapted ecosystems. Land managers are faced with tough questions regarding where to restore a natural fire regime.

FIRE COMPONENTS

A fire regime describes the pattern that fire follows in a particular ecosystem. Its 'severity' is a term that ecologists use to refer to the impact that a fire has on an ecosystem. Ecologists can define this in many ways, but one way is through an estimate of plant mortality. Fire can burn at three levels. Ground fires will burn through soil that is rich in organic matter. Surface fires will burn through dead plant material that is lying on the ground. Crown fires will burn in the tops of shrubs and trees. Ecosystems may experience predominantly one of these fire regimes or a mix of all three. Fires will often break out during a dry season, but in some areas wildfires may also commonly occur during a time of year when lightning is prevalent. The frequency over a span of years at which fire will occur at a particular location is a measure of how common wildfires are in a given ecosystem. It is either defined as the average interval between fires at a given site, or the average interval between fires in an equivalent specified area.

Abiotic Responses

Fire has important effects on the abiotic (non-living) components of an ecosystem, particularly the soil. Fire can affect the soil by direct contact with it and by its effects on the plant community associated with it. By removing overhead vegetation, fire can lead to increased solar radiation on the soil surface by day, resulting in greater warming, and to greater cooling through the loss of radiative heat at night. Fewer leaves left to intercept rain will allow more moisture to reach the soil surface. In addition, plant transpiration (the process by which water travels through plants and evaporates through pores in the leaves) will be reduced following a fire, allowing the soil to retain more moisture. Exposure to sunlight, wind and evaporation, however, will work in the other way, to dry the soil. The fire may have created an impermeable crust at the soil surface, if organic matter on the ground was heated by the fire into a waxy residue, and if this has happened, it may lead to increased soil erosion through surface run-off.

Fire may cause nutrient loss through a variety of mechanisms, including oxidation, volatilisation, and increased erosion and leaching by water. Temperatures must be very high, however, to cause a significant loss of nutrients, which are often replaced by organic matter left behind in the fire. Charcoal is able to counteract some nutrient and water loss because of its absorptive properties.

Overall, soils become more basic (higher pH) following fires because of acid combustion. By driving novel chemical reactions at high temperatures, fire can even alter the texture and structure of soils by affecting the clay content and the soil's porosity.

Biotic Responses and Adaptations

Plants

Plants have evolved many adaptations to cope with fire. In chaparral communities in Southern California, for example, some plants have leaves coated in flammable oils that encourage an intense fire. This heat causes their fire-activated seeds to germinate and the young plants can then capitalise on the lack of competition in a burnt landscape. Other plants have smoke-activated seeds, or fire-activated buds. The serotinous cones of the Lodgepole pine (*Pinus contorta*) are sealed with a resin that a fire melts away, releasing the seeds. Many plant species, including the shade-intolerant giant sequoia (*Sequoiadendron giganteum*), require fire to make gaps in the vegetation canopy that will let in light, allowing their seedlings to compete with the more shade-tolerant seedlings of other species, and so establish themselves. Because their stationary nature precludes any fire avoidance, plant species may only be fire-intolerant, fire-tolerant or fire-resistant.

Fire intolerance

Fire-intolerant plant species tend to be highly inflammable and are destroyed completely by fire. Some of these plants and their seeds may simply fade from the community after a fire and not return, others have adapted to ensure that their offspring survives into the next generation. Obligate seeders are plants with large, fire-activated seed banks that germinate, grow, and mature rapidly following a fire, in order to reproduce and renew the seed bank before the next fire.

Fire tolerance

Fire-tolerant species are able to withstand a degree of burning and continue growing despite damage from fire. These plants are sometimes referred to as 'resprouters'. Ecologists have shown that some species of resprouters store extra energy in their roots to aid recovery and re-growth following a fire.

Fire resistance

Fire-resistant plants suffer little damage during a characteristic fire regime. These include large trees whose flammable parts are high above surface fires. Mature Ponderosa Pine (*Pinus ponderosa*) is an example of a tree species that suffers virtually no crown damage under a naturally mild fire regime, because it sheds its lower, vulnerable branches as it matures.

Animals, birds and microbes

Like plants, animals display a range of abilities to cope with fire, but they differ from plants in that they must avoid the actual fire to survive. Although birds are vulnerable when nesting, they are generally able to escape a fire; indeed they often profit from being able to take prey fleeing from a fire and to recolonise burned areas quickly afterwards. Mammals are often capable of fleeing a fire, or seeking cover if they can burrow. Amphibians and reptiles may avoid flames by burrowing into the ground or using the burrows of other animals. Amphibians in particular are able to take refuge in water or very wet mud. Some arthropods also take shelter during a fire, although the heat and smoke may actually attract some of them, to their peril. Microbial organisms in the soil vary in their heat tolerance but are more likely to be able to survive a fire the deeper they are in the soil. A low fire intensity, a quick passing of the flames and a dry soil will also help. An increase in available nutrients after the fire has passed may result in larger microbial communities than before the fire.

LONG TERM IMPACTS

Fire behaviour is different in every ecosystem and the organisms in those ecosystems have adapted accordingly. One sweeping generality is that in all ecosystems, fire creates a mosaic of different habitat patches, with areas ranging from those having just been burned to those that have been untouched by fire for many years. This is a form of ecological succession in which a freshly burned site will progress through continuous and directional phases of colonisation following the destruction caused by the fire. Ecologists usually characterise succession through the changes in vegetation that successively arise. After a fire, the first species to re-colonise will be those whose seeds are already present in the ground, or those whose seeds are able to travel into the burned area quickly. These are generally fast-growing herbaceous plants that need lots of light and are poor competitors in crowded areas. As time passes, more slowly growing, shade-tolerant, and competitive, woody species will crowd out the herbaceous plants. These woody plants may be shrubs or trees. Different species of plants, animals, and microbes specialise in exploiting different stages in this process of succession, and by creating these different types of patches, fire allows a greater number of species to exist within a landscape. Soil characteristics will be a factor in determining the specific nature of a fire-adapted ecosystem, as will climate and topography.

Forests

Mild to moderate fires burn in the forest understory, removing small trees and herbaceous groundcover. Only high-intensity fires will burn into the crowns of the tallest trees. Crown fires may require support from ground fuels to maintain the fire in the forest canopy (passive crown fires), or the fire may burn in the canopy independently of any ground fuel support (an active crown fire). Fires used in the management of woodlands will typically aim for low to moderate intensity, whereas wildfires can evolve into crown fires. When a forest burns frequently and thus has less plant litter build-up, below-ground soil temperatures rise only slightly and will not be lethal to roots that lie deep in the soil. Although other characteristics of a forest will influence the impact of fire upon it, factors such as climate and topography play an important role in determining fire severity and fire extent. Fires spread most widely during drought years, are most severe on upper slopes and are influenced by the type of vegetation that is growing.

Fire Suppression

Fire serves many important functions within fire-adapted ecosystems. Fire plays an important role in nutrient cycling, diversity maintenance and habitat structure. The suppression of fire can lead to unforeseen changes in ecosystems that often adversely affect the plants, animals and humans that depend upon that habitat. Wildfires that deviate from a historical fire regime because of fire suppression are called 'uncharacteristic fires'.

Chaparral communities

In 2003, southern California witnessed powerful chaparral wildfires. Hundreds of homes and hundreds of thousands of acres of land went up in flames. Extreme fire weather (low humidity, low fuel moisture and high winds) and the accumulation of dead plant material from 8 years of drought, contributed to a catastrophic outcome. Although some have maintained that fire suppression contributed to an unnatural buildup of fuel loads, a detailed analysis of historical fire data has showed that this may not have been the case. Fire suppression activities had failed to exclude fire from the southern California chaparral. Research showing differences in fire size and frequency between southern California and Baja has been used to imply that the larger fires north of the border are the result of fire suppression, but this opinion has been challenged by numerous investigators and is no longer supported by the majority of fire ecologists.

Fish impacts

The Boise National Forest is a US national forest located north and east of the city of Boise, Idaho. Following several uncharacteristically large wildfires, an immediately negative impact on fish populations was observed, posing particular danger to small and isolated fish populations. In the long-term, however, fire appears to rejuvenate fish habitats by causing hydraulic changes that increase flooding and lead to silt removal and the deposition of a favourable habitat substrate. This leads to larger post-fire populations of the fish that are able to recolonise these improved areas. But although fire generally appears favourable for fish populations in these ecosystems, the more intense effects of uncharacteristic wildfires, in combination with the fragmentation of populations by human barriers to dispersal such as weirs and dams, will pose a threat to fish populations.

Ponderosa pine forests

Ponderosa pine forests now face severe damage under harsher fire regimes brought on by fire suppression and aggravated by natural drought cycles. Fires in these forests now result in crown fires that cause extensive tree-mortality. Fire suppression may also lead to increased defoliation of the trees by herbivorous insects, whose populations might otherwise be controlled by more regular outbreaks of wildfire.

Fire as a Management Tool

Restoration ecology is the name given to an attempt to reverse or mitigate some of the changes that humans have caused to an ecosystem. Controlled burning is one tool that is currently receiving considerable attention as a means of restoration and management. Applying fire to an ecosystem may create habitats for species that have been negatively impacted by fire suppression, or fire may be used as a way of controlling invasive species without resorting to herbicides or pesticides. But what should managers aim to restore their ecosystems to? Does 'natural' mean pre-human? Pre-European? Native American use of fire, not natural fires, historically maintained the diversity of the savannas of North America. When, how, and where managers should use fire as a management tool is a subject of debate.

EFFECTS OF FIRE ON PLANTS AND ANIMALS: INDIVIDUAL LEVEL

Plants

Obviously plants can't move away when a fire comes through an area, so how are they able to survive and persist?

Most vegetative survival involves the protection of tissue from heat which would otherwise destroy it. Fire resistance and tolerance is exhibited through: bark thickness, other vegetative insulation, above-ground resprouting, underground roots and stems.

Bark thickness: Thick bark insulates and protects the cambium from heat and damage.

Vegetative insulation: Some protection is afforded by leaf sheaths. Grasses have meristems at leaf base so are protected from heat and damage in this way. *Pandanus* also receives some protection from leaf sheaths.

Fire acts as a generalist herbivore removing plant material above the ground surface, thus enabling new herbaceous growth.

Above ground re-sprouting: While many trees are killed by total defoliation following a fire, some can re-sprout from epicormic buds, which are buds positioned beneath the bark. *Eucalyptus* trees are known for their ability to vegetatively regenerate branches along their trunks from buds. This is because epicormic buds of *Eucalyptus* trees are more protected than on other tree species because they are set much deeper at maximum bark thickness. The ability to survive and re-sprout depends on tree height, scorch and char heights, but also tree species, age, size (height) and the severity of the fire.

Ladder fires running up the trunks of these *Melaleuca* trees might look impressive but they can stimulate regrowth from epicormic buds and do not pose a threat to the trees.

Below-ground roots and underground stems: Because soil is a good insulator, buds underground are well protected. Plants can survive fires by re-sprouting from basal stems, and also from roots and horizontal rhizomes.

Root suckering can result in large clonal populations post-fire. Other plants have woody swellings at the base of their stems called lignotubers. They contain latent buds which are released from dormancy post-fire. Most *Eucalyptus* species have lignotubers. Geophytes survive burning because the storage organs are below ground protected from burning. They persist vegetatively between fires, and are stimulated to flower following a fire.

FIRE ECOLOGY

Fire is a natural component of many ecosystems, which include plants and animals that interact with one another and with their physical environment. Fire ecology examines the role of fire in ecosystems. Fire ecologists study the origins of fire, what influences spread and intensity, fire's relationship with ecosystems, and how controlled fires can be used to maintain ecosystem health.

Physical and Chemical Nature of Fire

For fuel to ignite it must be heated in the presence of oxygen to the ignition point or kindling temperature. Wood must reach about 800°F to burst into flame. As the wood is being heated to this point it dries as water, oils, and resins are boiled away. The chemical structure of the fuel is broken down and flammable gases are produced. The ignition of these flammable gases is known as flaming combustion. Flaming combustion transforms the surface of the wood to charcoal. At cooler temperatures, glowing combustion consumes charcoal, producing ash, water, and carbon dioxide. Many factors such as fuel, weather, topography, and fire history influence the probability of ignition and combustion.

Fire Behaviour

Fire behaviour is most often described by intensity and spread. Many factors influence this behaviour. Five factors that influence intensity are available fuel, moisture and temperature, fuel composition, wind, and topography. Available fuel is quantified by size and arrangement. The more available fuel, the more intense the fire and cool, moist fuels combust more slowly than hot, dry fuels. Fuel composition can make a fire more or less intense. Oils and resins increase the heat yield of the reaction and cause a fire to burn intensely whereas other chemical factors, such as high concentrations of minerals, can reduce flammability. Wind increases oxygen supply, convects heat and can produce 'spot fires' from fragments that blow down-wind. Finally, topography effects intensity. A fire ignited on the top of a slope is likely to spread slowly as it burns downhill, whereas a fire at the bottom of a slope will start rapidly and gain momentum as it burns uphill because warm air rises and preheats uphill fuels. Many of the factors that affect intensity also affect the rate of spread. For example, fires in dry, windy conditions with abundant fuel spread rapidly. Fuel continuity and topography also play a role in spread. Topographic features such as streams and lakes can create firebreaks, thus influencing the distribution of burns across landscapes. Finally, the composition of plant communities affects spread, as some species are more flammable than others.

Effects of Fire on Ecosystems

There is much yet to be learned about how wildland fire affects ecosystems. This is in part because each fire and each ecosystem has unique properties. However, some generalities can be made.

Mosaic patterns

Wildland fires create a mixture of totally burned, partially burned, and unburned sections called a burn mosaic. The varying degrees of burn are a result of many factors including wind shifts, daily temperature changes, moisture levels, and varying chemical composition of the vegetation. The burn mosaic results in varied regrowth rates that creates a vegetation mosaic.

Soil conditions

Wildland fires can be both a detriment and a benefit to soil. The soil can become more nutrient-rich after a fire due to the high mineral content of the ash and charcoal and also due to the warm, moist conditions that increase microbial activity. The intense heat can also cause soil particles to become water-repellant, causing rainwater to run off. As the water runs off it can carry soil particles with it and lead to erosion.

Animal populations

Some animals will perish in wildland fires, especially small animals, insects, and older and weaker individuals. However, fire has a greater affect on habitat than on individuals. While the vast majority of large mammals are able to flee fires, populations often suffer substantial losses in the months following a fire due to a loss of food sources. Food sources are scarce because of the fire itself and also because most natural fires occur shortly before winter. These habitat changes allow other animals to thrive. Scavenging animals find an increased abundance of food sources and predatory animals may benefit from reduced forest cover which makes prey more visible. Nutrient-rich new growth also benefits many animals and animals such as deer will even eat the nutrient-rich charcoal and ashes. Birds also thrive on increased seed availability and nesting sites in snags.

Plant populations

Vegetation composition is one factor that determines how a fire behaves. The fire behaviour in turn determines the extent to which the plant populations are affected. The more intense the fire, the more vegetation is killed. The initial vegetation losses may look harsh, but the reduced number of trees and shrubs minimise competition among the surviving individuals.

The organisms that survive the fire gain more access to nutrients, light, and water. Plants may exhibit increased growth, benefiting from the additional minerals in the soil as a result of the fire. Fire may also rid some plants of their parasites, increasing plant health. For example, a high-intensity fire kills dwarf mistletoe, a parasitic plant of the lodgepole. Some plant species have adaptations that allow them to survive, thrive, and even require fire for survival. The giant sequoia can produce bark that is 2 feet thick as protection from fire. Other plants such as the chaparral snowbush require the heat of wildland fires to crack their seed coats.

Fire regimes

Fire regimes are the patterns of wildland fires that include factors such as frequency, extent, intensity, type, and season. Regimes vary by ecosystem because each ecosystem has a different composition and structure determined by climate conditions, vegetation types, and ignition sources. Humans have altered many aspects of natural fire regimes over time. Currently, ecologists are studying evidence to try to determine historical fire records or natural fire regimes. Techniques include sampling fire scars on trees for evidence of a sequence of fires in the growth rings, sampling lake and reservoir sediments for extreme or unusual runoff events, using written and oral histories, and extrapolating from current patterns of weather, fuel build-up and lightning fires. Understanding natural fire regimes should lead to the most appropriate resource management policies. The variety of ecosystems and regimes dictates that there should be a variety of techniques and practices in any comprehensive management policy. One solution does not fit all.

Lodgepole pine

The Lodgepole Pine (Pinus contorta) is a dominant tree of the northern United States and Canada. A lodgepole stand may live 250 to 400 years. During the first century of the stand's existence surface fires are unable to climb into the forest canopy because the lower branches of a lodgepole die and drop-off as the tree grows taller. Slowly shade-tolerant spruces and firs start to grow on the dark floor of the stand, blocking out the sun and preventing lodgepoles from sprouting. Eventually spruces and firs would dominate if there were no forest fires. However lightning usually ignites a fire that destroys the stand. After a fire, lodgepole pines are often the first trees to reappear because the bare, sunlit soil that remains after a fire is ideal for lodgepole seedling growth. Lodgepoles also have special serotinous cones that are coated with a hard waxy substance. Fire melts this coating, allowing thousands of seeds to be released. The Lodgepole Pine is just one of many species that is dependent upon fire.

Human influence on wildland fire

The way a fire operates is largely determined by a region's wildlife. Altering these biotic components impacts the fire's effects. Humans have had one of the greatest influences on the biota of ecosystems. Native Americans and early settlers used fire extensively in their land management practices. Today, we clear vegetation for farming, homes, commercial buildings, and roads. We introduce nonnative species. We use forests to harvest timber. Our influences are countless. As a result, it is impossible to fully understand the extent to which humans have altered natural fire regimes. This makes fire management a complex and often controversial topic.

WILDFIRE

A wildfire is any uncontrolled fire in combustible vegetation that occurs in the countryside or a wilderness area. Other names such as brush fire, bushfire, forest fire, grass fire, hill fire, peat fire, vegetation fire, veldfire, and wildland fire may be used to describe the same phenomenon depending on the type of vegetation being burned. A wildfire differs from other fires by its extensive size, the speed at which it can spread out from its original source, its potential to change direction unexpectedly, and its ability to jump gaps such as roads, rivers and fire breaks. Wildfires are characterised in terms of the cause of ignition, their physical properties such as speed of propagation, the combustible material present, and the effect of weather on the fire.

Wildfires occur on every continent except Antarctica. Fossil records and human history contain accounts of wildfires, as wildfires can occur in periodic intervals. Wildfires can cause extensive damage, both to property and human life, but they also have various beneficial effects on wilderness areas. Some plant species depend on the effects of fire for growth and reproduction, although large wildfires may also have negative ecological effects.

Strategies of wildfire prevention, detection, and suppression have varied over the years, and international wildfire management experts encourage further development of technology and research. One of the more controversial techniques is controlled burning: permitting or even igniting smaller fires to minimise the amount of flammable material available for a potential wildfire. While some wildfires burn in remote forested regions, they can cause extensive destruction of homes and other property located in the wildland-urban interface: a zone of transition between developed areas and undeveloped wilderness.

Characteristics

The name wildfire was once a synonym for Greek fire but now refers to any large or destructive conflagration. Wildfires differ from other fires in that they take place outdoors in areas of grassland, woodlands, bushland, scrubland, peatland, and other wooded areas that act as a source of fuel, or combustible material. Buildings may become involved if a wildfire spreads to adjacent communities. While the causes of wildfires vary and the outcomes are always unique, all wildfires can be characterised in terms of their physical properties, their fuel type, and the effect that weather has on the fire.

Wildfire behaviour and severity result from the combination of factors such as available fuels, physical setting, and weather. While wildfires can be large, uncontrolled disasters that burn through 0.4 to 400 square kilometers (100 to 1,00,000 acres) or more, they can also be as small as 0.0010 square kilometers (0.25 acre) or less. Although smaller events may be included in wildfire modelling, most do not earn press attention. This can be problematic because public fire policies, which relate to fires of all sizes, are influenced more by the way the media portrays catastrophic wildfires than by small fires.

Causes

The four major natural causes of wildfire ignitions are lightning, volcanic eruption, sparks from rockfalls, and spontaneous combustion. The thousands of coal seam fires that are burning around the world, such as those in Centralia, Burning Mountain, and several coal-sustained fires in China, can also flare up and ignite nearby flammable material. However, many wildfires are attributed to human sources such as arson, discarded cigarettes, sparks from equipment, and power line arcs (as detected by arc mapping). In societies experiencing shifting cultivation where land is cleared quickly and farmed until the soil loses fertility, slash and burn clearing is often considered the least expensive way to prepare land for future use. Forested areas cleared by logging encourage the dominance of flammable grasses, and abandoned logging roads overgrown by vegetation may act as fire corridors. Annual grassland fires in Southern Vietnam can be attributed in part to the destruction of forested areas by herbicides, explosives, and mechanical land clearing and burning operations during the Vietnam War.

The most common cause of wildfires varies throughout the world. In the United States, Canada, and Northwest China, for example, lightning is the major source of ignition. In other parts of the world, human involvement is a major contributor. In Mexico, Central America, South America, Africa, Southeast Asia, Fiji, and New Zealand, wildfires can be attributed to human activities such as animal husbandry, agriculture, and land-conversion burning. Human carelessness is a major cause of wildfires in China and in the Mediterranean Basin. In Australia, the source of wildfires can be traced to both lightning strikes and human activities such as machinery sparks and cast-away cigarette butts.

Fuel type

The spread of wildfires varies based on the flammable material present and its vertical arrangement. For example, fuels uphill from a fire are more readily dried and warmed by the fire than those downhill, yet burning logs can roll downhill from the fire to ignite other fuels. Fuel arrangement and density is governed in part by topography, as land shape determines factors such as available sunlight and water for plant growth. Overall, fire types can be generally characterised by their fuels as follows:

1. Ground fires are fed by subterranean roots, duff and other buried organic matter. This fuel type is especially susceptible to ignition due to spotting. Ground fires typically burn by smoldering, and can burn slowly for days to months, such as peat fires in Kalimantan and Eastern Sumatra, Indonesia, which resulted from a riceland creation project that unintentionally drained and dried the peat.
2. Crawling or surface fires are fueled by low-lying vegetation such as leaf and timber litter, debris, grass, and low-lying shrubbery.
3. Ladder fires consume material between low-level vegetation and tree canopies, such as small trees, downed logs, and vines. Kudzu, Old World climbing fern, and other invasive plants that scale trees may also encourage ladder fires.
4. Crown, canopy, or aerial fires burn suspended material at the canopy level, such as tall trees, vines, and mosses. The ignition of a crown fire, termed crowning, is dependent on the density of the suspended material, canopy height, canopy continuity, and sufficient surface and ladder fires in order to reach the tree crowns. For example, ground-clearing fires lit by humans can spread into the Amazon rain forest, damaging ecosystems not particularly suited for heat or arid conditions.

Physical properties

Wildfires occur when all of the necessary elements of a fire triangle come together in a susceptible area: an ignition source is brought into contact with a combustible material such as vegetation, that is subjected to sufficient heat and has an adequate supply of oxygen from the ambient air. A high moisture content usually prevents ignition and slows propagation, because higher temperatures are required to evaporate any water within the material and heat the material to its fire point. Dense forests usually provide more shade, resulting in lower ambient temperatures and greater humidity, and are therefore less susceptible to wildfires. Less dense material such as grasses and leaves are easier to ignite because they contain less water than denser material such as branches and trunks. Plants continuously lose water by evapotranspiration, but water loss is usually balanced by water absorbed from the soil, humidity, or rain. When this balance is not maintained, plants dry out and are therefore more flammable, often a consequence of droughts.

A wildfire front is the portion sustaining continuous flaming combustion, where unburned material meets active flames, or the smoldering transition between unburned and burned material. As the front approaches, the fire heats both the surrounding air and woody material through convection and thermal radiation. First, wood is dried as water is vapourised at a temperature of 100°C (212°F). Next, the pyrolysis of wood at 230°C (450°F) releases flammable gases. Finally, wood can smolder at 380°C (720°F) or, when heated sufficiently, ignite at 590°C (1000°F). Even before the flames of a wildfire arrive at a particular location, heat transfer from the wildfire front warms the air to 800°C (1470°F), which pre-heats and dries flammable materials, causing materials to ignite faster and allowing the fire to spread faster. High-temperature and long-duration surface wildfires may encourage flashover or torching: the drying of tree canopies and their subsequent ignition from below.

Wildfires have a rapid forward rate of spread (FROS) when burning through dense, uninterrupted fuels. They can move as fast as 10.8 kilometers per hour (6.7 mph) in forests and 22 km per hour (14 mph) in grasslands. Wildfires can advance tangential to the main front to form a flanking front, or burn in the opposite direction of the main front by backing. They may also spread by jumping or spotting as winds and vertical convection columns carry firebrands (hot wood embers) and other burning materials through the air over roads, rivers, and other barriers that may otherwise act as firebreaks. Torching and fires in tree canopies encourage spotting, and dry ground fuels that surround a wildfire are especially vulnerable to ignition from firebrands. Spotting can create spot fires as hot embers and firebrands ignite fuels downwind from the fire. In Australian bushfires, spot fires are known to occur as far as 10 km from the fire front.

Especially large wildfires may affect air currents in their immediate vicinities by the stack effect: air rises as it is heated, and large wildfires create powerful updrafts that will draw in new, cooler air from surrounding areas in thermal columns. Great vertical differences in temperature and humidity encourage pyrocumulus clouds, strong winds, and fire whirls with the force of tornadoes at speeds of more than 80 km per hour (50 mph). Rapid rates of spread, prolific crowning or spotting, the presence of fire whirls, and strong convection columns signify extreme conditions.

Effect of weather

Heat waves, droughts, cyclical climate changes such as El Niño, and regional weather patterns such as high-pressure ridges can increase the risk and alter the behaviour of wildfires dramatically. Years of precipitation followed by warm periods can encourage more widespread fires and longer fire seasons. Since the mid-1980s, earlier snowmelt and associated warming has also been associated with an increase in length and severity of the wildfire season in the Western United States. However, one individual element does not always cause an increase in wildfire activity. For example, wildfires will not occur during a drought unless accompanied by other factors, such as lightning (ignition source) and strong winds (mechanism for rapid spread). Fire intensity also increases during day time hours. Burn rates of smoldering logs are up to five times greater during the day due to lower humidity, increased temperatures, and increased wind speeds. Sunlight warms the ground during the day which creates air currents that travel uphill. At night the land cools, creating air currents that travel downhill. Wildfires are fanned by these winds and often follow the air currents over hills and through valleys. Fires in Europe occur frequently during the hours of 12:00 pm and 2:00 pm. Wildfire suppression operations in the United States revolve around a 24 hour fire day that begins at 10:00 am due to the predictable increase in intensity resulting from the daytime warmth.

Ecology

Wildfires are common in climates that are sufficiently moist to allow the growth of vegetation but feature extended dry, hot periods. Such places include the vegetated areas of Australia and Southeast Asia, the veld in southern Africa, the fynbos in the Western Cape of South Africa, the forested areas of the United States and Canada, and the Mediterranean Basin. Fires can be particularly intense during days of strong winds, periods of drought, and during warm summer months. Global warming may increase the intensity and frequency of droughts in many areas, creating more intense and frequent wildfires.

Although some ecosystems rely on naturally occurring fires to regulate growth, many ecosystems suffer from too much fire, such as the chaparral in southern California and lower elevation deserts in the American Southwest. The increased fire frequency in these ordinarily fire-dependent areas has upset natural cycles, destroyed native plant communities, and encouraged the growth of fire-intolerant vegetation and non-native weeds. Invasive species, such as *Lygodium microphyllum* and *Bromus tectorum*, can grow rapidly in areas that were damaged by fires. Because they are highly flammable, they can increase the future risk of fire, creating a positive feedback loop that increases fire frequency and further destroys native growth.

In the Amazon Rainforest, drought, logging, cattle ranching practices, and slash-and-burn agriculture damage fire-resistant forests and promote the growth of flammable brush, creating a cycle that encourages more burning. Fires in the rainforest threaten its collection of diverse species and produce large amounts of CO_2. Also, fires in the rainforest, along with drought and human involvement, could damage or destroy more than half of the Amazon rainforest by the year 2030. Wildfires generate ash, destroy available organic nutrients, and cause an increase in water runoff, eroding away other nutrients and creating flash flood conditions. Wildfires can also have an effect on climate change, increasing the amount of carbon released into the atmosphere and inhibiting vegetation growth, which affects overall carbon uptake by plants.

Plant adaptation

Plants in wildfire-prone ecosystems often survive through adaptations to their local fire regime. Such adaptations include physical protection against heat, increased growth after a fire event, and flammable materials that encourage fire and may eliminate competition. For example, plants of the genus *Eucalyptus* contain flammable oils that encourage fire and hard sclerophyll leaves to resist heat and drought, ensuring their dominance over less fire-tolerant species. Dense bark, shedding lower branches, and high water content in external structures may also protect trees from rising temperatures. Fire-resistant seeds and reserve shoots that sprout after a fire encourage species preservation, as embodied by pioneer species. Smoke, charred wood, and heat can stimulate the germination of seeds in a process called serotiny. Exposure to smoke from burning plants promotes germination in other types of plants by inducing the production of the orange butenolide.

Atmospheric effects

Most of the earth's weather and air pollution reside in the troposphere, the part of the atmosphere that extends from the surface of the planet to a height of about 10 km. The vertical lift of a severe thunderstorm or pyrocumulonimbus can be enhanced in the area of a large wildfire, which can propel smoke, soot, and other particulate matter as high as the lower stratosphere. Previously, prevailing scientific theory held that most particles in the stratosphere came from volcanoes, but smoke and other wildfire emissions have been detected from the lower stratosphere. Pyrocumulus clouds can reach 6100 metres (20,000 ft) over wildfires. Increased fire by-products in the stratosphere can increase ozone concentration beyond safe levels.

Human involvement

The human use of fire for agricultural and hunting purposes during the Paleolithic and Mesolithic ages altered the preexisting landscapes and fire regimes. Woodlands were gradually replaced by smaller vegetation that facilitated travel, hunting, seed-gathering and planting.

Prevention

Wildfire prevention refers to the preemptive methods of reducing the risk of fires as well as lessening its severity and spread. Effective prevention techniques allow supervising agencies to manage air quality, maintain ecological balances, protect resources, and to limit the effects of future uncontrolled fires. North American firefighting policies may permit naturally caused fires to burn to maintain their ecological role, so long as the risks of escape into high-value areas are mitigated. However, prevention policies must consider the role that humans play in wildfires, since, for example, 95 per cent of forest fires in Europe are related to human involvement. Sources of human-caused fire may include arson, accidental ignition, or the uncontrolled use of fire in land-clearing and agriculture such as the slash-and-burn farming in Southeast Asia.

Wildfires are caused by a combination of natural factors such as topography, fuels, and weather. Other than reducing human infractions, only fuels may be altered to affect future fire risk and behaviour. Wildfire prevention programs around the world may employ techniques such as wildland fire use and prescribed or controlled burns. Wildland fire use refers to any fire of natural causes that is monitored but allowed to burn. Controlled burns are fires ignited by government agencies under less dangerous weather conditions.

Vegetation may be burned periodically to maintain high species diversity, and frequent burning of surface fuels limits fuel accumulation, thereby reducing the risk of crown fires. Using strategic cuts of trees, fuels may also be removed by handcrews in order to clean and clear the forest, prevent fuel build-up, and create access into forested areas. Chain saws and large equipment can be used to thin out ladder fuels and shred trees and vegetation to a mulch. Multiple fuel treatments are often needed to influence future fire risks, and wildfire models may be used to predict and compare the benefits of different fuel treatments on future wildfire spread.

However, controlled burns are reportedly 'the most effective treatment for reducing a fire's rate of spread, fireline intensity, flame length, and heat per unit of area' according to Jan Van Wagtendonk, a biologist at the Yellowstone Field Station. Additionally, while fuel treatments are typically limited to smaller areas, effective fire management requires the administration of fuels across large landscapes in order to reduce future fire size and severity.

Building codes in fire-prone areas typically require that structures be built of flame-resistant materials and a defensible space be maintained by clearing flammable materials within a prescribed distance from the structure. Communities in the Philippines also maintain fire lines 5 to 10 meters (16 to 33 ft) wide between the forest and their village, and patrol these lines during summer months or seasons of dry weather. Fuel buildup can result in costly, devastating fires as new homes, ranches, and other development are built adjacent to wilderness areas. Continued growth in fire-prone areas and rebuilding structures destroyed by fires has been met with criticism.

However, the population growth along the wildland-urban interface discourages the use of current fuel management techniques. Smoke is an irritant and attempts to thin out the fuel load is met with opposition due to desirability of forested areas, in addition to other wilderness goals such as endangered species protection and habitat preservation. The ecological benefits of fire are often overridden by the economic and safety benefits of protecting structures and human life. For example, while fuel treatments decrease the risk of crown fires, these techniques destroy the habitats of various plant and animal species. Additionally, government policies that cover the wilderness usually differ from local and state policies that govern urban lands. A new and ecologically evolutionary practice, termed 'Hydro-Pyrogeography', promises to bound wildfire from passing through any such wildland-urban interface anywhere on earth that the practice is put into place, and thereby diminishing, even eliminating the above-referred oppositions and concerns to traditional fuel management techniques.

ECONOMICS OF FIRE MANAGEMENT POLICY

Similar to that of military operations, fire management is often very expensive in the US. Today, it is not uncommon for suppression operations for a single wildfire to exceed costs of $1 million in just a few days. Although fire suppression offers many benefits to society, other options for fire management exist. While these options can't completely replace fire suppression as a fire management tool, other options can play an important role in overall fire management and can therefore affect the costs of fire suppression.

The application of fire management tools requires making certa in tradeoffs. Below is a sample of some costs and benefits associated with the tools currently used in fire management. Current approaches to fire management are an almost complete turnaround compared to historic approaches. In fact, it is commonly accepted that past fire suppression, along with other factors, has resulted in larger, more intense wildfire events which are seen today. In economic terms, expenditures used for wildfire suppression in the early 20th century have contributed to increased suppression costs which are being realised today. As is the case with many public policy issues, costs and benefits associated with particular fire management tools are difficult to accurately quantify. Ultimately, costs and benefits should be weighed against one another on a case-by-case basis in planning wildland fire management operations.

Depending on the tradeoffs that a land manager is willing to make, a combination of the following fire management tools could be used. For instance, prescribed fire and/or mechanical fuels reduction could be used to help prevent or lessen the intensity of a wildfire thereby reducing or eliminating suppression costs. In addition, prescribed fire and/or mechanical fuels reduction could be used to improve soil conditions in fields or in forests to the benefit of wildlife or natural resources. On the other hand, the use of prescribed fire requires much advanced planning and can have negative impacts on human health in nearby communities (Table 8.1).

Table 8.1. Cost and benefits of wildland fire management tools.

	Costs	Benefits
Suppression	Labour intensive	Can reduce human health impacts
	Requires high level of planning	Can protect forest resources and agricultural
	Can be very expensive	resources
	Particular strategies can be very inefficient (i.e. aerial retardant drops)	Can save private dwellings and commercial buildings
	Can increase intensity and likelihood of future wildfires	

(Contd ...)

	Costs	*Benefits*
	Inhibits natural ecological processes in many cases	
Prescribed fire	Can be expensive to implement	Can provide habitate for wildlife
	Requires skilled workforce to implement Requires high level of planning	Can improve forest resources and agricultural resources
	Can impact human health (e.g. smoke and its effect on those with asthma or allergies)	Can reduce hazardous fuel loading
		Mimics natural processes but under more controlled circumstances
Mechanical fuels reduction	Requires use of heavy machinery (resulting in fossil fuel consumption, soil compaction, etc.)	Can provide habitat for wildlife
		Can improve forest resources and agricultural resources
	Can be expensive to implement	Can reduce hazardous fuel loading
	Does not mimic natural processes	Does not produce large amounts of smoke

Detection

Fast and effective detection is a key factor in wildfire fighting. Early detection efforts were focused on early response, accurate results in both daytime and night time, and the ability to prioritise fire danger. Currently, public hotlines, fire lookouts in towers, and ground and aerial patrols can be used as a means of early detection of forest fires. However, accurate human observation may be limited by operator fatigue, time of day, time of year, and geographic location. Electronic systems have gained popularity in recent years as a possible resolution to human operator error. These systems may be semi- or fully-automated and employ systems based on the risk area and degree of human presence, as suggested by GIS data analyses. An integrated approach of multiple systems can be used to merge satellite data, aerial imagery, and personnel position via global positioning system (GPS) into a collective whole for near-realtime use by wireless Incident Command Centers. A small, high risk area that features thick vegetation, a strong human presence, or is close to a critical urban area can be monitored using a local sensor network. Detection systems may include wireless sensor networks that act as automated weather systems: detecting temperature, humidity, and smoke. These may be battery-powered, solar-powered or tree-rechargeable: able to recharge their battery systems using the small electrical currents in plant material. Larger, medium-risk areas can be monitored by scanning towers that incorporate fixed cameras and sensors to detect smoke or additional factors such as the infrared signature of carbon dioxide produced by fires. Additional capabilities such as night vision, brightness detection, and colour change detection may also be incorporated into sensor arrays.

Satellite and aerial monitoring can provide a wider view and may be sufficient to monitor very large, low risk areas. These more sophisticated systems employ GPS and aircraft-mounted infrared or high-resolution visible cameras to identify and target wildfires. Satellite-mounted sensors such as Envisat's Advanced Along Track Scanning Radiometer and European Remote-Sensing Satellite's Along-Track Scanning Radiometer can measure infrared radiation emitted by fires, identifying hot spots greater than 39°C (102°F). The National Oceanic and Atmospheric Administration's Hazard Mapping System combines remote-sensing data from satellite sources such as geostationary operational environmental satellite (GOES), moderate-resolution imaging spectroradiometer (MODIS), and advanced very high resolution radiometer (AVHRR) for detection of fire and smoke plume location.

Wildfire Modelling

Wildfire modelling is concerned with numerical simulation of wildfires in order to comprehend and predict fire behaviour. Wildfire modelling can ultimately aid wildfire suppression, increase the safety of firefighters and the public, and minimise damage. Using computational science, wildfire modelling involves the statistical analysis of past fire events to predict spotting risks and front behaviour. Various wildfire propagation models have been proposed in the past, including simple ellipses and egg- and fan-shaped models. Early attempts to determine wildfire behaviour assumed terrain and vegetation uniformity. However, the exact behaviour of a wildfire's front is dependent on a variety of factors, including windspeed and slope steepness.

Modern growth models utilise a combination of past ellipsoidal descriptions and Huygens' Principle to simulate fire growth as a continuously expanding polygon. Extreme value theory may also be used to predict the size of large wildfires. However, large fires that exceed suppression capabilities are often regarded as statistical outliers in standard analyses, even though fire policies are more influenced by catastrophic wildfires than by small fires.

VOLCANOES

A volcano is an opening or rupture, in a planet's surface or crust, which allows hot magma, volcanic ash and gases to escape from below the surface. Volcanoes are generally found where tectonic plates are diverging or converging. A mid-oceanic ridge, for example the Mid-Atlantic Ridge, has examples of volcanoes caused by divergent tectonic plates pulling apart; the Pacific Ring of Fire has examples of volcanoes caused by convergent tectonic plates coming together. By contrast, volcanoes are usually not created where two tectonic plates slide past one another. Volcanoes can also form where there is stretching and thinning of the earth's crust (called 'non-hotspot intraplate volcanism'), such as in the East African Rift, the Wells Gray-Clearwater volcanic field and the Rio Grande Rift in North America. Volcanoes can be caused by mantle plumes. These so-called hotspots, for example at Hawaii, can occur far from plate boundaries. Hotspot volcanoes are also found elsewhere in the solar system, especially on rocky planets and moons.

Plate Tectonics and Hotspots

Divergent plate boundaries

Divergent plate boundaries are locations where plates are moving away from one another. This occurs above rising convection currents. The rising current pushes up on the bottom of the lithosphere, lifting it and flowing laterally beneath it. This lateral flow causes the plate material above to be dragged along in the direction of flow. At the crest of the uplift, the overlying plate is stretched thin, breaks and pulls apart (Fig. 8.1).

At the mid-oceanic ridges, two tectonic plates diverge from one another. New oceanic crust is being formed by hot molten rock slowly cooling and solidifying. The crust is very thin at mid-oceanic ridges due to the pull of the tectonic plates. The release of pressure due to the thinning of the crust leads to adiabatic expansion, and the partial melting of the mantle causing volcanism and creating new oceanic crust. Most divergent plate boundaries are at the bottom of the oceans, therefore most volcanic activity is submarine, forming new seafloor. Black smokers or deep sea vents are an example of this kind of volcanic activity. Where the mid-oceanic ridge is above sea-level, volcanic islands are formed, for example, Iceland.

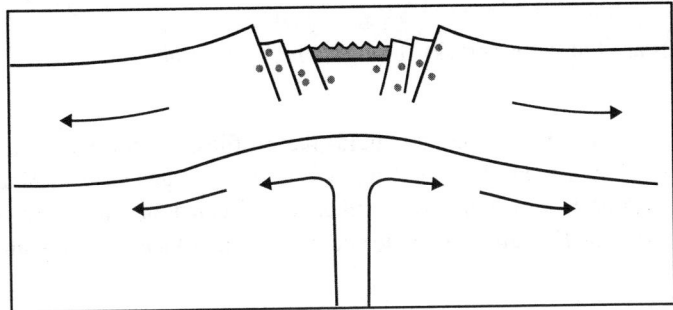

Fig. 8.1. Divergent plate boundaries.

Convergent plate boundaries

Subduction zones are places where two plates, usually an oceanic plate and a continental plate, collide. In this case, the oceanic plate subducts, or submerges under the continental plate forming a deep ocean trench just offshore. Water released from the subducting plate lowers the melting temperature of the overlying mantle wedge, creating magma. This magma tends to be very viscous due to its high silica content, so often does not reach the surface and cools at depth. When it does reach the surface, a volcano is formed. Typical examples for this kind of volcano are Mount Etna and the volcanoes in the Pacific Ring of Fire.

Hotspots

Hotspots are not usually located on the ridges of tectonic plates, but above mantle plumes, where the convection of the earth's mantle creates a column of hot material that rises until it reaches the crust, which tends to be thinner than in other areas of the earth. The temperature of the plume causes the crust to melt and form pipes, which can vent magma. Because the tectonic plates move whereas the mantle plume remains in the same place, each volcano becomes dormant after a while and a new volcano is then formed as the plate shifts over the hotspot. The Hawaiian Islands are thought to be formed in such a manner, as well as the Snake River Plain, with the Yellowstone Caldera being the part of the North American plate currently above the hot spot.

Volcanic features

The most common perception of a volcano is of a conical mountain, spewing lava and poisonous gases from a crater at its summit. This describes just one of many types of volcano, and the features of volcanoes are much more complicated. The structure and behaviour of volcanoes depends on a number of factors. Some volcanoes have rugged peaks formed by lava domes rather than a summit crater, whereas others present landscape features such as massive plateaus. Vents that issue volcanic material (lava, which is what magma is called once it has escaped to the surface, and ash) and gases (mainly steam and magmatic gases) can be located anywhere on the landform.

Other types of volcano include cryovolcanoes (or ice volcanoes), particularly on some moons of Jupiter, Saturn and Neptune; and mud volcanoes, which are formations often not associated with known magmatic activity.

Active mud volcanoes tend to involve temperatures much lower than those of igneous volcanoes, except when a mud volcano is actually a vent of an igneous volcano.

Fissure vents

Volcanic fissure vents are flat, linear cracks through which lava emerges.

Shield volcanoes

Shield volcanoes, so named for their broad, shield-like profiles, are formed by the eruption of low-viscosity lava that can flow a great distance from a vent, but not generally explode catastrophically. Since low-viscosity magma is typically low in silica, shield volcanoes are more common in oceanic than continental settings. The Hawaiian volcanic chain is a series of shield cones, and they are common in Iceland, as well.

Lava domes

Lava domes are built by slow eruptions of highly viscous lavas. They are sometimes formed within the crater of a previous volcanic eruption (as in Mount Saint Helens), but can also form independently, as in the case of Lassen Peak. Like stratovolcanoes, they can produce violent, explosive eruptions, but their lavas generally do not flow far from the originating vent.

Cryptodomes

Cryptodomes are formed when viscous lava forces its way up and causes a bulge. The 1980 eruption of Mount St. Helens was an example. Lava was under great pressure and forced a bulge in the mountain, which was unstable and slid down the north side.

Volcanic cones

Volcanic cones or cinder cones are the result from eruptions that erupt mostly small pieces of scoria and pyroclastics (both resemble cinders, hence the name of this volcano type) that build up around the vent. These can be relatively short-lived eruptions that produce a cone-shaped hill perhaps 30 to 400 meters high. Most cinder cones erupt only once. Cinder cones may form as flank vents on larger volcanoes, or occur on their own. Parícutin in Mexico and Sunset Crater in Arizona are examples of cinder cones. In New Mexico, Caja del Rio is a volcanic field of over 60 cinder cones.

Stratovolcanoes

Stratovolcanoes or composite volcanoes are tall conical mountains composed of lava flows and other ejecta in alternate layers, the strata that give rise to the name. Stratovolcanoes are also known as composite volcanoes, created from several structures during different kinds of eruptions. Strato/composite volcanoes are made of cinders, ash and lava. Cinders and ash pile on top of each other, lava flows on top of the ash, where it cools and hardens, and then the process begins again. Classic examples include Mt. Fuji in Japan, Mayon Volcano in the Philippines, and Mount Vesuvius and Stromboli in Italy.

In recorded history, explosive eruptions by stratovolcanoes have posed the greatest hazard to civilisations, as ash is produced by an explosive eruption. No supervolcano erupted in recorded history. Shield volcanoes have not an enormous pressure build up from the lava flow. Fissure vents and monogenetic volcanic fields (volcanic cones) have not powerful explosive eruptions, as they are many times under extension. Stratovolcanoes ($30°–35°$) are steeper than shield volcanoes (generally $5°–10°$), their lose tephra are material for dangerous lahars.

Supervolcanoes

A supervolcano is a large volcano that usually has a large caldera and can potentially produce devastation on an enormous, sometimes continental, scale. Such eruptions would be able to cause severe cooling of global temperatures for many years afterwards because of the huge volumes of sulphur and ash erupted. They are the most dangerous type of volcano. Examples include Yellowstone Caldera in Yellowstone National Park and Valles Caldera in New Mexico (both western United States), Lake Taupo in New Zealand, Lake Toba in Sumatra, Indonesia and Ngorogoro Crater in Tanzania, Krakatoa near Java and Sumatra, Indonesia. Supervolcanoes are hard to identify centuries later, given the enormous areas they cover. Large igneous provinces are also considered supervolcanoes because of the vast amount of basalt lava erupted, but are non-explosive.

Submarine volcanoes

Submarine volcanoes are common features on the ocean floor. Some are active and, in shallow water, disclose their presence by blasting steam and rocky debris high above the surface of the sea. Many others lie at such great depths that the tremendous weight of the water above them prevents the explosive release of steam and gases, although they can be detected by hydrophones and discolouration of water because of volcanic gases. Pumice rafts may also appear. Even large submarine eruptions may not disturb the ocean surface.

Because of the rapid cooling effect of water as compared to air, and increased buoyancy, submarine volcanoes often form rather steep pillars over their volcanic vents as compared to above-surface volcanoes. They may become so large that they break the ocean surface as new islands. Pillow lava is a common eruptive product of submarine volcanoes. Hydrothermal vents are common near these volcanoes, and some support peculiar ecosystems based on dissolved minerals.

Subglacial volcanoes

Subglacial volcanoes develop underneath icecaps. They are made up of flat lava which flows at the top of extensive pillow lavas and palagonite. When the icecap melts, the lavas on the top collapse, leaving a flat-topped mountain.

These volcanoes are also called table mountains, tuyas or (uncommonly) mobergs. Very good examples of this type of volcano can be seen in Iceland, however, there are also tuyas in British Columbia. The origin of the term comes from Tuya Butte, which is one of the several tuyas in the area of the Tuya River and Tuya Range in northern British Columbia.

Tuya Butte was the first such landform analysed and so its name has entered the geological literature for this kind of volcanic formation. The Tuya Mountains Provincial Park was recently established to protect this unusual landscape, which lies north of Tuya Lake and south of the Jennings River near the boundary with the Yukon Territory.

Mud volcanoes

Mud volcanoes or mud domes are formations created by geo-excreted liquids and gases, although there are several different processes which may cause such activity. The largest structures are 10 kilometres in diameter and reach 700 meters high.

Effects of Volcanoes

There are many different types of volcanic eruptions and associated activity: phreatic eruptions (steam-generated eruptions), explosive eruption of high-silica lava (e.g. rhyolite), effusive eruption of low-silica lava (e.g. basalt), pyroclastic flows, lahars (debris flow) and carbon dioxide emission. All of these activities can pose a hazard to humans. Earthquakes, hot springs, fumaroles, mud pots and geysers often accompany volcanic activity (Fig. 8.2).

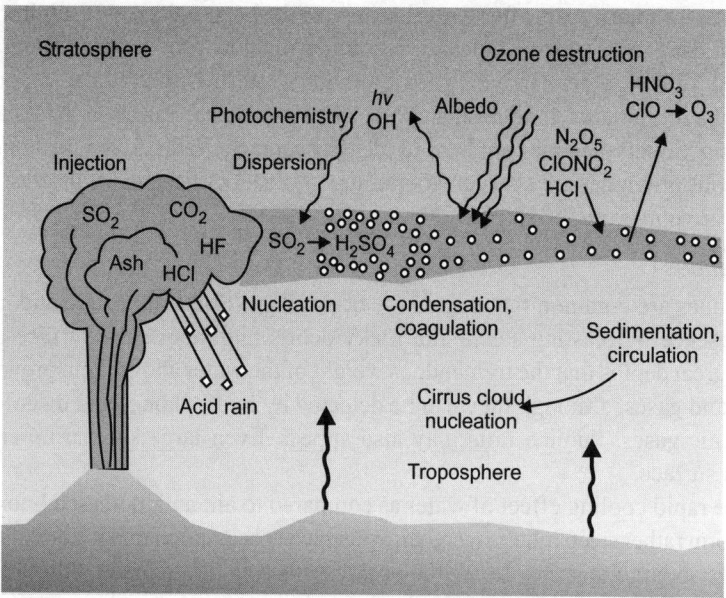

Fig. 8.2. Volcanic 'injection'.

The concentrations of different volcanic gases can vary considerably from one volcano to the next. Water vapour is typically the most abundant volcanic gas, followed by carbon dioxide and sulphur dioxide.

SECTION III

Community Ecology

Chapter 9

Community Structure and Classification

INTRODUCTION

A community is the set of all populations that inhabit a certain area. Communities can have different sizes and boundaries. These are often identified with some difficulty. An ecosystem is a higher level of organisation of the community plus its physical environment. Ecosystems include both the biological and physical components affecting the community/ecosystem. We can study ecosystems from a structural view of population distribution or from a functional view of energy flow and other processes.

Ecologists find that within a community many populations are not randomly distributed. This recognition that there was a pattern and process of spatial distribution of species was a major accomplishment of ecology. Two of the most important patterns are open community structure and the relative rarity of species within a community. Do species within a community have similar geographic range and density peaks? If they do, the community is said to be a closed community, a discrete unit with sharp boundaries known as ecotones. An open community, however, has its populations without ecotones and distributed more or less randomly.

In a forest, where we find an open community structure, there is a gradient of soil moisture. Plants have different tolerances to this gradient and occur at different places along the continuum. Where the physical environment has abrupt transitions, we find sharp boundaries developing between populations. For example, an ecotone develops at a beach separating water and land.

Open structure provides some protection for the community. Lacking boundaries, it is harder for a community to be destroyed in an all or nothing fashion. Species can come and go within communities over time, yet the community as a whole persists. In general, communities are less fragile and more flexible than some earlier concepts would suggest. Most species in a community are far less abundant than the dominant species that provide a community its name: for example oak-hickory, pine, etc. Populations of just a few species are dominant within a community, no matter what community we examine. Resource partitioning is thought to be the main cause for this distribution.

CLASSIFICATION OF COMMUNITIES

There are two basic categories of communities: terrestrial (land) and aquatic (water). These two basic types of community contain eight smaller units known as biomes. A biome is a large-scale category containing many communities of a similar nature, whose distribution is largely controlled by climate.

1. Terrestrial biomes: Tundra, grassland, desert, taiga, temperate forest, tropical forest.
2. Aquatic biomes: Marine, freshwater.

Terrestrial Biomes

Tundra and desert

The tundra and desert biomes occupy the most extreme environments, with little or no moisture and extremes of temperature acting as harsh selective agents on organisms that occupy these areas. These two biomes have the fewest numbers of species due to the stringent environmental conditions. In other words, not everyone can live there due to the specialised adaptations required by the environment.

Tropical rain forests

Tropical rain forests occur in regions near the equator. The climate is always warm (between 20° and 25°C) with plenty of rainfall (at least 190 cm/year). The rain forest is probably the richest biome, both in diversity and in total biomass. The tropical rain forest has a complex structure, with many levels of life. More than half of all terrestrial species live in this biome. While diversity is high, dominance by a particular species is low.

While some animals live on the ground, most rain forest animals live in the trees. Many of these animals spend their entire life in the forest canopy. Insects are so abundant in tropical rain forests that the majority have not yet been identified. Charles Darwin noted the number of species found on a single tree, and suggested the richness of the rain forest would stagger the future systematist with the size of the catalogue of animal species found there. Termites are critical in the decomposition and nutrient cycling of wood. Birds tend to be brightly coloured, often making them sought after as exotic pets. Amphibians and reptiles are well represented. Lemurs, sloths, and monkeys feed on fruits in tropical rain forest trees. The largest carnivores are the cats (jaguars in South America and leopards in Africa and Asia). Encroachment and destruction of habitat put all these animals and plants at risk.

Epiphytes are plants that grow on other plants. These epiphytes have their own roots to absorb moisture and minerals, and use the other plant more as an aid to grow taller. Some tropical forests in India, Southeast Asia, West Africa, Central and South American are seasonal and have trees that shed leaves in dry season. The warm, moist climate supports high productivity as well as rapid decomposition of detritus. With its year long growing season, tropical forests have a rapid cycling of nutrients. Soils in tropical rain forests tend to have very little organic matter since most of the organic carbon is tied up in the standing biomass of the plants.

These tropical soils, termed laterites, make poor agricultural soils after the forest has been cleared. About 17 million hectares of rain forest are destroyed each year (an area equal in size to Washington state). Estimates indicate the forests will be destroyed (along with a great part of the earth's diversity) within 100 years. Rainfall and climate patterns could change as a result.

Temperate forests

The temperate forest biome occurs south of the taiga in eastern North America, eastern Asia, and much of Europe. Rainfall is abundant (30–80 inches/year; 75–150 cm) and there is a well-defined growing season of between 140 and 300 days. The eastern United States and Canada are covered (or rather were once covered) by this biome's natural vegetation, the eastern deciduous forest. Dominant plants include beech, maple, oak; and other deciduous hardwood trees. Trees of a deciduous forest have broad leaves, which they lose in the fall and grow again in the spring.

Sufficient sunlight penetrates the canopy to support a well-developed understory composed of shrubs, a layer of herbaceous plants, and then often a ground cover of mosses and ferns. This stratification beneath the canopy provides a numerous habitats for a variety of insects and birds. The deciduous forest also contains many members of the rodent family, which serve as a food source for bobcats, wolves, and foxes. This area also is a home for deer and black bears. Winters are not as cold as in the taiga, so many amphibian and reptiles are able to survive.

Shrubland (Chaparral)

The shrubland biome is dominated by shrubs with small but thick evergreen leaves that are often coated with a thick, waxy cuticle, and with thick underground stems that survive the dry summers and frequent fires. Shrublands occur in parts of South America, western Australia, central Chile, and around the Mediterranean Sea. Dense shrubland in California, where the summers are hot and very dry, is known as chaparral. This Mediterranean-type shrubland lacks an understory and ground litter, and is also highly flammable. The seeds of many species require the heat and scarring action of fire to induce germination.

Grasslands

Grasslands occur in temperate and tropical areas with reduced rainfall (10–30 inches per year) or prolonged dry seasons. Grasslands occur in the Americas, Africa, Asia, and Australia. Soils in this region are deep and rich and are excellent for agriculture. Grasslands are almost entirely devoid of trees, and can support large herds of grazing animals. Natural grasslands once covered over 40 per cent of the earth's land surface. In temperate areas where rainfall is between 10 and 30 inches a year, grassland is the climax community because it is too wet for desert and too dry for forests.

Most grasslands have now been utilised to grow crops, especially wheat and corn. Grasses are the dominant plants, while grazing and burrowing species are the dominant animals. The extensive root systems of grasses allows them to recover quickly from grazing, flooding, drought, and sometimes fire.

Temperate grasslands include the Russian steppes, the South American pampas, and North American prairies. A tall-grass prairie occurs where moisture is not quite sufficient to support trees. A short-grass-prairie, survives on less moisture and occurs between a tall-grass prairie and desert.

Animal life includes mice, prairie dogs, rabbits, and animals that feed on them (hawks and snakes). Prairies once contained large herds of buffalo and pronghorn antelope, but with human activity these once great herds have dwindled.

The savanna is a tropical grassland that contains some trees. The savanna contains the greatest variety and numbers of herbivores (antelopes, zebras, and wildebeests, among others). This environment supports a large population of carnivores (lions, cheetahs, hyenas, and leopards). Any plant litter not consumed by grazers is attacked by termites and other decomposers. Once again, human activities are threatening this biome, reducing the range for herbivores and carnivores. Will extinction of the great cats be a result?

Deserts

Deserts are characterised by dry conditions (usually less than 10 inches per year; 25 cm) and a wide temperature range. The dry air leads to wide daily temperature fluctuations from freezing at night to over 120 degrees during the day. Most deserts occur at latitudes of 30° N or S where descending air masses are dry. Some deserts occur in the rainshadow of tall mountain ranges or in coastal areas near cold offshore currents. Plants in this biome have developed a series of adaptations (such as succulent stems, and small, spiny, or absent leaves) to conserve water and deal with these temperature extremes. Photosynthetic modifications (CAM) are another strategy to life in the drylands.

The Sahara and a few other deserts have almost no vegetation. Most deserts, however, are home to a variety of plants, all adapted to heat and lack of abundant water (succulents and cacti). Animal life of the Sonoran desert includes arthropods (especially insects and spiders), reptiles (lizards and snakes), running birds (the roadrunner of the American southwest and Warner Brothers cartoon fame), rodents (kangaroo rat and pack rat), and a few larger birds and mammals (hawks, owls, and coyotes).

The taiga (pronounced 'tie-guh') is a coniferous forest extending across most of the northern area of northern Eurasia and North America. This forest belt also occurs in a few other areas, where it has different names: the montane coniferous forest when near mountain tops; and the temperate rain forest along the Pacific Coast as far south as California. The taiga receives between 10 and 40 inches of rain per year and has a short growing season. Winters are cold and short, while summers tend to be cool. The taiga is noted for its great stands of spruce, fir, hemlock, and pine. These trees have thick protective leaves and bark, as well as needlelike (evergreen) leaves can withstand the weight of accumulated snow. Taiga forests have a limited understory of plants, and a forest floor covered by low-lying mosses and lichens. Conifers, alders, birch and willow are common plants; wolves, grizzly bears, moose, and caribou are common animals. Dominance of a few species is pronounced, but diversity is low when compared to temperate and tropical biomes.

Tundra

The tundra, shown in Fig. 9.1, covers the northernmost regions of North America and Eurasia, about 20 per cent of the earth's land area. This biome receives about 20 cm (8–10 inches) of rainfall annually. Snow melt makes water plentiful during summer months. Winters are long and dark, followed by very short summers. Water is frozen most of the time, producing frozen soil, permafrost. Vegetation includes no trees, but rather patches of grass and shrubs; grazing musk ox, reindeer, and caribou exist along with wolves, lynx, and rodents. A few animals highly adapted to cold live in the tundra year-round (lemming, ptarmigan). During the summer the tundra hosts numerous insects and migratory animals. The ground is nearly completely covered with sedges and short grasses during the short summer. There are also plenty of patches of lichens and mosses. Dwarf woody shrubs flower and produce seeds quickly during the short growing season. The alpine tundra occurs above the timberline on mountain ranges, and may contain many of the same plants as the arctic tundra.

Fig. 9.1. Left: View of the tundra, locality unknown. Right: Caribou, an animal characteristic of the tundra.

Climate, altitude and terrestrial biomes

Climate controls biome distribution by an altitudinal gradient and a latitudinal gradient. With increases of either altitude or latitude, cooler and drier conditions occur. Cooler conditions can cause aridity since cooler air can hold less water vapour than can warmer air. This is shown in Fig. 9.2.

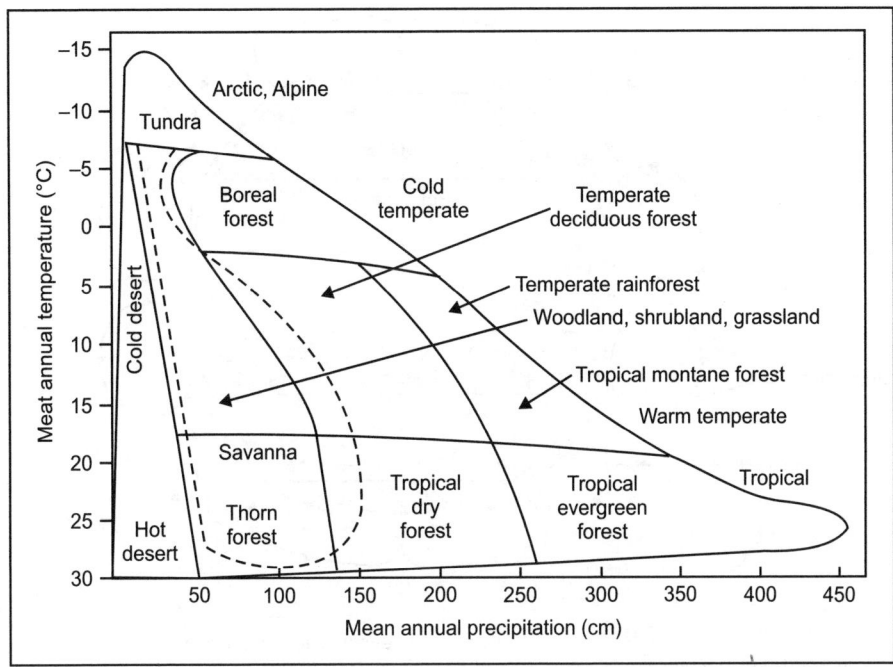

Fig. 9.2. Effect of temperature on precipitation.

Deserts can occur in warm areas due to a blockage of air circulation patterns that form a rain shadow, or from atmospheric circulation patterns as shown in Fig. 9.3. Warm air rises, producing low pressure areas. Cooler air sinks, producing high pressure areas. The tropics tend to be atmospheric low pressure zones the arctic areas atmospheric highs. Relative humidity is a measure of how much water an air mass at a given temperature can hold. In short, warm air can hold more moisture than can cold air. This basic physical feature of air helps explain the distribution of some of the world's great deserts.

The warm, moist air masses in the tropics rise upward in the atmosphere as they heat. The pressure of air rising forces air in the upper atmosphere to flow away north and south. This air at higher elevations is cooler and loses much of its moisture as rainfall. When the air masses begin to descend they heat up and begin to draw moisture from the lands they descend upon, at 30 degrees north and south of the equator. Many of the world's deserts are at approximately 30 degrees latitude, as shown in Fig. 9.3.

Rain shadow deserts also form when cool, dry air masses descend after passing over a tall mountain range, such as the Coast Range and Sierras in California. The Sonoran desert in Arizona is a doubly caused desert, being at 30 degrees latitude as well as in the rain shadow of California mountains.

Aquatic biomes

Conditions in water are generally less harsh than those on land. Aquatic organisms are buoyed by water support, and do not usually have to deal with desiccation. Despite covering 71 per cent of the earth's surface, areas of the open ocean are a vast aquatic desert containing few nutrients and very little life, as shown by Fig. 9.4. Clearcut biome distinctions in water, like those on land, are difficult to make. Dissolved nutrients controls many local aquatic distributions. Aquatic communities are classified into: freshwater (inland) communities and marine (saltwater or oceanic) communities.

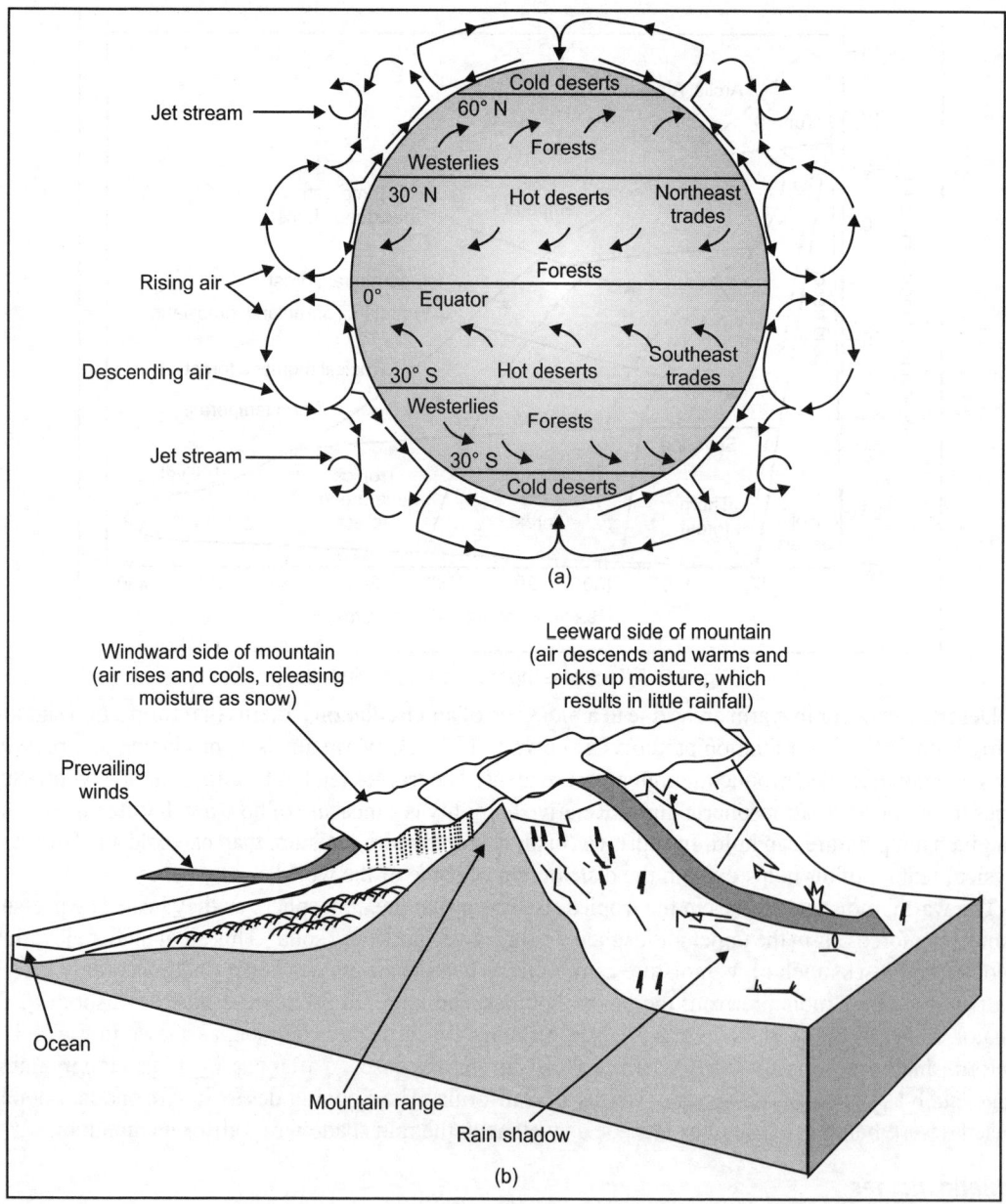

Fig. 9.3. (a) Air circulation patterns and the global distribution of wet and dry areas, (b) rainshadows and deserts.

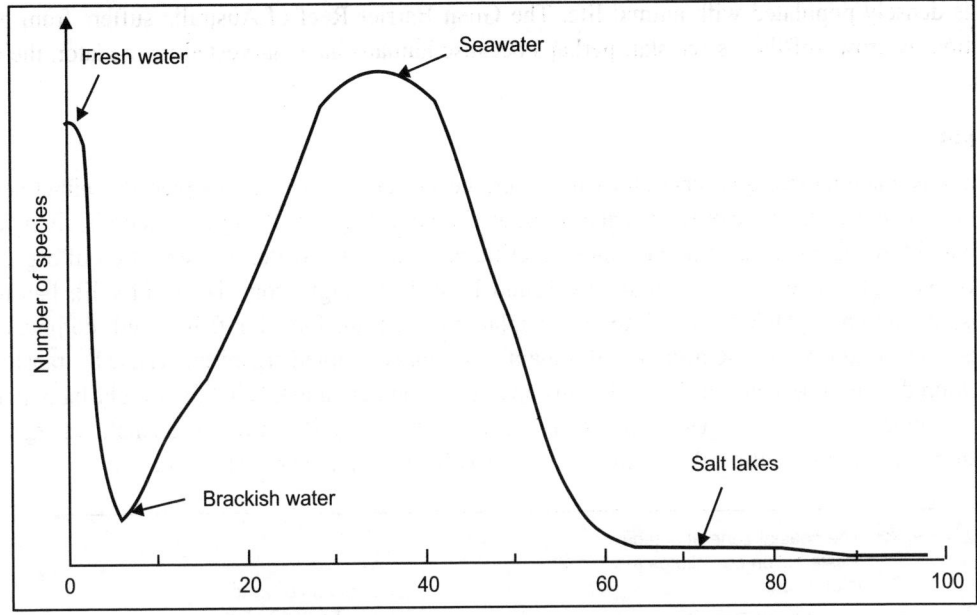

Fig. 9.4. Species diversity and salt concentration.

Marine Biome

The marine biome contains more dissolved minerals than the freshwater biome. Over 70 per cent of the earth's surface is covered in water, by far the vast majority of that being saltwater. There are two basic categories to this biome: benthic and pelagic. Benthic communities (bottom dwellers) are subdivided by depth: the shore/shelf and deep sea. Pelagic communities (swimmers or floaters suspended in the water column) include planktonic (floating) and nektonic (swimming) organisms. The upper 200 meters of the water column is the euphotic zone to which light can penetrate.

Coastal communities

Estuaries are bays where rivers empty into the sea. Erosion brings down nutrients and tides wash in salt water; forms nutrient trap. Estuaries have high production for organisms that can tolerate changing salinity. Estuaries are called 'nurseries of the sea' because many young marine fish develop in this protected environment before moving as adults into the wide open seas.

Seashores

Rocky shorelines offer anchorage for sessile organisms. Seaweeds are main photosynthesisers and use holdfasts to anchor. Barnacles glue themselves to stone. Oysters and mussels attach themselves by threads. Limpets and periwinkles either hide in crevices or fasten flat to rocks.

Sandy beaches and shores are shifting strata. Permanent residents therefore burrow underground. Worms live permanently in tubes. Amphipods and ghost crabs burrow above high tide and feed at night.

Coral reefs

Areas of biological abundance in shallow, warm tropical waters. Stony corals have calcium carbonate exoskeleton and may include algae. Most form colonies; may associate with zooxanthellae dinoflagellates.

Reef is densely populated with animal life. The Great Barrier Reef of Australia suffers from heavy predation by crown-of-thorns sea star, perhaps because humans have harvested its predator, the giant triton.

Oceans

Oceans cover about three-quarters of the earth's surface. Oceanic organisms are placed in either pelagic (open water) or benthic (ocean floor) categories, as shown in Fig. 9.5. Pelagic division is divided into neritic and three levels of pelagic provinces. Neritic province has greater concentration of organisms because sunlight penetrates; nutrients are found here. Epipelagic zone is brightly lit, has much photosynthetic phytoplankton, that support zooplankton that are food for fish, squid, dolphins, and whales. Mesopelagic zone is semi-dark and contains carnivores; adapted organisms tend to be translucent, red coloured, or luminescent; for example: shrimps, squids, lantern and hatchet fishes. The bathypelagic zone is completely dark and largest in size; it has strange-looking fish. Benthic division includes organisms on continental shelf (sublittoral), continental slope (bathyal), and the abyssal plain.

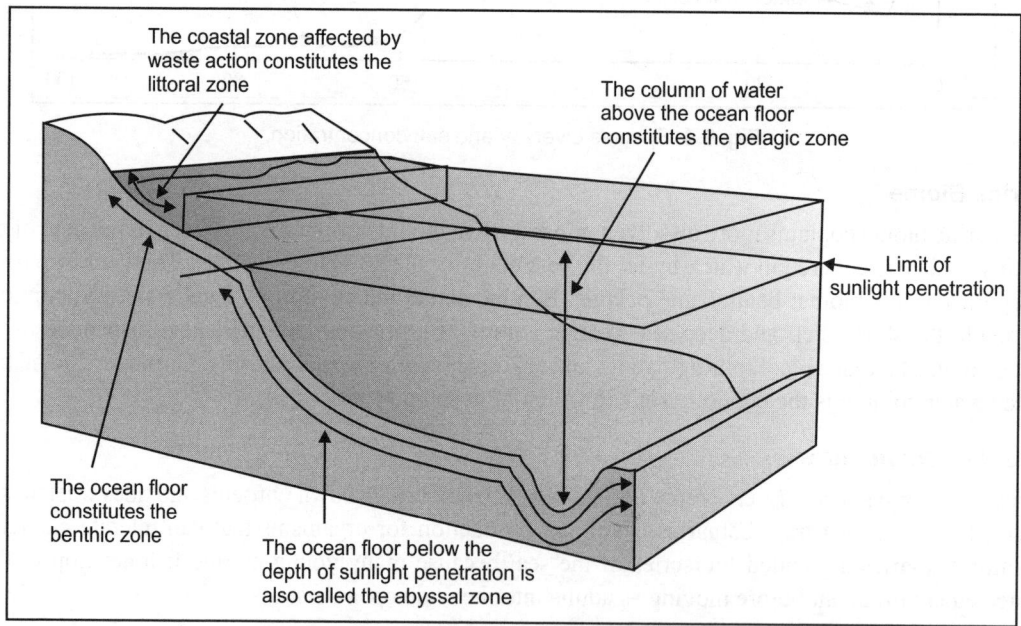

Fig. 9.5. Zones within the marine biome.

Sublittoral zone harbours seaweed that becomes sparse where deeper; most dependent on slow rain of plankton and detritus from sunlit water above. Bathyal zone continues with thinning of sublittoral organisms. Abyssal zone is mainly animals at soil-water interface of dark abyssal plain; inspite of high pressure, darkness and coldness, many invertebrates thrive here among sea urchins and tubeworms.

Thermal vents along oceanic ridges form a very unique community. Molten magma heats seawater to 350°C, reacting with sulphate to form hydrogen sulphide (H_2S). Chemosynthetic bacteria obtain energy by oxidising hydrogen sulphide. The resulting food chain supports a community of tubeworms and clams.

Freshwater Biome

The freshwater biome is subdivided into two zones: running waters and standing waters. Larger bodies of freshwater are less prone to stratification (where oxygen decreases with depth). The upper layers have abundant oxygen, the lowermost layers are oxygen-poor. Mixing between upper and lower layers in a pond or lake occurs during seasonal changes known as spring and fall overturn.

Lakes are larger than ponds, and are stratified in summer and winter, as shown in Fig. 9.6. The epilimnion is the upper surface layer. It is warm in summer. The hypolimnion is the cold lower layer. A sudden drop in temperature occurs at the middle of the thermocline. Layering prevents mixing between the lower hypolimnion (rich in nutrients) and the upper epilimnion (which has oxygen absorbed from its surface). The epilimnion warms in spring and cools in fall, causing a temporary mixing. As a consequence, phytoplankton become more abundant due to the increased amounts of nutrients.

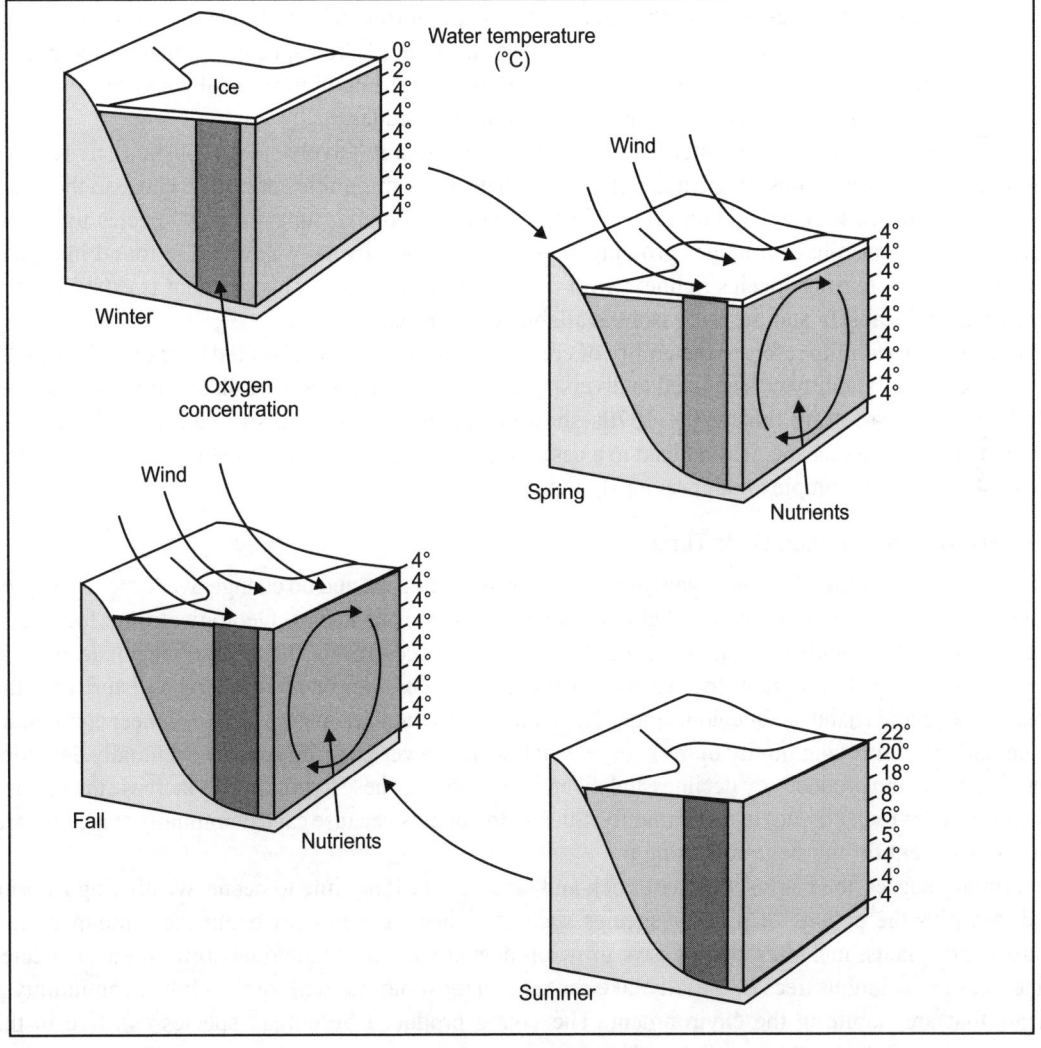

Fig. 9.6. Lake overturn.

Life zones also exist in lakes and ponds. The littoral zone is closest to shore. The limnetic zone is the sunlit body of the lake. Below the level of sunlight penetration is the dark profundal zone. At the soil-water interface we find the benthic zone. The term benthos is applied to animals and other organisms that live on or in the benthic zone.

Rapidly flowing, bubbling streams have insects and fish adapted to oxygen-rich water. Slow moving streams have aquatic life more similar to lake and pond life.

Community Density and Stability

Communities are made up of species adapted to the conditions of that community. Diversity and stability help define a community and are important in environmental studies. Species diversity decreases as we move away from the tropics. Species diversity is a measure of the different types of organisms in a community (also referred to as species richness). Latitudinal diversity gradient refers to species richness decreasing steadily going away from the equator. A hectare of tropical rain forest contains 40–100 tree species, while a hectare of temperate zone forest contains 10–30 tree species. In marked contrast, a hectare of taiga contains only a paltry 1–5 species. Habitat destruction in tropical countries will cause many more extinctions per hectare than it would in higher latitudes.

Environmental stability is greater in tropical areas, where a relatively stable/constant environment allows more different kinds of species to thrive. Equatorial communities are older because they have been less disturbed by glaciers and other climate changes, allowing time for new species to evolve. Equatorial areas also have a longer growing season. The depth diversity gradient is found in aquatic communities. Increasing species richness with increasing water depth. This gradient is established by environmental stability and the increasing availability of nutrients.

Community stability refers to the ability of communities to remain unchanged over time. During the 1950s and 1960s, stability was equated to diversity: diverse communities were also stable communities. Mathematical modelling during the 1970s showed that increased diversity can actually increase interdependence among species and lead to a cascade effect when a keystone species is removed. Thus, the relation is more complex than previously thought.

Change in Communities Over Time

Biological communities, like the organisms that comprise them, can and do change over time. Ecological time focuses on community events that occur over decades or centuries. Geological time focuses on events lasting thousands of years or more. Community succession is the sequential replacement of species by immigration of new species and local extinction of older ones following a disturbance that creates unoccupied habitats for colonisation. The initial rapid coloniser species are the pioneer community. Eventually a climax community of more or less stable but slower growing species eventually develops. During succession productivity declines and diversity increases. These trends tend to increase the biomass (total weight of living tissue) in a community. Succession occurs because each community stage prepares the environment for the stage following it.

Primary succession begins with bare rock and takes a very long time to occur. Weathering by wind and rain plus the actions of pioneer species such as lichens and mosses begin the buildup of soil. Herbaceous plants, including the grasses, grow on deeper soil and shade out shorter pioneer species. Pine trees or deciduous trees eventually take root and in most biomes will form a climax community of plants that are stabile in the environment. The young produced by climax species can live in that environment, unlike the young produced by successional species.

Secondary succession occurs when an environment has been disturbed, such as by fire, geological activity, or human intervention (farming or deforestation in most cases). This form of succession often begins in an abandoned field with soil layers already in place. Compared to primary succession, which must take long periods of time to build or accumulate soil, secondary succession occurs rapidly. The herbaceous pioneering plants give way to pines, which in turn may give way to a hardwood deciduous forest (in the classical old field succession models developed in the eastern deciduous forest biome). Early researchers assumed climax communities were determined for each environment. Today we recognise the outcome of competition among whatever species are present as establishing the climax community. Climax communities tend to be more stable than successional communities. Early stages of succession show the most growth and are most productive. Pioneer communities lack diversity, make poor use of inputs, and lose heat and nutrients. As succession proceeds, species variety increases and nutrients are recycled more. Climax communities make fuller use of inputs and maintain themselves, thus, they are more stable. Human activity (such as clearing a climax forest community to establish a farm field consisting of a cultivated pioneering species, say corn or wheat) replaces climax communities with simpler communities. Communities are composed of species that evolve, so the community must also evolve. Comparing marine communities of 500 million years ago with modern communities shows modern communities composed of quite different organisms. Modern communities also tend to be more complex, although this may be a reflection of the nature of the fossil record as well as differences between biological and fossil species.

Disturbance of a Community

The basic effect of human activity on communities is community simplification, an overall reduction of species diversity. Agriculture is a purposeful human intervention in which we create a monoculture of a single favoured (crop) species such as corn. Most of the agricultural species are derived from pioneering communities. Inadvertent human intervention can simplify communities and produce stressed communities that have fewer species as well as a superabundance of some species. Disturbances favour early successional (pioneer) species that can grow and reproduce rapidly.

ECOSYSTEMS AND COMMUNITIES

Ecosystems include both living and nonliving components. These living, or biotic, components include habitats and niches occupied by organisms. Nonliving, or abiotic, components include soil, water, light, inorganic nutrients, and weather. An organism's place of residence, where it can be found, is its habitat. A niche is often viewed as the role of that organism in the community, factors limiting its life, and how it acquires food. Producers, a major niche in all ecosystems, are autotrophic, usually photosynthetic, organisms. In terrestrial ecosystems, producers are usually green plants. Freshwater and marine ecosystems frequently have algae as the dominant producers.

Consumers are heterotrophic organisms that eat food produced by another organism. Herbivores are a type of consumer that feeds directly on green plants (or another type of autotroph). Since herbivores take their food directly from the producer level, we refer to them as primary consumers. Carnivores feed on other animals (or another type of consumer) and are secondary or tertiary consumers. Omnivores, the feeding method used by humans, feed on both plants and animals. Decomposers are organisms, mostly bacteria and fungi that recycle nutrients from decaying organic material. Decomposers break down detritus, nonliving organic matter, into inorganic matter. Small soil organisms are critical in helping bacteria and fungi shred leaf litter and form rich soil. Even if communities do differ in structure, they have some common uniting processes such as energy flow and matter cycling, shown in Fig. 9.7. Energy flows move through feeding relationships. The term ecological niche refers to how an organism functions in an ecosystem. Food webs, food chains, and food pyramids are three ways of representing energy flow.

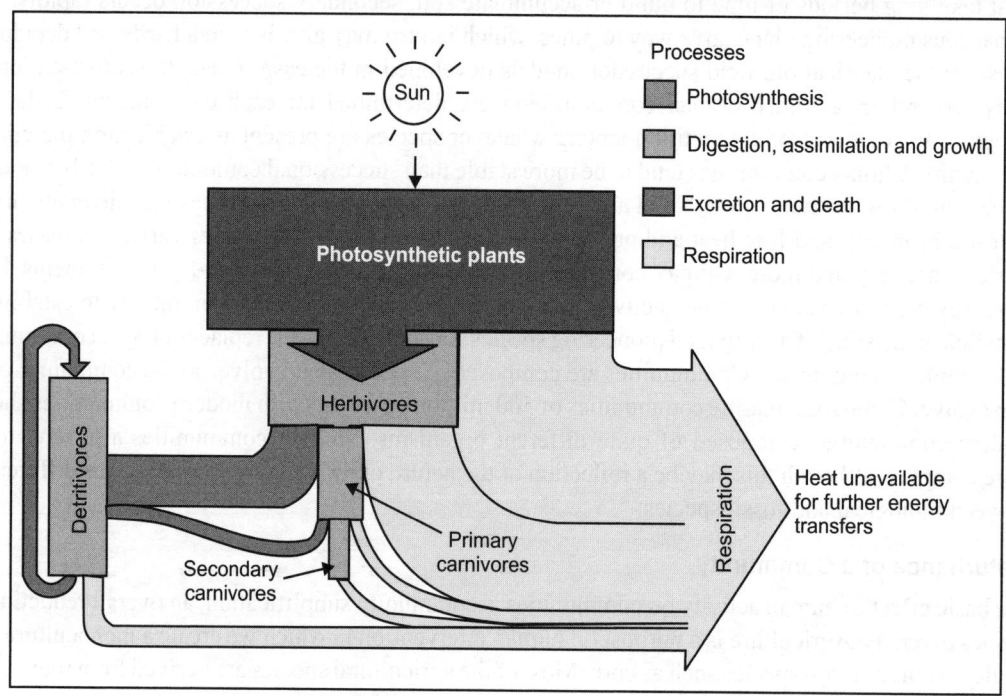

Fig. 9.7. The flow of energy through an ecosystem.

Producers absorb solar energy and convert it to chemical bonds from inorganic nutrients taken from environment. Energy content of organic food passes up food chain; eventually all energy is lost as heat, therefore requiring continual input. Original inorganic elements are mostly returned to soil and producers; can be used again by producers and no new input is required.

Energy flow in ecosystems, as with all other energy, must follow the two laws of thermodynamics. Recall that the first law states that energy is neither created nor destroyed, but instead changes from one form to another (potential to kinetic). The second law mandates that when energy is transformed from one form to another, some usable energy is lost as heat. Thus, in any food chain, some energy must be lost as we move up the chain. The ultimate source of energy for nearly all life is the Sun. Recently, scientists discovered an exception to this once unchallenged truism: communities of organisms around ocean vents where food chain begins with chemosynthetic bacteria that oxidise hydrogen sulphide generated by inorganic chemical reactions inside the earth's crust. In this special case, the source of energy is the internal heat engine of the earth. Food chains indicate who eats whom in an ecosystem and represent one path of energy flow through an ecosystem. Natural ecosystems have numerous interconnected food chains. Each level of producer and consumers is a trophic level. Some primary consumers feed on plants and make grazing food chains; others feed on detritus. The population size in an undisturbed ecosystem is limited by the food supply, competition, predation, and parasitism. Food webs help determine consequences of perturbations: if titmice and vireos fed on beetles and earthworms, insecticides that killed beetles would increase competition between birds and probably increase predation of earthworms, etc. The trophic structure of an ecosystem forms an ecological pyramid. The base of this pyramid represents the producer trophic level. At the apex is the highest level consumer, the top predator. Other pyramids can be recognised in an ecosystem. A pyramid of numbers is based on how many organisms occupy each trophic level. The pyramid of biomass is calculated by multiplying the average weight for organisms times the number of organisms at each trophic level. An energy pyramid illustrates the amounts of energy available at each successive trophic level. The energy pyramid always shows a decrease moving up trophic levels because:

1. Only a certain amount of food is captured and eaten by organisms on the next trophic level.
2. Some of food that is eaten cannot be digested and exits digestive tract as undigested waste.
3. Only a portion of digested food becomes part of the organism's body; rest is used as source of energy.
4. Substantial portion of food energy goes to build up temporary ATP in mitochondria that is then used to synthesise proteins, lipids, carbohydrates, fuel contraction of muscles, nerve conduction, and other functions.
5. Only about 10 per cent of the energy available at a particular trophic level is incorporated into tissues at the next level. Thus, a larger population can be sustained by eating grain than by eating grain-fed animals since 100 kg of grain would result in 10 human kg but if fed to cattle, the result, by the time that reaches the human is a paltry 1 human kg.

A food chain is a series of organisms each feeding on the one preceding it. There are two types of food chain: decomposer and grazer. Grazer food chains begin with algae and plants and end in a carnivore. Decomposer chains are composed of waste and decomposing organisms such as fungi and bacteria. This is shown in Fig. 9.8.

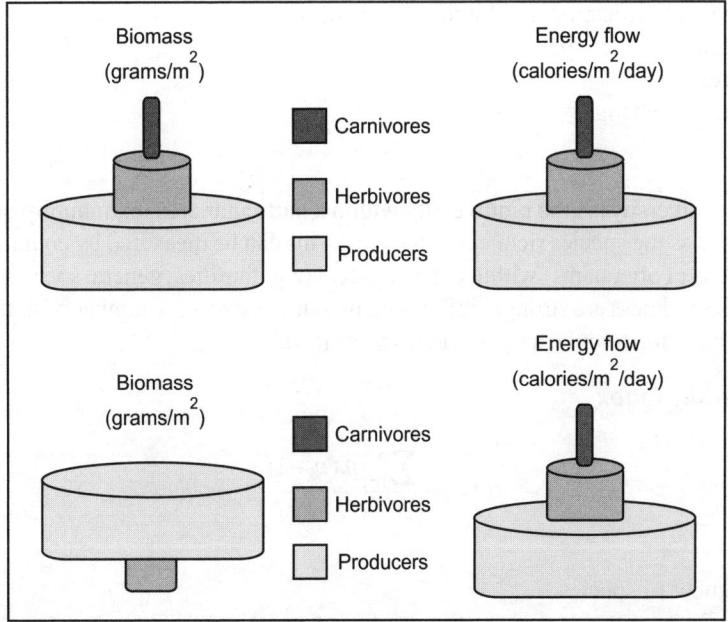

Fig. 9.8. Energy flow and the relative proportions of various levels in the food chain.

Food chains are simplifications of complex relationships. A food web is a more realistic and accurate depiction of energy flow. Food webs are networks of feeding interactions among species. The food pyramid provides a detailed view of energy flow in an ecosystem. The first level consists of the producers (usually plants). All higher levels are consumers. The shorter the food chain the more energy is available to organisms. Most humans occupy a top carnivore role, about 2 per cent of all calories available from producers ever reach the tissues of top carnivores. Leakage of energy occurs between each feeding level. Most natural ecosystems therefore do not have more than five levels to their food pyramids. Large carnivores are rare because there is so little energy available to them atop the pyramid.

Food generation by producers varies greatly between ecosystems. Net primary productivity (NPP) is the rate at which producer biomass is formed. Tropical forests and swamps are the most productive terrestrial ecosystems. Reefs and estuaries are the most productive aquatic ecosystems. All of these productive areas are in danger from human activity. Humans redirect nearly 40 per cent of the net primary productivity and directly or indirectly use nearly 40 per cent of all the land food pyramid. This energy is not available to natural populations.

SPECIES DIVERSITY

Species diversity is an index that incorporates the number of species in an area and also their relative abundance. It is a more comprehensive value than species richness. The most common index of species diversity is a family of equations called Simpson's Diversity Index. Here is one such example:

$$D = (n/N)^2$$

where, n is the total number of organisms of a particular species and N is the total number of organisms of all species. D is the value of diversity. It can range between 0 and 1, where, 0 is infinite diversity, and 1 is the least diverse an ecosystem can possibly be (i.e. only one species present).

Humans have a huge effect on species diversity; the main reasons are:
1. Destruction, modification and fragmentation of habitat.
2. Introduction of exotic species.
3. Overharvesting.
4. Global climate change.

Alpha Diversity

Alpha diversity (α-diversity) is the biodiversity within a particular area, community or ecosystem, and is usually expressed as the species richness of the area. This can be measured by counting the number of taxa (distinct groups of organisms) within the ecosystem (e.g. families, genera, species). However, such estimates of species richness are strongly influenced by sample size, so a number of statistical techniques can be used to correct for sample size to get comparable values.

Simpson's diversity index

$$D = 1 - \frac{\sum_{i=1}^{s} n_i(n_i - 1)}{N(N-1)}$$

where,

S is the number of species.

N is the total percentage cover or total number of organisms.

n_i is the percentage cover of a species or number of organisms of species i.

Shannon index

$$H' = -\sum_{i=1}^{s} p_i \ln p_i$$

where,

S is the number of species. Also called species richness.

p_i is the relative abundance of each species, calculated as the proportion of individuals of a given species to the total number of individuals in the community: n_i/N.

n_i is the number of individuals in each species; the abundance of each species.

N is the total number of all individuals.

Fisher's alpha

Rarefaction takes hypothetical subsamples of n organisms from the more-sampled region, and calculates the average number of species in such subsamples. This average can be compared to the number of species actually found in the less-sampled region

Examples: Alpha diversity has been measured both in extant ecosystems and in extinct communities. It is an especially useful measure to track earth's past biodiversity because if it is used correctly, diversity can be measured independently of the fossil record which is not complete and therefore subject to errors. Examples include:

1. A study of the alpha diversity of Triassic tetrapod families found that recovery after the Permo-Triassic extinction event took 30 million years.
2. A study on trilobites revealed that following the end-Ordovician extinction, trilobite alpha diversity was comparable to those of the Late Cambrian communities.

Beta Diversity

Beta diversity (β-diversity) is a measure of biodiversity which works by comparing the species diversity between ecosystems or along environmental gradients. This involves comparing the number of taxa that are unique to each of the ecosystems.

It is the rate of change in species composition across habitats or among communities. It gives a quantitative measure of diversity of communities that experience changing environments.

Absolute value

At its simplest, beta diversity is the total number of species that are unique between communities. This can be represented by the following equation:

$$\beta = (S_1 - c) + (S_2 - c)$$

where, S_1 = the total number of species recorded in the first community, S_2 = the total number of species recorded in the second community, and c = the number of species common to both communities.

Indices

Sørensen's similarity index

$$\beta = \frac{2c}{S_1 + S_2}$$

where, S_1 = the total number of species recorded in the first community, S_2 = the total number of species recorded in the second community, and c = the number of species common to both communities. The Sørensen index is a very simple measure of beta diversity, ranging from a value of 0, where there is no species overlap between the communities, to a value of 1 when exactly the same species are found in both communities.

Whittaker's measure

$$\beta = \frac{S}{\overline{\alpha}} \quad \text{or} \quad \beta = \frac{S}{\overline{\alpha}} - 1$$

where, S = the total number of species recorded in both communities, $\overline{\alpha}$ = average number of species found within the communities.

Gamma Diversity

Gamma diversity (γ-diversity) is a measure of biodiversity. It refers to the total species richness over a large area or region. It is the product of the α diversity of component ecosystems and the β diversity between component ecosystems.

According to Whittaker, gamma diversity is the richness in species of a range of habitats in a geographic area (e.g. a landscape, an island) and it is consequent on the alpha diversity of the individual communities and the range of differentiation or beta diversity among them. Like alpha diversity, it is a quality which simply has magnitude, not direction and can be represented by a single number (a scalar).

Gamma diversity can be expressed in terms of the species richness of component communities as follows:

$$\gamma = S_1 + S_2 - c$$

where, S_1 = the total number of species recorded in the first community, S_2 = the total number of species recorded in the second community, and c = the number of species common to both communities.

The internal relationship between alpha, beta and gamma diversity can be represented as:

$$\beta = \gamma/\alpha$$

Diversity Index

A diversity index is a statistic which is intended to measure the local members of a set consisting of various types of objects. Diversity indices can be used in many fields of study to assess the diversity of any population in which each member belongs to a unique group, type or species. For instance, it is used in ecology to measure biodiversity in an ecosystem, in demography to measure the distribution of population of various demographic groups, in economics to measure the distribution over sectors of economic activity in a region, and in information science to describe the complexity of a set of information.

In measuring human diversity, the diversity index measures the probability that any two residents, chosen at random, would be of different ethnicities. If all residents are of the same ethnic group it's zero. If half are from one group and half from another it's 0.50. Below, a series of diversity indices is discussed.

Species richness

The species richness S is simply the number of species present in an ecosystem. This index makes no use of relative abundances. In practice, measuring the total species richness in an ecosystem is impossible, except in very depauperate systems. The observed number of species in the system is a biased estimator of the true species richness in the system, and the observed species number increases non-linearly with sampling effort. Thus S, if indicating the observed species richness in an ecosystem, is usually referred to as species density.

LAKE STRATIFICATION AND MIXING

Many of our Illinois lakes and reservoirs are deep enough to stratify or form 'layers' of water with different temperatures. Such thermal stratification occurs because of the large differences in density (weight) between warm and cold waters. Density depends on temperature: water is most dense (heaviest) at about 39°F, and less dense (lighter) at temperatures warmer and colder than 39°F.

Stratification Process

In the fall, chilly air temperatures cool the lake's surface. As the surface water cools, it becomes more dense and sinks to the bottom. Eventually the entire lake reaches about 39°F (4°C). As the surface water cools even more, it becomes less dense and 'floats' on top of the denser 39°F water, forming ice at 32°F (0°C). The lake water below the ice remains near 39°F. This situation is referred to as winter stratification. Winter stratification remains stable because the ice cover prevents wind from mixing the water. Come spring, the ice melts and the surface water begins to warm above 32°F. The increasing density of the warming water along with wind action cause this surface water to sink and mix with the deeper water— a process called spring turnover. During this time period, most of the lake water is at the same temperature, and surface and bottom waters mix freely. Lakes with a small surface area, especially if protected from the wind, typically completely mix for only a brief time in the spring—usually just a few days. In comparison, large lakes often circulate for weeks. As the sun continues to warm the lake surface through late spring and early summer, the temperature differences increase between the surface and deeper waters. In lake areas deeper than about 10 to 12 feet, the temperature differences eventually create a physical the metalimnion and extending to the lake bottom is the colder (heavier), usually dark, and relatively undisturbed hypolimnion.

The most important actions causing lake mixing are wind, inflowing water, and outflowing water. While wind influences the surface waters of all lakes, its ability to mix the entire water volume in summer-stratified lakes is greatly reduced. This is because the rapid change in temperature and density within the metalimnion acts like a physical barrier between the epilimnion and hypolimnion. Though not an absolute barrier, it takes a lot of energy to disrupt it.

The stability of a lake's stratification depends on many factors, most importantly the lake's depth, shape, and size. Also playing a role are climate, orientation of the lake to the wind, and inflow/outflow. As noted earlier, in shallow lakes (less than about 10 to 12 feet deep) wind forces are usually strong enough to mix the water from top to bottom and thereby thwart summer stratification. Lakes with a lot of water flowing through them (i.e. a short water residence time) also do not develop persistent thermal stratification. While a temperature gradient from warmer surface to cooler bottom waters may exist in such lakes, a true metalimnion is not typically formed.

Summer stratification continues until fall when surface waters begin to cool and sink. The metalimnion begins to 'erode' and weaken, and continues to do so as the lake cools. Wind energy helps mix the lake deeper and deeper. When the whole lake reaches a similar temperature, wind forces are again able to mix the lake from top to bottom in a process called fall turnover. The transition from summer stratification to fall turnover can occur within just a few hours, especially if accompanied by strong winds.

Effects of Stratification

Stratification has important implications for fisheries management, phytoplankton (algae) populations, and water supply quality. A discussion of a few stratification impacts follows.

Dissolved oxygen

Just after summer stratification is established, the hypolimnion is rich in dissolved oxygen from the early spring mixing of the lake. However, because the metalimnion acts as a barrier between the epilimnion and hypolimnion, the hypolimnion is essentially cut off from oxygen exchange with the atmosphere and is often too dark for plants and algae to grow and produce oxygen by photosynthesis. In a eutrophic (nutrient-rich) lake, the hypolimnion can become anoxic (without oxygen, or anaerobic) as the summer progresses. This occurs as its supply of oxygen is consumed by bacteria and other bottom-dwelling organisms. A lack of dissolved oxygen can have serious consequences.

Phosphorus and nitrogen: In anoxic conditions, the nutrients phosphorus and ammonia-nitrogen become more soluble (dissolvable) and are released from the bottom sediments into the hypolimnion. During the summer, stratified lakes can sometimes partially mix (such as with the passing of a cold front accompanied by strong winds and cold rains), allowing some of these nutrients to 'escape' into the epilimnion and potentially stimulate an algal bloom. For similar reasons, algal blooms often are seen at fall turnover as nutrient-rich bottom water is brought to the lake surface where there is ample sunlight to support algae growth. Ammonia-nitrogen also can have an impact on fish. Fish are sensitive to ammonia and are repelled by high levels in the water.

Metals and other compounds: Some metals and other elements—notably iron, manganese, and sulphur (as hydrogen sulphide)—also become increasingly soluble and are released from anoxic bottom sediments. These compounds cause taste and odour problems—a potentially serious concern in drinking water supply reservoirs. Additionally, hydrogen sulphide concentrations above 1 mg/l are lethal to many gamefish as well as some zooplankton (microscopic animals that are an important fish food).

Fish: Low oxygen levels may restrict where fish can go in a lake and limit the types and numbers of fish in the hypolimnion. Warmwater fish (e.g. bass and bluegill) need at least 5 mg/l of dissolved oxygen to survive, while coldwater fish (e.g. trout) require 6–7 mg/l. In eutrophic lakes, as summer progresses and dissolved oxygen levels become too low in the hypolimnion, fish are confined to the epilimnion and a portion of the metalimnion. As ice covers a lake in early winter, there usually is adequate oxygen in the water to sustain fish and other aquatic organisms. You may be surprised to learn that certain algae and rooted aquatic plants grow right through the winter and photosynthesise, producing oxygen. However, bacterial decomposition of organic matter on the lake bottom can consume more oxygen than photosynthesis can replace, causing a decline in dissolved oxygen levels as the winter season progresses. If enough snow covers the ice or if the ice is opaque, sunlight may be inadequate or unable to penetrate and photosynthesis will stop. If the lake's supply of oxygen falls too low before ice-out, a partial or total fishkill can occur.

Temperature

In summer-stratified lakes, water temperatures decrease from the surface to the bottom. As discussed above, a warm surface layer (the epilimnion) 'floats' on a colder layer (the hypolimnion). Different fish species prefer different water temperatures. Hence, a lake's temperature variations are important in influencing what types and how many fish will live and reproduce in that lake. If the colder, deeper waters of the hypolimnion have enough oxygen, then that area will provide a refuge for fish species that prefer, or require, cold water temperatures. However, if dissolved oxygen levels become too low in the hypolimnion and fish are forced into the warmer surface waters, coldwater fish species may not be able to survive.

QUADRAT METHOD

Many questions come to mind as we set out to study the importance of a particular species of plant in a community: How widely distributed is this species? How many plants of this species are present in the community? How much of the total available space does this species occupy? If you can answer these questions, you can conduct some interesting and important studies. For example, you can compare the vegetation in one region with that in another: you can study seasonal changes in vegetation within a given area: you can determine the relationships between plant populations and abiotic factors: you can investigate ecological plant succession. If you require only a rough idea of the contribution of each plant species to a community, a quantitative study is often sufficient. For example, to estimate the abundance of each species you simply prepare a list of the species present and categorise each as abundant, frequent, occasional or rare. It is difficult, however, to remain objective when using such an approach. The observer may list a species as abundant because of its height or colour make sit quite easy to spot when, in actual fact, other less obvious species are more abundant. Thus, for precise work, quantitative methods must be used. Very accurate quantitative information could be obtained if a team of observers went into the study area and identified, counted, and measured every plant. In large areas this would obviously be impractical, if not impossible. Besides, it is unnecessary, since ecologists have developed sampling techniques that give equally valid results in much shorter time periods. The most commonly used techniques are described here. Although these studies yield interesting information on there own, they are best performed in conjunction with such things as animal population studies, soil studies, and measurement of the appropriate physical factors. The quadrat method is one of the most widely used means of attaining quantitative information about the compositions and structure of plant communities. In, principle, the method appears quite simple. You merely sample the study area at several sites using quadrats. You then assume these sample plots give a reliable picture of the vegetation overt the total study area. This assumption is true only if you have picked the proper size and shape of quadrat and if you use a suitable number of and arrangement of quadrats.

TRANSECT METHOD

It is impossible to count every living and non-living thing in an ecosystem. So a great method for finding out what things exist in an ecosystem is by establishing transects. A transect is a defined area in which sample population counts of plants and animals can be taken. The defined area has to be large enough to truly characterise the biotic and abiotic factors of the ecosystem. The size of a transect is determined by the biotic factors in the area chosen to investigate. If the area is in a grassland, where plants are plentiful and close together, you may choose a one-meter square in which to count all species of plants and animals. If the area is in a sparsely vegetated area such as a desert, you may choose a 10-metre or even a 100-metre square in which to count your populations.

Transects can be established in many ways. Students can measure off a square in a predetermined size by marking corners with pencils in the ground and running string between pencils. You can build a transect square out of PVC pipe that can be disassembled and assembled quickly. You can use a hula-hoop. The size, transect shape and materials for making the transect just need to be determined ahead of time so that all groups of students use the same transect in the same ecosystem so the results can be averaged fairly. In order to establish exactly where you begin the first point of your transect square or center of your transect circle, one can throw a rock over their shoulder. Wherever it lands is either the first point or the center.

Abiotic factors such as soil temperature, type and percolation rate and weather data need to also be collected at the transect site. Soil characteristics and climate are the determining factors for the representative plant and animal populations. Going to different ecosystems and collecting the same data for comparative purpose could be a valuable and exciting investigative experience that brings students to a true understanding of what makes an ecosystem what it is. In addition, it is what real scientists do. It is a real life meaningful endeavour.

PINPOINT METHOD

The pinpoint method (or point-intercept method) is used for non-destructive measurements of plant cover and plant biomass. In a pinpoint analysis, a frame (or a transect) with a fixed grid pattern is placed above the vegetation. A pin that is inserted vertically through one of the grid points into the vegetation will typically touch a number of plants, and the number of times the pin touches different plant species is recorded. This procedure is repeated at each grid point.

VEGETATION CLASSIFICATION METHODS

Vegetation is often chosen as the basis for the classification of terrestrial ecosystems because it generally integrates the ecological processes acting on a site or landscape more measurably than any other factor or set of factors. Because patterns of co-occurring plant species are easily measured, they have received more attention than those other components, such as fauna. Vegetation is a critical component of energy flow in ecosystems and provides habitat for many organisms. In addition, vegetation is often used to infer soil and climate patterns. For these reasons, a classification of terrestrial ecological communities based on vegetation can serve to describe many facets of ecological patterns across the landscape.

Vegetation Concepts

The structure of plant communities was widely debated throughout much of the 20th century. Essentially, two general models were proposed: the community as discrete unit, and the continuum. The community-unit hypothesis formulated by Clements states that communities are highly structured, repeatable and identifiable associations of species controlled by climate. The alternative continuum model of Whittaker and Curtis states that plant communities change gradually along complex environmental gradients, such that no discrint associations of species can be identified.

Continua of independent species distributions revealed in gradient analyses have generally been interpreted as evidence for Gleason's concept of individualistic species assemblages and this concept has been organised into the individualistic-continuum theory. However, while the continuum model grew out of Gleason's essays on the individualistic distribution of species they should not be considered synonymous. The individualistic hypothesis is a species-scale phenomenon involving the tolerance of individuals of different species to local environmental conditions, which may include interspecific interactions. In contrast, the continuum model is a community-level construct of the collective distributions and abundance of species along environmental gradients. It is therefore possible, that individualistic distribution of species gives rise to discrete communities as well as to continuum.

Although most ecologists and vegetation scientists now accept the continuum model to be correct, the debate concerning the validity of these models still continues. Westman suggested the debate endures because empirical evidence exists that supports both points of view. On the other hand, Shipley and Keddy determined that neither model applied to species distributions along complex environmental gradients in wetlands. Roberts suggested that both the community-unit and continuum models were consistent with a mechanistic view of vegetation development. From a hierarchical perspective, the two models are not competitive; rather, they reflect differences in scale of perception.

ECOTONE

An ecotone is a transition area between two adjacent but different patches of landscape, such as forest and grassland. It may be narrow or wide, and it may be local (the zone between a field and forest) or regional (the transition between forest and grassland ecosystems). An ecotone may appear on the ground as a gradual blending of the two communities across a broad area, or it may manifest itself as a sharp boundary line.

Formation of Ecotone

Changes in the physical environment may produce a sharp boundary, as in the example of the interface between areas of forest and cleared land (Krummholz). Elsewhere, a more gradually blended interface area will be found, where species from each community will be found together as well as unique local species. Mountain ranges often create such ecotones, due to the wide variety of climatic conditions experienced on their slopes. They may also provide a boundary between species due to the obstructive nature of their terrain. Mont Ventoux in France is a good example, marking the boundary between the flora and fauna of northern and southern France. Most wetlands are ecotones.

Plants in competition extend themselves on one side of the ecotone as far as their ability to maintain themselves allows. Beyond this competitors of the adjacent community take over. As a result the ecotone represents a shift in dominance. Ecotones are particularly significant for mobile animals, as they can exploit more than one set of habitats within a short distance. The ecotone contains not only species common to the communities on both sides; it may also include a number of highly adaptable species that tend to colonise such transitional areas. This can produce an edge effect along the boundary line, with the area displaying a greater than usual diversity of species.

The phenomenon of increased variety of plants as well as animals at the community junction is called the 'edge effect' and is essentially due to a locally broader range of suitable environmental conditions or ecological niches.

Ecotones and Ecoclines

An ecotone is often associated with an ecocline: a 'physical transition zone' between two systems. The ecotone and ecocline concepts are sometimes confused: an ecocline can signal an ecotone chemically (ex: pH or salinity gradient), or microclimatically (hydrothermal gradient) between two ecosystems.

In contrast:

1. An ecocline is a variation of the physicochemical environment dependent of one or two physico-chemical factors of life, and thus presence/absence of certain species. An ecocline can be a thermocline, chemocline (chemical gradient), halocline (salinity gradient) or pycnocline (variations in density of water induced by temperature or salinity).

2. An ecotone describes a variation in species prevalence and is often not strictly dependent a major physical factor separating an ecosystem from another, with resulting habitat variability. An ecotone is often unobtrusive and harder to measure.

because empirical evidence exists that supports both inferences [E.C. Or, flex: flex: bone, Schaffer, and Ecology: assumed that neither model inabilities to species climates from climate community. Commander edit species mechanism. Reflects suggested that both the observation and the coefficient analysis were empirical collisions combine level of vegetation developments have a different. A process that the fact were vegetation diverse influence may reflect differences in some

BUT ONE

pop measure where two provides adjacent the different database. Furthermore, at and for to open seed and all the community is the big of mass. Furthermore, range the property at site level at open and the site analysis used. For a commune may topics in the central thing furthermore, range the property at site level at open and

RANGE Tile

commune of Empire

Chapter 10

Community Dynamics

INTRODUCTION

Ecological succession, a fundamental concept in ecology, refers to more or less predictable and orderly changes in the composition or structure of an ecological community. Succession may be initiated either by formation of new, unoccupied habitat (e.g. a lava flow or a severe landslide) or by some form of disturbance (e.g. fire, severe windthrow, logging) of an existing community. Succession that begins in areas where no soil is initially present is called primary succession, whereas succession that begins in areas where soil is already present is called secondary succession.

The trajectory of ecological change can be influenced by site conditions, by the interactions of the species present, and by more stochastic factors such as availability of colonists or seeds or weather conditions at the time of disturbance. Some of these factors contribute to predictability of succession dynamics; others add more probabilistic elements. In general, communities in early succession will be dominated by fast-growing, well-dispersed species (opportunist, fugitive or *r*-selected life-histories). As succession proceeds, these species will tend to be replaced by more competitive (*k*-selected) species.

Trends in ecosystem and community properties in succession have been suggested, but few appear to be general. For example, species diversity almost necessarily increases during early succession as new species arrive, but may decline in later succession as competition eliminates opportunistic species and leads to dominance by locally superior competitors. Net Primary Productivity, biomass, and trophic level properties all show variable patterns over succession, depending on the particular system and site.

Ecological succession was formerly seen as having a stable end-stage called the climax, sometimes referred to as the 'potential vegetation' of a site, shaped primarily by the local climate. This idea has been largely abandoned by modern ecologists in favour of nonequilibrium ideas of how ecosystems function. Most natural ecosystems experience disturbance at a rate that makes a 'climax' community unattainable. Climate change often occurs at a rate and frequency sufficient to prevent arrival at a climax state. Additions to available species pools through range expansions and introductions can also continually reshape communities.

The development of some ecosystem attributes, such as pedogenesis and nutrient cycles, are both influenced by community properties, and in turn, influence further community development. This process may occur only over centuries or millennia. Coupled with the stochastic nature of disturbance events and other long-term (e.g. climatic) changes, such dynamics make it doubtful whether the 'climax' concept ever applies or is particularly useful in considering actual vegetation.

TYPES OF SUCCESSION

Primary and Secondary Succession

If the development begins on an area that has not been previously occupied by a community, such as a newly exposed rock or sand surface, a lava flow, glacial tills or a newly formed lake, the process is known as primary succession.

If the community development is proceeding in an area from which a community was removed it is called secondary succession. Secondary succession arises on sites where the vegetation cover has been disturbed by humans or other animals (an abandoned crop field or cut-over forest or natural forces such as water, wind storms, and floods). Secondary succession is usually more rapid as the colonising area is rich in leftover soil, organic matter and seeds of the previous vegetation, whereas in primary succession the soil itself must be formed, and seeds and other living things must come from outside the area.

Seasonal and Cyclic Succession

Unlike secondary succession, these types of vegetation change are not dependent on disturbance but are periodic changes arising from fluctuating species interactions or recurring events. These models propose a modification to the climax concept towards one of dynamic states.

Causes of Plant Succession

Autogenic succession can be brought by changes in the soil caused by the organisms there. These changes include accumulation of organic matter in litter or humic layer, alteration of soil nutrients, change in pH of soil by plants growing there. The structure of the plants themselves can also alter the community. For example, when larger species like trees mature, they produce shade on to the developing forest floor that tends to exclude light-requiring species. Shade-tolerant species will invade the area.

Allogenic succession is caused by external environmental influences and not by the vegetation. For example soil changes due to erosion, leaching or the deposition of silt and clays can alter the nutrient content and water relationships in the ecosystems. Animals also play an important role in allogenic changes as they are pollinators, seed dispersers and herbivores. They can also increase nutrient content of the soil in certain areas or shift soil about (as termites, ants, and moles do) creating patches in the habitat. This may create regeneration sites that favour certain species.

Climatic factors may be very important, but on a much longer time-scale than any other. Changes in temperature and rainfall patterns will promote changes in communities. As the climate warmed at the end of each ice age, great successional changes took place. The tundra vegetation and bare glacial till deposits underwent succession to mixed deciduous forest. The greenhouse effect resulting in increase in temperature is likely to bring profound Allogenic changes in the next century. Geological and climatic catastrophes such as volcanic eruptions, earthquakes, avalanches, meteors, floods, fires, and high wind also bring allogenic changes.

CLEMENT'S THEORY OF SUCCESSION/MECHANISMS OF SUCCESSION

F.E. Clement developed a descriptive theory of succession and advanced it as a general ecological concept. His theory of succession had a powerful influence on ecological thought. Clement's concept is usually termed classical ecological theory. According to Clement, succession is a process involving several phases:

1. Nudation: Succession begins with the development of a bare site, called Nudation (disturbance).
2. Migration: It refers to arrival of propagules.

3. Ecesis: It involves establishment and initial growth of vegetation.
4. Competition: As vegetation became well established, grew, and spread, various species began to compete for space, light and nutrients. This phase is called competition.
5. Reaction: During this phase autogenic changes affect the habitat resulting in replacement of one plant community by another.
6. Stabilisation: Reaction phase leads to development of a climax community.

SERAL COMMUNITIES

A seral community is an intermediate stage found in an ecosystem advancing towards its climax community. In many cases more than one seral stage evolves until climax conditions are attained. A prisere is a collection of seres making up the development of an area from non-vegetated surfaces to a climax community. Depending on the substratum and climate, a seral community can be one of the following:
1. Hydrosere: Community in freshwater.
2. Lithosere: Community on rock.
3. Psammosere: Community on sand.
4. Xerosere: Community in dry area.
5. Halosere: Community in saline body (e.g. a marsh).

CHANGES IN ANIMAL LIFE

Animal life also exhibit changes with changing communities. In lichen stage the fauna is sparse. It comprises few mites, ants and spiders living in the cracks and crevices. The fauna undergoes a qualitative increase during herb grass stage. The animals found during this stage include nematodes, insects larvae, ants, spiders, mites, etc. The animal population increases and diversifies with the development of forest climax community. The fauna consists of invertebrates like slugs, snails, worms, millipedes, centipedes, ants, bugs; and vertebrates such as squirrels, foxes, mouse, moles, snakes, various birds, salamanders and frogs.

MICROSUCCESSION/SERULE

Succession of micro-organisms like fungi, bacteria, etc. occurring within a microhabitat is known as microsuccession or serule. This type of succession occurs within communities, for example in dead trees, animal droppings, etc. Microbial communities may also change due to products secreted by the bacteria present. Changes of pH in a habitat could provide ideal conditions for a new species to inhabit the area. In some cases the new species may outcompete the present ones for nutrients leading to the primary species demise. Changes can also occur by microbial succession with variations in water availability and temperature.

CLIMAX CONCEPT

According to classical ecological theory, succession stops when the sere has arrived at an equilibrium or steady state with the physical and biotic environment. Barring major disturbances, it will persist indefinitely. This end point of succession is called climax.

Climax Community

The final or stable community in a sere is the climax community or climatic vegetation. It is self-perpetuating and in equilibrium with the physical habitat. There is no net annual accumulation of organic matter in a climax community mostly. The annual production and use of energy is balanced in such a community.

Characteristics of Climax

1. The vegetation is tolerant of environmental conditions.
2. It has a wide diversity of species, a well-drained spatial structure, and complex food chains.
3. The climax ecosystem is balanced. There is equilibrium between gross primary production and total respiration, between energy used from sunlight and energy released by decomposition, between uptake of nutrients from the soil and the return of nutrient by litter fall to the soil.
4. Individuals in the climax stage are replaced by others of the same kind. Thus the species composition maintains equilibrium.
5. It is an index of the climate of the area. The life or growth forms indicate the climatic type.

Types of Climax

1. Climatic climax: If there is only a single climax and the development of climax community is controlled by the climate of the region, it is termed as climatic climax. For example, development of Maple-beech climax community over moist soil. Climatic climax is theoretical and develops where physical conditions of the substrate are not so extreme as to modify the effects of the prevailing regional climate.
2. Edaphic climax: When there are more than one climax communities in the region, modified by local conditions of the substrate such as soil moisture, soil nutrients, topography, slope exposure, fire, and animal activity, it is called edaphic climax. Succession ends in an edaphic climax where topography, soil, water, fire or other disturbances are such that a climatic climax cannot develop.
3. Catastrophic climax: Climax vegetation vulnerable to a catastrophic event such as a wildfire. For example, in California, chaparral vegetation is the final vegetation. The wildfire removes the mature vegetation and decomposers. A rapid development of herbaceous vegetation follows until the shrub dominance is re-established. This is known as catastrophic climax.
4. Disclimax: When a stable community, which is not the climatic or edaphic climax for the given site, is maintained by man or his domestic animals, it is designated as Disclimax (disturbance climax) or anthropogenic subclimax (man-generated). For example, overgrazing by stock may produce a desert community of bushes and cacti where the local climate actually would allow grassland to maintain itself.
5. Subclimax: The prolonged stage in succession just preceding the climatic climax is subclimax.
6. Preclimax and postclimax: In certain areas different climax communities develop under similar climatic conditions. If the community has life forms lower than those in the expected climatic climax, it is called preclimax; a community that has life forms higher than those in the expected climatic climax is postclimax. Preclimax strips develop in less moist and hotter areas, whereas Postclimax strands develop in more moist and cooler areas than that of surrounding climate.

Theories Regarding Nature of Climax

There are three schools of interpretations explaining the climax concept:

1. **Monoclimax** or **climatic climax theory** was advanced by Clements and recognises only one climax whose characteristics are determined solely by climate (climatic climax). The processes of succession and modification of environment overcome the effects of differences in topography, parent material of the soil, and other factors. The whole area would be covered with uniform plant community. Communities other than the climax are related to it, and are recognised as subclimax, postclimax and disclimax.

2. **Polyclimax theory** was advanced by Tansley. It proposes that the climax vegetation of a region consists of more than one vegetation climaxes controlled by soil moisture, soil nutrients, topography, slope exposure, fire, and animal activity.

3. **Climax pattern theory** was proposed by Whittaker. The climax pattern theory recognises a variety of climaxes governed by responses of species populations to biotic and abiotic conditions. According to this theory the total environment of the ecosystem determines the composition, species structure, and balance of a climax community. The environment includes the species responses to moisture, temperature, and nutrients, their biotic relationships, availability of flora and fauna to colonise the area, chance dispersal of seeds and animals, soils, climate, and disturbance such as fire and wind. The nature of climax vegetation will change as the environment changes. The climax community represents a pattern of populations that corresponds to and changes with the pattern of environment. The central and most widespread community is the climatic climax.

More recently another possible idea has been put forward called the theory of alternative stable states which suggests that there is not one end point but many which transition between each other over ecological time.

FOREST SUCCESSION

The forests, being an ecological system are subject to the species succession process. There are 'opportunistic' or 'pioneer' species that produce great quantity of seeds that are disseminated by the wind, and therefore can colonise big empty extensions, and they are capable to germinate and grow under direct sun exposition. Once they have produced a closed canopy, the lack of direct sun radiation at soil makes it difficult for their own seedlings to develop. It is then the opportunity for shade 'tolerant' species to get established under the protection of pioneer. When these pioneers will die, the shade tolerants will replace them. The shade tolerant species are capable of growing under the canopy, and therefore, in the absence of catastrophes, will stay. For this reason it is said than the stand has reached its climax. When an important catastrophe will arrive, the opportunity for the pioneers will be open again, provided they are not absent at a reasonable range.

An example of pioneer species, in forests of northeastern North America are Betula alleghaniensis (Yellow birch) and Prunus serotina (Black cherry), that are particularly well-adapted to exploit large gaps in forest canopies, but are intolerant of shade and are eventually replaced by other (shade-tolerant) species in the absence of disturbances that create such gaps.

Things in nature are usually neither white nor black, and there are intermediates. It is therefore normal that between the two extremes light/shade there is a gradation, and there are species that may act as pioneer or tolerant, depending on circumstances. It is of paramount importance to know the tolerance of species in order to practice an effective silviculture.

HYDROSERE

A hydrosere is a plant succession which occurs in a freshwater lake. In time, an area of open freshwater will naturally dry out, ultimately becoming woodland. During this change, a range of different landtypes such as swamp and marsh will succeed each other. The succession from open water to climax woodland is likely to take at least two hundred years. Some intermediate stages will last a shorter time than others. For example, swamp may change to marsh within a decade or less. How long it takes will depend largely on the amount of siltation occurring in the area of open water.

Stages

Hydrosere is the primary succession sequence which develops in aquatic environments such as lakes and ponds. It results in conversion of water body and its community into a land community. The early changes are allogenic as inorganic particles such as sand and clay are washed from catchment areas and begin filling the basin of the water body. Later, remains of dead plants also fill up these bodies and contribute to further changes in the environment.

If water body is large and very deep, a strong wave action is at work, therefore in these bodies a noticeable change cannot easily be observed. However, in smaller water body such as a pond the succession is easily recognisable. Different plant communities occupy different zones in a water body and exhibit concentric zonation. The edges of the water body are occupied by rooted species, submerged species are found in the littoral zone and plankton and floating species occupy the open water zone.

Phytoplankton stage

Unicellular floating algal plants such as diatoms are pioneer species of a bare water body, such as a pond. Their spores are carried by air to the pond. The phytoplankton are followed by zooplankton. They settle down to the bottom of the pond after death, and decay into humus that mixes with silt and clay particles brought into the basin by run off water and wave action and form soil. As soil build up, the pond becomes shallower and further environmental changes follow.

Submerged stage

As the water body becomes shallower, more submerged rooted species are able to become established due to increasing light penetration in the shallower water. This is suitable for growth of rooted submerged species such as Myriophyllum, Vallisneria, Elodea, Hydrilla, and Ceratophyllum. These plants root themselves in mud. Once submerged species colonise the successional changes are more rapid and are mainly autogenic as organic matter accumulates. Inorganic sediment is still entering the lake and is trapped more quickly by the net of plant roots and rhizomes growing on the pond floor. The pond becomes sufficiently shallow (2–5 ft) for floating species and less suitable for rooted submerged plants.

Floating stage

The floating plants are rooted in the mud, but some or all their leaves float on the surface of the water. These include species like Nymphaea, Nelumbo and Potamogeton. Some free-floating species also become associated with root plants. The large and broad leaves of floating plants shade the water surface and conditions become unsuitable for growth of submerged species which start disappearing. The plants decay to form organic mud which makes the pond more shallow yet (1–3 ft).

Reed swamp stage

The pond is now invaded by emergent plants such as Phragmites (reed-grasses), Typha (cattail), and Zizania (wild rice) to form a reed-swamp (in North American usage, this habitat is called a marsh). These plants have creeping rhizomes which knit the mud together to produce large quantities of leaf litter. This litter is resistant to decay and reed peat builds up, accelerating the autogenic change. The surface of the pond is converted into water-saturated marshy land.

Sedge-meadow stage

Successive decreases in water level and changes in substratum help members of Cyperaceae and Graminae such as Carex spp. and Juncus to establish themselves. They form a mat of vegetation extending towards the centre of the pond. Their rhizomes knit the soil further. The above water leaves transpire water to lower the water level further and add additional leaf litter to the soil. Eventually the sedge peat accumulates above the water level and soil is no longer totally waterlogged. The habitat becomes suitable for invasion of herbs (secondary species) such as Mentha, Caltha, Iris, and Galium which grow luxuriantly and bring further changes to the environment. Mesic conditions develop and marshy vegetation begins to disappear.

Woodland stage

The soil now remains drier for most of the year and becomes suitable for development of wet woodland. It is invaded by shrubs and trees such as Salix (willow), Alnus (alders), and Populus. These plants react upon the habitat by producing shade, lower the water table still further by transpiration, build up the soil, and lead to the accumulation of humus with associated micro-organisms. This type of wet woodland is also known as carr.

Climax stage

Finally a self perpetuating climax community develops. It may be a forest if the climate is humid, grassland in case of sub-humid environment or a desert in arid and semi-arid conditions. A forest is characterised by presence of all types of vegetation including herbs, shrubs, mosses, shade-loving plants and trees. Decomposers are frequent in climax vegetation. The overall changes taking place during development of successional communities are building up of substratum, shallowing of water, addition of humus and minerals, soil building and aeration of soil. As the water body fills in with sediment, the area of open water decreases and the vegetation types moves inwards as the water becomes shallower. Many of the above mentioned communities can be seen growing together in a water body. The center is occupied by floating and submerged plants with reeds nearer the shores, followed by sedges and rushes growing at the edges. Still further are shrubs and trees occupying the dry land.

Examples: An example is a small kettle lake called Sweetmere, in Shropshire, UK. Sweetmere is one of many small kettle lakes which formed at the end of the last glacial period when the temperatures began to increase. The ice began to melt and retreat approximately 10,000 years ago.

As the climate slowly began to warm this allowed algae, water lillies and floating aquatic plants to begin to colonise the lake. These, in essence, were the pioneer species. Once these began to die it provided organic matter to the lake bed sediment and therefore increased fertility and reduced depth. As a result this allowed deeper rooted species to develop such as reed, bulrush and reedmace. At this point there is a growing floating raft of thick organic matter within the lake. Because the bulrushes and reeds have relatively deep roots, this encouraged bioconstruction which traps more sediment, allowing sedges, willow and alder to become established. This process further decreased the water depth and raised the lakebed thus making it drier.

Drier conditions now meant that a wider range of species could inhabit the area. Birch and alder came into dominance. All species which have grown have occurred because of seed transfer either by animals, birds, wind or water transfer. Water level is further reduced as a result of further bioconstruction and also due to increasing temperatures there is increased evaporation from the lake.

Underneath the birch canopy developed terrestrial shrubs and grasses. This then increased the acidity which increased the rates of nutrient exchange. The area has been artificially drained and this allowed the oak and ash community to develop. This is the seral stage.

The lake is now being managed by cutting down certain species in order to stop the whole lake becoming dried up and dominated by the oak and ash woodland.

Another example of a hydrosere is Loch a' Mhuilin, located on the Isle of Arran, Scotland. This small lake lies behind a ridge of material deposited towards the end of the last Ice Age. The lake exhibits characteristic features of a hydrosere, the succession from a fresh water surface with small pioneer plant species to a sub-climax vegetation of alder and willow. The climax vegetation of oak and beech woodland has not been achieved due to the impact of human activities of clearing grazing land, as well as grazing by red deer and rabbits.

Lithosere

A lithosere (a sere originating on rock) is a plant succession that begins life on a newly exposed rock surface, such as one left bare as a result of glacial retreat, tectonic uplift as in the formation of a raised beach or volcanic eruptions. For example, the lava fields of Eldgjá in Iceland where Laki and Katla fissures erupted in the year 935 and the solidified lava has, over time, begun to form a lithosere.

Pioneer species are the first organisms that colonise an area, of which lithoseres are an example. They will typically be very hardy (i.e. they will be xerophytes, wind-resistant or cold-resistant). In the case of a lithosere the pioneer species will be cyanobacteria and algae, which create their own food and water, i.e. they are autotrophic and so do not require any external nutrition (except sunlight). For example, the first lithosere observed after the volcanic explosion of Krakatau was algae. Other examples of lithoseres include communities of mosses and lichens, as they are extremely resilient and are capable of surviving in areas without soil.

As more mosses and lichens colonise the area, they, along with natural elements such as wind and frost shattering, begin to weather the rock down. This over time creates more soil, leading to increased water retention. Early on, when there is little water, lichens dominate as they are more suited to a lack of water; but as water retention increases, mosses become more dominant as they are faster growing, and these further break the rocks down. The amount of soil is also increased by the decaying mosses and lichens. This improves the fertility of the soil as humus is increased, allowing grasses and ferns to colonise. Over time, flowering plants will emerge, followed by shrubs. As the soil gets progressively deeper, larger and more advanced plants are able to grow. This is the case in Surtsey, a 'new', small volcanic island located off the south coast of Iceland. Surtsey was 'created' in the 1960s and currently its plant succession has reached the stage where ferns and grasses have begun to start growing in the south of the island where the lava cooled first.

As the plant succession develops further, trees start to appear. The first trees (or pioneer trees) that appear are typically fast growing trees such as birch, willow or rowan. In turn these will be replaced by slow growing, larger trees such as ash and oak. This is the climax community on a lithosere, defined as the point where a plant succession does not develop any further — it reaches a delicate equilibrium with the environment, in particular the climate.

In the off chance of a phenomenon which effectively removes most of the lifeforms in these areas, the resultant landscape is considered to be a disclimax, where there is a loss of the previous climax community. In most cases, should the area be left to regenerate as normal, the area eventually becomes a climax community again.

Chapter 11

Population Dynamics

INTRODUCTION

A population is all the organisms that both belong to the same species and live in the same geographical area. The area that is used to define the population is such that inter-breeding is possible between any pair within the area and more probable than cross-breeding with individuals from other areas. Normally breeding is substantially more common within the area than across the border. In sociology, population refers to a collection of human beings. Demography is a sociological discipline which entails the statistical study of human populations.

POPULATION ECOLOGY

Population ecology is a major sub-field of ecology that deals with the dynamics of species populations and how these populations interact with the environment. It is the study of how the population sizes of species living together in groups change over time and space. The development of population ecology owes much to demography and actuarial life tables. Population ecology is important in conservation biology, especially in the development of population viability analysis (PVA) which makes it possible to predict the long-term probability of a species persisting in a given habitat patch, such as a national park. Although population ecology is a subfield of biology, it provides interesting problems for mathematicians and statisticians who work in population dynamics.

Fundamentals of Population Ecology

One of the first laws of population ecology is the Thomas Malthus' exponential law of population growth. A population will grow (or decline) exponentially as long as the environment experienced by all individuals in the population remains constant. This principle in population ecology provides the basis for formulating predictive theories and tests that follow. Simplified population models usually start with four key variables including death, birth, immigration, and emigration. Mathematical models used to calculate changes in population demographics and evolution hold the assumption (or null hypothesis) of no external influence. Models can be more mathematically complex where 'several competing hypotheses are simultaneously confronted with the data'. For example, in a closed system where immigration and emigration does not take place, the per capita rates of change in a population can be described as:

$$\frac{dN}{dT} = B - D = bN - dN = (b - d)N = rN,$$

where, N is the total number of individuals in the population, B is the number of births, D is the number of deaths, b and d are the per capita rates of birth and death respectively, and r is the per capita rate of population change. This formula can be read as the rate of change in the population (dN/dT) is equal to births minus deaths (B–D).

Using these techniques, Malthus population principle of growth was later transformed into a mathematical model known as the logistic equation:

$$\frac{dN}{dT} = aN\left(1 - \frac{N}{K}\right),$$

where, N is the biomass density, a is the maximum per-capita rate of change, and K is the carrying capacity of the population. The formula can be read as follows:

The rate of change in the population (dN/dT) is equal to growth (aN) that is limited by carrying capacity ($1 - N/K$).

From these basic mathematical principles the discipline of population ecology expands into a field of investigation that queries the demographics of real populations and tests these results against the statistical models. The field of population ecology often uses data on life history and matrix algebra to develop projection matrices on fecundity and survivorship. This information is used for managing wildlife stocks and setting harvest quotas.

r/K selection

An important concept in population ecology is the r/K selection theory. The first variable is r (the intrinsic rate of natural increase in population size, density independent) and the second variable is K (the carrying capacity of a population, density dependent). An r-selected species (e.g. many kinds of insects, such as aphids) is one that has high rates of fecundity, low levels of parental investment in the young, and high rates of mortality before individuals reach maturity. Evolution favours productivity in r-selected species. In contrast, a K-selected species (such as humans) has low rates of fecundity, high levels of parental investment in the young, and low rates of mortality as individuals mature. Evolution in K-selected species favours efficiency in the conversion of more resources into fewer offspring.

Metapopulation

Populations are also studied and conceptualised through the 'metapopulation' concept. The metapopulation concept was introduced in 1969: 'As a population of populations which go extinct locally and recolonise'.

Metapopulation ecology is a simplified model of the landscape into patches of varying levels of quality. Patches are either occupied or they are not. Migrants moving among the patches are structured into metapopulations either as sources or sinks. Source patches are productive sites that generate a seasonal supply of migrants to other patch locations. Sink patches are unproductive sites that only receive migrants. In metapopulation terminology there are emigrants (individuals that leave a patch) and immigrants (individuals that move into a patch). Metapopulation models examine patch dynamics over time to answer questions about spatial and demographic ecology. An important concept in metapopulation ecology is the rescue effect, where small patches of lower quality (i.e. sinks) are maintained by a seasonal influx of new immigrants. Metapopulation structure evolves from year to year, where some patches are sinks, such as dry years, and become sources when conditions are more favourable. Ecologists utilise a mixture of computer models and field studies to explain metapopulation structure.

POPULATION STRUCTURE

Population structure may refer to many aspects of population ecology:

1. Population stratification.

2. Population pyramid.
3. Age class structure.
4. F-statistics.
5. Population density.
6. Population distribution.
7. Population dynamics andpopulation growth.
8. Population genetics.
9. Population size.

Population Stratification

Population stratification is the presence of a systematic difference in allele frequencies between subpopulations in a population possibly due to different ancestry, especially in the context of association studies. Population stratification is also referred to as population structure, in this context.

Causes of Population Stratification

The basic cause of population stratification is nonrandom mating between groups, often due to their physical separation (e.g. for populations of African and European descent) followed by genetic drift of allele frequencies in each group. In some contemporary populations there has been recent admixture between individuals from different populations, leading to populations in which ancestry is variable (as in African Americans). Over tens of generations, random mating can eliminate this type of stratification. In some parts of the globe (e.g. in Europe), population structure is best modelled by isolation-by-distance, in which allele frequencies tend to vary smoothly with location.

Population Stratification and Association Studies

Population stratification can be a problem for association studies, such as case-control studies, where the association found could be due to the underlying structure of the population and not a disease associated locus. Also the real disease causing locus might not be found in the study if the locus is less prevalent in the population where the case subjects are chosen. For this reason, it was common in the 1990s to use family-based data where the effect of population stratification can easily be controlled for using methods such as the TDT. But if the structure is known or a putative structure is found, there are a number of possible ways to implement this structure in the association studies and thus compensate for any population bias. Most contemporary genome-wide association studies take the view that the problem of population stratification is manageable, and that the logistic advantages of using unrelated cases and controls make these studies preferable to family-based association studies.

The two most widely used approaches to this problem include genomic control, which is a relatively nonparametric method for controlling the inflation of test statistics, and structured association methods, which use genetic information to estimate and control for population structure. Currently, the most-widely used structured association method is Eigenstrat, developed by Alkes Price and colleagues.

Population Pyramid

A population pyramid, also called an age structure diagram, is a graphical illustration that shows the distribution of various age groups in a human population (typically that of a country or region of the world), which ideally forms the shape of a pyramid when the region is healthy. It is also used in ecology to determine the overall age distribution of a population; an indication of the reproductive capabilities and likelihood of the continuation of a species (Fig. 11.1).

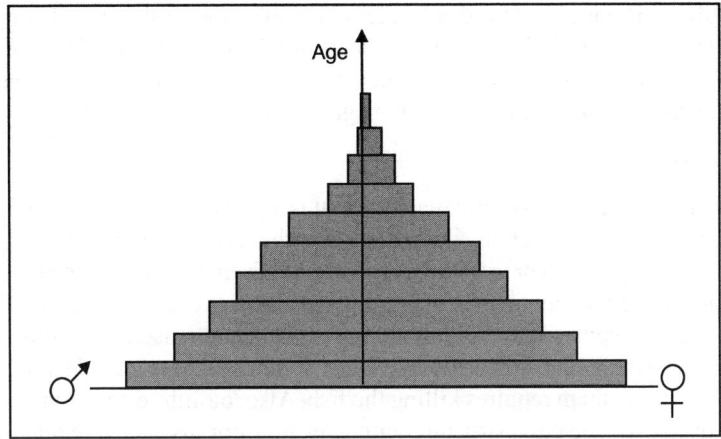

Fig. 11.1. This distribution is named for the frequently pyramidal shape of its graph.

Population pyramids are often viewed as the most effective way to graphically depict the age and sex distribution of a population, partly because of the very clear image these pyramids present.

Types of population pyramid

While all countries' population pyramids differ, four general types have been identified by the fertility and mortality rates of a country (Fig. 11.2).

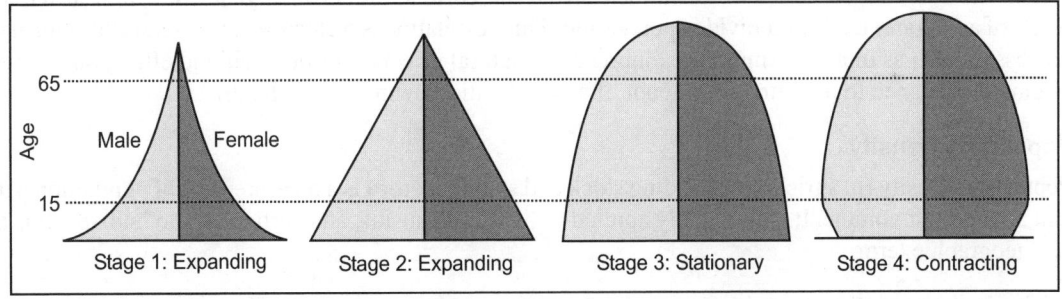

Fig. 11.2. Population pyramids for 4 stages of the demographic transition model.

Stable pyramid: A population pyramid showing an unchanging pattern of fertility and mortality.

Stationary pyramid: A population pyramid typical of countries with low fertility and low mortality, very similar to a constrictive pyramid.

Expansive pyramid: A population pyramid showing a broad base, indicating a high proportion of children, a rapid rate of population growth, and a low proportion of older people. This wide base indicates a large number of children. A steady upwards narrowing shows that more people die at each higher age band. This type of pyramid indicates a population in which there is a high birth rate, a high death rate and a short life expectancy. This is the typical pattern for less economically developed countries, due to little access to and incentive to use birth control, negative environmental factors and poor access to health care.

Constrictive pyramid: A population pyramid showing lower numbers or percentages of younger people. The country will have a greying population which means that people are generally older, as the country has long life expectancy, a low death rate, but also a low birth rate. This pyramid has been occurring more frequently, especially when immigrants are factored out, and is often a typical pattern for a very developed country, a high over-all education and easy access and incentive to use birth control, good health care and a low number to no negative environmental factors.

Age Class Structure

Age class structure in fisheries and wildlife management is a part of population assessment. Age can be determined by counting growth rings in fish scales, otoliths, cross-sections of fin spines for species with thick spines such as triggerfish or teeth for a few species. Each method has its merits and drawbacks. Fish scales are easiest to obtain, but may be unreliable if scales have fallen off of the fish and new ones grown in their places. Fin spines may be unreliable for the same reason, and most fish do not have spines of sufficient thickness for clear rings to be visible. Otoliths will have stayed with the fish throughout its life history, but obtaining them requires killing the fish. Also, otoliths often require more preparation before ageing can occur. An age class structure with gaps in it, for instance a regular bell curve for the population of 1–5 year-old fish, excepting a very low population for the 3-year olds, implies a bad spawning year 3 years ago in that species. Often fish in younger age class structures have very low numbers because they were small enough to slip through the sampling nets, and may in fact have a very healthy population.

F-Statistics

In population genetics, F-statistics describe the level of heterozygosity in a population; more specifically the degree of (usually) a reduction in heterozygosity when compared to Hardy–Weinberg expectation. F-statistics can also be thought of as a measure of the correlation between genes drawn at different levels of a (hierarchically) subdivided population. This correlation is influenced by several evolutionary processes, such as mutation, migration, inbreeding, natural selection or the Wahlund effect, but it was originally designed to measure the amount of allelic fixation owing to genetic drift.

Population Density

Population density (in agriculture standing stock and standing crop) is a measurement of population per unit area or unit volume. It is frequently applied to living organisms, and particularly to humans. It is a key geographic term.

Biological population densities

Population density is population divided by total land area. Low densities may cause an extinction vortex and lead to further reduced fertility. This is called the Allee effect after the scientist who identified it. Examples of the causes in low population densities include:
1. Increased problems with locating mates.
2. Increased inbreeding.

Different species have different expected densities. R-selected species commonly have high population densities, while K-selected species may have lower densities. Low densities may be associated with specialised mate location adaptations such as specialised pollinators, as found in the orchid family (Orchidaceae).

Human population density

For humans, population density is the number of people per unit of area usually per square kilometer or mile (which may include or exclude cultivated or potentially productive area). Commonly this may be calculated for a county, city, country, another territory or the entire world.

The world's population is 6.8 billion, and earth's total area (including land and water) is 510 million square kilometers (197 million square miles). Therefore the worldwide human population density is 6.8 billion ÷ 510 million = 13.3 per km² (34.5 per sq. mile). If only the earth's land area of 150 million km² (58 million sq. miles) is taken into account, then human population density increases to 45.3 per km² (117.2 per sq. mile). This calculation includes all continental and island land area, including Antarctica. If Antarctica is also excluded, then population density rises to 50 people per km² (129.28 per sq. mile). Considering that over half of the earth's land mass consists of areas inhospitable to human inhabitation, such as deserts and high mountains, and that population tends to cluster around seaports and fresh water sources, this number by itself does not give any meaningful measurement of human population density.

Several of the most densely-populated territories in the world are city-states, microstates, micronations or dependencies. These territories share a relatively small area and a high urbanisation level, with an economically specialised city population drawing also on rural resources outside the area, illustrating the difference between high population density and overpopulation.

Cities with high population densities are, by some, considered to be overpopulated, though the extent to which this is the case depends on factors like quality of housing and infrastructure and access to resources. Most of the most densely-populated cities are in southern and eastern Asia, though Cairo and Lagos in Africa also fall into this category.

Methods of measurement

While arithmetic density is the most common way of measuring population density, several other methods have been developed which aim to provide a more accurate measure of population density over a specific area.

1. Arithmetic density: The total number of people/area of land (measured in km² or sq. miles).
2. Physiological density: The total population/area of arable land.
3. Agricultural density: The total rural population/area of arable land.
4. Residential density: The number of people living in an urban area/area of residential land.
5. Urban density: The number of people inhabiting an urban area/total area of urban land.
6. Ecological optimum: The density of population which can be supported by the natural resources.

Population Distribution

In biology, the range or distribution of a species is the geographical area within which that species can be found. Within that range, dispersion is variation in local density. The term is often qualified:

1. Sometimes a distinction is made between a species' native range and the places to which it has been introduced by human agency (deliberately or accidentally), as well as where it has been re-introduced following extirpation.
2. For species which are found in different regions at different times of year, terms such as summer range and winter range are often employed.
3. For species where only part of their range is used for breeding activity, the terms breeding range and non-breeding range are not used.
4. When discussing mobile animals, the species' natural range is often discussed, as opposed to areas where it occurs as a vagrant.
5. Geographic or temporal qualifiers are often added, e.g. British range or pre-1950 range.

There are at least five types of distribution patterns:

1. Scattered/random (random placement).

2. Clustered/grouped (the majority are placed in one area).
3. Linear (their placements form a line).
4. Radial (placements form a '*x*' shape).
5. Regular/ordered (they are not random at all, but follow a set placement. Much like a grid).

BIOLOGICAL DISPERSAL

Biological dispersal refers to species movement away from an existing population or away from the parent organism. Through simply moving from one habitat patch to another, the dispersal of an individual has consequences not only for individual fitness, but also for population dynamics, population genetics, and species distribution. Understanding dispersal and the consequences both for evolutionary strategies at a species level, and for processes at an ecosystem level, requires understanding on the type of dispersal, the dispersal range of a given species, and the dispersal mechanisms involved.

Types of Dispersal

In general there are two basic types of dispersal:
1. Density independent dispersal: Organisms have evolved adaptations for dispersal that take advantage of various forms of kinetic energy occurring naturally in the environment. This is referred to as density independent or passive dispersal and operates on many groups of organisms (some invertebrates, fish, insects and sessile organisms such as plants) that depend on animal vectors, wind, gravity or current for dispersal.
2. Density dependent dispersal: Density dependent or active dispersal for many animals largely depends on factors such as local population size, resource competition, habitat quality, and habitat size. Due to population density, dispersal may relieve pressure for resources in an ecosystem, and competition for these resources may be a selection factor for dispersal mechanisms.

Dispersal of organisms is a critical process for understanding both geographic isolation in evolution through gene flow and the broad patterns of current geographic distributions (biogeography).

At some time during its life, an organism, whether animal or plant, moves or is moved, so that it or its offspring do not die exactly where they were born. Such movement is called dispersal. Some organisms are motile throughout their lives, but others are adapted to move or be moved at precise, limited phases of their life cycles. This commonly is called the Dispersive phase of the life cycle. The strategies of organisms' entire life cycles often are predicated on the nature and circumstances of their dispersive phases.

Dispersal range

'Dispersal range' refers to the distance a species can move from an existing population or the parent organism. An ecosystem depends critically on the ability of individuals and populations to disperse from one habitat patch to another. Therefore, biological dispersal is critical to the stability of ecosystems.

Environmental constraints

Few species are ever evenly or randomly distributed within or across landscapes. In general, species significantly vary across the landscape in association with environmental features that influence their reproductive success and population persistence. Spatial patterns in environmental features (e.g. resources) permit individuals to escape unfavourable conditions and seek out new locations. This allows the organism to 'test' new environments for their suitability, provided they are within animal's geographic range. In addition, the ability of a species to disperse over a gradually changing environment could enable a population to survive extreme conditions (i.e. climate change).

Dispersal barriers

A dispersal barrier may mean that the dispersal range of a species is much smaller than the species distribution. An artificial example is habitat fragmentation due to human land use. Natural barriers to dispersal that limit species distribution include mountain ranges and rivers. An example is the separation of the ranges of the two species of chimpanzee by the Congo River. On the other hand, human activities may also expand the dispersal range of a species by providing new dispersal methods (e.g. ships).

Dispersal mechanisms

Most animals are capable of locomotion and the basic mechanism of dispersal is movement from one place to another. Locomotion allows the organism to 'test' new environments for their suitability, provided they are within the animal's range. Movements are usually guided by inherited behaviours.

The formation of barriers to dispersal or gene flow between adjacent areas can isolate populations on either side of the emerging divide. The geographic separation and subsequent genetic isolation of portions of an ancestral population can result in speciation. Geodispersal occurs with the erosion of these barriers to dispersal or gene flow, thereby allowing the mixing of previously isolated populations.

Plant dispersal mechanisms

Seed dispersal is the movement or transport of seeds away from the parent plant. Plants have limited mobility and consequently rely upon a variety of dispersal vectors to transport their propagules, including both abiotic and biotic vectors. Seeds can be dispersed away from the parent plant individually or collectively, as well as dispersed in both space and time. The patterns of seed dispersal are determined in large part by the dispersal mechanism and this has important implications for the demographic and genetic structure of plant populations, as well as migration patterns and species interactions. There are five main modes of seed dispersal: gravity, wind, ballistic, water and by animals and the sky.

Animal dispersal mechanisms

Non-motile animals

There are numerous animal forms that are non-motile, such as sponges, Bryozoans, Tunicates, sea anemones, corals, and oysters. In common, they are all either marine or aquatic. It may seem curious that plants have been so successful at stationary life on land, while animals have not, but the answer lies in the food supply. Plants produce their own food from sunlight and carbon dioxide — both generally more abundant on land than in water. Animals fixed in place must rely on the surrounding medium to bring food at least close enough to grab, and this occurs in the three-dimensional water environment, but with much less abundance in the atmosphere.

All of the marine and aquatic invertebrates whose lives are spent fixed to the bottom (more or less; anemones are capable of getting up and moving to a new location if conditions warrant) produce dispersal units. These may be specialised 'buds' or motile sexual reproduction products or even a sort of alteration of generations as in certain cnidaria.

Corals provide a good example of how sedentary species achieve dispersion. Corals reproduce by releasing sperm and eggs directly into the water. These release events are coordinated by lunar phase in certain warm months, such that all corals of one or many species on a given reef will release on the same single or several consecutive nights. The released eggs are fertilised, and the resulting zygote develops quickly into a multicellular planula. This motile stage then attempts to find a suitable substratum for settlement. Most are unsuccessful and die or are fed upon by zooplankton and bottom dwelling predators such as anemones and other corals. However, untold millions are produced, and a few do succeed in locating spots of bare limestone, where they settle and transform by growth into a polyp. All things being favourable, the single polyp grows into a coral head by budding off new polyps to form a colony.

Motile animals

The majority of all animals are motile. Although motile animals can, in theory, disperse themselves by their spontaneous and independent locomotive powers, a great many species utilise the existing kinetic energies in the environment, resulting in passive movement. Dispersal by water currents is especially associated with the physically small inhabitants of marine waters known as zooplankton. The term plankton comes from the Greek, meaning 'wanderer' or 'drifter'.

Dispersal by dormant stages

Many animal species, especially freshwater invertebrates, are able to disperse by wind or by transfer with an aid of larger animals (birds, mammals or fishes) as dormant eggs, dormant embryos or, in some cases, dormant adult stages. Tardigrades, some rotifers and some copepods are able to withstand desiccation as adult dormant stages. Many other taxa (Cladocera, Bryozoa, Hydra, Copepoda and so on) can disperse as dormant eggs or embryos. Freshwater sponges usually have special dormant propagules called gemmulae for such a dispersal. Many kinds of dispersal dormant stages are able to withstand not only desiccation and low and high temperature, but also action of digestive enzymes during their transfer through digestive tracts of birds and other animals, high concentration of salts and many kinds of toxicants. Such dormant-resistant stages made possible the long-distance dispersal from one water body to another and broad distribution ranges of many freshwater animals.

Population Dynamics and Population Growth

Population dynamics

Population dynamics is the branch of life sciences that studies short- and long-term changes in the size and age composition of populations, and the biological and environmental processes influencing those changes. Population dynamics deals with the way populations are affected by birth and death rates, and by immigration and emigration, and studies topics such as ageing populations or population decline.

The mathematical model often viewed as the best to govern the population dynamics of any given species is called the exponential model. With the exponential model, the rate of change of any given population is proportional to the already existing population.

Fisheries and wildlife management

In fisheries and wildlife management, population is affected by three dynamic rate functions.
1. Natality or birth rate, often recruitment, which means reaching a certain size or reproductive stage. Usually refers to the age a fish can be caught and counted in nets.
2. Population growth rate, which measures the growth of individuals in size and length. More important in fisheries, where population is often measured in biomass.
3. Mortality, which includes harvest mortality and natural mortality. Natural mortality includes non-human predation, disease and old age.

If N_1 is the number of individuals at time 1 then:

$$N_1 = N_0 + B - D + I - E$$

where, N_0 is the number of individuals at time 0, B is the number of individuals born, D the number that died, I the number that immigrated, and E the number that emigrated between time 0 and time 1.

If we measure these rates over many time intervals, we can determine how a population's density changes over time. Immigration and emigration are present, but are usually not measured.

All of these are measured to determine the harvestable surplus, which is the number of individuals that can be harvested from a population without affecting long-term stability or average population size. The harvest within the harvestable surplus is considered compensatory mortality, where the harvest deaths are substituting for the deaths that would occur naturally. It started in Europe. Harvest beyond that is additive mortality, harvest in addition to all the animals that would have died naturally. These terms are not the universal good and evil of population management, for example, in deer, the DNR are trying to reduce deer population size overall to an extent, since hunters have reduced buck competition and increased deer population unnaturally.

Intrinsic rate of increase

The rate at which a population increases in size if there are no density-dependent forces regulating the population is known as the intrinsic rate of increase:

$$(dN/dt)(1/N) = r$$

Where, (dN/dt) is the rate of increase of the population and N is the population size, r is the intrinsic rate of increase. This is therefore the theoretical maximum rate of increase of a population per individual The concept is commonly used in insect population biology to determine how environmental factors affect the rate at which pest populations increase.

Population growth

Population growth is the change in a population over time, and can be quantified as the change in the number of individuals of any species in a population using 'per unit time' for measurement. In biology, the term population growth is likely to refer to any known organism, but this article deals mostly with the application of the term to human populations in demography.

In demography, population growth is used informally for the more specific term population growth rate, and is often used to refer specifically to the growth of the human population of the world. Simple models of population growth include the malthusian growth model and the logistic model.

Determinants of Population growth

Population growth is determined by four factors, births (B), deaths (D), immigrants (I), and emigrants (E). Using a formula expressed as:

$$\Delta P \equiv B - D + I - E$$

In other words, the population growth of a period can be calculated in two parts, natural growth of population ($B - D$) and mechanical growth of population ($I - E$), in which Mechanical growth of population is mainly affected by social factors, e.g. the advanced economies are growing faster while the backward economies are growing slowly even with negative growth.

Population growth rate

In demographics and ecology, population growth rate (PGR) is the fractional rate at which the number of individuals in a population increases. Specifically, PGR ordinarily refers to the change in population over a unit time period, often expressed as a percentage of the number of individuals in the population at the beginning of that period. This can be written as the formula:

$$\text{Growth rate} = \frac{\left(\text{Population at end of period} - \text{Population at beginning of period}\right)}{\text{Population at beginning of period}}$$

(In the limit of a sufficiently small time period.)

The above formula can be expanded to: Growth rate = crude birth rate — crude death rate + net immigration rate or $\Delta P/P = (B/P) - (D/P) + (I/P) - (E/P)$, where P is the total population, B is the number of births, D is the number of deaths, I is the number of immigrants, and E is the number of emigrants.

This formula allows for the identification of the source of population growth, whether due to natural increase or an increase in the net immigration rate. Natural increase is an increase in the native-born population, stemming from either a higher birth rate, a lower death rate or a combination of the two. Net immigration rate is the difference between the number of immigrants and the number of emigrants.

The most common way to express population growth is as a ratio, not as a rate. The change in population over a unit time period is expressed as a percentage of the population at the beginning of the time period. That is:

Growth ratio = Growth rate × 100%.

A positive growth ratio (or rate) indicates that the population is increasing, while a negative growth ratio indicates the population is decreasing. A growth ratio of zero indicates that there were the same number of people at the two times — net difference between births, deaths and migration is zero. However, a growth rate may be zero even when there are significant changes in the birth rates, death rates, immigration rates, and age distribution between the two times. Equivalently, per cent death rate = the average number of deaths in a year for every 100 people in the total population.

A related measure is the net reproduction rate. In the absence of migration, a net reproduction rate of more than one indicates that the population of women is increasing, while a net reproduction rate less than one (sub-replacement fertility) indicates that the population of women is decreasing.

Excessive growth and decline

Population exceeding the carrying capacity of an area or environment is called overpopulation. It may be caused by growth in population or by reduction in capacity. Spikes in human population can cause problems such as pollution and traffic congestion, these might be resolved or worsened by technological and economic changes. Conversely, such areas may be considered 'underpopulated' if the population is not large enough to maintain an economic system. Between these two extremes sits the notion of the optimum population.

Human population growth rate

Globally, the growth rate of the human population has been declining since peaking in 1962 and 1963 at 2.20 per cent per annum. In 2009, the estimated annual growth rate was 1.1 per cent. The CIA World Factbook gives the world annual birthrate, mortality rate, and growth rate as 1.915 per cent, 0.812 per cent, and 1.092 per cent respectively. The last one hundred years have seen a rapid increase in population due to medical advances and massive increase in agricultural productivity made possible by the Green Revolution.

The actual annual growth in the number of humans fell from its peak of 88.0 million in 1989, to a low of 73.9 million in 2003, after which it rose again to 75.2 million in 2006. Since then, annual growth has declined. In 2009, the human population increased by 74.6 million, which is projected to fall steadily to about 41 million per annum in 2050, at which time the population will have increased to about

9.2 billion. Each region of the globe has seen great reductions in growth rate in recent decades, though growth rates remain above 2 per cent in some countries of the Middle East and Sub-Saharan Africa, and also in South Asia, Southeast Asia, and Latin America.

Some countries experience negative population growth, especially in Eastern Europe mainly due to low fertility rates, high death rates and emigration. In Southern Africa, growth is slowing due to the high number of HIV-related deaths. Some Western Europe countries might also encounter negative population growth. Japan's population began decreasing in 2005.

Population Genetics

Population genetics is the study of allele frequency distribution and change under the influence of the four main evolutionary processes: natural selection, genetic drift, mutation and gene flow. It also takes into account the factors of recombination, population subdivision and population structure. It attempts to explain such phenomena as adaptation and speciation.

Population genetics is the study of the frequency and interaction of alleles and genes in populations. A population is a set of organisms in which any pair of members can breed together. This implies that all members belong to the same species and live near each other.

For example, all of the moths of the same species living in an isolated forest are a population. A gene in this population may have several alternate forms, which account for variations between the phenotypes of the organisms. An example might be a gene for coloration in moths that has two alleles: black and white. A gene pool is the complete set of alleles for a gene in a single population; the allele frequency for an allele is the fraction of the genes in the pool that is composed of that allele (for example, what fraction of moth coloration genes are the black allele). Evolution occurs when there are changes in the frequencies of alleles within a population; for example, the allele for black colour in a population of moths becoming more common.

Hardy–Weinberg principle

To understand the mechanisms that cause a population to evolve, it is useful to consider what conditions are required for a population not to evolve. The Hardy-Weinberg principle states that the frequencies of alleles (variations in a gene) in a sufficiently large population will remain constant if the only forces acting on that population are the random reshuffling of alleles during the formation of the sperm or egg, and random combination of the alleles in these sex cells during fertilisation. Such a population is said to be in Hardy–Weinberg equilibrium as it is not evolving. Hardy–Weinberg equilibrium is impossible in nature. Genetic equilibrium is an ideal state that provides a baseline to measure genetic change against (Fig. 11.3).

Allele frequencies in a population remain static across generations, provided the following conditions are at hand: random mating, no mutation (the alleles do not change), no migration or emigration (no exchange of alleles between populations), infinitely large population size, and no selective pressure for or against any traits.

In the simplest case of a single locus with two alleles: the dominant allele is denoted A and the recessive a and their frequencies are denoted by p and q; freq(A) = p; freq (a) = q; $p + q = 1$. If the population is in equilibrium, then we will have freq(AA) = p^2 for the AA homozygotes in the population, freq(aa) = q^2 for the aa homozygotes, and freq(Aa) = $2pq$ for the heterozygotes.

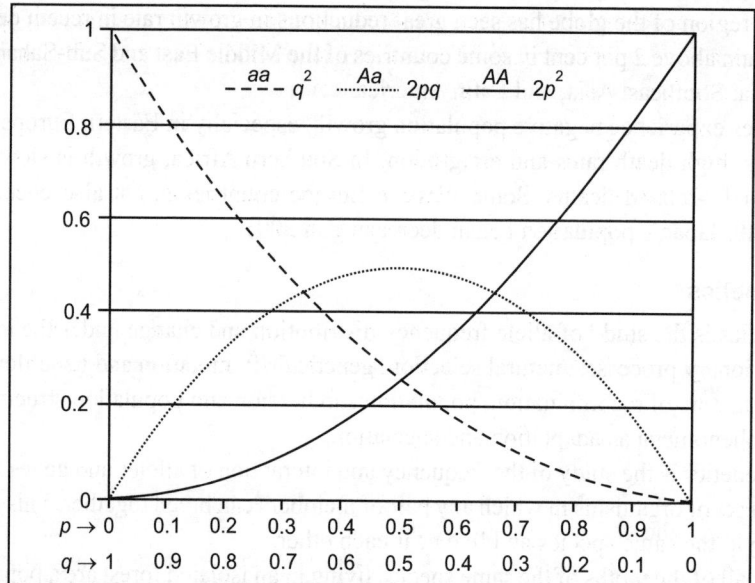

Fig. 11.3. Hardy–Weinberg principle for two alleles: The horizontal axis shows the two allele frequencies p and q and the vertical axis shows the genotype frequencies. Each graph shows one of the three possible genotypes.

Scope and theoretical considerations

The mathematics of population genetics were originally developed as part of the modern evolutionary synthesis. According to Beatty, it defines the core of the modern synthesis.

According to Lewontin, the theoretical task for population genetics is a process in two spaces: a 'genotypic space' and a 'phenotypic space'. The challenge of a complete theory of population genetics is to provide a set of laws that predictably map a population of genotypes (G_1) to a phenotype space (P_1), where selection takes place, and another set of laws that map the resulting population (P_2) back to genotype space (G_2) where Mendelian genetics can predict the next generation of genotypes, thus completing the cycle. Even leaving aside for the moment the non-Mendelian aspects of molecular genetics, this is clearly a gargantuan task. Visualising this transformation schematically:

$$G_1 \xrightarrow{T_1} P_1 \xrightarrow{T_2} P_2 \xrightarrow{T_3} G_2 \xrightarrow{T_4} G_2' \to \ldots$$

T_1 represents the genetic and epigenetic laws, the aspects of functional biology or development, that transform a genotype into phenotype. We will refer to this as the 'genotype-phenotype map'. T_2 is the transformation due to natural selection, T_3 are epigenetic relations that predict genotypes based on the selected phenotypes and finally T_4 the rules of Mendelian genetics.

In practice, there are two bodies of evolutionary theory that exist in parallel, traditional population genetics operating in the genotype space and the biometric theory used in plant and animal breeding, operating in phenotype space. The missing part is the mapping between the genotype and phenotype space. This leads to a 'sleight of hand' (as Lewontin terms it) whereby variables in the equations of one domain, are considered parameters or *constants*, where, in a full-treatment they would be transformed themselves by the evolutionary process and are in reality functions of the state variables in the other domain. The 'sleight of hand' is assuming that we know this mapping. Proceeding as if we do understand it is enough to analyse many cases of interest. For example, if the phenotype is almost one-to-one with genotype (sickle-cell disease) or the time-scale is sufficiently short, the 'constants' can be treated as such; however, there are many situations where it is inaccurate.

The four processes

Natural selection

Natural selection is the process by which heritable traits that make it more likely for an organism to survive and successfully reproduce become more common in a population over successive generations.

The natural genetic variation within a population of organisms means that some individuals will survive more successfully than others in their current environment. Factors which affect reproductive success are also important, an issue which Charles Darwin developed in his ideas on sexual selection. Natural selection acts on the phenotype or the observable characteristics of an organism, but the genetic (heritable) basis of any phenotype which gives a reproductive advantage will become more common in a population. Over time, this process can result in adaptations that specialise organisms for particular ecological niches and may eventually result in the emergence of new species.

Natural selection is one of the cornerstones of modern biology. The term was introduced by Darwin in his groundbreaking 1859 book 'On the Origin of Species', in which natural selection was described by analogy to artificial selection, a process by which animals and plants with traits considered desirable by human breeders are systematically favoured for reproduction. The concept of natural selection was originally developed in the absence of a valid theory of heredity; at the time of Darwin's writing, nothing was known of modern genetics. The union of traditional Darwinian evolution with subsequent discoveries in classical and molecular genetics is termed the modern evolutionary synthesis. Natural selection remains the primary explanation for adaptive evolution.

Genetic drift

Genetic drift is the change in the relative frequency in which a gene variant (allele) occurs in a population due to random sampling and chance. That is, the alleles in the offspring in the population are a random sample of those in the parents. A population's allele frequency is the fraction or percentage of its gene copies compared to the total number of gene alleles that share a particular form.

Genetic drift is an evolutionary process which leads to changes in allele frequencies over time. It may cause gene variants to disappear completely, and thereby reduce genetic variability. In contrast to natural selection, which makes gene variants more common or less common depending on their reproductive success, the changes due to genetic drift are not driven by environmental or adaptive pressures, and may be beneficial, neutral or detrimental to reproductive success.

The effect of genetic drift is larger in small populations, and smaller in large populations. Vigorous debates wage among scientists over the relative importance of genetic drift compared with natural selection. Ronald Fisher held the view that genetic drift plays at the most a minor role in evolution, and this remained the dominant view for several decades. In 1968 Motoo Kimura rekindled the debate with his neutral theory of molecular evolution which claims that most of the changes in the genetic material are caused by genetic drift.

Mutation

Mutations are changes in the DNA sequence of a cell's genome and are caused by radiation, viruses, transposons and mutagenic chemicals, as well as errors that occur during meiosis or DNA replication. Errors are introduced particularly often in the process of DNA replication, in the polymerisation of the second strand. These errors can also be induced by the organism itself, by cellular processes such as hypermutation. Mutations can have an impact on the phenotype of an organism, especially if they occur within the protein coding sequence of a gene. Error rates are usually very low (1 error in every 10 million–100 million bases) due to the 'proofreading' ability of DNA polymerases. Without proofreading, error rates are a thousandfold higher. Chemical damage to DNA occurs naturally as well, and cells use DNA repair mechanisms to repair mismatches and breaks in DNA. Nevertheless, the repair sometimes fails to return the DNA to its original sequence.

In organisms that use chromosomal crossover to exchange DNA and recombine genes, errors in alignment during meiosis can also cause mutations. Errors in crossover are especially likely when similar sequences cause partner chromosomes to adopt a mistaken alignment; this makes some regions in genomes more prone to mutating in this way. These errors create large structural changes in DNA sequence — duplications, inversions or deletions of entire regions or the accidental exchanging of whole parts between different chromosomes (called translocation).

Mutation can result in several different types of change in DNA sequences; these can either have no effect, alter the product of a gene or prevent the gene from functioning. Studies in the fly Drosophila melanogaster suggest that if a mutation changes a protein produced by a gene, this will probably be harmful, with about 70 per cent of these mutations having damaging effects, and the remainder being either neutral or weakly beneficial. Due to the damaging effects that mutations can have on cells, organisms have evolved mechanisms such as DNA repair to remove mutations. Therefore, the optimal mutation rate for a species is a trade-off between costs of a high mutation rate, such as deleterious mutations, and the metabolic costs of maintaining systems to reduce the mutation rate, such as DNA repair enzymes. Viruses that use RNA as their genetic material have rapid mutation rates, which can be an advantage since these viruses will evolve constantly and rapidly, and thus evade the defensive responses of, e.g. the human immune system.

Mutations can involve large sections of DNA becoming duplicated, usually through genetic recombination. These duplications are a major source of raw material for evolving new genes, with tens to hundreds of genes duplicated in animal genomes every million years. Most genes belong to larger families of genes of shared ancestry. Novel genes are produced by several methods, commonly through the duplication and mutation of an ancestral gene or by recombining parts of different genes to form new combinations with new functions.

Here, domains act as modules, each with a particular and independent function, that can be mixed together to produce genes encoding new proteins with novel properties. For example, the human eye uses four genes to make structures that sense light: three for colour vision and one for night vision; all four arose from a single ancestral gene. Another advantage of duplicating a gene (or even an entire genome) is that this increases redundancy; this allows one gene in the pair to acquire a new function while the other copy performs the original function. Other types of mutation occasionally create new genes from previously noncoding DNA.

Gene flow

Gene flow is the exchange of genes between populations, which are usually of the same species. Examples of gene flow within a species include the migration and then breeding of organisms or the exchange of pollen. Gene transfer between species includes the formation of hybrid organisms and horizontal gene transfer.

Migration into or out of a population can change allele frequencies, as well as introducing genetic variation into a population. Immigration may add new genetic material to the established gene pool of a population. Conversely, emigration may remove genetic material. As barriers to reproduction between two diverging populations are required for the populations to become new species, gene flow may slow this process by spreading genetic differences between the populations. Gene flow is hindered by mountain ranges, oceans and deserts or even man-made structures such as the Great Wall of China, which has hindered the flow of plant genes.

Depending on how far two species have diverged since their most recent common ancestor, it may still be possible for them to produce offspring, as with horses and donkeys mating to produce mules. Such hybrids are generally infertile, due to the two different sets of chromosomes being unable to pair up during meiosis. In this case, closely related species may regularly interbreed, but hybrids will be selected against and the species will remain distinct. However, viable hybrids are occasionally formed and these new species can either have properties intermediate between their parent species or possess a totally new phenotype. The importance of hybridisation in creating new species of animals is unclear, although cases have been seen in many types of animals, with the gray tree frog being a particularly well-studied example.

Hybridisation is, however, an important means of speciation in plants, since polyploidy (having more than two copies of each chromosome) is tolerated in plants more readily than in animals. Polyploidy is important in hybrids as it allows reproduction, with the two different sets of chromosomes each being able to pair with an identical partner during meiosis. Polyploids also have more genetic diversity, which allows them to avoid inbreeding depression in small populations.

Horizontal gene transfer is the transfer of genetic material from one organism to another organism that is not its offspring; this is most common among bacteria. In medicine, this contributes to the spread of antibiotic resistance, as when one bacteria acquires resistance genes it can rapidly transfer them to other species. Horizontal transfer of genes from bacteria to eukaryotes such as the yeast Saccharomyces cerevisiae and the adzuki bean beetle Callosobruchus chinensis may also have occurred. An example of larger-scale transfers are the eukaryotic bdelloid rotifers, which appear to have received a range of genes from bacteria, fungi, and plants. Viruses can also carry DNA between organisms, allowing transfer of genes even across biological domains. Large-scale gene transfer has also occurred between the ancestors of eukaryotic cells and prokaryotes, during the acquisition of chloroplasts and mitochondria.

Gene flow is the transfer of alleles from one population to another. Migration into or out of a population may be responsible for a marked change in allele frequencies. Immigration may also result in the addition of new genetic variants to the established gene pool of a particular species or population.

There are a number of factors that affect the rate of gene flow between different populations. One of the most significant factors is mobility, as greater mobility of an individual tends to give it greater migratory potential. Animals tend to be more mobile than plants, although pollen and seeds may be carried great distances by animals or wind.

Maintained gene flow between two populations can also lead to a combination of the two gene pools, reducing the genetic variation between the two groups. It is for this reason that gene flow strongly acts against speciation, by recombining the gene pools of the groups, and thus, repairing the developing differences in genetic variation that would have led to full speciation and creation of daughter species.

For example, if a species of grass grows on both sides of a highway, pollen is likely to be transported from one side to the other and vice versa. If this pollen is able to fertilise the plant where it ends up and produce viable offspring, then the alleles in the pollen have effectively been able to move from the population on one side of the highway to the other.

Recombination

Basic models of population genetics consider only one gene locus at a time. Because of epistasis, the phenotypic effect of an allele at one locus may depend on which alleles are present at many other loci. If all genes are in linkage equilibrium, the effect of an allele at one locus can be averaged across the gene pool at other loci. In reality, one allele is frequently found in linkage disequilibrium with genes at other loci, especially with genes located nearby on the same chromosome. Recombination breaks up this linkage disequilibrium too slowly to avoid genetic hitchhiking, where an allele at one locus rises to high frequency because it is linked to an allele under selection at a nearby locus. This is a problem for population genetic models that treat one gene locus at a time. It can, however, be exploited as a method for detecting the action of natural selection via selective sweeps. In the extreme case of primarily asexual populations, different population genetic equations can be derived and solved, and these behave quite differently to the sexual case.

Genetic structure

Because of physical barriers to migration, along with limited vagility, and natal philopatry, natural populations are rarely panmictic. There is usually a geographic range within which individuals are more closely related to one another than those randomly selected from the general population. This is described as the extent to which a population is genetically structured. Genetic structuring can be caused by migration due to historical climate change, species range expansion or current availability of habitat.

Microbial population genetics

Microbial population genetics is a rapidly advancing field of investigation with relevance to many other theoretical and applied areas of scientific investigations. The population genetics of micro-organisms lays the foundations for tracking the origin and evolution of antibiotic resistance and deadly infectious pathogens. Population genetics of micro-organisms is also an essential factor for devising strategies for the conservation and better utilisation of beneficial microbes.

Population Size

In population genetics and population ecology, population size (usually denoted N) is the number of individual organisms in a population. The effective population size (N_e) is defined as 'the number of breeding individuals in an idealised population that would show the same amount of dispersion of allele frequencies under random genetic drift or the same amount of inbreeding as the population under consideration'. N_e is usually less than N (the absolute population size) and this has important applications in conservation genetics.

Small population size results in increased genetic drift. Population bottlenecks are when population size reduces for a short period of time. Overpopulation may indicate any case in which the population of any species of animal may exceed the carrying capacity of its ecological niche.

Birth Rate

Crude birth rate is the nativity or childbirths per 1000 people per year (in estimation review points). According to the United Nations' World Population Prospects: The 2008 Revision Population Database, crude birth rate is the number of births over a given period divided by the person-years lived by the population over that period. It is expressed as number of births per 1000 population. CBR = (births in a period/population of person-years over that period).

Another indicator of fertility that is frequently used is the total fertility rate, which is the average number of children born to each woman over the course of her life. In general, the total fertility rate is a better indicator of (current) fertility rates because, unlike the crude birth rate, it is not affected by the age distribution of the population. Fertility rates tend to be higher in less economically developed countries and lower in more economically developed countries.

Birth rate and the demographic transition model

The demographic transition model describes how population mortality and fertility decline as social and economic development occurs through time. The two major factors in the demographic transition model are crude birth rate (CBR) and crude death rate (CDR). There are four stages to the demographic model. In the first and second stages, CBR remains high because people are still in agrarian cultures and need more labour to work on farms. In addition, the chances of children dying are high because medicine is not as advanced during that phase. In the third stage, CBR starts to decline due to women's increasing participation in society and the reduced need for families to have many children to work on farms. In the fourth stage, CBR is sustained at a very low level, with some countries having rates that are below replacement levels in other countries.

Mortality Rate

Mortality rate is a measure of the number of deaths (in general or due to a specific cause) in some population, scaled to the size of that population, per unit time. Mortality rate is typically expressed in units of deaths per 1000 individuals per year; thus, a mortality rate of 9.5 in a population of 1,00,000 would mean 950 deaths per year in that entire population or 0.95 per cent out of the total. It is distinct from morbidity rate, which refers to the number of individuals in poor health during a given time period (the prevalence rate) or the number of newly appearing cases of the disease per unit of time (incidence rate).

One distinguishes:

1. The crude death rate, the total number of deaths per year per 1000 people. As of July 2009 the crude death rate for the whole world is about 8.37 per 1000 per year according to the current CIA World Factbook.
2. The perinatal mortality rate, the sum of neonatal deaths and fetal deaths (stillbirths) per 1000 births.
3. The maternal mortality rate, the number of maternal deaths per 1,00,000 women of reproductive age in same time period.
4. The infant mortality rate, the number of deaths of children less than 1 year old per 1000 live births.
5. The child mortality rate, the number of deaths of children less than 5 years old per 1000 live births.
6. The standardised mortality ratio (SMR)—This represents a proportional comparison to the numbers of deaths that would have been expected if the population had been of a standard composition in terms of age, gender, etc.
7. The age-specific mortality rate (ASMR)— This refers to the total number of deaths per year per 1000 people of a given age.

In regard to the success or failure of medical treatment or procedures, one would also distinguish:

1. The early mortality rate, the total number of deaths in the early stages of an ongoing treatment or in the period immediately following an acute treatment.
2. The late mortality rate, the total number of deaths in the late stages of an ongoing treatment or a significant length of time after an acute treatment.

Note that the crude death rate as defined above and applied to a whole population can give a misleading impression. The crude death rate depends on the age (and gender) specific mortality rates and the age (and gender) distribution of the population. The number of deaths per 1000 people can be higher for developed nations than in less-developed countries, despite life expectancy being higher in developed countries due to standards of health being better. This happens because developed countries typically have a completely different population age distribution, with a much higher proportion of older people, due to both lower recent birth rates and lower mortality rates. A more complete picture of mortality is given by a life table which shows the mortality rate separately for each age. A life table is necessary to give a good estimate of life expectancy.

Biotic Potential

Biotic potential is the maximum reproductive capacity of a population if resources are unlimited. Full expression of the biotic potential of an organism is restricted by environmental resistance, any condition that inhibits the increase in number of the population. It is generally only reached when environmental conditions are very favourable. A species reaching its biotic potential would exhibit exponential population growth and be said to have a high fertility, that is, how many offspring are produced per mother.

Biotic Potential is a fundamental species characteristic, defined by Chapman as 'the inherent power of organisms to reproduce and survive'. In 1931, Chapman redescribed it as: 'It is a sort of algebraic sum of the number of young produced at each reproduction, number of reproductions over a period of time, sex ratio of the species, and their general ability to survive under given physical conditions'.

Chapman relates to a 'vital index':

$$\text{Vital index} = (\text{Number of births}/\text{Number of deaths}) \times 100$$

Biotic potential is the highest possible vital index of a species; therefore, when the species has its highest birthrate and lowest mortality rate.

Significance of biotic potential

If the potential value of population increase can be determined, the impact of the environment upon the population also can be determined. Compute the biotic potential (potential increase) and subtract the actual or observed value of decrease; this difference represents how effective the environment is in preventing the species from attaining its full potential.

Components of biotic potential

1. Reproductive potential — potential natality: It is the upper limit to biotic potential (in the absence of mortality).
2. Survival potential: Because reproductive potential does not account for the number of gametes surviving, survival potential is a necessary component of biotic potential; it is the reciprocal of mortality (in the absence of mortality, biotic potential = reproductive potential).

Life Table

In actuarial science, a life table is a table which shows, for each age, what the probability is that a person of that age will die before his or her next birthday. From this starting point, a number of inferences can be derived.

1. The probability of surviving any particular year of age.
2. Remaining life expectancy for people at different ages.

Life tables can be constructed using projections of future mortality rates, but more often they are a snapshot of age-specific mortality rates in the recent past, and do not purport to be projections. For various reasons, such as advances in medicine, age-specific mortality rates vary over time. Life tables are usually constructed separately for men and for women because of their substantially different mortality rates. Other characteristics can also be used to distinguish different risks, such as smoking status, occupation, and socio-economic class.

Life tables can be extended to include other information in addition to mortality, for instance health information to calculate health expectancy. Health expectancies, of which disability-free life expectancy (DFLE) and healthy life years (HLY) are the best-known examples, are the remaining number of years a person can expect to live in a specific health state, such as free of disability. Two types of life tables are used to divide the life expectancy into life spent in various states: (i) multi-state life tables (also known as increment-decrement life tables) based on transition rates in and out of the different states and to death, and (ii) prevalence-based life tables (also known as the Sullivan method) based on external information on the proportion in each state.

Life tables can also be extended to show life expectancies in different labour force states or marital status states. Life tables are also used extensively in biology and epidemiology. The concept is also of importance in product life cycle management.

To do this, actuaries develop mathematical models of the rates and timing of the events. They do this by studying the incidence of these events in the recent past, and sometimes developing expectations of how these past events will change over time (for example, whether the progressive reductions in mortality rates in the past will continue) and deriving expected rates of such events in the future, usually based on the age or other relevant characteristics of the population. These are called mortality tables if they show death rates, and morbidity tables if they show various types of sickness or disability rates.

The availability of computers and the proliferation of data gathering about individuals has made possible calculations that are more voluminous and intensive than those used in the past (i.e. they crunch more numbers) and it is more common to attempt to provide different tables for different uses, and to factor in a range of non-traditional behaviours (e.g. gambling, debt load) into specialised calculations utilised by some institutions for evaluating risk. This is particularly the case in non-life insurance (e.g. the pricing of motor insurance can allow for a large number of risk factors, which requires a correspondingly complex table of expected claim rates). However the expression 'life table' normally refers to human survival rates and is not relevant to non-life assurance.

Mathematics used in life tables

The basic algebra used in life tables is as follows:

1. q_x: The probability that someone aged exactly x will die before reaching age $(x + 1)$.
2. p_x: The probability that someone aged exactly x will survive to age $(x + 1)$:

$$p_x = 1 - q_x$$

3. l_x: The number of people who survive to age x. Note that this is based on a starting point of l_0 lives, typically taken as 1,00,000:

$$l_{x+1} = l_x \cdot (1 - q_x) = l_x \cdot p_x$$

$$\frac{l_{x+1}}{l_x} = p_x$$

4. d_x: The number of people who die aged x last birthday:

$$d_x = l_x - l_{x+1} = l_x \cdot (1 - p_x) = l_x \cdot q_x$$

5. $_t p_x$: The probability that someone aged exactly x will survive for t more years, i.e. live up to at least age $x = t$ years.

$$_t p_x = \frac{l_x + t}{l_x}$$

6. $_{t|k} q_x$: The probability that someone aged exactly x will survive for t more years, then die within the following k years.

$$_{t|k} q_x = {_t p_x} \cdot kq_x + t = \frac{l_{x+t} - l_{x+t+k}}{l_x}$$

7. μ_x : The force of mortality, i.e. the instantaneous mortality rate at age x, i.e. the number of people dying in a short interval starting at age x, divided by l_x and also divided by the length of the interval. Unlike q_x, the instantaneous mortality rate, μ_x, may exceed 1.

Another common variable is:

$$m_x$$

This symbol refers to central rate of mortality. It is approximately equal to the average force of mortality, averaged over the year of age.

POPULATION REGULATION

Population regulation is the control of the size of a population. This regulation implies a tendency of the population to achieve or return to a size at equilibrium or in harmony with the surrounding environment. If a population tends to remain about the same size, then it is said to be stable. There are basically two different types of population regulation—classified according to the types of factors that control the size of the population.

Density-dependent Control

A density-dependent factor is one where the effect of the factor on the size of the population depends upon the original density or size of the population. A disease is a good example of a density-dependent factor. If a population is dense and the individuals live close together, then each individual will have a higher probability of catching the disease than if the individuals had been living farther apart. Not only will a greater number of individuals be affected, but, more importantly, a greater proportion of the population will be affected if they are living close together. For example, bird populations are often regulated more by this type of regulation.

Characteristic of the factor

In general, density-dependent factors are biological factors, such as diseases, parasites, competition, and predation.

Characteristics of populations controlled primarily by density-dependent factors

A population being controlled primarily by density-dependent factors will have the first growth form which is called the 'self-limiting' growth form. Furthermore, in populations being controlled by density-dependent factors, growth rates are usually inversely proportional to population density. For example, if the population density is high, the growth rate is low. Conversely, if the density is low, the growth rate is high.

Characteristics of ecosystems having populations controlled by density-dependent factors

Because of the nature of these biological factors, such as disease and competition, this type of regulation will usually occur in.

1. Ecosystems where the communities have many species, i.e. where many biological interactions are taking place.
2. Ecosystems not usually stressed periodically by physical factors, i.e. ecosystems that are usually more stable.

Density-independent Control

A density-independent factor is one where the effect of the factor on the size of the population is independent of and does not depend upon the original density or size of the population. The effect of weather is an example of a density-independent factor. A severe storm and flood coming through an area can just as easily wipe out a large population as a small one. Another example would be a harmful pollutant put into the environment, e.g. a stream. The probability of that harmful substance at some concentration killing an individual would not change depending on the size of the population. For example, populations of small mammals are often regulated more by this type of regulation.

Characteristic of the factor

In general, density-independent factors are physical factors, such as weather factors (e.g. severe winter) or the presence of harmful chemicals.

Characteristic of populations controlled primarily by density-independent factors

Many populations controlled by density-independent factors have the second growth form, the 'resource-limited' type. There is much less biological control and the control is a more haphazard, physical control. The population size often goes over the carrying capacity before some other physical factor decreases the population size. Unlike the case for density-dependent factors, in populations being controlled by density-independent factors, growth rates do not seem to show any trend at all relative to population density.

Characteristics of ecosystems having populations controlled by density-independent factors

This type of regulation will usually occur in.

1. Ecosystems where the communities have few species, i.e. where fewer biological interactions are taking place,
2. Ecosystems are usually stressed periodically by physical factors (such as periodic flooding through a flood plain).

Note: Population regulation factors, then, can be classified as above into two types, but what usually happens in nature is that a population is actually controlled by a combination of density-dependent and density-independent factors. Some populations will be primarily controlled by one type and other populations will be controlled primarily by the other type. Note that both types of factors are external forces on the population.

SECTION IV

Plant Ecology

Chapter 12

Nature and Structure of Plant Community

INTRODUCTION

Community ecology is the study of the interactions between populations of co-existing species. In ecology, a community is an assemblage of two or more populations of different species occupying the same geographical area. The term community has a variety of uses. In its simplest form it refers to groups of organisms in a specific place and/or time, for example, 'the fish community of Lake Ontario before industrialisation'.

A plant community is a recognisable and complex assemblage of plant species which interact with each other as well as with the elements of their environment and is distinct from adjacent assemblages. A plant community is not a static entity: rather it may vary in appearance and species composition from location to location and also over time. What makes each of these communities distinguishable to us is its general physiognomy or physical structure. This overall appearance is created by the particular species present, as well as their size, abundance, and distribution relative to one another. Dominant species, those whose presence most influences the community environment and composition, are often the largest or the most abundant and may be a single species or several codominant species. Dominance may also be sociologic, expressed in the form of allelopathogens, chemical compounds manufactured by some plants that inhibit the growth and development of other species and/or seedlings of the same species within a certain distance. Community structure and distribution are dictated by the delicate balance of environmental factors: soils, climate, topography, geography, fire, time, and humans and other living beings.

Plant communities may occur as relatively obvious ribbons across a landscape, such as the lush green path of a river through the desert. More often, however, adjacent communities interdigitate, their boundaries less distinguishable. Ecotones are areas where adjacent plant communities overlap with, transition, and grade into one another and have a unique set of characteristics which are defined spatially, temporally, and by diverse interactions among the adjacent communities. Earlier literature referred to these conspicuous components of the landscape as 'edge' or 'margin' habitats, and numerous current studies focus on the 'edge effect' and 'landscape boundaries'. Ecotones vary in size and species composition, containing elements of each of the bordering communities. They may be predictable and recognisable entities whose physiognomy and even species composition may be similar from one geographical area to another. In other instances, the boundary between communities may be a band of barren soil caused by animal browsing or by phytotoxins (plant-produced allelopathogens), for example, between the grassland and the black sage (*Salvia mellifera*)-dominated coastal scrub on several hillsides of Poly Canyon. Where chaparral and coastal scrub adjoin in the Canyon, the ecotones vary in proportion, probably most directly as a result of varying soils and topography. Another example is on steep slopes above Brizzolara Creek. Where coastal live oak woodland and riparian communities meet, ecotones can be quite broad. This makes it very difficult to tell exactly where one community ends and the other begins. Regardless of their scale, 'ecotones have important characteristics, and play an integral part in the behaviour of the landscape as a whole'.

Communities vary over time. Fires, floods, grazing, and plowing are some disturbances which quickly change a community. As the vegetation returns to a disturbed area, it may represent the same species that were there before or other species may invade the area. This change in communities is called succession. Natural ecological succession tends to proceed at a relatively gradual pace, sometimes taking hundreds of years. During this time, communities evolve from one to another, through a series of seral (temporary, non-climax) stages. Eventually, in the absence of further disturbance, a climax community develops, one which is at equilibrium with the environment. In Poly Canyon, there are areas whose probable climax community is chaparral, but which now are dominated by black sage (*Salvia mellifera*). Black sage is primarily a component of coastal scrub, but it invades areas previously dominated by chaparral which have burned or have otherwise been disturbed. Just as individual plants and animals are named and sorted into groups of similar types, plant communities are also classified. There are many ways of looking at communities. One can focus on the characteristics of the habitat, such as alpine meadow, where the habitat remains the same although the species composition and physiognomy or community structure may vary from site to site. Another focus is the physiognomy of a community, for example chaparral: dense vegetation of woody shrubs with evergreen hardened leaves and relatively little understory or leaf litter. The species composition varies from locality to locality, but the overall appearance of the community remains the same. A third focus of community classification is the species composition. For example, the sole dominant species of a coastal live oak woodland is the coast live oak (*Quercus agrifolia*). Numerous plant community classification systems have been formulated for California, each based on distinct experiences and with differing goals. Four of these most commonly used locally include: Holland and Keil, Barbour and Major, Natural Diversity Data Base (NDDB). Natural Communities Program, and California Native Plant Society. The first two are more didactic and provide a basic understanding of the California flora for the academic community and general public. The second two share the objective of recognition of all natural plant communities and ecosystems and their inclusion in the state data base for the purpose of legally pursuing their protection. With this more holistic approach, the protection of whole ecosytems would reduce the number of state and federal listings while increasing the number of species protected. The plant communities (per Holland and Keil) found in Poly Canyon are chaparral, coastal scrub, coastal live oak woodland, riparian, grassland, serpentinite, rock outcrop, and anthropogenic (i.e. man-caused, disturbed or ruderal areas such as roadsides and pastures). Some characteristics of plant communities are shown in Table 12.1.

PLANT COMMUNITY CONCEPTS

Vegetation and Plant Communities

1. Attributes of vegetation:
    (a) Physiognomy.
    (b) Floristic composition.
    (c) Structure (spatial arrangement and sizes of plants and plant parts) as characterised by abundances, density, heights, stem diameters, cover, biomass, spatial pattern, etc.

2. The debate over community concepts: What is the nature of a plant community? Are there general models of spatial patterns of biotic communities? In the debate over 'plant community concepts' there are two fundamental questions:
    (a) Can discrete (separate) units of vegetation be clearly defined? Or, is there continuous, gradual variation over space in the attributes of vegetation? This is the question of 'spatial variation'.
    (b) How integrated is a plant community? Does a plant community function as a coordinated, harmonious whole? This is the question of interdependence or functional integrity.

Table 12.1. Characteristics of plant communities.

Some characteristic of plant communities	
Species composition	Change over time
Typical species	Adaptation to disturbance
Relative importance (cover, density)	Succession
Accidental and ubiquitous species	Stability
	Response to climatic change
Species patterns	Creation of, and control over a microenvironment
Spatial	Shading
Niche breadth and overlap	
Species diversity	Nutrient cycling
Richness	Nutrient demand
Evenness	Storage capacity
Diversity within stands and between stands	Rate of nutrient return to the soil
	Nutrient retention efficiency of the nutrient cycles
Structure (physiognomy)	Productivity
Life forms (herbaceous, shrub, tree)	Biomass
Cover	Annual net productivity
Phenology (blooming, setting seed, dormat)	

Traditional Concepts of Plant Communities

Organismal view — Frederick Clements

1. High degree of integration in a plant community.
2. A plant community is an organic entity (e.g. a complex organism).

Implications of the organismal view:

1. Discrete, repeatable vegetation units can be recognised and classified.
2. Succession constitutes the series of life history stages of the complex organism.

Quasi-organismal view — Alex Tansley

A plant community differs from a true organism in the following ways:

1. Lack of clear delimitation.
2. Lack of genetic unity.
3. Same community type may have different 'life histories' (i.e. successional pathways).
4. Lack of coordinated reproduction.
5. Lack of structural integrity.

However, Tansley saw a strong enough similarity between an organism and a community so that an analogy is valid.

Individualistic concept of the plant association — Henry Gleason

Based on three simple premises:

1. Dispersal of propagules occur at different rates and therefore different sites arrive at different times at a bare site.
2. The site (operational environment) acts as a filter so that only certain species can survive at a particular site.
3. Sites vary over space and time.

Therefore, species composition at any particular site will be unique because of chance dispersal and the independent distribution of each species.

Implications

1. Vegetation varies gradually in space and discrete boundaries between different vegetation units are rare.
2. Due to the uniqueness of associations, vegetation cannot be perfectly classified.

Supporting evidence for validity of Gleason's concept:

1. Quantitative studies of floristic composition and vegetation gradients.
2. Independent migration of species as shown by paleoecological studies.

Continuum/Gradient concept

1. Continuum concept—J.T. Curtis and R.P. McIntosh, 1951.
2. Gradient concept—R.H. Whittaker, 1951.

This viewpoint stresses the idea that plant species populations are independently distributed along environmental gradients (e.g. elevation, moisture) so that plant communities change gradually in their species composition. This contrasts with the organismal idea of a plant community in which discrete (clearly demarcated) plant communities are postulated. The continuum and gradient concepts emphasise the continuous nature of spatial variation in species composition.

RIPARIAN PLANT COMMUNITY STRUCTURE

This section describes the equipment, sampling protocols, sampling frequency, and site selection considerations for monitoring of the riparian zone plant community. Parameters include species abundance, composition, per cent cover, stem density, and basal area. These parameters describe the riparian plant community and identify changes in the riparian zone that may occur over time.

Minimum Equipment

Riparian vegetation monitoring requires relatively limited equipment but should be conducted by persons trained in botany, field plant identification, and the use of systematic keys for plant identification. Field equipment should include, at a minimum:

1. Laminated, scaled maps (e.g. NWI, soil survey maps) or aerial photographs, protected in clear, resealable plastic bag or folder, depicting the stream/river reach, riparian study area, and bordering uplands
2. Field notebook, pencils, waterproof permanent markers, and clipboard.
3. Data sheets.

4. Tape measure (100 ft or 300 ft open-reel fibreglass tape).
5. Meter stick for measuring plant heights.
6. Rebar or wooden survey stakes to serve as permanent monuments/markers to identify transect endpoints and plot locations.
7. Diameter at breast height (DBH) tape to measure trees.
8. Camera and photo monitoring data sheets.
9. Resealable plastic bags for plant specimens.
10. Hand lens for keying-out plants.
11. Plant identification keys.

Monitoring Design

A plant community is an association of plant species in a given place. Community structure is inclusive of all plants that occur in the tree, sapling and shrub, and ground cover (vine/liana and herbs) vegetative layers. The composition and per cent areal cover of plants, as well as their general condition with respect to both native and non-native species, describes riparian plant communities. Collecting and analysing plant community data following well-recognised methods, such as the step-by-step protocol listed below, provides the basis for documenting these communities for purposes of their protection, conservation, and/or restoration.

Sampling Protocol

The goal of this protocol is to characterise riparian vegetation by sampling permanent vegetation monitoring stations along permanent transects. Transects should be established within each of the three reaches: (i) the reference reach upstream of the barrier's influence, (ii) the impoundment or reach affected by the barrier, and (iii) the reach downstream of the barrier. Along each transect, sampling stations are selected to characterise the vegetation within general cover types such as floating emergent and submerged aquatic vegetation; emergent wetlands; scrub/shrub wetlands; and forested wetlands. The instructions below describe where to locate the transects and sampling stations and how to sample the vegetation at each station.

1. Identify vegetation cover types present at the barrier removal monitoring site: Using aerial photos and field visits, identify cover types (tree, shrub, vine, liana, and herbaceous layers), wetland versus adjacent upland community types, and their condition.
2. Establish the transects: Establish a minimum of three transects in the reference reach, three transects in the impoundment reach, and two transects in the downstream reach. Transects should adequately represent the plant cover types identified in step 1. Additional transects should be established if time and resources allow, particularly for extended riparian zone/stream reaches that may be affected by the barrier removal.
3. Install rebar/wooden survey stakes at each transect's start and end point: End points should be located upgradient of the wetland-upland boundary or outside the area that is expected to change with barrier removal. Label/number each transect stake. Transects may also be referenced to monumented cross-sections or offset a known distance from the cross sections. Transects should not be co-located with the monumented cross sections because surveying will trample vegetation.
4. Mark and label the transect start and end points: Mark them on a scaled site map or aerial photo. Record the GPS coordinates of these locations.
5. Establish at least one sampling station within each distinct vegetation cover type along each transect: Sampling stations should be chosen to characterise each of the cover types identified

in step 1. Ideally, sampling stations will be established via a systematic random approach, where the vegetation units are first identified in step 1, and station locations are then randomly selected along the transect within the identified cover types (e.g. herbaceous plants sampled every 5 metres, shrubs every 15 metres, and trees every 30 metres). Supplemental, post-restoration sampling stations may need to be established to accommodate changing cover types, particularly where deepwater impoundment drawdown results in a vegetation community.

6. Mark each station with a stake or other permanent monument: Use GPS to determine station location, and record the distance along the transect from the starting point to the station. Document whether a station stake is the center point or a corner point for each plot, particularly if a station differs from the layout used for the remainder of the monitoring area stations.

7. At each station, estimate species cover: Estimate cover within all of the layers that are present: herbaceous; sapling and shrub; and tree. Herbaceous vegetation is sampled using a 1 m^2 (10.8 ft^2) quadrat. The sapling/shrub layer is sampled within a 5 m (approx. 16 ft) radius of the sampling station. Trees are sampled within a 9 m (approx. 30 ft) radius of the station.

The herbaceous layer includes all non-woody, emergent species of all heights (including bryophytes) and woody-stemmed plants <3 ft (approx. 1 m) in height. The monitoring quadrat should be 1 m^2 (10.8 ft^2) and can be defined using an increment-calibrated, 1-meterby-1-meter frame made of PVC-pipe or a similar method. Estimate cover of the vertical plant shoots' aerial projections lying only inside the plot as a percentage of the plot area. Total cover in a plot may exceed 100 per cent, as plant projections often overlap one another. When the project area has high stem density of herbaceous plants but relatively low species (< 5) diversity, a 0.5 m^2 (5.4 ft^2) quadrat may be used. When monitoring prior to barrier removal with plots located inside or near the edge of the impoundment, identify floating or submerged plants that are present. Identify each species, and record each species percent cover within the plot. Also estimate and record per cent of both barren ground and dead plant cover. If time allows, also count the number of stems in a 0.25 m^2 (2.7 ft^2) or 0.5 m^2 (5.4 ft^2) quadrat. The shrub and sapling layer includes all woody stemmed plants that are more than 3 ft (approx. 1 m) but no taller than 20 ft (approx. 6 m) tall and that have a diameter at breast height (DBH) between 0.4 inch (1 cm) and 5.0 inches (approx. 13 cm). DBH is measured at 4.5 ft (approx. 1.5 m) above ground level. For the shrub and sapling layer, monitor within a 5 m (approx. 16 ft) radius of the sampling station point. Identify the species of each plant, and record species per cent cover within plot. Note the number of dead standing shrubs. If time allows, randomly sub-sample the plot by counting woody stems in a 1 m^2 (10.8 ft^2) quadrat.

The tree layer includes all woody plants that are taller than 20 ft (approx. 6 m) and have a DBH greater than 5 inches (approx. 13 cm). Monitor within a 9 m (approx. 30 ft) radius of the sampling station point. Identify the species of each plant, and use a DBH measuring tape to obtain individual DBHs, which will be used later to calculate basal area [$A = \pi (d)^2/4$, where, $\pi = 3.14$ and $d =$ DBH] of each species within each plot. Also note the number of dead standing trees within the sample area.

When estimating per cent cover, values should be recorded as whole integers that can be categorised according to a standardised, commonly used Braun–Blanquet cover class scale (Table 12.2). These cover class categories can be used to expedite field sampling; the mid-point values are used in place of the actual corresponding field estimate values to minimise the variability of results that can arise when multiple people estimate cover. Once field assessments are completed, cover-abundance scores can be used to calculate plant species cover for assessment sites. Refer to the Analysis and Calculations section below for database management and calculations used for generating results.

Table 12.2. Braun-Blanquet cover class scale and mean values to estimate cover class.

Category	Per cent cover	Mid-point
T	<1	None
1	1–5	3
2	6–15	10.5
3	16–25	20.5
4	26–50	38
5	51–75	63
6	76–95	85.5
7	96–100	98

Sampling Frequency

It is optimal to monitor vegetation during the peak of the vascular plant growing season. For the northeastern United States, this period is generally between July 15 and August 31. Some riparian plant species flower in spring or early summer, so the monitoring team may want to consider a site assessment during spring, if time allows. Monitoring is conducted at least once annually for each monitoring year, and all stations should be monitored during each monitoring period. The project and reference sites should be monitored within the same time period and as close in time to one another as possible. Vegetation monitoring should include a minimum of one year of pre-restoration and three years of post-restoration assessment, and preferably over a longer period (such as once every 3 to 5 years) for post-restoration assessment. This is particularly important for reforesting sites and if a goal of the restoration is to document ecological succession of the riparian zone.

Baseline versus post-removal monitoring

Ideally, monitoring plots are monitored at least once prior to removal of the stream barrier to define a baseline condition. When funding and time allow, it may be beneficial to monitor two or more years prior to a barrier removal because this better accounts for environmental variability. Some removals result in very little change in riparian shoreline locations, whereas others can result in substantial change. If changes in shoreline vegetation are expected with impoundment draw-down, then baseline vegetation transects should include areas of impoundment habitat where vegetation and substrate conditions are documented.

Analysis and Calculations

After field assessments are completed using standardised field data sheets and handwritten data are checked for clarity and legibility, vegetation data should be entered into an excel spreadsheet where it can be manipulated for statistical analysis. In the spreadsheet, columns should represent species, and rows should represent the sampling plots. First create columns for all species found at the project site(s). Then enter per cent cover data. Alternatively, the data can be entered using the Braun-Blanquet cover scale, which uses a ranking system that facilitates similarity testing and ordination procedures. Tree plot basal areas are then totaled to derive per cent cover of tree species within the plot. Non-parametric tests can be used to evaluate differences in vegetation communities between sites (e.g. project restoration reach versus reference reach) or site conditions between sampling years.

Chapter 13

Plant Succession

INTRODUCTION

Succession is a directional non-seasonal cumulative change in the types of plant species that occupy a given area through time. It involves the processes of colonisation, establishment, and extinction which act on the participating plant species. Most successions contain a number of stages that can be recognised by the collection of species that dominate at that point in the succession. Succession begin when an area is made partially or completely devoid of vegetation because of a disturbance. Some common mechanisms of disturbance are fires, wind storms, volcanic eruptions, logging, climate change, severe flooding, disease, and pest infestation. Succession stops when species composition changes no longer occur with time, and this community is said to be a climax community.

The concept of a climax community assumes that the plants colonising and establishing themselves in a given region can achieve stable equilibrium. The idea that succession ends in the development of a climax community has had a long history in the fields of biogeography and ecology. One of the earliest proponents of this idea was Frederic Clements who studied succession at the beginning of the 20th century. However, beginning in the 1920s scientists began refuting the notion of a climax state. By 1950, many scientists began viewing succession as a phenomenon that rarely attains equilibrium. The reason why equilibrium is not reached is related to the nature of disturbance. Disturbance acts on communities at a variety of spatial and temporal scales. Further, the effect of disturbance is not always 100 per cent. Many disturbances remove only a part of the previous plant community. As a result of these new ideas, plant communities are now generally seen as being composed of numerous patches of various size at different stages of successional development.

TYPES OF SUCCESSION

Various types of plant succession are discussed below:
1. Primary succession: Primary succession is the establishment of plants on land that has not been previously vegetated-Mount Saint Helens. Begins with colonisation and establishment of pioneer species.
2. Secondary succession: Secondary succession is the invasion of a habitat by plants on land that was previously vegetated. Removal of past vegetation may be caused by natural or human disturbances such as fire, logging, cultivation or hurricanes.
3. Allogenic succession: Allogenic succession is caused by a change in environmental conditions which in turn influences the composition of the plant community. In Cornwall England,

observations on the estuary of the Fal river suggest that the deposition of silt may be causing an allogenic succession from salt marsh to woodland. Measurements indicate sedimentation rates of about 1 cm per year on the mud flats that are found 15 km (9 miles) into the estuary. Over the last 100 years, this salt marsh has increased its elevation and has extended itself seaward by 800 meters (2600 feet). The adjacent woodland has followed the salt marsh by invading its landward limit.

4. Autogenic succession: Autogenic succession is a succession where both the plant community and environment change, and this change is caused by the activities of the plants over time. Mt. St. Helens after the last volcanic eruption.

5. Progressive succession: Progressive succession is a succession where the community becomes complex and contains more species and biomass over time.

6. Retrogressive succession: Retrogressive succession is a succession where the community becomes simplistic and contains fewer species and less biomass over time. Some retrogressive successions are allogenic in nature. For example, the introduction of grazing animals result in degenerated rangeland. Table 13.1 describes some of the plant, community, and ecosystem attributes that change with succession.

Table 13.1. Comparison of plant, community, and ecosystem characteristics between early and late stages of succession.

Attribute	Early stages of succession	Late stages of succession
Plant biomass	Small	Large
Plant longevity	Short	Long
Seed dispersal characteristics of dominant plants	Well dispersed	Poorly dispersed
Plant morphology and physiology	Simple	Complex
Photosynthetic efficiency of dominant plants at low light	Low	High
Rate of soil nutrient resource consumption by plants	Fast	Slow
Plant recovery rate from resource limitation	Fast	Slow
Plant leaf canopy structure	Multilayered	Monolayer
Site of nutrient storage	Litter and soil	Living biomass and litter
Role of decomposers in cycling nutrients to plants	Minor	Great
Biogeochemical cycling	Open and rapid	Closed and slow
Rate of net primary productivity	High	Low
Community site characteristics	Extreme	Moderate (mesic)
Importance of macroenvironment on plant success	Great	Moderate
Ecosystem stability	Low	High
Plant species diversity	Low	High
Life-history type	r	K
Seed longevity	Long	Short

SUCCESSION MECHANISMS

An overview of the mechanisms of succession has been produced by Connell and Slatyer. Connell and Slatyer propose three models, of which the first (facilitation) is the classical explanation most often invoked in the past, while the other two (tolerance and inhibition) may be equally important but have frequently been overlooked.

The essential feature of facilitation succession, in contrast with either the tolerance or inhibition models, is that changes in the abiotic environment are imposed by the developing plant community. Thus, the entry and growth of the later species depends on earlier species preparing the ground.

The tolerance model suggests that a predictable sequence is produced because different species have different strategies for exploiting resources. Later species are able to tolerate lower resource levels due to competition and can grow to maturity in the presence of early species, eventually out competing them.

The inhibition model applies when all species resist invasions of competitors. Later species gradually accumulate by replacing early individuals when they die. An important distinction between models is the cause of death of the early colonists. In the case of facilitation and tolerance, they are killed in competition for resources, notably light and nutrients. In the case of the inhibition model, however, the early species are killed by very local disturbances caused by extreme physical conditions or the action of predators.

PROCESS OF SUCCESSION

Alluvial deposit: Layer of broken rocky matter or sediment, formed from material that has been carried in suspension by a river or stream and dropped as the velocity of the current decreases. River plains and deltas are made entirely of alluvial deposits, but smaller pockets can be found in the beds of upland torrents. Alluvial deposits can consist of a whole range of particle sizes, from boulders down through cobbles, pebbles, gravel, sand, silt, and clay. The raw materials are the rocks and soils of upland areas that are loosened by erosion and washed away by mountain streams. Much of the world's richest farmland lies on alluvial deposits. These deposits can also provide an economic source of minerals. River currents produce a sorting action, with particles of heavy material deposited first while lighter materials are washed downstream. Hence heavy minerals such as gold and tin, present in the original rocks in small amounts, can be concentrated and deposited on stream beds in commercial quantities. Such deposits are called 'placer ores'.

Estuarine deposits: An estuary is a wide opening at the mouth of a river into which the sea has penetrated by the depression of the land. In such bodies of water the tide often scours with much force. Estuaries abound along our Atlantic coast, Delaware and Chesapeake Bays and the mouth of the Hudson being excellent examples of such. The water in them is brackish, and unfavourable to abundant aquatic life, for only a limited number of marine animals, and fewer fresh-water ones, flourish in brackish water. Estuarine deposits are, in general, much like those of the sea, except that they are apt to be of a finer grain for a given depth of water; muds are abundantly laid down, especially in the more sheltered nooks and bays, with fine and coarse sands and gravels in the more exposed situations. The sands are apt to show a confused stratification from the conflicting currents and eddies in which they are deposited, but with horizontal layers formed at slack water. Extensive mud-flats often surround an estuary, especially if the rise and fall of the tide be great.

Dune: In physical geography, a dune is a hill of sand built by aeolian processes. Dunes occur in different forms and sizes, formed by interaction with the wind. Most kinds of dunes are longer on the windward side where the sand is pushed up the dune and have a shorter 'slip face' in the lee of the wind. The valley or trough between dunes is called a slack. A 'dune field' is an area covered by extensive sand dunes. Large dune fields are known as ergs.

Dune habitats provide niches for highly specialised plants and animals, including numerous rare species and some endangered species. Due to widespread human population expansion, dunes face destruction through land development and recreational usages, as well as alteration to prevent the encroachment of sand onto inhabited areas. Some countries, notably the United States, Australia, Canada, New Zealand, the United Kingdom, and Netherlands, have developed significant programs of dune protection through the use of sand dune stabilisation. In the UK a Biodiversity Action Plan has been developed to assess dunes loss and to prevent future dunes destruction.

Landslips: Landslips large portions of land which from some cause have become detached from their original position, and slid down to a lower level. They are especially common in volcanic districts, where the trembling of the earth that frequently accompanies the eruption of a volcano is sufficient to split off large portions of mountains, which slide down to the plains below. Water is another great agent in producing landslips. It operates in various ways. The most common method is when water insinuates itself into minute cracks, which are widened and deepened by its freezing in winter. When the fissure becomes sufficiently deep, on the melting of the ice, a landslip is produced.

Scree: Scree, also called talus, is a term given to an accumulation of broken rock fragments at the base of crags, mountain cliffs or valley shoulders. Landforms associated with these materials are sometimes called scree slopes or talus piles. These deposits typically have a concave upwards form, while the maximum inclination of such deposits corresponds to the angle of repose of the mean debris size.

The terms scree and talus are often used interchangeably, though scree commonly refers to smaller material like mixed gravel and loose dirt, talus to rocks larger than scree. Talus is usually the preferred term in scientific writing.

Formation of scree or talus deposits results from physical and chemical weathering and erosional processes acting on a rock face. The predominant processes that degrade a rock slope depend largely on the regional climate (temperature, amount of rainfall, etc.). Examples include:

1. Mechanical weathering by ice.
2. Chemical weathering by mineral hydration and salt deposition.
3. Thermal stresses.
4. Topographic stresses.
5. Biotic processes.

Erosion: Erosion is the process of weathering and transport of solids (sediment, soil, rock and other particles) in the natural environment or their source and deposits them elsewhere. It usually occurs due to transport by wind, water or ice; by down-slope creep of soil and other material under the force of gravity; or by living organisms, such as burrowing animals, in the case of bioerosion.

A certain amount of erosion is natural and, in fact, healthy for the ecosystem. For example, gravels continuously move downstream in watercourses. Excessive erosion, however, causes serious problems, such as receiving water sedimentation, ecosystem damage and outright loss of soil.

Erosion is distinguished from weathering, which is the process of chemical or physical breakdown of the minerals in the rocks, although the two processes may occur concurrently.

Pacific ring of fire: The pacific ring of fire (or sometimes just the ring of fire) is an area where large numbers of earthquakes and volcanic eruptions occur in the basin of the Pacific Ocean. In a 40,000 km (25,000 mi) horseshoe shape, it is associated with a nearly continuous series of oceanic trenches, volcanic arcs, and volcanic belts and/or plate movements.

Migration: All plants possess the effective organs of reproduction which are shed and dispersed. These may be the spores, of various forms in lower non-vascular or vascular plants or seeds and other organs of vegetative reproduction in seed bearing plants. The fruits as such in indehiscent types or seeds in dehiscent types are adapted to the effective dispersal. The agencies which bring about dispersal are wind, water, animals including man.

Ecesis: It is the process by which a plant or animal becomes established in a new habitat.

Competition: This phenomenon involves struggle for existence between two or more individuals growing in an area, that make successive demands, that are similar in nature, on the soil. This leads to a struggle. Such as struggle is usually between trees or shrubs or herbs, but never between a tree and a lichen or a tree and a herb.

Reaction: It is not only the environment which affects plants. Plants also produce their effects on the environment which is called reaction. The first affects are on the substratum. This leads to the formation of soil from rocks, addition of materials to soil. Plants change the structure and texture of soil in course of time by addition of humus. The soil water and air composition is also affected.

Climax community: An ecological community in which populations of plants or animals remain stable and exist in balance with each other and their environment. A climax community is the final stage of succession, remaining relatively unchanged until destroyed by an event such as fire or human interference. Climax community is discussed in detail in the end of this chapter.

CHRONOSEQUENCE AND TOPOSEQUENCE

Chronosequence

A chronosequence (in forest sciences) is a set of forested sites that share similar attributes but are of different ages. Since many processes in forest ecology take a long time (decades or centuries) to develop, chronosequence methods are used to represent and study the time-dependent development of a forest. Field data from a chronosequence can be collected in a short period of several months. For example, chronosequences are often used to study the changes in plant communities during succession. Importance of chronosequence is summarised below:

1. Chronosequences and associated space-for-time substitutions are an important and often necessary tool for studying temporal dynamics of plant communities and soil development across multiple time-scales. However, they are often used inappropriately, leading to false conclusions about ecological patterns and processes, which has prompted recent strong criticism of the approach. Here, we evaluate when chronosequences may or may not be appropriate for studying community and ecosystem development.

2. Chronosequences are appropriate to study plant succession at decadal to millennial time-scales when there is evidence that sites of different ages are following the same trajectory. They can also be reliably used to study aspects of soil development that occur between temporally linked sites over time-scales of centuries to millennia, sometimes independently of their application to shorter-term plant and soil biological communities.

3. Some characteristics of changing plant and soil biological communities (e.g. species richness, plant cover, vegetation structure, soil organic matter accumulation) are more likely to be related in a predictable and temporally linear manner than are other characteristics (e.g. species composition and abundance) and are therefore more reliably studied using a chronosequence approach.

4. Chronosequences are most appropriate for studying communities that are following convergent successional trajectories and have low biodiversity, rapid species turnover and low frequency and severity of disturbance. Chronosequences are least suitable for studying successional

trajectories that are divergent, species-rich, highly disturbed or arrested in time because then there are often major difficulties in determining temporal linkages between stages.

5. Synthesis: We conclude that, when successional trajectories exceed the life span of investigators and the experimental and observational studies that they perform, temporal change can be successfully explored through the judicious use of chronosequences.

Toposequence

Adjacent soils that show differing profile characteristics reflecting the influence of local topography are called toposequences. As a general rule, soil profiles on the convex upper slopes in a toposequence are more shallow and have less distinct subsurface horizons than soils at the summit or on lower, concave-upward slopes.

Toposequence variations in soil properties are characterised and related to variations in populations of total isolatable bacteria and arthrobacters. Increases in soil NO_3-N, available phosphorous, NO_3-N-producing power, *Arthrobacter* counts, and the percentage of the total counts represented by arthrobacters are correlated with decreases in soil acidity. The total bacterial counts are not correlated with soil acidity but are associated with percentage of soil organic matter and percentage of clay. The percentage of the total counts represented by arthrobacters is lowest at the summit position and increased downslope to the highest value in the toeslope position.

CYCLIC SUCCESSION

Cyclic succession is a pattern of vegetation change in which in a small number of species tend to replace each other over time in the absence of large-scale disturbance. Observations of cyclic replacement have provided evidence against traditional Clementsian views of an end-state climax community with stable species compositions. Cyclic succession is one of several kinds of ecological succession, a concept in community ecology (Fig. 13.1).

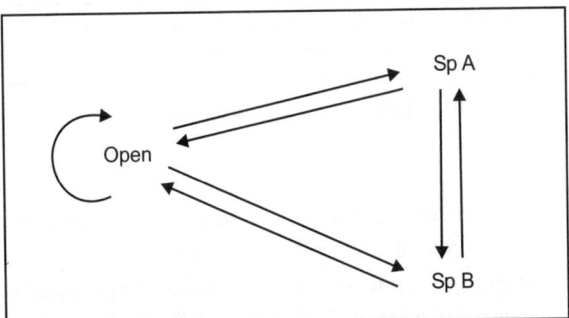

Fig. 13.1. Graphic model of cyclic succession.

When used narrowly, 'cyclic succession' refers to processes not initiated by wholesale exogenous disturbances or long-term physical changes in the environment. However, broader cyclic processes can also be observed in cases of secondary succession in which regular disturbances such as insect outbreaks can 'reset' an entire community to a previous stage.

These examples differ from the classic cases of cyclic succession discussed below in that entire species groups are exchanged, as opposed to one species for another. On geologic time scales, climate cycles can result in cyclic vegetation changes by directly altering the physical environment.

The cyclic model of succession was proposed in 1947 by British ecologist Alexander Watt. In a seminal paper on vegetation patterns in grass, heath, and bog communities, Watt describes the plant community is a regenerating entity consisting of a 'space-time mosaic' of species, whose cyclic behaviour can be characterised by patch dynamics. Based on the current composition and its corresponding stage of succession, he explains, a community can either be in an 'upgrade' phase toward late-successional shrubs or 'downgrade' degenerate phase toward grasses. These phases would occur in a predictable cycle. Watt's study has since become a classic example frequently cited in scientific ecology.

Modelling Cyclic Succession

The cyclic model of succession can be displayed in terms of a transition matrix. Based on the Markov chain, the matrix describes the likelihood of future states based on the milieu of present states. The three states in the simplest cyclic model are open substrate (usually a bare patch of land), Species A dominance, and Species B dominance. With respect to facilitation, inhibition, and tolerance models of succession, the key feature of the cyclic model is that A and B are not autosuccessional—that is, they do not facilitate their own growth. Rather, A will either facilitate the succession of B or be eliminated (through mortality) such that the patch occupied becomes open substrate. Likewise, B will either facilitate the succession of A or be eliminated. Open substrate can remain open or become occupied by either A or B. This configuration results in a cyclic scheme of species dominance (Fig.13.2).

Future state	Present state		
	Open	Sp A	Sp B
Open	+	+	+
Sp A	+	0	+
Sp B	+	+	0

Where + indicates a transition from present to future
0 indicates no transition

Fig. 13.2. Cyclic succession matrix.

Mechanisms

Cyclic succession is a descriptive phenomenon that can be accounted for in several ways. In Watt's bog system, he suggested that factors endogenous to the plant species were at play. He writes, 'Each patch in this space-time mosaic is dependent on its neighbours and develops under conditions partly imposed by them.' In other words, species life history characteristics fluctuate cyclically under the influence of surrounding species. These periodic shifts in life history properties produce observable changes in community composition. In the system Watt observed, phasic development was specifically responsible for changes in growth and mortality rate.

As a result of changes in survival and growth ability, the balance of species dominance shifts, thus marking discrete stages. If the milieu of interspecific relationships satisfies the conditions described in the model above, a cyclic pattern of succession is observed. Exogenous factors, such as depredation by herbivores, can also be indirect drivers for cyclic succession if they differentially modulate plant life history properties over time. Density-dependent root gnawing by rodents is proposed as one such mechanism in the Larrea-Opuntia system. Watt noted that cyclic fluctuations in mortality rate could also be produced through differential response to seasonal conditions like frost.

It is important to note that patterns cyclic succession cannot be readily linked to any single species, as Watt's *Calluna* bushes have been observed in non-cyclic systems. Rather, it is the aggregate composition of species that gives rise to the cyclic process.

Additional empirical evidence

Another salient example of cyclic replacement occurs in a two-species plant community in the Sonoran Desert. Even though water availability is limiting such that only one species would be predicted to survive, *Larrea tridentata* and *Opuntia leptocaulis* are observed to replace each other in the absence of environmental disturbance.

XEROSERE

Xerosere is a plant succession which is limited by water availability. It includes the different stages in a *xerarch succession*. Xerarch succession of ecological communities originated in extremely dry situation such as sand deserts, sand dunes, salt deserts, rock deserts, etc. A xerosere may include lithoseres (on rock) and psammoseres.

Stages of Xerosere

Bare rocks

Bare rocks are produced when glaciers recede or volcanoes erupt. Erosion of these rocks is brought by rain water and wind loaded with soil particles. The rain water combines with atmospheric carbon dioxide that corrodes the surface of the rocks and produce crevices. Water enters these crevices, freezes and expands to separate boulders. These boulders move down under the influence of gravity and wear particles from the rocks. Also when the wind loaded with soil particles strikes against the rocks, it removes soil particles. All these processes lead to formation of a little soil at the surface of these bare rocks. Animals such as spiders which can hide between boulders or stones invade these rocks. These animals live by feeding on insects which have been blown in or flown in. Algal and fungal spores reach these rocks by air from the surrounding areas. These spores grow and form symbiotic association, the lichen, which act as pioneer species of bare rocks. The process of succession starts when autotrophic organisms start living in the rocks.

Crustose lichen stage

A bare rock consists of solid surface or very large boulders and there is no place for rooting plants to colonise. The thalli of crustose lichens can adhere to the surface of rock and absorb moisture from atmosphere; therefore, these colonise the bare surfaces of rocks first. The propagules of these lichens are brought by air from the surrounding areas. These lichens produce acids which corrode the rock and their thalli collect wind blown soil particles among them that help in formation of a thin film of soil. When these lichens die their thalli are decomposed to add humus. This promotes soil building and the environment becomes suitable for growth of foliose and fruticose type of lichens.

Foliose and fruticose lichen stage

Foliose lichens have leaf-like thalli, while the fruticose lichens are like small bushes. They are attached to the substratum at one point only, therefore, do not cover the soil completely. They can absorb and retain more water and are able to accumulate more dust particles. Their dead remains are decomposed to humus which mixes with soil particles and help building substratum and improving soil moisture contents further. The shallow depressions in the rocks and crevices become filled with soil and topsoil layer increases further. These autogenic changes favour growth and establishment of mosses.

Moss stage

The spores of xerophytic mosses, such as Polytrichum, Tortula, and Grimmia, are brought to the rock where they succeed lichens. Their rhizoids penetrate soil among the crevices, secrete acids and corrode the rocks. The bodies of mosses are rich in organic and inorganic compounds. When these die they add these compounds to the soil, increasing the fertility of the soil. As mosses develop in patches they catch soil particles from the air and help increase the amount of substratum. The changing environment leads to migration of lichens and helps invasion of herbaceous vegetation that can out-compete mosses.

Herb stage

Herbaceous weeds, mostly annuals such as asters, evening primroses, and milk weeds, invade the rock. Their roots penetrate deep down, secrete acids and enhance the process of weathering. Leaf litter and death of herbs add humus to the soil. Shading of soil results in decrease in evaporation and there is a slight increase in temperature. As a result the xeric conditions begin to change and biennial and perennial herbs and xeric grasses such as Aristida, Festuca, and Poa, begin to inhabit. These climatic conditions favour growth of bacterial and fungal populations, resulting in increase in decomposition activity.

Shrub stage

The herb and grass mixture is invaded by shrub species, such as Rhus and Phytocarpus. Shrub consists of densely packed bushes with growth stunted by want of water and high transpiration rate. Early invasion of shrub is slow, but once a few bushes have become established, birds invade the area and help disperse scrub seeds. This results in dense scrub growth shading the soil and making conditions unfavourable for the growth of herbs, which then begin to migrate. The soil formation continues and its moisture content increases. The environment becomes mesic (moderately moist).

Tree stage

Change in environment favours colonisation of tree species. The tree saplings begin to grow among the scrubs and establish themselves. The kind of tree species inhabiting the area depends upon the nature of the soil. In poorly drained soils oaks establish themselves. The trees form canopy and shade the area. Shade-loving scrubs continue to grow as secondary vegetation. Leaf litter and decaying roots weather the soil further and add humus to it making the habitat more favourable for growth to trees. Mosses and ferns make their appearance and fungi population grows abundantly.

Climax stage

The succession culminates in a climax community, the forest. Many intermediate tree stages develop prior to establishment of a climax community. The forest type depends upon climatic conditions. The climax forest may be:

Oak-Hickory Climax Forest: In dry habitat oaks and hickories are climax vegetation. There is only one tree stage and forests are characterised by presence of scrubs, herbs, ferns, and mosses.

Beech-Hemlock Climax Forest: These climax forests develop in mesic climates. The dominant vegetation is Beech and Hemlock. There are many intermediate tree stages. The other vegetation types include herbs, ferns, and mosses.

American Beech-Sugar Maple Climax Forest: These climax forests develop in mesic climates in the Northeastern United States. The dominant vegetation is American Beech and Sugar Maple.

Spruce-Alpine Fir Climax Forest: At high altitudes in Rocky Mountains the climax forest is dominated by spruces and alpine firs.

PSAMMOSERE

A psammosere is a seral community, an ecological succession that began life on newly exposed coastal sand. Most common psammoseres are sand dune systems.

In a psammosere, the organisms closest to the sea will be pioneer species: salt-tolerant species such as littoral algae and glasswort with marram grass stabilising the dunes. Progressing inland many characteristic features change and help determine the natural succession of the dunes. For instance, the drainage slows down as the land becomes more compact and has better soils, and the pH drops as the proportion of seashell fragments reduces and the amount of humus increases. Sea purslane, sea lavender, meadow grass and heather eventually grade into a typical non-maritime terrestrial eco-system. The first trees (or pioneer trees) that appear are typically fast-growing trees such as birch, willow or rowan. In turn these will be replaced by slow-growing, larger trees such as ash and oak. This is the climax community, defined as the point where a plant succession does not develop any further because it has reached equilibrium with the environment, in particular the climate.

In an idealised coastal psammosere model, at the seaward edge of the sand dune the pH of the soil is typically alkaline/neutral with a pH of 7.0/8.0 particularly where shell fragments provide a significant component of the sand. Tracking inland across the dunes a podsol develops with a pH of 5.0/4.0 followed by mature podsols at the climax with a pH of 3.5–4.5.

FUNGAL SUCCESSION ON DUNG

Although many people shun dung as a subject fit for study, it is a marvelous resource that is constantly being produced and deposited in large quantities at convenient locations. Even though it has passed through an animal's digestive tract, dung retains many nutrients. Thus, it attracts its own fauna and flora consisting of bacteria, fungi, protozoa, platyhelminths, nematodes, annelids, and arthropods. Coprophilous (dung-loving) fungi are uniquely adapted to herbivore dung. They are deposited with dung, and they grow and reproduce there. They disperse their spores from the heap to a location from which they will be consumed by a herbivore, pass through its gut, and again be deposited with the dung heap.

Dung as a Home

Dung consists of the macerated and undigested remains of plant food plus vast quantities of bacteria (mostly dead) as well as animal waste products, such as broken-down red blood cells and bile pigments. The nature of herbivore dung depends on the efficiency of the digestive tract, which, in turn, depends on the animal's digestive anatomy and its microflora. Ruminants produce fine-textured dung of fibrous plant material whereas horses, with a less efficient system, produce much coarser dung. Dung decomposes rapidly because the macerated material has a high nitrogen content, available aeration, and a high water content that is protected from fluctuations.

Although there is little available protein in dung, many other undigested food components are present. Dung is rich in water-soluble vitamins, growth factors, and mineral ions, some of which are metabolic by-products of the microbes in a herbivore's gut. For example, coprogen, an organo-iron compound found in dung, is necessary for the growth and reproduction of the fungus Pilobolus crystallinus. Dung also contains a large amount of readily available carbohydrates.

Fungal Dung Inhabitants

Considering the great variation in the feeding habits, habitats, and digestive systems of herbivores, it is surprising how universal coprophilous fungi are. All classes of the Kingdom Fungi are found on dung, with the Zygomycetes usually appearing first, followed by the Ascomycetes, and finally the Basidiomycetes. Their distribution is influenced locally by the number of herbivores in an area. Some species are restricted to a particular herbivore; for example, *Lasiobolus cainii* is found only on porcupine dung. However, many coprophilous fungi grow indiscriminately on any herbivore dung. The greatest variety of fungi have been reported on cow, rabbit, and horse dung, but this could be because the majority of research has focused on these animals.

Adaptations of Fungi to Dung

Coprophilous fungi are highly specialised for growth on dung, and some never occur elsewhere. While some dung fungi show few modifications peculiar to their habitat, most do have some unique features. Many exhibit some very specialised structures to ensure survival in their unique habitat.

Herbivores do not graze near their own dung; therefore, the spores must be propelled beyond this 'zone of repugnance'. Thus, the spores or spore masses are relatively large and heavy. In the Zygomycete Pilobolus, for instance, the entire sporangium is discharged as a unit. In the bird's nest fungus, *Cyathus stercoreus*, the peridioles (the 'eggs') containing many spores are violently discharged when a raindrop hits the peridium (the 'nest'). The spores/masses, because of their weight, do not remain in the air long, but follow a parabolic trajectory landing on nearby grass without the aid of air currents. The sporangium and sporangiophore of Pilobolus measure about 0.5–1.0 cm, yet the sporangium has been propelled as much as 1.8 m vertically and 2.1 m horizontally.

Some coprophilous fungi exhibit a phototropic response that determines the direction the spore mass will be projected and ensures that the spores clear the substrate. Spore discharge is always during the day. The entire sporangium of Pilobolus, for example, grows toward the light source. To demonstrate this phototropic response, place the mature culture of Pilobolus inside a container with a hole punched into one side of the container top. Wrap the container inside aluminum foil and punch another hole in the foil aligned over the container hole. Place transparent tape over the hole, and set the container in a window. The following day, remove the foil and container. The majority of the ejected sporangia will be found stuck to the tape or around the light source in the container top.

The spores are dark to protect them from ultraviolet light until they are consumed by a herbivoure. In some fungi, melanin is present in the spore walls; in others, a dark membrane covers the spore mass. *Coprinus comatus*, a mushroom that fruits on dung, exhibits these dark spores.

The spores/mass are often mucilagenous so that they stick to vegetation upon impact, and the mucilage, when dry, cements firmly. In *Cyathus*, the peridiole has a sticky piece of hypha, the funiculus, which attaches to vegetation upon impact and wraps the peridiole firmly around it. Other coprophilous fungi, such as *Mucor hiemalis*, form a sticky droplet around their spores. When an insect visits the dung, the spores stick to the insect's body. If the insect rests again on other vegetation or another dung heap, the spores rub off and adhere to the new environment. Many of the coprophilous fungal spores will not germinate until after passing through an herbivore's digestive tract: They must be heated inside the gut, digested by the gut enzymes and/or bacteria or stimulated by the higher pH of dung.

Fungal Dung Succession

Some observers have noted a true ecological succession on dung in that first the Zygomycetes, then the Ascomycetes, and finally the Basidiomycetes appear. Early researchers suggested that this succession was a nutritional one. They postulated that the Zygomycetes or sugar fungi, appear first, because their spores germinate quickly and their mycelium grows rapidly, exploiting the fresh substrate; that is, they utilise the simple sugars and hemicelluloses present in dung. The Zygomycetes are not capable of utilising the cellulose and lignin found in dung. When the simpler carbon sources are metabolised, the Zygomycetes disappear and are replaced by the Ascomycetes, which can utilise cellulose. These are then replaced by the Basidiomycetes, which can utilise both the cellulose and the lignin.

Although this nutritional hypothesis of dung succession was an attractive one that seemed to fit the observations, it ignored some important ecological and physiological facts. For example, some spores had already germinated or had their dormancy broken in the herbivore gut. Also, no scientist had yet considered the interference/competition/enhancement effects of the various dung inhabitants on each other.

The nutritional hypothesis was based solely on the order of appearance of fruiting structures. Mycelial development and growth may not have been proceeding at the same rate, so a second hypothesis, the reproduction hypothesis of dung succession, was developed. This hypothesis was based on the time it takes each kind of fungus to fruit.

The simple sporangia of the Zygomycetes could develop much more quickly and require much less energy than the more complex Ascomycete fruiting body. The even larger Basidiomycete fruitification would require the largest amount of energy expenditure and would, therefore, take the longest amount of time to appear. This hypothesis still neglects the interrelationships between fungi themselves and between fungi and other dung inhabitants.

The dung heap is not inhabited exclusively by fungi. As mentioned previously, vast populations of bacteria, protozoa, platyhelminths, nematodes, annelids, and arthropods coexist with the fungi. These compete for resources, and some parasitise or consume the fungi while others provide substrates for them. Also, these organisms can deplete the dung of nutritionally necessary compounds, such as nitrogenous ones, and can produce waste products that enhance or retard the growth and reproduction of some fungi. For instance, ammonia, a waste product of the bacterial degradation of proteins, stimulates sporangial production of Pilobolus, which grows better in the presence of other microbes.

Coprophilous bacteria enhance the growth of some of the coprophilous fungi, but retard the growth of others. Also, the number of fly larvae, Lycoriella mali that survive to pupate increases as fungal competition increases. It could be that as fungal growth is inhibited, more resources are available to the fly larvae or that larvae may be favored by the enzymes produced by fungi to inhibit other fungi!

There is evidence that some fungi excrete products that inhibit fungal competitors. A few weeks after dung is deposited, the Zygomycetes and Ascomycetes disappear, but the Basidiomycetes continue fruiting for months.

Obviously, the substrate has not been depleted. Some Basidiomycetes suppress some of the Zygomycetes and Ascomycetes by hyphal interference. Within minutes of contact, the sensitive species' hyphae undergo vacuolisation and lose turgor. This is followed by a drastic alteration of cell membrane permeability and subsequent death of the hyphae.

In the classroom, dung succession is easy to observe and is sure to promote the interest of students. Fresh or weathered dung can be placed in empty containers. If the dung is dry, add a small amount of distilled water. Plastic wrap secured with a rubber band will allow observation and prevent evaporation of the necessary moisture. The dung culture should be observed over a period of three to four weeks to study fungal succession. The success of coprophilous fungi, as measured by their ability to produce and maintain a fruiting body, is not simply a matter of competing against other fungi. Instead, it is a complex web of interactions, some combative, some inhibitive, some mutually or exclusively beneficial between and among the bacteria, fungi, protozoans, platyhelminths, nematodes, annelids, and arthropods. It's a jungle in there!

CLIMAX COMMUNITY

In ecology, a climax community or climatic climax community, is a biological community of plants and animals which, through the process of ecological succession—the development of vegetation in an area over time—has reached a steady state. This equilibrium occurs because the climax community is composed of species best adapted to average conditions in that area. The term is sometimes also applied in soil development.

Frederic Clements's Use of Climax

Clements described the successional development of an ecological communities as comparable to the ontogenetic development of individual organisms. Clements suggested only comparisons to very simple organisms. Later ecologists developed this idea that the ecological community is a 'superorganism' and even sometimes claimed that communities could be homologous to complex organisms.

Clements's theory sought to define a single climax-type for each area. Arthur Tansley developed this idea with the 'polyclimax'—multiple steady-state end-points, determined by edaphic factors, in a given climatic zone. Clements had called these end-points other tems, not climaxes, and had thought they were not stable, because by definition climax vegetation is best-adapted to the climate of a given area. Henry Gleason's early challenges to Clements's organism simile, and other of his strategies for describing vegetation, were largely disregarded for several decades until substantially vindicated by research in the 1950s and 1960s. Meanwhile, climax theory was deeply incorporated in both theoretical ecology and in vegetation management. Clements's terms such as pre-climax, post-climax, plagioclimax and disclimax continued to be used to describe the many communities which persist in states that diverge from the climax ideal for a particular area.

Though the views are sometimes attributed to him, Clements never argued that climax communities must always occur or that the dominant cause of vegetation is climate or that the different species in an ecological community are tightly integrated physiologically or that plant communities have sharp boundaries in time or space. Rather, he employed the idea of a climax community—of the form of vegetation best adapted to some idealised set of environmental conditions—as a conceptual starting point for describing the vegetation in a given area.

There are good reasons to believe that the species best adapted to some conditions might appear there, when those conditions occur. But much of Clements's work was devoted to characterising what happens when those ideal conditions do not occur. In those circumstances, vegetation other than the ideal climax will often occur instead. But those different kinds of vegetation can still be described as deviations from the climax ideal. Therefore, Clements developed a very large vocabulary of theoretical terms describing the various possible causes of vegetation, and various non-climax states vegetation adopts as a consequence. His method of dealing with ecological complexity was to define an ideal form of vegetation—the climax community and describe other forms of vegetation as deviations from that ideal.

Rejection of Climax Theory

Support among ecologists for the climax theory declined, because they found the theory with its many coined terms difficult to apply, because they were dissatisfied how it compared to observed individual organisms, and because better theories developed.

Although Clements recognised that vegetation follows gradients rather than being tightly bound, his rhetorical comparisons of ecological communities to organisms fostered the impression that communities, including the climax, have distinct edges in space and time. Yet Robert Whittaker's research demonstrated plant species distribute themselves along nutrient and other environmental gradients. Many ecologists saw this as a major reason to stop using the climax concept.

More recent palynological studies showed that modern species assemblages are ephemeral; vegetation in eastern North America since the last glacial maximum has consisted of several different species assemblages, many of which have no analogues in modern 'climax' communities. That would mean, at least, that the climax types for those areas could not be stable to the degree clements believed they were.

Ultimately, even if succession tends towards a steady state, the time required to achieve this state is unrealistically long; in most cases, external disturbances and environmental change occur so frequently that the realisation of a climax community is unlikely, and therefore it has come to be regarded as a less useful concept. Long-term vegetation dynamics are now more often characterised as resulting from the action of stochastic factors.

Continuing Usage of 'Climax'

Despite the overall abandonment of climax theory, during the 1990s use of climax concepts again became more popular among some theoretical ecologists. Many authors and nature-enthusiasts continue to use the term 'climax' in a diluted form to refer to what might otherwise be called mature or old-growth communities. The term 'climax' has also been adopted as description for a late successional stage for marine macroinvertebrate communities.

A heterotroph is a creature that must ingest biomass to obtain its energy and nutrition. In direct contrast, autotrophs are capable of assimilating diffuse, inorganic energy and materials and using these to synthesise biochemicals. Green plants, for example, use sunlight and simple inorganic molecules to photosynthesise organic matter. All heterotrophs have an absolute dependence on the biological products of autotrophs for their sustenance — they have no other source of nourishment.

All animals are heterotrophs, as are most micro-organisms (the major exceptions being microscopic algae and blue-green bacteria). Heterotrophs can be classified according to the sorts of biomass that they eat. Animals that eat living plants are known as herbivores, while those that eat other animals are known as carnivores.

Many animals eat both plants and animals, and these are known as omnivores. Animal parasites are a special type of carnivore that are usually much smaller than their prey, and do not usually kill the animals that they feed upon. Heterotrophic micro-organisms mostly feed upon dead plants and animals, and are known as decomposers. Some animals also specialise on feeding on dead organic matter, and are known as scavengers or detritivores. Even a few vascular plants are heterotrophic, parasitising the roots of other plants and thereby obtaining their own nourishment. These plants, which often lack chlorophyll, are known as saprophytes.

Humans, of course, are heterotrophs. This means that humans can only sustain themselves by eating plants or by eating animals that have themselves grown by eating plants. All of these foods must be specifically grown for human consumption in agricultural ecosystems or be gathered from natural ecosystems. If humans and their societies are to be sustained over the long term, it can only be through the wise use of the species and ecosystems that sustain them. This is a fact, and a consequence of the inextricable links of humans with other species and with the products and services of ecosystems. The intimate dependency of humans on other creatures is a biological and ecological relationship that can be difficult for modern people to remember as they purchase their food in stores, and do not directly participate in its growth, harvesting, and processing.

Chapter 14

Biotic Factors of Ecology

INTRODUCTION

Biotic components are the living things that shape an ecosystem. A biotic factor is any living component that affects another organism, including animals that consume the organism in question, and the living food that the organism consumes. Biotic factors include human influence. Biotic components are contrasted to abiotic components, which are non-living components of an organism's environment, such as temperature, light, moisture, air currents, etc. Biotic components usually include:

1. Producers, i.e. autotrophs, e.g. plants; they convert the energy (from the sun or other sources such as hydrothermal vents) into food.
2. Consumers, i.e. heterotrophs, e.g. animals; they depend upon producers for food.
3. Decomposers, i.e. detritivores, e.g. fungi and bacteria; they break down chemicals from producers and consumers into simpler form which can be reused.

Odum classified the biotic interactions into two broad catogories:

1. Positive interactions: One interacting species helps the other either one way or on reciprocal terms, e.g. commensalism (one benefitted, other not harmed), cooperation (on benefitted, other not harmed), cooperation (favourable in reciprocal but not obligatory) and mutualism (favourable in reciprocal and obligatory).
2. Negative interactions: One interacting species is harmed. The other may be harmed, benefitted or neutral, e.g. competition (for a common requirement), predation (on eating the other), antibiosis (inhibiting development of the other by chemicals).

MUTUALISM

Mutualism is the way two organisms biologically interact where each individual derives a fitness benefit (i.e. increased reproductive output). Similar interactions within a species are known as co-operation. It can be contrasted with interspecific competition, in which each species experiences reduced fitness, and exploitation or parasitism, in which one species benefits at the expense of the other. Mutualism and symbiosis are sometimes used as if they are synonymous, but this is strictly incorrect: symbiosis is a broad category, defined to include relationships which are mutualistic, parasitic or commensal. Mutualism is only one type.

A well known example of mutualism is the relationship between ungulates (such as cows) and bacteria within their intestines. The ungulates benefit from the cellulase produced by the bacteria, which facilitates digestion; the bacteria benefit from having a stable supply of nutrients in the host environment.

Mutualism plays a key part in ecology. For example, mutualistic interactions are vital for terrestrial ecosystem function as more than 48 per cent of land plants rely on mycorrhizal relationships with fungi to provide them with inorganic compounds and trace elements. In addition, mutualism is thought to have driven the evolution of much of the biological diversity we see, such as flower forms (important for pollination mutualisms) and co-evolution between groups of species. However mutualism has historically received less attention than other interactions such as predation and parasitism.

Measuring the exact fitness benefit to the individuals is not always straightforward, particularly when the individuals can receive benefits from a range of species, for example most plant-pollinator mutualisms. It is therefore common to categorise mutualisms according to the closeness of the association, using terms such as obligate versus facultative. Defining 'closeness', however, is also problematic. It can refer to mutual dependency (the species cannot live without one another) or the biological intimacy of the relationship in relation to physical closeness (e.g. one species living within the tissues of the other species).

Types of Relationships

Mutualistic transversals can be thought of as a form of 'biological barter' in which species trade resources (for example carbohydrates or inorganic compounds) or services such as gamete, offspring dispersal or protection from predators.

Resource-resource relationships

Resource-resource interactions, in which one type of resource is traded for a different resource, are probably the most common form of mutualism; for example mycorrhizal associations between plant roots and fungi, with the plant providing carbohydrates to the fungus in return for primarily phosphate but also nitrogenous compounds. Other examples include rhizobia bacteria which fix nitrogen for leguminous plants (family Fabaceae) in return for energy-containing carbohydrates.

Service-resource relationships

Service-resource relationships are also common. Pollination in which nectar or pollen (food resources) are traded for pollen dispersal (a service) or ant protection of aphids, where the aphids trade sugar-rich honeydew (a by-product of their mode of feeding on plant sap) in return for defense against predators such as ladybird beetles. Phagophiles feed (resource) on ectoparasites, thereby providing anti-pest service. Zoochory is an example where animals disperse the seeds of plants. This is similar to pollination in that the plant produces food resources (for example, fleshy fruit, overabundance of seeds) for animals that disperse the seeds (service).

Service-service relationships

Strict service-service interactions are very rare, for reasons that are far from clear. One example is the relationship between sea anemones and anemonefish in the family Pomacentridae: the anemones provide the fish with protection from predators (which cannot tolerate the stings of the anemone's tentacles) and the fish defend the anemones against butterflyfish (family Chaetodontidae) which eat anemones.

Humans and Mutualism

Humans also engage in mutualisms with other species, including their gut flora (without which they would not be able to digest food efficiently) and domesticated animals such as horses, which provide transportation in return for food and shelter. In traditional agriculture, many plants will function mutualistically as companion plants, providing each other with shelter, soil fertility and the repelling of pests. For example, beans may grow up cornstalks as a trellis, while fixing nitrogen in the soil for the corn, as exploited in the Three Sisters gardening technique.

SYNTROPHY

Syntrophy, cross-feeding or cross feeding is the phenomenon that one species lives off the products of another species. For example house dust mites live off human skin flakes, of which a healthy human being produces about 1 gram per day. These mites can also produce chemicals that stimulate the production of skin flakes, and people can become allergic to these compounds.

Another example are the many organisms that feast on faeces or dung. A cow eats a lot of grass, the cellulose of which is transformed into lipids by micro-organisms in the cow's large intestine. These micro-organisms cannot use the lipids because of lack of dioxygen in the intestine, so the cow does not take up all lipids produced. When the processed grass leaves the intestine as dung and comes into open air, many organisms, such as the dung beetle, feast on it.

Yet another example is the community of micro-organisms in soil that live off leaf litter. Leaves typically last one year and are then replaced by new ones. These micro-organisms mineralise the discarded leaves and release nutrients that are taken up by the plant. Such relationships are called reciprocal syntrophy because the plant lives off the products of micro-organisms. Many symbiotic relationships are based on syntrophy. Finally, anaerobic fermentation/methanogenesis is an example of a syntrophic relationship between different groups of micro-organism. Although fermentative bacteria are not strictly dependent on syntrophyic relationships, they still gain profit from the activities of the hydrogen-scavenging organisms, as the fermentative bacteria gain maximum energy yield when protons are used as electron acceptor with concurrent H_2 production. Also, acetogenic bacteria and methanogenic archea are the two groups of micro-organisms living in syntrophy during the methanogenesis. Some fermentation products such as fatty acids longer than two carbon atoms, alcohols longer than one carbon atom, and branched-chain and aromatic fatty acids, cannot directly be used in methanogenesis. In acetogenesis process, these products are oxidised to acetate and H_2 by obligated proton reducing bacteria in syntrophic relationship with methanogenic archaea as low H_2 partial pressure is essential for acetogenic reactions to be thermodynamically favourable ($\Delta G < 0$). Syntrophic interactions are very important in all living communities, and are important to the Dynamic Energy Budget theory.

Mycorrhiza

A mycorrhiza is a symbiotic (generally mutualistic, but occasionally weakly pathogenic) association between a fungus and the roots of a vascular plant. In a mycorrhizal association, the fungus colonises the host plant's roots, either intracellularly as in arbuscular mycorrhizal fungi (AMF) or extracellularly as in ectomycorrhizal fungi. They are an important component of soil life and soil chemistry.

The mechanisms of increased absorption are both physical and chemical. Mycorrhizal mycelia are much smaller in diameter than the smallest root, and thus can explore a greater volume of soil, providing a larger surface area for absorption. Also, the cell membrane chemistry of fungi is different from that of plants (including organic acid excretion which aids in ion displacement). Mycorrhizas are especially beneficial for the plant partner in nutrient-poor soils.

Types of Mycorrhiza

Mycorrhizas are commonly divided into ectomycorrhizas and endomycorrhizas. The two types are differentiated by the fact that the hyphae of ectomycorrhizal fungi do not penetrate individual cells within the root, while the hyphae of endomycorrhizal fungi penetrate the cell wall and invaginate the cell membrane. Additionally, many plants in the order Ericales form a third type, ericoid mycorrhizas, while some members of the Ericales form arbutoid and monotropoid mycorrhizas.

Endomycorrhiza

Endomycorrhizas are variable and have been further classified as arbuscular, ericoid, arbutoid, monotropoid, and orchid mycorrhizas. Arbuscular mycorrhizas or AM (formerly known as vesicular-arbuscular mycorrhizas or VAM), are mycorrhizas whose hyphae enter into the plant cells, producing structures that are either balloon-like (vesicles) or dichotomously-branching invaginations (arbuscules). The fungal hyphae do not in fact penetrate the protoplast (i.e. the interior of the cell), but invaginate the cell membrane. The structure of the arbuscules greatly increases the contact surface area between the hypha and the cell cytoplasm to facilitate the transfer of nutrients between them.

Ectomycorrhiza

Ectomycorrhizas or EcM, are typically formed between the roots of around 10 per cent of plant families, mostly woody plants including the birch, dipterocarp, eucalyptus, oak, pine, and rose families and fungi belonging to the Basidiomycota, Ascomycota, and Zygomycota. Some EcM fungi, such as many *Leccinum* and *Suillus*, are symbiotic with only one particular genus of plant, while other fungi, such as the *Amanita*, are generalists that form mycorrhizas with many different plants.

Ectomycorrhizas consist of a hyphal sheath or mantle, covering the root tip and a hartig net of hyphae surrounding the plant cells within the root cortex. In some cases the hyphae may also penetrate the plant cells, in which case the mycorrhiza is called an ectendomycorrhiza. Outside the root, the fungal mycelium forms an extensive network within the soil and leaf litter.

Rhizobium and legume symbiosis

Symbiosis between rhizobia and leguminous plants leads to the formation of N_2-fixing root nodules. The interaction of rhizobia and plants shows a high degree of host specificity based on the exchange of chemical signals between the symbiotic partners. The plant signals, flavonoids exuded by the roots, activate the expression of nodulation genes, resulting in the production of the rhizobial lipochitooligosaccharide signals (Nod factors). Nod factors act as morphogens that, under conditions of nitrogen limitation, induce cells within the root cortex to divide and to develop into nodule primordia.

Root Nodule

Root nodules occur on the roots of plants (primarily Fabaceae) that associate with symbiotic nitrogen-fixing bacteria. Under nitrogen-limiting conditions, capable plants form a symbiotic relationship with a host-specific strain of bacteria known as rhizobia. This process has evolved multiple times within the Fabaceae, as well as in other species found within the Rosid clade.

Within legume nodules, nitrogen gas from the atmosphere is converted into ammonia, which is then assimilated into amino acids (the building blocks of proteins), nucleotides (the building blocks of DNA and RNA as well as the important energy molecule ATP), and other cellular constituents such as vitamins, flavones, and hormones. Their ability to fix gaseous nitrogen makes legumes an ideal agricultural organism as their requirement for nitrogen fertilizer is reduced. Indeed high nitrogen content blocks nodule development as there is no benefit for the plant of forming the symbiosis. The energy for splitting the nitrogen gas in the nodule comes from sugar that is translocated from the leaf (a product of photosynthesis). Malate as a breakdown product of sucrose is the direct carbon source for the bacteroid. Nitrogen fixation in the nodule is very oxygen sensitive. Legume nodules harbor an iron containing protein called leghaemoglobin, closely related to animal myoglobin, to facilitate the conversion of nitrogen gas to ammonia.

Two main types of nodule have been described: determinate and indeterminate.

1. Determinate nodules are found on tropical (sub) legumes, such as those of the genera *Glycine* (soyabean), *Phaseolus* (common bean), *Lotus*, and *Vigna*. Determinate nodules lose meristematic activity shortly after initiation, thus growth is due to cell expansion resulting in mature nodules which are spherical in shape.
2. Indeterminate nodules are found on temperate legumes like *Pisum* (pea), *Medicago* (alfalfa), *Trifolium* (clover), and *Vicia* (vetch). They earned the moniker 'indeterminate' because they maintain an active apical meristem that produces new cells for growth over the life of the nodule. This results in the nodule having a generally cylindrical shape.

Nodulation

Legumes release compounds called flavonoids from their roots, which trigger the production of nod factors by the bacteria. When the nod factor is sensed by the root, a number of biochemical and morphological changes happen: cell division is triggered in the root to create the nodule, and the root hair growth is redirected to wind around the bacteria multiple times until it fully encapsulates one or more bacteria. The bacteria encapsulated divide multiple times, forming a microcolony. From this microcolony, the bacteria enter the developing nodule through a structure called an infection thread, which grows through the root hair into the basal part of the epidermis cell, and onwards into the root cortex; they are then surrounded by a plant-derived membrane and differentiate into bacteroids that fix nitrogen. Nodulation is controlled by a variety of processes, both external (heat, acidic soils, drought, nitrate) and internal (autoregulation of nodulation, ethylene). Autoregulation of nodulation controls nodule numbers per plant through a systemic process involving the leaf. Leaf tissue senses the early nodulation events in the root through an unknown chemical signal, then restricts further nodule development in newly developing root tissue.

Connection to root structure

Root nodules apparently have evolved three times within the Fabaceae but are rare outside that family. The propensity of these plants to develop root nodules seems to relate to their root structure. In particular, a tendency to develop lateral roots in response to abscisic acid may enable the later evolution of root nodules.

In other species

Root nodules that occur on non-legume genera like Parasponia in association with Rhizobium bacteria, and those that arise from symbiotic interactions with Actinobacteria *Frankia* in some plant genera such as Alnus, vary significantly from those formed in the legume-rhizobia symbiosis. In these symbioses the bacteria are never released from the infection thread.

COMMENSALISM

In ecology, commensalism is a class of relationship between two organisms where one organism benefits but the other is neutral (there is no harm or benefit). There are three other types of association: mutualism (where both organisms benefit), competition (where both organisms are harmed) and parasitism (one organism benefits and the other one is harmed).

Commensalism is harder to demonstrate than parasitism and mutualism, for it is easier to show a single instance whereby the host is affected, than it is to prove or disprove that possibility. Often, a detailed investigation will show that the host indeed has become affected by the relationship.

Protocooperation

Protocooperation is where two species interact with each other beneficially; they have no need to interact with each other they interact purely for the gain that they receive from doing this. It is not at all necessary for protocooperation to occur; growth and survival is possible in the absence of the interaction. The interaction that occurs can be between different kingdoms.

Protocooperation is a form of mutualism, but they do not depend on each other for survival. An example of protocooperation happens between soil bacteria or fungi, and the plants that occur growing in the soil. None of the species rely on the relationship for survival, but all of the fungi, bacteria and higher plants take part in shaping soil composition and fertility. Soil bacteria and fungi interrelate with each other, forming nutrients essential to the plants survival. The plants obtain nutrients from root nodules and decomposing organic substance. Plants benefit by getting essential mineral nutrients and carbon dioxide.

Examples

Ants and aphids: A further example of protocooperation is the connection between ants and aphids. The ant searches for food on trees and shrubs that are hosts to honeydew-secreting species such as aphids, mealybugs, and some scales. The ant gathers the sugary substance and takes it to its nest as food for its offspring. It has been known for the ant to stimulate the aphid to secrete honeydew straight into its mouth. Some ant species even look after the honeydew producers from natural predators. In areas where the ant inhabits the same ecosystem as the plant the plants normally suffer from a higher presence of aphids which is detrimental to the plant but not to the two species protocooperating.

Flowers and insects: The flowers of plants that are pollinated by insects and birds benefit from protocooperation. The plants, particularly those with large bright colourful flowers bearing nectar glands, experience cross pollination because of the insects activities. This is beneficial to the insect that has got the food supply of pollen and nectar required for its survival.

Birds: Protocooperation can occur in birds. The Egyptian plover removes insect pests from the backs of buffalo, antelope, giraffes, and rhinos. The cattle egret in America as well does the same task of removing the unwanted insects and parasites.

Fish: Certain fish perform the task of cleaning other fish, by removing ectoparasites, cleaning wounded flesh, and getting rid of dead flesh. Even predatory fish rely on cleansing symbionts, and adopt a placid state while they are cleansed. The fish that do the cleansing are often concentrated around specific sites where the other fish come to be cleansed these are known as cleansing stations.

COMPETITION (ECOLOGY)

In ecology, the interaction between two or more organisms or groups of organisms, that use a common resource in short supply. There can be competition between members of the same species and competition between members of different species. Competition invariably results in a reduction in the numbers of one or both competitors, and in evolution contributes both to the decline of certain species and to the evolution of adaptations.

The resources in short supply for which organisms compete may be obvious things, such as mineral salts for animals and plants or light for plants. However, there are less obvious resources. For example, competition for suitable nesting sites is important in some species of birds. Competition results in a reduction in breeding success for one or other organism(s). Because of this it is one of the most important aspects of natural selection, which may result in evolutionary change if the environment is changing. Competition also results in the distribution of organisms we see in habitats. It is believed that organisms tend to occur where the pressures of competition are not as great as in other areas. In agriculture cultivation methods are designed to reduce competition. For example, a crop of wheat is sown at a density that minimises competition within the same species. The plants are grown far enough apart to reduce competition between the roots of neighbouring wheat plants for soil mineral nutrients. The spraying of the ground to kill weeds reduces competition between the wheat and weed plants. Some weeds would grow taller than the wheat and deprive it of light.

Competition can have both beneficial and detrimental effects. Many evolutionary biologists view inter-species and intra-species competition as the driving force of adaptation, and ultimately of evolution. However, some biologists, most famously Richard Dawkins, prefer to think of evolution in terms of competition between single genes, which have the welfare of the organism 'in mind' only insofar as that welfare furthers their own selfish drives for replication. Some social darwinists claim that competition also serves as a mechanism for determining the best-suited group; politically, economically and ecologically. On the negative side, competition can cause injury and loss to the organisms involved, and drain valuable resources and energy. Human competition can be expensive, as is the case with political elections, international sports competitions, advertising wars and arms races.

AMENSALISM

In amensalism, one microbial population growing on a substrate is inhibitory to the other population. This relationship is based on the production of certain microbicidal chemicals (allelopathic substances) or antibiotics. Amensalism leads to pre-emptive colonisation of a habitat, i.e. once an organism establishes itself within a habitat, it prevents other population from surviving in that habitat. For example, lactic acid bacteria prevent other microbial population from surviving in its substrate by producing large amounts of acids that prove detrimental to the other microbial population. Similarly, *E. coli* cannot survive in the rumen due to the production of volatile fatty acids produced by the already existing anaerobes.

Fatty acids produced by microbes on the skin surface prevent the colonisation of these surfaces by other organisms like yeasts. Similarly, acids produced by micro-organisms in the vaginal tract are responsible for preventing infection by Candida albicans. Some other examples of amensalism are: oxidation of sulphur by Thiobacillus thiooxidans which produces sulphuric acid which lowers aquatic pH thereby inhibiting many other microbes. Production of oxygen by algae may alter the habitat that can prove detrimental to obligate anaerobes.

Zymogenous populations grow under condition of high organic matter which permit the production of antibiotics. For example, *Cephalosporium graminerum* is a wheat pathogen that grows in dead wheat tissue and produces antifungal antibiotics to prevent attack by other fungi.

Similarly *Trichophyton mentagrophytes* (a dermatophytic fungi) in the skin of New Zealand hedgehog produces penicillin (an antibiotic) on the skin and prevents the growth of the penicillin sensitive *Staphylococcus aureus* which is otherwise a common inhabitant of the skin. Bacteriocins are similar to antibiotics but their action is restricted to micro-organisms very closely related to each other. They are peptides coded by plasmids and they are produced by one population to inhibit other microbial populations. For example, *Lactobacillus* species produce NISIN a bacteriocin that preserves food material like dairy products from spoilage by other bacteria.

ALLELOPATHY

Allelopathy is a biological phenomenon by which an organism produces one or more biochemicals that influence the growth, survival, and reproduction of other organisms. These biochemicals are known as allelochemicals and can have beneficial (positive allelopathy) or detrimental (negative allelopathy) effects on the target organisms. Allelochemicals are a subset of secondary metabolites, which are not required for metabolism (i.e. growth, development and reproduction) of the allelopathic organism. Allelochemicals with negative allelopathic effects are an important part of plant defense against herbivory.

Allelopathy is characteristic of certain plants, algae, bacteria, coral, and fungi. Allelopathic interactions are an important factor in determining species distribution and abundance within plant communities, and are also thought to be important in the success of many invasive plants.

The process by which a plant acquires more of the available resources (such as nutrients, water or light) from the environment without any chemical action on the surrounding plants is called resource competition. This process is not negative allelopathy, although both processes can act together to enhance the survival rate of the plant species.

Examples of allelopathy. One of the most studied aspects of allelopathy is the role of allelopathy in agriculture. Current research is focused on the effects of weeds on crops, on weeds, and crops on crops. This research furthers the possibility of using allelochemicals as growth regulators and natural herbicides, to promote sustainable agriculture.

A famous case of purported allelopathy is in desert shrubs. One of the most widely known early examples was *Salvia leucophylla*. Bare zones around the shrubs were hypothesised to be caused by volatile terpenes emitted by the shrubs. However, like many allelopathy studies, it was based on artificial lab experiments and unwarranted extrapolations to natural ecosystems. Allelopathy has been shown to play a crucial role in forests, influencing the composition of the vegetation growth, and also provides an explanation for the patterns of forest regeneration. The black walnut (*Juglans nigra*) produces the allelochemical juglone, which affects some species greatly while others not at all.

PREDATION

In ecology, predation describes a biological interaction where a predator (an organism that is hunting) feeds on its prey (the organism that is attacked). Predators may or may not kill their prey prior to feeding on them, but the act of predation always results in the death of its prey and the eventual absorption of the prey's tissue through consumption. Other categories of consumption are herbivory (eating parts of plants) and detritivory, the consumption of dead organic material (detritus). All these consumption categories fall under the rubric of consumer-resource systems. It can often be difficult to separate our various types of feeding behaviours. For example, parasitic species prey on a host organism and then lay their eggs on it for their offspring to feed on it while it continues to live or on its decaying corpse after it has died. The key characteristic of predation however is the predator's direct impact on the prey population. On the other hand, detritivores simply eat dead organic material arising from the decay of dead individuals and have no direct impact on the 'donor' organism(s). Selective pressures imposed on one another has led to an evolutionary arms race between prey and predator, resulting in various antipredator adaptations.

Functional Classification

Classification of predators by the extent to which they feed on and interact with their prey is one way ecologists may wish to categorise the different types of predation. Instead of focusing on what they eat, this system classifies predators by the way in which they eat, and the general nature of the interaction between predator and prey species. Two factors are considered here: How close the predator and prey are physically (in the latter two cases the term prey may be replaced with host). Additionally, whether or not the prey are directly killed by the predator is considered, with true predation and parasitoidism involving certain death.

True predation

A true predator can commonly be known as one which kills and eats another organism. Whereas other types of predator all harm their prey in some way, this form certainly kills them. Predators may hunt actively for prey or sit and wait for prey to approach within striking distance, as in ambush predators. Some predators kill large prey and dismember or chew it prior to eating it, such as a jaguar; others may eat their (usually much smaller) prey whole, as does a bottlenose dolphin swallowing a fish or a snake or duck or stork swallowing a frog.

Grazing

Grazing organisms may also kill their prey species, but this is seldom the case. While some herbivores like zooplankton live on unicellular phytoplankton and have no choice but to kill their prey, many only eat a small part of the plant. Grazing livestock may pull some grass out at the roots, but most is simply grazed upon, allowing the plant to regrow once again. Kelp is frequently grazed in subtidal kelp forests, but regrows at the base of the blade continuously to cope with browsing pressure. Animals may also be 'grazed' upon; female mosquitos land on hosts briefly to gain sufficient proteins for the development of their offspring. Starfish may be grazed on, being capable of regenerating lost arms.

Parasitism

Parasites can at times be difficult to distinguish from grazers. Their feeding behaviour is similar in many ways, however they are noted for their close association with their host species. While a grazing species such as an elephant may travel many kilometers in a single day, grazing on many plants in the process, parasites form very close associations with their hosts, usually having only one or at most a few in their lifetime. This close living arrangement may be described by the term symbiosis, 'living together', but unlike mutualism the association significantly reduces the fitness of the host. Parasitic organisms range from the macroscopic mistletoe, a parasitic plant, to microscopic internal parasites such as cholera. Some species however have more loose associations with their hosts. Lepidoptera (butterfly and moth) larvae may feed parasitically on only a single plant or they may graze on several nearby plants. It is therefore wise to treat this classification system as a continuum rather than four isolated forms.

Parasitoidism

Parasitoids are organisms living in or on their host and feeding directly upon it, eventually leading to its death. They are much like parasites in their close symbiotic relationship with their host or hosts. Like the previous two classifications parasitoid predators do not kill their hosts instantly. However, unlike parasites, they are very similar to true predators in that the fate of their prey is quite inevitably death. A well known example of a parasitoids are the ichneumon wasps, solitary insects living a free life as an adult, then laying eggs on or in another species such as a caterpillar. Its larva(e) feed on the growing host causing it little harm at first, but soon devouring the internal organs until finally destroying the nervous system resulting in prey death. By this stage the young wasp(s) are developed sufficiently to move to the next stage in their life cycle. Though limited mainly to the insect order Hymenoptera, Diptera and Coleoptera parasitoids make up as much as 10 per cent of all insect species.

Degree of specialisation

Among predators there is a large degree of specialisation. Many predators specialise in hunting only one species of prey. Others are more opportunistic and will kill and eat almost anything (examples: humans, leopards, and dogs).

Ecological Role

Predators may increase the biodiversity of communities by preventing a single species from becoming dominant. Such predators are known as keystone species and may have a profound influence on the balance of organisms in a particular ecosystem. Introduction or removal of this predator or changes in its population density, can have drastic cascading effects on the equilibrium of many other populations in the ecosystem. For example, grazers of a grassland may prevent a single dominant species from taking over.

The elimination of wolves from Yellowstone National Park had profound impacts on the trophic pyramid. Without predation, herbivores began to over-graze many woody browse species, affecting the area's plant populations. Additionally, wolves often kept animals from grazing in riparian areas, which protected beavers from having their food sources encroached upon. The removal of wolves had a direct effect on beaver populations, as their habitat became territory for grazing. Furthermore, predation keeps hydrological features such as creeks and streams in normal working order. Increased browsing on willows lenr and conifers along Blacktail Creek due to a lack of predation resulted in channel incision because those species helped slow the water down and hold the soil in place.

Adaptations and Behaviour

The act of predation can be broken down into a maximum of four stages: Detection of prey, attack, capture and finally consumption. The relationship between predator and prey is one which is typically beneficial to the predator, and detrimental to the prey species. Sometimes, however, predation has indirect benefits to the prey species, though the individuals preyed upon themselves do not benefit. This means that, at each applicable stage, predator and prey species are in an evolutionary arms race to maximise their respective abilities to obtain food or avoid being eaten. This interaction has resulted in a vast array of adaptations in both groups.

CARNIVOROUS PLANT

Carnivorous plants are plants that derive some or most of their nutrients (but not energy) from trapping and consuming animals or protozoans, typically insects and other arthropods. Carnivorous plants appear adapted to grow in places where the soil is thin or poor in nutrients, especially nitrogen, such as acidic bogs and rock outcroppings. Charles Darwin wrote Insectivorous Plants, the first well-known treatise on carnivorous plants, in 1875.

True carnivory is thought to have evolved independently six times in five different orders of flowering plants, and these are now represented by more than a dozen genera. These include about 630 species that attract and trap prey, produce digestive enzymes, and absorb the resulting available nutrients. Additionally, over 300 protocarnivorous plant species in several genera show some but not all these characteristics.

Trapping Mechanisms

Five basic trapping mechanisms are found in carnivorous plants.
1. Pitfall traps (pitcher plants) trap prey in a rolled leaf that contains a pool of digestive enzymes or bacteria.
2. Flypaper traps use a sticky mucilage.
3. Snap traps utilise rapid leaf movements.
4. Bladder traps suck in prey with a bladder that generates an internal vacuum.
5. Lobster-pot traps force prey to move towards a digestive organ with inward-pointing hairs.

These traps may be active or passive, depending on whether movement aids the capture of prey. For example, *Triphyophyllum* is a passive flypaper that secretes mucilage, but whose leaves do not grow or move in response to prey capture. Meanwhile, sundews are active flypaper traps whose leaves undergo rapid acid growth, which is an expansion of individual cells as opposed to cell division. The rapid acid growth allows the sundew tentacles to bend, aiding in the retention and digestion of prey.

Pitfall traps

Pitfall traps are thought to have evolved independently on at least four occasions. The simplest ones are probably those of Heliamphora, the marsh pitcher plant. In this genus, the traps are clearly derived evolutionarily from a simple rolled leaf whose margins have sealed together. These plants live in areas of high rainfall in South America such as Mount Roraima and consequently have a problem ensuring their pitchers do not overflow. To counteract this problem, natural selection has favoured the evolution of an overflow similar to that of a bathroom sink—a small gap in the zipped-up leaf margins allows excess water to flow out of the pitcher.

Flypaper traps

The flypaper trap is based on a sticky mucilage or glue. The leaf of flypaper traps is studded with mucilage-secreting glands, which may be short and nondescript (like those of the butterworts) or long and mobile (like those of many sundews). Flypapers have evolved independently at least five times.

In the genus Pinguicula, the mucilage glands are quite short (sessile), and the leaf, while shiny (giving the genus its common name of butterwort), does not appear carnivorous. However, this belies the fact that the leaf is an extremely effective trap of small flying insects (such as fungus gnats), and its surface responds to prey by relatively rapid growth. This thigmotropic growth may involve rolling of the leaf blade (to prevent rain from splashing the prey off the leaf surface) or dishing of the surface under the prey to form a shallow digestive pit.

Snap traps

The only two active snap traps—the Venus flytrap (*Dionaea muscipula*) and the waterwheel plant (*Aldrovanda vesiculosa*)—are believed to have had a common ancestor with similar adaptations. Their trapping mechanism has also been described as a 'mouse trap', 'bear trap' or 'man trap', based on their shape and rapid movement. However, the term snap trap is preferred as other designations are misleading, particularly with respect to the intended prey. *Aldrovanda* is aquatic and specialised in catching small invertebrates; *Dionaea* is terrestrial and catches a variety of arthropods, including spiders.

Bladder traps

Bladder traps are exclusive to the genus *Utricularia* or bladderworts. The bladders (vesicula) pump ions out of their interiors. Water follows by osmosis, generating a partial vacuum inside the bladder. The bladder has a small opening, sealed by a hinged door. In aquatic species, the door has a pair of long trigger hairs. Aquatic invertebrates such as *Daphnia* touch these hairs and deform the door by lever action, releasing the vacuum. The invertebrate is sucked into the bladder, where it is digested. Many species of *Utricularia* (such as *U. sandersonii*) are terrestrial, growing on waterlogged soil, and their trapping mechanism is triggered in a slightly different manner. Bladderworts lack roots, but terrestrial species have anchoring stems that resemble them. Temperate aquatic bladderworts generally die back to a resting turion during the winter months, and *U. macrorhiza* appears to regulate the number of bladders it bears in response to the prevailing nutrient content of its habitat.

Lobster-pot traps

A lobster-pot trap is a chamber that is easy to enter, and whose exit is either difficult to find or obstructed by inward-pointing bristles. Lobster pots are the trapping mechanism in Genlisea, the corkscrew plants. These plants appear to specialise in aquatic protozoa. A Y-shaped modified leaf allows prey to enter but not exit. Inward-pointing hairs force the prey to move in a particular direction. Prey entering the spiral entrance that coils around the upper two arms of the Y are forced to move inexorably towards a stomach in the lower arm of the Y, where they are digested. Prey movement is also thought to be encouraged by water movement through the trap, produced in a similar way to the vacuum in bladder traps, and probably evolutionarily related to it.

Outside of *Genlisea*, features reminiscent of lobster-pot traps can be seen in *Sarracenia psittacina*, *Darlingtonia californica*, and, some horticulturalists argue, *Nepenthes aristolochioides*.

Borderline carnivores

To be a fully fledged carnivore, a plant must attract, kill, and digest prey; and it must benefit from absorbing the products of the digestion (mostly amino acids and ammonium ions). To many horticulturalists, these distinctions are a matter of taste. There is a spectrum of carnivory found in plants: from completely non-carnivorous plants like cabbages, to borderline carnivores, to unspecialised and simple traps, like *Heliamphora*, to extremely specialised and complex traps, like that of the Venus flytrap.

Evolution

The evolution of carnivorous plants is obscured by the paucity of their fossil record. Very few fossils have been found, and then usually only as seed or pollen. Carnivorous plants are generally herbs, and their traps primary growth. They generally do not form readily fossilisable structures such as thick bark or wood. The traps themselves would probably not be preserved in any case.

Pitfall traps are quite clearly derived from rolled leaves. The vascular tissues of *Sarracenia* is a case in point. The keel along the front of the trap contains a mixture of leftward and rightward-facing vascular bundles, as would be predicted from the fusion of the edges of an adaxial (stem-facing) leaf surface. Flypapers also show a simple evolutionary gradient from sticky, non-carnivorous leaves, through passive flypapers to active forms.

The pitfall traps may have evolved simply by selection pressure for the production of more deeply cupped leaves, followed by 'zipping up' of the margins and subsequent loss of most of the hairs, except at the bottom, where they help retain prey.

The lobster-pot traps of *Genlisea* are difficult to interpret. They may have developed from bifurcated pitchers that later specialised on ground-dwelling prey or, perhaps, the prey-guiding protrusions of bladder traps became more substantial than the net-like funnel found in most aquatic bladderworts. Whatever their origin, the helical shape of the lobster pot is an adaptation that displays as much trapping surface as possible in all directions when buried in moss.

Ecology and Modelling of Carnivory

Carnivorous plants are widespread but rather rare. They are almost entirely restricted to habitats such as bogs, where soil nutrients are extremely limiting, but where sunlight and water are readily available. Only under such extreme conditions is carnivory favoured to an extent that makes the adaptations obvious.

The archetypal carnivore, the Venus flytrap, grows in soils with almost immeasurable nitrate and calcium levels. Plants need nitrogen for protein synthesis, calcium for cell wall stiffening, phosphate for nucleic acid synthesis, and iron for chlorophyll synthesis. The soil is often waterlogged, which favours the production of toxic ions such as ammonium, and its pH is an acidic 4 to 5. Ammonium can be used as a source of nitrogen by plants, but its high toxicity means that concentrations high enough to fertilise are also high enough to cause damage.

However, the habitat is warm, sunny, constantly moist, and the plant experiences relatively little competition from low growing *Sphagnum* moss. Still, carnivores are also found in very atypical habitats. *Drosophyllum lusitanicum* is found around desert edges and *Pinguicula valisneriifolia* on limestone (calcium-rich) cliffs.

In all the studied cases, carnivory allows plants to grow and reproduce using animals as a source of nitrogen, phosphorus and possibly potassium. However, there is a spectrum of dependency on animal prey. Pygmy sundews are unable to use nitrate from soil because they lack the necessary enzymes (nitrate reductase in particular). Common butterworts (*Pinguicula vulgaris*) can use inorganic sources of nitrogen better than organic sources, but a mixture of both is preferred. European bladderworts seem to use both sources equally well. Animal prey makes up for differing deficiencies in soil nutrients.

Plants use their leaves to intercept sunlight. The energy is used to reduce carbon dioxide from the air with electrons from water to make sugars (and other biomass) and a waste product, oxygen, in the process of photosynthesis. Leaves also respire, in a similar way to animals, by burning their biomass to generate chemical energy. This energy is temporarily stored in the form of ATP (adenosine triphosphate), which acts as an energy currency for metabolism in all living things. As a waste product, respiration produces carbon dioxide.

For a plant to grow, it must photosynthesise more than it respires. Otherwise, it will eventually exhaust its biomass and die. The potential for plant growth is net photosynthesis, the total gross gain of biomass by photosynthesis, minus the biomass lost by respiration. Understanding carnivory requires a cost-benefit analysis of these factors.

In carnivorous plants, the leaf is not just used to photosynthesise, but also as a trap. Changing the leaf shape to make it a better trap generally makes it less efficient at photosynthesis. For example, pitchers have to be held upright, so that only their opercula directly intercept light. The plant also has to expend extra energy on non-photosynthetic structures like glands, hairs, glue and digestive enzymes. To produce such structures, the plant requires ATP and respires more of its biomass. Hence, a carnivorous plant will have both decreased photosynthesis and increased respiration, making the potential for growth small and the cost of carnivory high.

Being carnivorous allows the plant to grow better when the soil contains little nitrate or phosphate. In particular, an increased supply of nitrogen and phosphorus makes photosynthesis more efficient, because photosynthesis depends on the plant being able to synthesise very large amounts of the nitrogen-rich enzyme RuBisCO (ribulose-1,5-*bis*-phosphate carboxylase/oxygenase), the most abundant protein on earth. It is intuitively clear that the Venus flytrap is more carnivorous than Triphyophyllum peltatum. The former is a full-time moving snap-trap; the latter is a part-time, non-moving flypaper. The energy 'wasted' by the plant in building and fuelling its trap is a suitable measure of the carnivory of the trap.

Pitcher Plant

Pitcher plants are carnivorous plants whose prey-trapping mechanism features a deep cavity filled with liquid known as a pitfall trap. It has been widely assumed that the various sorts of pitfall trap evolved from rolled leaves, with selection pressure favouring more deeply cupped leaves over evolutionary time. However, some pitcher plant genera (such as *Nepenthes*) are placed within clades consisting mostly of flypaper traps: this indicates that this view may be too simplistic, and some pitchers may have evolved from the common ancestors of today's flypaper traps by loss of mucilage.

Whatever their evolutionary origins, foraging, flying or crawling insects such as flies are attracted to the cavity formed by the cupped leaf, often by visual lures such as anthocyanin pigments, and nectar bribes. The sides of the pitcher are slippery and may be grooved in such a way so as to ensure that the insects cannot climb out. The small bodies of liquid contained within the pitcher traps are called phytotelmata. They drown the insect, and the body of it is gradually dissolved. This may occur by bacterial action (the bacteria being washed into the pitcher by rainfall) or by enzymes secreted by the plant itself. Furthermore, some pitcher plants contain mutualistic insect larvae, which feed on trapped prey, and whose excreta the plant absorbs. Whatever the mechanism of digestion, the prey items are converted into a solution of amino acids, peptides, phosphates, ammonium and urea, from which the plant obtains its mineral nutrition (particularly nitrogen and phosphorus). Like all carnivorous plants, they grow in locations where the soil is too poor in minerals and/or too acidic for most plants to survive.

Types of pitcher plants

The families Nepenthaceae and Sarraceniaceae are the best-known and largest groups of pitcher plants. The Nepenthaceae contains a single genus, Nepenthes, containing about 120 species and numerous hybrids and cultivars. In these Old World pitcher plants, the pitchers are borne at the end of tendrils that extend from the midrib of an otherwise unexceptional leaf. The plants themselves are often climbers, accessing the canopy of their habitats using the aforementioned tendrils, although others are found on the ground in forest clearings or as epiphytes on trees.

In contrast, the New World pitcher plants (Sarraceniaceae), which comprise three genera, are ground-dwelling herbs whose pitchers arise from a horizontal rhizome. In this family, the entire leaf forms the pitcher, whereas in the Nepenthaceae, the pitcher arises from the terminal portion of the leaf. The species of Heliamphora, which are popularly known as marsh pitchers (or erroneously as sun pitchers), have a simple rolled-leaf pitcher, at the tip of which is a spoon-like structure that secretes nectar.

MYRMECOPHILY

Myrmecophily is the term applied to positive interspecies associations between ants and a variety of other organisms such as plants, arthropods, and fungi. Myrmecophily refers to mutualistic associations with ants, though in its more general use the term may also refer to commensal or even parasitic interactions.

The term myrmecophile, is used mainly for animals that associate with ants. There are an estimated 10,000 species of ants (Formicidae), with a higher diversity in the tropics. In most terrestrial ecosystems ants are ecologically and numerically dominant, being the main invertebrate predators. As a result, ants play a key role in controlling arthropod richness, abundance, and community structure. There is evidence that the evolution of myrmecophilous interactions has contributed to the abundance and ecological success of ants, by ensuring a dependable and energy-rich food supply and thus providing a competitive advantage for ants over other invertebrate predators. Most myrmecophilous associations are opportunistic, unspecialised, and facultative (meaning both species are capable of surviving without the interaction), though obligate mutualisms (those in which one or both species are dependent on the interaction for survival) have also been observed for many species.

A myrmecophile is an organism that lives in association with ants. Myrmecophiles may have various roles in their host ant colony. Many consume waste materials in the nests, such as dead ants, dead larvae or fungi growing in the nest. Some myrmecophiles, however, feed on the stored food supplies of ants, and a few are predatory on ant eggs, larvae or pupae. Others benefit the ants by providing a food source for them. Many myrmecophilous relationships are obligate, meaning one or the other participant requires the relationship for survival. Some associations are facultative, benefiting one or both participants but not being necessary to their survival.

Significance of Myrmecophily in Ecology

Mutualisms are geographically ubiquitous, found in all organismic kingdoms, and play a major role in all ecosystems. Combined with the fact that ants are one of the most dominant lifeforms on earth, it is clear that myrmecophily plays a significant role in the evolution and ecology of diverse organisms, and in the community structure of many terrestrial ecosystems.

Evolution of positive interactions

Questions of how and why species coevolve are of great interest and significance. In many myrmecophilous organisms it is clear that ant associations have been influential in the ecological success, diversity, and persistence of species. Analyses of phylogenetic information for myrmecophilous organisms as well as ant lineages have demonstrated that myrmecophily has arisen independently in most groups multiple times. Because there have been multiple gains (and perhaps losses) of myrmecophilous adaptations, the evolutionary sequence of events in most lineages is unknown. Exactly how these associations evolve also remains unclear.

Species coexistence

In addition to leading to coevolution, mutualisms also play an important role in structuring communities. One of the most obvious ways in which myrmecophily influences community structure is by allowing for the coexistence of species which might otherwise be antagonists or competitors. For many myrmecophiles, engaging in ant associations is first and foremost a method of avoiding predation by ants. For example, the caterpillars of lycaenid butterflies are an ideal source of food for ants: they are slow-moving, soft-bodied, and highly nutritious, yet they have evolved complex structures to not only appease ant aggression but to elicit protective services from the ants. In order to explain why ants cooperate with other species as opposed to predating on them, two related hypotheses have been proposed: cooperation either provides ants with resources that are otherwise difficult to find or it ensures the long-term availability of those resources.

SAPROPHYTES

Saprophytes are the organisms that act as the rainforests decomposers, competing with the heavy rainfall which constantly washes away nutrients on the forest floors. Some fungi, called mycorrhizals, are examples of plant life that carry out this function. Decomposers work extremely efficiently and, together with the warmth and wetness which helps accelerate decomposition, can often break down dead animals and vegetation within 24 hours. Decomposition in montane forests, which are colder and less humid, however, can sometimes take up to six weeks.

Many saprotrophs are so small, called microbes, that they cannot be seen with the naked eye. Other decomposers, which include insects, grubs, snails, slugs, beetles and ants, aid in recycling valuable nutrients from dead organic matter which is then released back into the soil to be reabsorbed rapidly by plants and trees. Decayed matter contains essential nutrients like iron, calcium, potassium and phosphorous all of which are necessary to promote healthy rainforest growth. Thus decomposers must work continuously to release these and other elements into the soil.

HERBIVORE

Herbivores are organisms that eat plants. Herbivory is a form of consumption in which an organism principally eats autotrophs such as plants, algae and photosynthesising bacteria. More generally, organisms that feed on autotrophs in general are known as primary consumers.

By strict interpretation of this definition, many fungi, some bacteria, many animals, some protists and a small number of parasitic plants might be considered herbivores. However, herbivory generally refers to animals eating plants. Fungi, bacteria and protists that feed on living plants are usually termed plant pathogens (plant diseases). Microbes that feed on dead plants are saprotrophs. Flowering plants that obtain nutrition from other living plants are usually termed parasitic plants.

Food Chain

Herbivores form an important link in the food chain as they consume plants in order to receive the carbohydrates produced by a plant from photosynthesis. Carnivores in turn consume herbivores for the same reason, while omnivores can obtain their nutrients from either plants or animals. Due to a herbivore's ability to survive solely on tough and fibrous plant matter, they are termed the primary consumers in the food cycle (chain). Herbivory, carnivory, and omnivory and call be regarded as special cases of Consumer-Resource Systems.

Predator–Prey Theory (Herbivore–Plant interactions)

According to the theory of predator-prey interactions, the relationship between herbivores and plants is cyclic. When prey (plants) are numerous their predators (herbivores) increase in numbers, reducing the prey population, which in turn causes predator number to decline. The prey population eventually recovers, starting a new cycle. This suggests that the population of the herbivore fluctuates around the carrying capacity of the food source, in this case the plant.

Several factors play into these fluctuating populations and help stabilise predator-prey dynamics. For example, spatial heterogeneity is maintained, which means there will always be pockets of plants not found by herbivores. This stabilising dynamic plays an especially important role for specialist herbivores that feed on one species of plant and prevents these specialists from wiping out their food source. Eating a second prey type helps herbivores' populations stabilise. Alternating between two or more plant types provides population stability for the herbivore, while the populations of the plants oscillate. This plays an important role for generalist herbivores that eat variety of plants. Keystone herbivores keep vegetation populations in check and allow for a greater diversity of both herbivores and plants. When an invasive herbivore or plant enters the system, the balance is thrown off and the diversity can collapse to a monotaxon system.

Plant Defense Against Herbivory

Plant defense against herbivory or host-plant resistance (HPR) describes a range of adaptations evolved by plants which improve their survival and reproduction by reducing the impact of herbivores. Plants use several strategies to defend against damage caused by herbivores. Many plants produce secondary metabolites, known as allelochemicals, that influence the behaviour, growth or survival of herbivores. These chemical defenses can act as repellents or toxins to herbivores or reduce plant digestibility.

Other defensive strategies used by plants include escaping or avoiding herbivores in time or in place, for example by growing in a location where plants are not easily found or accessed by herbivores or by changing seasonal growth patterns. Another approach diverts herbivores toward eating non-essential parts or enhances the ability of a plant to recover from the damage caused by herbivory. Some plants encourage the presence of natural enemies of herbivores, which in turn protect the plant. Each type of defense can be either constitutive (always present in the plant) or induced (produced in reaction to damage or stress caused by herbivores).

Grazing and Browsing: How Plants are Affected

Grazing can have a neutral, positive or negative effect on rangeland plants, depending on how it is managed. Land owners and managers can better protect rangeland plants, and, in turn, other rangeland resources, if they understand:

1. The effects of grazing and browsing (eating the leaves and young twigs of trees and shrubs) on individual plants and plant populations.
2. The indicators that show which plants are in danger of overuse by grazing and browsing animals.
3. The grazing management practices that help preserve the rangeland resource.

Understanding these factors and knowing the available management options allows landowners and managers to make better decisions about which actions are best for a particular site and when to take action. Timely action can preserve the long-term health of the rangeland as well as the viability of livestock and wildlife operations.

Interactions between range plants and range animals

Rangelands are ecosystems that have adapted to withstand such disturbances as drought, flood, fire, and grazing. All disturbances affect plants to some extent, either directly or indirectly, depending on the timing, intensity, and frequency of the disturbance. Generally, the more diverse the vegetation, the better rangeland can withstand disturbance.

Rangeland plants provide nutrients—proteins, starches and sugars—to grazing and browsing livestock and wildlife. These nutrients or plant foods, are produced by photosynthesis. Because photosynthesis occurs only in green plant tissue and mostly in the leaves, a plant becomes less able to produce food, at least temporarily, when its leaves are removed (defoliation) by grazing and browsing animals. Products of photosynthesis are just as important to plants as they are to animals. Like all other living things, plants need food to survive and grow. The food that plants make for themselves through photosynthesis is used for major plant functions such as surviving dormancy, growing new roots, growing new leaves in the spring, and replacing leaves lost to grazing or browsing. Most native rangelands evolved under grazing. Therefore, rangeland plants have developed the ability to withstand a certain level of grazing or browsing. Although grazing animals do disturb rangeland, research has shown that rangelands gain few benefits when livestock are totally excluded for long periods.

What happens to a plant after grazing or browsing?

Grazing affects not just the leaves, but also other parts and functions of plants, including the root system, food production after defoliation, and the destination of food products within the plant after defoliation.

Food reserves and the root system

When a plant's leaves are removed, its roots are also affected. Excessive defoliation makes the root system smaller.

Food production after defoliation

Grazing and browsing decrease, at least temporarily, a plant's food production by reducing the amount of green plant material available to produce food. Other factors affecting food production after grazing or browsing include the amount, kind, and age of plant material (leaf, sheath, stem) remaining on the plant. In many plant species, including some grasses, the leaves on grazed or browsed plants produce food at higher rates than leaves of the same age on plants that have not been grazed or browsed.

Destination of food products after defoliation

Plants use the foods they produce for growth and maintenance. Any excess food is sent from the food-producing plant parts to other parts both above and below ground, where it is stored.

Once a plant has been defoliated, it may change the destination of its food products. The destination of that food varies with plant species. In some species, more food is sent to growing shoots and less to roots. This process occurs for a few days until the food-producing tissues can be reestablished. In some grass species, more food products may even be sent to the more active food-producing leaf blades rather than to less active leaf sheaths. A plant's ability to send food products to new shoots after defoliation can help it quickly reestablish its food-producing parts. Plant species that have this ability are better able to tolerate grazing.

How do plants cope with grazing and browsing?

The ability of plants to survive grazing or browsing is called grazing or browsing resistance. The most grazing resistant plants are grasses, followed by forbs (herbaceous plants other than grass), deciduous shrubs and trees, and evergreen shrubs and trees.

When a grass seedling develops, it produces a primary tiller or shoot. This primary tiller has both a main growing point and secondary growing points located at or below ground level. Additional tillers can develop from secondary growing points at the base of a tiller. Tillers can also develop from buds at the nodes of stolons (above-ground lateral stems, such as in buffalograss) or rhizomes (below-ground lateral stems, such as in Johnsongrass) of grasses with these structures.

Cool-season grasses begin growth in the fall, maintain some live basal leaves through winter, and continue growth in the spring. Tillers produced in the fall are exposed to cold and can produce seedheads in spring. Tillers initiated in the spring usually do not produce seedheads. In comparison, warm-season grasses produce new tillers in late summer and early fall. Although these young tillers die back when exposed to frost, their buds will produce new tillers the following spring. Tillers of most grasses live only 1 to 2 years. Individual leaves usually live less than a year and most only a few months.

A plant can produce leaves only at an intact growing point. As long as that growing point is close to the ground, it is protected from being eaten (Fig 14.1). At some point, most grasses elevate at least some of their growing points to produce tillers or shoots, that have seedheads.

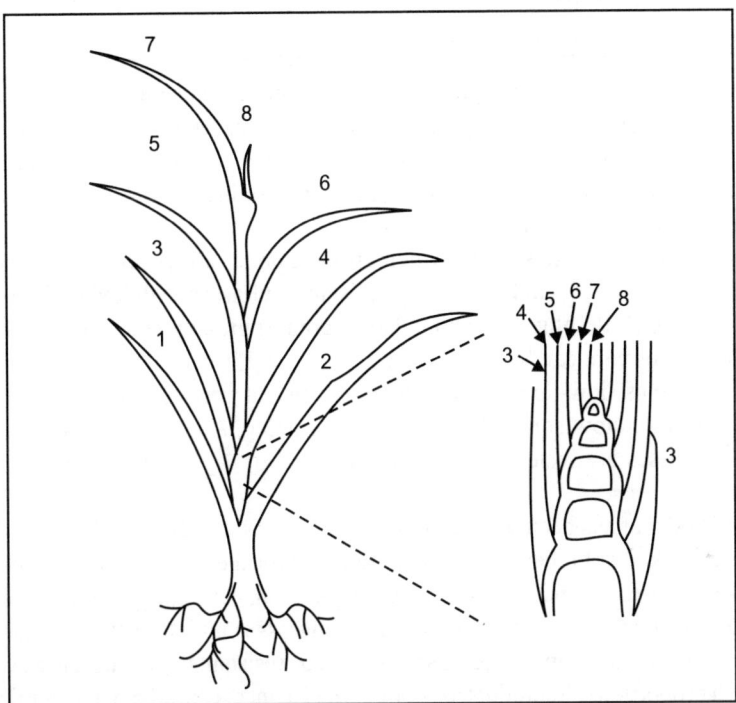

Fig. 14.1. This illustration represents a grass tiller (or shoot) and its main growing point. On the left are the grass tiller and eight leaves, numbered 1 to 8. On the right is an enlargement of the area near the base of this tiller where the main growing point is located. All the leaves shown have developed from this growing point. As long as the growing point is close to the ground as shown here, it is safe from being eaten and can continue to produce leaves for the life of the tiller (1 to 2 years).

Tillers stop producing new leaves when a seedhead develops from the growing point or when the growing point is eaten. Plants then must depend on other tillers to continue producing new leaves or wait until basal buds produce new tillers.

Excessive grazing of a grass plant when its growing points are elevated reduces new leaf production, and therefore, the ability of the plant to produce food and tolerate grazing. Destruction of the growing point also prevents seed production and production of new seedlings. Grasses should be rested from grazing periodically to allow them to produce leaf material to feed the plant and to allow seed production.

Timing of growing point elevation varies among grass species (Table 14.1). For example, growing points of buffalograss and other sod-forming grasses remain close to the ground, giving these grasses high grazing resistance.

Table 14.1. Examples of growing point elevation and grazing resistance for some common range grasses.

Grass species	Growing point elevation/reproductive	Tiller ratio grazing resistance
Buffalograss	Remain close to ground	High
Little bluestem	Elevation late w/large number reproductive tillers	Moderate
Sideoats grama	Elevation late w/large number reproductive tillers	Moderate
Switchgrass	Elevation early	Low
Yellow indiangrass	Elevation early	Low
Johnsongrass	High proportion of reproductive tillers	Low

Little bluestem and sideoats grama keep their growing points close to the ground until just before seedheads emerge. Although this strategy protects growing points from being eaten for a longer period, these two grasses produce many tillers with seedheads, which means that many growing points are exposed. The combined effect of delayed elevation and the production of many tillers with seedheads gives these two grasses moderate grazing resistance.

Yellow indiangrass and switchgrass elevate their growing points above ground level soon after growth begins. This early elevation results in low grazing resistance. Grasses with low (yellow indiangrass and switchgrass) to moderate (little bluestem and sideoats grama) grazing resistance require more care in grazing management. This care can be accomplished in several ways.

One way to manage these low- to moderate-grazing resistant grasses is to lower grazing pressure by stocking fewer animals to allow some plants to escape grazing. Another method is to make sure that pastures with these grasses are rested from grazing every 3 or 4 years during the growing season to allow the plants to produce seed.

Still another method that has been used successfully is intensive-early stocking. With this approach, grazing animals are stocked at higher than normal numbers for the first part of the growing season and then removed from pastures for the rest of the growing season. This approach has typically been used with stocker (young steer and heifer) operations. Johnsongrass is an interesting contradiction. Because it produces strong rhizomes (underground stems), it should be resistant to grazing. However, Johnsongrass also produces a high proportion of reproductive stems, which cancels the advantage of rhizome production and results in lower grazing resistance.

The growing points of forbs, like those of grasses, remain close to the ground early in the growing season. Forb species that elevate growing points early are less resistant to grazing. For woody plants, growing points are elevated above ground and, therefore, are easily accessible to browsing animals. If these growing points are removed, lateral buds are stimulated to sprout and produce leaves. However, woody plants replace leaves relatively slowly.

Grazing avoidance and grazing tolerance

Grazing resistance can be divided into avoidance and tolerance (Fig. 14.2). Grazing avoidance mechanisms decrease the chance that a plant will be grazed or browsed. Grazing tolerance mechanisms promote growth after grazing or browsing.

Grazing resistance factors can be related to plant anatomy, plant chemistry or plant physiology:
1. Anatomical features that help plants resist being grazed include leaf accessibility (leaf angle, leaf length), awns or spines, leaf hair and/or wax, tough leaves, grass species with more vegetative stems (fewer growing points exposed) than reproductive stems, and the ability to replace leaves, which depends on growing points.

2. Chemical factors of grazing resistance include those compounds that make plants taste bad, toxic or hard to digest.
3. Physiological factors include sending new food products to new leaves, water-use efficiency, and root growth and function.

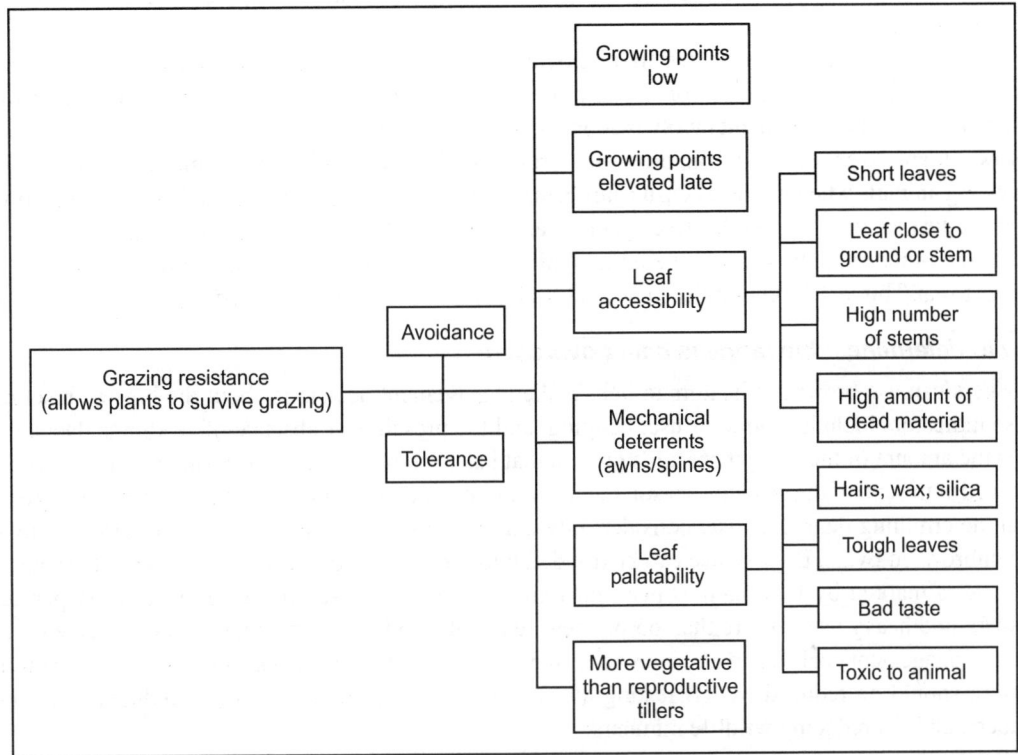

Fig. 14.2. Examples of plant grazing-resistance mechanisms.

Competition and grazing

Competition from neighboring plants for soil nutrients and water affects plant response to defoliation. Studies have shown that when competition is reduced, leaf growth in defoliated plants can be similar to that in nondefoliated plants. Competition can be reduced by: (i) lowering grazing pressure by stocking fewer animals, and (ii) resting plants from grazing. If competition is not reduced, new leaf growth may not occur because of a lack of available nutrients to grow new leaves. Therefore, plants that are grazed severely while neighbouring plants are not grazed or grazed less severely are at a competitive disadvantage.

Do plants benefit from grazing?

It is not clear if plants benefit from being grazed. Certain species may benefit from grazing but not necessarily from being grazed. For example, plants may benefit indirectly from removal of competition or from the creation of a favourable environment for seed germination or directly from removal of self-shading or removal of inactive leaves. Some grazed plants experience compensatory photosynthesis (food production). However, this response does not mean that the plants benefit from being grazed, only that they have ways to cope with grazing.

Browse management considerations

Browsing animals such as goats and deer prefer certain browse species. Preferred species vary with natural regions (such as the Edwards Plateau, Rio Grande Plain, trans Pecos, etc.) of Texas. However, Texas kidneywood and Texas or Spanish oak are examples of highly preferred species; live oak represents a moderately preferred species; and ashe juniper (blueberry cedar) and mesquite are examples of low-preference species.

Without proper management, the more desirable browse species can disappear because of these preferences, while less desirable or undesirable species become more abundant. From a livestock perspective, proper management involves controlling browsing livestock numbers and controlling access to browse plants to provide rest from browsing. From a wildlife standpoint, proper management involves harvesting animals when wildlife census numbers and browse use signs indicate a danger to the browse resource. Just as with grasses, browse species can be managed to promote and maintain key species, that is, the preferred plants that make up a significant part of the production of browse available for animals to eat. This task is accomplished by controlling animal numbers and providing rest from browsing.

How to determine if the range is being overused

Managers can use browse indicators to help make management decisions about the browse resource. These indicators include degree of use, hedging, and the presence or absence of seedlings. Degree of use is the amount of the current season's growth that has been removed by browsing animals. It is best observed at the end of the growing season in late fall for deciduous plants and late winter for evergreens. When determining degree of use, consider only current season growth by comparing browsed twigs with unbrowsed twigs. Browse use can be divided into three levels of current season growth removal: light use is marked by less than 40 per cent removal; moderate use ranges from 40 to 65 per cent removal; and heavy use is more than 65 per cent removal. Moderate use on key browse species is the correct management goal. When use approaches the upper limit of moderate use for key species, browsing pressure should be reduced by: (i) resting areas from browsing livestock use or reducing livestock numbers, and (ii) reducing wildlife numbers.

Hedging is a plant response to browsing marked by twigs that have many lateral branches. A moderate degree of hedging is acceptable because it keeps browse material within easy reach of animals and stimulates leaf and twig growth. However, excessive hedging produces short twigs with smaller than normal leaves and twigs. Eventually, entire plants can die from excessive hedging. Another indicator of excess browsing pressure is the hedging of low-preference plants such as agarita. When animals consume plants they do not normally eat, it usually means that not enough of their preferred food is available.

To provide forage, browse plants must be within reach of browsing animals. As hedging increases, the lower branches disappear and a browse line develops. A browse line is the height on trees or shrubs below which there is little or no browse and above which browse cannot be reached by animals. Areas where trees or shrubs have a highly developed browse line have a park-like appearance. In the early development of a browse line, light begins to show through the lower vegetation. With continued browsing pressure, a distinct browse line develops. Development of browse lines on low-preference plants such as ashe juniper (blueberry cedar) also indicates excessive use of the range. The height of browse lines depends on browsing animal species. For example, white-tailed deer usually browse to about 3 to 4 feet, goats to about 4 to 5 feet, and exotic wildlife species to 6 feet and more. To keep woody plant populations healthy, plants must be allowed to reproduce. Therefore, the presence of seedlings of desirable browse plants is another indicator that managers can use to check for range overuse.

Chapter 15

Halophytes and Other Type of Plants

INTRODUCTION

A halophyte is a plant that grows where it is affected by salinity in the root area or by salt spray, such as in saline semi-deserts, mangrove swamps, marshes and sloughs, and seashores. An example of a halophyte is the salt marsh grass *Spartina alterniflora* (smooth cordgrass) shown in Fig. 15.1. Relatively few plant species are halophytes—perhaps only 2 per cent of all plant species. The large majority of plant species are 'glycophytes', and are damaged fairly easily by salinity.

One quantitative measure of salt tolerance is the 'total dissolved solids' in irrigation water that a plant can tolerate. Sea water typically contains 40 grams per litre (g/l) of dissolved salts (mostly sodium chloride). Beans and rice can tolerate about 1–3 g/l, and are considered glycophytes (as are most crop plants). At the other extreme, *Salicornia bigelovii* (dwarf glasswort) grows well at 70 g/l of dissolved solids, and is a promising halophyte for use as a crop. Plants such as barley (*Hordeum vulgare*) and the date palm (*Phoenix dactylifera*) can tolerate about 5 g/l, and can be considered as marginal halophytes.

Adaptation to saline environments by halophytes may take the form of salt tolerance or salt avoidance. Plants that avoid the effects of high salt even though they live in a saline environment may be referred to as facultative halophytes rather than 'true' or obligatory, halophytes. For example, a short-lived plant species that completes its reproductive life cycle during periods (such as a rainy season) when the salt concentration is low would be avoiding salt rather than tolerating it. Or a plant species may maintain a 'normal' internal salt concentration by excreting excess salts through its leaves or by concentrating salts in leaves that later die and drop off.

MANGROVE

Mangroves are various kinds of trees up to medium height and shrubs that grow in saline coastal sediment habitats in the tropics and subtropics — mainly between latitudes 25°N and 25°S. The word is used in at least three senses: (i) most broadly to refer to the habitat and entire plant assemblage or mangal, for which the terms mangrove forest biome, mangrove swamp and mangrove forest are also used, (ii) to refer to all trees and large shrubs in the mangal, and (iii) narrowly to refer to the mangrove family of plants, the Rhizophoraceae or even more specifically just to mangrove trees of the genus Rhizophora.

The mangrove biome or mangel, is a distinct saline woodland or shrubland habitat characterised by a depositional coastal environments, where fine sediments (often with high organic content) collect in areas protected from high-energy wave action. Mangroves dominate three quarters of tropical coastlines.

247

The saline conditions tolerated by various mangrove species range from brackish water, through pure seawater (30 to 40 ppt), to water concentrated by evaporation to over twice the salinity of ocean seawater (up to 90 ppt).

Fig. 15.1. *Spartina alterniflora* (cordgrass), a halophyte.

Ecology of Mangrove

Mangroves are found in tropical and subtropical tidal areas. Areas where mangals occur include estuaries and marine shorelines. The intertidal existence to which these trees are adapted repesents the major limitation to the number of species able to thrive in their habitat. High tide brings in salt water, and when the tide recedes, solar evaporation of the seawater in the soil leads to further increases in salinity. The return of tide can flush out these soils, bringing them back to salinity levels comparable to that of seawater. At low tide, organisms are also exposed to increases in temperature and desiccation, and are then cooled and flooded by the tide. Thus, in order for a plant to survive in this environment, it must tolerate broad ranges of salinity, temperature, and moisture, as well as a number of other key environmental factors. It is unsurprising, perhaps, that only a select few species make up the mangrove tree community.

About 110 species are considered mangroves. However, a given mangrove typically features only a small number of tree species. It is not uncommon for a mangrove forest in the Caribbean to feature only three or four tree species. For comparison, the tropical rainforest biome contains thousands of tree species. That is not to say that mangrove forests lack diversity. Though the trees themselves are few in species, the ecosystem that these trees create provides a home for a great variety of other organisms.

Mangroves require a number of physiological adaptations to overcome the problems of anoxia, high salinity and frequent tidal inundation. Each species has its own solutions to these problems; this may be the primary reason why, on some shorelines, mangrove tree species show distinct zonation. Small environmental variations within a mangal may lead to greatly differing methods for coping with the environment. Therefore, the mix of species is partly determined by the tolerances of individual species to physical conditions, like tidal inundation and salinity, but may also be influenced by other factors such as predation of plant seedlings by crabs.

Once established, mangrove roots provide an oyster habitat and slow water flow, thereby enhancing sediment deposition in areas where it is already occurring. The fine, anoxic sediments under mangroves act as sinks for a variety of heavy (trace) metals which colloidal particles in the sediments scavenged from the water. Mangrove removal disturbs these underlying sediments, often creating problems of trace metal contamination of seawater and biota.

Mangroves protect coastal areas from erosion, storm surge (especially during hurricanes), and tsunamis. The mangroves' massive root systems are efficient at dissipating wave energy. Likewise, they slow down tidal water enough that its sediment is deposited as the tide comes in, leaving all except fine particles when the tide ebbs. In this way, mangroves build their own environment. Because of the uniqueness of mangrove ecosystems and the protection against erosion they provide, they are often the object of conservation programs, including national biodiversity action plans.

However, mangroves' protective value is sometimes overstated. Wave energy is typically low in areas where mangroves grow, so their effect on erosion can only be measured over long periods. Their capacity to limit high-energy wave erosion is limited to events such as storm surges and tsunamis. Erosion often occurs on the outer sides of bends in river channels that wind through mangroves, while new stands of mangroves are appearing on the inner sides where sediment is accruing.

The unique ecosystem found in the intricate mesh of mangrove roots offers a quiet marine region for young organisms. In areas where roots are permanently submerged, the organisms they host include algae, barnacles, oysters, sponges, and bryozoans, which all require a hard surface for anchoring while they filter feed. Shrimps and mud lobsters use the muddy bottoms as their home. Mangrove crabs mulch the mangrove leaves, adding nutrients to the mangal muds for other bottom feeders. In at least some cases, export of carbon fixed in mangroves is important in coastal food webs.

Mangrove plantations in Vietnam, Thailand, the Philippines and India host several commercially important species of fish and crustaceans. Despite restoration efforts, developers and others have removed over half of the world's mangroves in recent times.

Biology

Of the recognised 110 mangrove species, only about 54 species in 20 genera from 16 families constitute the 'true mangroves', species that occur almost exclusively in mangrove habitats. Demonstrating convergent evolution, many of these species found similar solutions to the tropical conditions of variable salinity, tidal range (inundation), anaerobic soils and intense sunlight. Plant biodiversity is generally low in a given mangal. This is especially true in higher latitudes and in the Americas. The greatest biodiversity occurs in the mangal of New Guinea, Indonesia and Malaysia.

Adaptations to low oxygen

Red mangroves, which can survive in the most inundated areas, prop themselves above the water level with stilt roots and can then absorb air through pores in their bark (lenticels). Black mangroves live on higher ground and make many pneumatophores (specialised root-like structures which stick up out of the soil like straws for breathing) which are also covered in lenticels.

Limiting salt intake

Red mangroves exclude salt by having significantly impermeable roots which are highly suberised, acting as an ultrafiltration mechanism to exclude sodium salts from the rest of the plant. Analysis of water inside mangroves has shown 90–97 per cent of salt has been excluded at the roots (Fig. 15.2).

Fig. 15.2. Salt crystals formed on grey mangrove leaf.

Limiting water loss

Because of the limited fresh water available in salty intertidal soils, mangroves limit the amount of water they lose through their leaves. They can restrict the opening of their stomata (pores on the leaf surfaces, which exchange carbon dioxide gas and water vapour during photosynthesis). They also vary the orientation of their leaves to avoid the harsh midday sun and so reduce evaporation from the leaves.

Nutrient uptake

The biggest problem that mangroves face is nutrient uptake. Because the soil is perpetually waterlogged, there is little free oxygen. Anaerobic bacteria liberate nitrogen gas, soluble iron, inorganic phosphates, sulphides, and methane, which makes the soil much less nutritious and contributes to mangroves' pungent odour.

Increasing survival of offspring

In this harsh environment, mangroves have evolved a special mechanism to help their offspring survive. Mangrove seeds are buoyant and therefore suited to water dispersal. Unlike most plants, whose seeds germinate in soil, many mangroves (e.g. red mangrove) are viviparous, whose seeds germinate while still attached to the parent tree.

SUNDARBANS

The sundarbans is the largest single block of tidal halophytic mangrove forest in the world. The forest lies in the vast delta on the Bay of Bengal formed by the super confluence of the Padma, Brahmaputra and Meghna rivers across Saiyan southern Bangladesh. The seasonally-flooded Sundarbans freshwater swamp forests lie inland from the mangrove forests on the coastal fringe. The forest covers 10,000 sq. km. of which about 6000 are in Bangladesh.

The Sundarbans is intersected by a complex network of tidal waterways, mudflats and small islands of salt-tolerant mangrove forests. The interconnected network of waterways makes almost every corner of the forest accessible by boat. The area is known for the eponymous Royal Bengal Tiger (*Panthera tigris tigris*), as well as numerous fauna including species of birds, spotted deer, crocodiles and snakes. The fertile soils of the delta have been subject to intensive human use for centuries, and the ecoregion has been mostly converted to intensive agriculture, with few enclaves of forest remaining. The remaining forests, pain together with the Sundarbans mangroves, are important habitat for the endangered tiger. Additionally, the Sundarbans serves a crucial function as a protective barrier for the millions of inhabitants in and around Khulna and Mongla against the floods that result from the cyclones. Sundarbans have also been enlisted amongst the finalist in the 'New Seven Wonders of Nature'.

Physiography

The mangrove-dominated Ganges Delta — the Sundarbans — is a complex ecosystem comprising one of the three largest single tract of mangrove forests of the world. Situated mostly in Bangladesh, India, has a small portion of it, estimated to be about 19 per cent, while the major part (81 per cent) is situated in the southwest corner of Bangladesh. To the south the forest meets the Bay of Bengal; to the east it is bordered by the Baleswar River and to the north there is a sharp interface with intensively cultivated land. The natural drainage in the upstream areas, other than the main river channels, is every where impeded by extensive embankments and polders.

The Sundarbans was originally measured (about 200 years ago) to be of about 16,700 km². Now it has dwindled to about 1/3 of the original size. The total land area today is 4143 km² (including exposed sandbars: 42 km²) and the remaining water area of 1874 km² encompasses rivers, small streams and canals. Rivers in the Sundarbans are meeting places of salt water and freshwater. Thus, it is a region of transition between the freshwater of the rivers originating from the Ganges and the saline water of the Bay of Bengal.

The Sundarbans along the Bay of Bengal has evolved over the millennia through natural deposition of upstream sediments accompanied by intertidal segregation. The physiography is dominated by deltaic formations that include innumerable drainage lines associated with surface and subaqueous levees, splays and tidal flats. There are also marginal marshes above mean tide level, tidal sandbars and islands with their networks of tidal channels, subaqueous distal bars and proto-delta clays and silt sediments. The Sundarbans' floor varies from 0.9 m to 2.11 m above sea level.

Biotic factors here play a significant role in physical coastal evolution and for wildlife a variety of habitats have developed including beaches, estuaries, permanent and semi-permanent swamps, tidal flats, tidal creeks, coastal dunes, back dunes and levees. The mangrove vegetation itself assists in the formation of new landmass and the intertidal vegetation plays an important role in swamp morphology. The activities of mangrove fauna in the intertidal mudflats develop micromorphological features that trap and hold sediments to create a substratum for mangrove seeds. The morphology and evolution of the eolian dunes is controlled by an abundance of xerophytic and halophytic plants. Creepers and grasses and sedges stabilises sand dunes and uncompacted sediments.

The Sunderbans mudflats are found at the estuary and on the deltaic islands where low velocity of river and tidal current occurs. The flats are exposed in low tides and submerged in high tides, thus being changed morphologically even in one tidal cycle. The interiorparts of the mudflats are magnificent home of luxuriant mangroves.

Ecoregions

Sundarbans features two ecoregions— 'Sundarbans freshwater swamp forests' (IM0162) and 'Sundarbans mangroves' (IM1406).

The Sundarbans freshwater swamp forests are a tropical moist broadleaf forest ecoregion of Bangladesh. It represents the brackish swamp forests that lie behind the Sundarbans Mangroves where the salinity is more pronounced. The freshwater ecoregion is an area where the water is only slightly brackish and becomes quite fresh during the rainy season, when the freshwater plumes from the Ganges and Brahmaputra rivers push the intruding salt water out and also bring a deposit of silt.

The Sundarbans Mangroves ecoregion on the coast forms the seaward fringe of the delta and is the world's largest mangrove ecosystem, with 20,400 square kilometers (7900 square miles) of area covered. The dominant mangrove species *Heritiera fomes*, locally known as sundri or sundari, is the tree for which the Sundarbans are thought to be named. Mangrove forests are not home to a great variety of plants. They have a thick canopy and the undergrowth is mostly seedlings of the mangrove trees.

Climate change impact

The physical development processes along the coast are influenced by a multitude of factors, comprising wave motions, micro and macro-tidal cycles and long shore currents typical to the coastal tract. The shore currents vary greatly along with the monsoon. These are also affected by cyclonic action. Erosion and accretion through these forces maintains varying levels, as yet not properly measured, of physiographic change whilst the mangrove vegetation itself provides a remarkable stability to the entire system. During each monsoon season almost all the Bengal Delta is submerged, much of it for half a year. The sediment of the lower delta plain is primarily advected inland by monsoonal coastal setup and cyclonic events. One of the greatest challenges people living on the Ganges Delta may face in coming years is the threat of rising sea levels caused mostly by subsidence in the region and partly by climate change.

Flora

The Sundarbans flora is characterised by the abundance of *Heritiera fomes*, *Excoecaria agallocha*, *Ceriops decandra* and *Sonneratia apetala*. A total 245 genera and 334 plant species were recorded by David Prain in 1903. Since Prain's report there have been considerable changes in the status of various mangrove species and taxonomic revision of the man-grove flora. However, very little exploration of the botanical nature of the Sundarbans has been made to keep up with these changes.

Whilst most of the mangroves in other parts of the world are characterised by members of the Rhizophoraceae, Avicenneaceae or Laganculariaceae, the mangroves of Bangladesh are dominated by the Sterculiaceae and Euphorbiaceae.

Fauna

The Sundarbans provide a unique ecosystem and a rich wildlife habitat. According to the 2011 tiger census, the Sundarbans have about 270 tigers. Although previous rough estimates had suggested much higher figures close to 300, the 2011 census provided the first ever scientific estimate of tigers from the area. Tiger attacks are frequent in the Sundarbans. Between 100 and 250 people are killed per year.

There is much more wildlife here than just the endangered Royal Bengal Tiger (*Panthera tigris*). Most importantly mangroves are a transition from the marine to freshwater and terrestrial systems and provide critical habitat for numerous species of small fish, crabs, fidler crabs, hermit crabs, shrimps and other crustaceans that are adapted to feed, shelter and reproduce among the tangled mass of roots, known as pneumatophores, that grow upward from the anaerobic mud to get the trees' supply of oxygen. Fishing Cats, Macaques, wild boar, Common Grey Mongoose, Fox, Jungle Cat, Flying Fox, Pangolin, and Chital are also found in abundance in the Sundarbans (Figs 15.3 and 15.4).

Fig. 15.3. Chital deer are widely seen in southern parts of Sundarban.

Fig. 15.4. A Royal Bengal tiger in Sundarban.

The management of wildlife is presently restricted to the protection of fauna from poaching and designation of some areas as wildlife sanctuaries where no extraction of forest produce is allowed and the wildlife face few disturbances. Although the fauna of Bangladesh have diminished in recent times and the Sundarbans has not been spared from this decline, the mangrove forest retains several good wildlife habitats and their associated fauna. Of these the tiger and dolphin are target species for planning wildlife management and tourism development. There are high profile and vulnerable mammals living in two contrasting environments and their uses and management are strong indicators of the general condition of wildlife and its management.

MESOPHYTE

Mesophytes are terrestrial plants which are adapted to neither a particularly dry nor particularly wet environment. An example of a mesophytic habitat would be a rural temperate meadow, which might contain Goldenrod, Clover, Oxeye Daisy, and *Rosa multiflora*. Mesophytes make up the largest ecological group of terrestrial plants, and usually grow under moderate to hot and humid climatic regions.

Mesophytes do not have any specific morphological adaptations, however they usually have broad, flat and green leaves. Mesophytes do not have any special internal structure. Epidermis is single layered usually with obvious stomata and large cortex.

Mesophytes generally require a more or less continuous water supply. They usually have larger, thinner leaves compared to xerophytes, sometimes with a greater number of stomata on the undersides of leaves. Because of their lack of particular xeromorphic adaptations, when they are exposed to extreme conditions they lose water rapidly, and are not tolerant of drought. Mesophytes are very intermediate in water use and needs. These plants are found in average conditions of temperature and moisture and grow in soil that has no water logging. The roots of mesophytes are well developed, branched and provided with a root cap. The shoot system is well organised .The stem is generally aerial, branched, straight, thick and hard. Leaves are thin, broad in middle, dark green and of variable shape and measurement.

For example, in hot weather they may overheat and suffer from temperature stress. They have no specific adaptations to overcome this, but, if there is enough water in the soil to allow this, they can increase their rate of transpiration by opening their stomata, thus meaning some heat is removed by the evaporating water. However these plants can only tolerate saturated soil for a certain amount of time without a warm temperature. In dry weather they may suffer from water stress (losing more water via transpiration than can be gained from the soil). Again they have no specific adaptations to overcome this, and can only respond by closing their stomata to prevent further transpiration. This does actually have some benefits as it reduces the surface area of the leaf exposed to the atmosphere, which reduces transpiration. Prolonged periods of dehydration, however, can lead to permanent wilting, cell plasmolysis, and subsequent death. Since mesophytes prefer moist, well drained soils, most crops are mesophytes. Some examples are: corn (maize), privet, lilac, goldenrod, clover, and oxeye daisy.

XEROPHYTE

A xerophyte or xerophytic organism (xero meaning dry, phyte meaning plant) is a plant which is able to survive in an environment with little available water or moisture, such as a desert. Xerophytic plants may have adaptations of their shape and form (morphology) or internal functions (physiology) that reduce their water loss or store water during periods of dryness. Plants with such morphological adaptations are called xeromorphic (Fig. 15.5).

Plants absorb water from the soil, which then evaporates from the surface of the plant; this process is known as evapotranspiration. In dry environments, a typical (mesophytic) plant would evaporate water faster than the rate at which water was replaced in the soil, leading to wilting. By contrast, xerophytic plants may exhibit a variety of specialised adaptations to survive in these conditions. Xerophytes may absorb water from their own storage, allocate water specifically to sites of new tissue growth or lose less water to the atmosphere and so convert a greater proportion of water in the soil to growth or have other adaptations to manage water supply and enable them to survive. The morphological consequences of these adaptations are collectively called xeromorphisms.

Plants like the cactus (plural- 'cacti', family-Cactaceae) and other succulents are typically found in deserts where low rainfall amounts are the norm. Other xerophytes, such as the bromeliads, can survive both extremely wet and extremely dry periods and can be found in seasonally moist habitats such as tropical forests, exploiting niches where water supplies are limited or too intermittent for mesophytic plants. Similarly, chaparral plants are adapted to Mediterranean climates with wet winters and dry summers. Plants that live under arctic conditions also have a need for xerophytic adaptations, since water is unavailable for uptake when the ground is frozen. Plants that survive lack of water during times of freezing by going dormant, with no other adaptations for times of low water, are not usually referred to as xerophytic.

Fig. 15.5. The Joshua tree is an example of a xerophyte.

Types of Xerophytic Plants

1. Succulent plants—typically store water in stems or leaves. They include the Cactaceae family which typically have stems that are round and store a lot of water. Often, as in cacti where the leaves are reduced to spines, their leaves are vestigial or they do not have leaves.
2. Bulbs—water is stored in their bulbs, at or below ground level. They may spend a period of dormancy during drought conditions underground, and are therefore known as drought evaders.

Importance of Water Conservation

If the water potential inside the leaf is higher than outside the leaf, the water vapour will diffuse out of the leaf down this gradient. This loss of water vapour from the leaves is called transpiration, and the water vapour diffuses through open stomata in the leaf. Although this is a normal and important process in all plants, it is vital that plants living in dry conditions have adaptations that decrease this water potential gradient, and decrease the size of open stomata, in order to reduce water loss from the plant. It is important for a plant living in these conditions to conserve water because without enough water, plant cells lose turgor and the plant tissue wilts. If the plant loses too much water, it will pass its permanent wilting point, where the plant will die.

Morphological Adaptations

Xerophytic plants may have similar shapes, forms, and structures, and look very similar, even if the plants are not very closely related, through a process called convergent evolution. For example, some species of cacti (members of the family Cactaceae), which evolved only in the Americas, may appear similar to Euphorbias, which are distributed worldwide. Unrelated species of caudiciforms, plants with swollen bases used to store water, may also display such similarities (Fig. 15.6).

Cereus peruvianus   *Euphorbia virosa*

Fig. 15.6. The cactus *Cereus peruvianus* looks superficially very similar to *Euphorbia virosa* because of convergent evolution.

Reduction of surface area

Xerophytic plants may have less surface area than other plants. This reduces the area exposed to the air, which reduces water loss by evaporation. Xerophytes may have smaller leaves and/or fewer branches than other plants. An example of leaf surface reduction are the spines of a cactus, which are modified leaves. An example of compaction and reduction of branching are the barrel cacti. Other xerophytes may have their leaves compacted at the base, as in a basal rosette, which may be smaller than the plant's flower. This adaptation is exhibited by some *Agaves* and *Eriogonums* growing near Death Valley.

Reduction in air flow

Some xerophytes have tiny hairs on their surface to provide a wind break and reduce air flow, thereby reducing the rate of evaporation. When a plant surface is covered with tiny hairs, it is called tomentose.

Reflectivity

The colour of a plant or of the waxes or hairs on its surface, may reflect sunlight and reduce evaporation. An extreme example is the white chalky wax (epicuticular wax) coating of *Dudleya brittonii*, which has the highest ultraviolet light (UV) reflectivity of any known naturally occurring biological substance.

Physiological Adaptations

Water storage

Plants may store water in root structures, trunk structures, stems, and leaves. Water storage in swollen parts of the plant is called 'succulence'. A swollen trunk or root at the ground level of a plant is called a caudex, and plants with swollen bases are called caudiciforms.

Stomata

Tiny pores on the surface of a xerophytic plant called stomata may open only at night so as to reduce evaporation.

Surface resins and waxes

Plants may secrete resins and waxes (epicuticular wax) on their surface which reduce evaporation. Examples are the heavily scented and flammable resins (volatile organic compounds) of some chaparral plants, such as *Malosma laurina* or the chalky wax of *Dudleya pulverulenta*.

Dropping leaves

Plants may drop their leaves in times of dryness (drought deciduous) or modify the leaves produced during such times such that the leaves are smaller.

Dormancy, change of photosynthetic mechanism, change of allocation of sugars produced

During dry times, xerophytic plants may stop growing and go dormant, change the kind of photosynthesis or change the allocation of the products of photosynthesis from growing new leaves to strengthening the roots.

Seed modification

Seeds may be modified to require an excessive amount of water before germinating, so as to ensure a sufficient water supply for the seedling's survival. An example of this is the California poppy whose seeds lie dormant during drought and then germinate, grow, flower, and form seeds within four weeks of rainfall.

Modification of environment

Plant material on the ground around a plant (leaf litter) may provide an evaporative barrier to prevent water loss. Waxes shed from the plant may coat the ground reducing evaporation from the ground in the immediate vicinity of the plant, as in the case of *Dudleya pulverulenta*.

Chapter 16

Phytogeography

INTRODUCTION

Phytogeography, also called geobotany, is the branch of biogeography that is concerned with the geographic distribution of plant species. Phytogeography is concerned with all aspects of plant distribution, from the controls on the distribution of individual species ranges (at both large and small scales) to the factors that govern the composition of entire communities and floras. The basic data elements of phytogeography are occurrence records (presence or absence of a species) with operational geographic units such as political units or geographical coordinates. These data are often used to construct phytogeographic provinces (floristic provinces) and elements.

The questions and approaches in phytogeography are largely shared with zoogeography, except zoogeography is concerned with animal distribution rather than plant distribution. The term phytogeography itself suggests a broad meaning.

Phytogeography is often divided into two main branches: Ecological phytogeography and Historical phytogeography. The former investigates the role of current day biotic and abiotic interactions in influencing plant distributions; the latter are concerned with historical reconstruction of the origin, dispersal, and extinction of taxa. Floristics is a study of the flora of some territory or area. Traditional phytogeography concerns itself largely with floristics and floristic classification.

PLANT DISTRIBUTION

Plant distribution is affected by a great number of different factors, which are discussed below.

Climate

Climate is generally accepted as being the dominant factor affecting plant geographic ranges. It has many different forms and affects different species in different ways, but ultimately the distribution of plant species on earth is controlled by the climate. The most important climatic factors are temperature and precipitation, temperature includes mean summer and winter temperatures, yearly maximum and minimum temperatures, the length of the growing season (above a certain temperature), frosts, soil freezing or the length of the thaw in cold environments. Precipitation principally limits distribution by its absence, i.e. a droughts, but can also have an effect in the form of snow cover, flooding, etc. It is likely that the distribution of most northern hemisphere plants is ultimately (i.e. if not held earlier by another factor), limited by cold temperatures in the north and summer drought in the south, though obviously different plants are better than others at coping with one or the other or both.

Other climatic variables include: Day length, light intensity, humidity, and wind speeds. Each may be required in different intensities by different species or act as a trigger for a process such as flowering that may never occur in some climates. For example, many plants use a set day length as a trigger for flowering, but this will occur at different times of the year at different latitudes.

Climate can have big effects in very small areas, the most dramatic indication of the effects of the climate are arctic and mountain treelines. Here the edge of a species range is clearly visible. Similar 'drought lines' can occur at the edges of deserts or 'wet lines' near a marsh, where the ground gets too wet for trees (or many other plants) to tolerate.

Example of climate as a range limiting factor: That the ranges of some species are determined by the climate is well known in botanical circles: Short leaved Lime trees, (*Tilia cordata*), reach a northern limit in Britain in the southern Lake District and Northumbria, approximately on the isotherm of 18°C average daily air temperature. This species is limited by the effect of low summer temperatures on the development of reproductive structures. Above this line it is too cold for successful pollen tube development, and thus seed cannot be fertilised. Researchers C.D. Piggott and J.P. Huntley, who carried out much of the work on climate and range limits, found that around 1 per cent of fruit produced at the range limit was fertile, as opposed to over 60 per cent further south. Most of the trees on the range edge are old, almost all the saplings date from the exceptionally warm summer of 1959, in a normal year seed production is too low to allow regeneration.

Land Use and Habitat

Land use and habitat are broadly the same thing and have an obvious effect on plant ranges. A species will not be found in unfavourable habitat, for example, agricultural weeds are found on agricultural land, Bluebells and understory shade plants in woodland, etc. The prescence or absence of habitat is usually fairly obvious, if a species range follows the available habitat closely then this is often apparent on maps. Habitat fragmentation may be less obvious. Good dispersal ability is important in avoiding the effects of Habitat fragmentation.

Soil and Nutrients

Edaphic or soil factors can influence plant distribution in a similar way to land use and habitat, *Brachypodium pinnatum* is a plant of calcareous chalk and limestone soils. Once again, this is usually fairly obvious, especially in the UK, where soil types have been well mapped.

Competition

Competition with other species can influence plant ranges. In the absence of competitors most species would be found over much wider areas. In fact competition interacts with all the other factors, most species ranges are limited when the climate, or soil conditions, becoming increasingly more adverse reach the point where the species is unable to thrive enough to stave off its competitors. A spot free of competition is very rare in the wild, but management in gardens allows many species to be grown far outside natural ranges and habitats.

Herbivory

Herbivory can be, like competition, an interaction with another factor, reducing a species ability to survive until it cannot cope with herbivory. Remove this or competitors and most ranges would shift outwards a fair bit. A mountain Herb of the Alps, *Arnica montana*, is restricted to higher altitudes by slug herbivory. On lower slopes the new shoots of the plant coincide with peak slug activity, higher up where it is colder the slug activity is reduced and the plant is better sable to survive. Note the indirect effect of climate. This effect may be common on mountain slopes, and Insect herbivores can also have a significant effect.

Dispersal Ability

Another major restriction on plant ranges is the species own ability to travel to new sites for colonisation. Plants have a great many methods of dispersal, but great mountain range or ocean will prevent dispersal completely for most species. A great any species are restricted to their continent of origin, lacking the means to leave it. Any area of unsuitable habitat too large to cross can act as a 'barrier to dispersal', e.g. an area of intense herbivory or competition. Many plants are dispersed by animals, and the ranges, dispersal ability, and any extinction of these will have a profound effect on dispersal.

The effect of climate, and many of the other factors affects dispersal ability first, in the two examples above it is reproductive structures that are affected, preventing the plant from dispersing further, and this has long been known to almost always be the case. These structures are sensitive, and take a second priority to survival under adverse conditions. Many plant species are self incompatible, meaning an individual is unable to fertilise its own seed, this can effectively stop dispersal except by vegetative growth, if there are not enough individuals.

Genetic Factors

Low genetic diversity can lead to inbreeding depression, the over expression of harmful deleterious mutations, as the number of individuals with two copies (homozygous) of these genes increases. The harmful effects of inbreeding can reduce a species ability to compete or cope with adverse conditions at a range edge, but for this to occur there needs to be almost no pollen or seed flow into these populations, and this is probably going to be due to something else. All populations get sparser towards range limits, as a species can only survive in progressively more ideal locations to cope with the stresses, making dispersal between them more difficult, rarer plants support less pollinators, etc.

Thus the principal factor may be climate, but it should be obvious by now that rarely does a factor work in isolation, a species range limit is a point on a continuum where the conditions become so adverse that the species cannot compete effectively or allocate enough resources to dispersal or a number of possible other combinations. Genetic problems can only exacerbate the problem.

SPECIES DISTRIBUTION

Species distribution is the manner in which a biological taxon is spatially arranged. Species distribution is not to be confused with dispersal, which is the movement of individuals away from their area of origin or from centers of high population density. A similar concept is the species range. A species range is often represented with a species range map. Biogeographers try to understand the factors determining a species' distribution. The pattern of distribution is not permanent for each species. Distribution patterns can change seasonally, in response to the availability of resources, and also depending on the scale at which they are viewed. Dispersion usually takes place at the time of reproduction. Populations within a species are translocated through many methods, including dispersal by people, wind, water and animals. People are one of the largest distributors due to the current trends in globalisation and the expanse of the transportation industry. For example, large tankers often fill their ballasts with water at one port and empty them in another, causing a wider distribution of aquatic species.

Species Distribution Model (SDM)

Species distribution can now be potentially predicted based on pattern of biodiversity at spatial scales. A general hierarchical model can integrate disturbance, dispersal and population dynamics. Based on factors of dispersal, disturbance, resources limiting climate, and other species distribution, predictions of species distribution can create a bioclimate range, or bioclimate envelope. The envelope can range from a local to a global scale or a density independence to density dependence. The hierarchical model takes into consideration of requirements and impacts or resources as well as local extinctions in disturbance factors. Models can integrate the dispersal/migration model, the disturbance model, and abundance model. SDM's can be used to assess climate change impacts and conservation management issues. Species distribution models include, presence/absence models, the dispersal/migration models, disturbance models, and abundance models. A prevalent way of creating predicted distribution maps for different species is to reclassify a land cover layer depending on whether or not the species in question would be predicted to habit each cover type. This simple SDM is often modified through the use of range data or ancillary information — such as elevation or water distance.

Recent studies have indicated that the grid size used can have an effect on the output of these species distribution models. The standard 50 × 50 km grid size can select up to 2.89 times more area than when modeled with a 1 × 1 km grid for the same specie. This has several effects on the species conservation planning under climate change predictions (global climate models — which are frequently used in the creation of species distribution models—usually consists of 50–100 km size grids) which could lead to over-prediction of future ranges in species distribution modelling. This can result in the misidentification of protected areas intended for a specie's future habitat.

Abiotic and Biotic Factors

The distribution of species into clumped, uniform, or random depends on different abiotic and biotic factors. Any non-living chemical or physical factor in the environment is considered an abiotic factor. There are three main types of abiotic factors: climatic factors consist of sunlight, atmosphere, humidity, temperature, and salinity; edaphic factors are abiotic factors regarding soil, such as the coarseness of soil, local geology, soil pH, and aeration; and social factors include land use and water availability. An example of the effects of abiotic factors on species distribution can be seen in drier areas, where most individuals of a species will gather around water sources, forming a clumped distribution.

Biotic factors, such as predation, disease, and competition for resources such as food, water, and mates, can also affect how a species is distributed. A biotic factor is any behaviour of an organism that affects another organism, such as a predator consuming its prey. For example, biotic factors in a quail's environment would include their prey (insects and seeds), competition from other quail, and their predators, such as the coyote. An advantage of a herd, community, or other clumped distribution allows a population to detect predators earlier, at a greater distance, and potentially mount an effective defense. Due to limited resources, populations may be evenly distributed to minimise competition, as is found in forests, where competition for sunlight produces an even distribution of trees.

Statistical determination of distribution patterns

There are various ways to determine the distribution pattern of species. The Clark-Evans nearest neighbour method can be used to determine if a distribution is clumped, uniform or random. To utilise the Clark-Evans nearest neighbour method, researchers examine a population of a single species. The distance of an individual to its nearest neighbour is recorded for each individual in the sample. For two individual that are each other's nearest neighbour, the distance is recorded twice, once for each individual. To receive accurate results, it is suggested that the number of distance measurements is at least 50. The average distance between nearest neighbours is compared to the expected distance in the case of random distribution to give the ratio:

$$\frac{\text{Mean distance}}{\frac{1}{2}\sqrt{\text{density}}}$$

If this ratio (R) is equal to 1, then the population is randomly dispersed. If R is significantly greater than 1, the population is evenly dispersed. Lastly, if R is significantly less than 1, the population is clumped. Statistical tests (such as t-test, chi squared, etc.) can then be used to determine whether R is significantly different from 1.

The Variance/Mean ratio method focuses mainly on determining whether a species fits a randomly spaced distribution, but can also be used as evidence for either an even or clumped distribution. To utilise the Variance/Mean ratio method, data is collected from several random samples of a given population. In this analysis, it is imperative that data from at least 50 sample plots is considered. The number of individuals present in each sample is compared to the expected counts in the case of random distribution. The expected distribution can be found using Poisson distribution. If the Variance/Mean ratio is equal to 1, the population is found to be randomly distributed. If it is significantly greater than 1, the population is found to be clumped distribution. Finally, if the ratio is significantly less than 1, the population is found to be evenly distributed. Typical statistical tests used to find the significance of the variance/mean ratio include *Student's t-test* and *chi squared*.

However, many researchers believe that species distribution models based on statistical analysis, without including ecological models and theories, are too incomplete for prediction. Instead of conclusions based on presence-absence data, probabilities that convey the likelihood a species will occupy a given area are more preferred because these models include an estimate of confidence in the likelihood of the species being present/absent. Additionally, they are also more valuable than data collected based on simple presence or absence because models based on probability allow the formation of spatial maps that indicates how likely a species is to be found in a particular area. Similar areas can then be compared to see how likely it is that a species will occur there also; this leads to a relationship between habitat suitability and species occurrence.

Species distribution data are useful for population monitoring, biodiversity mapping, and conservation management. Suitable habitats have been described for only a small percentage of species; consequently, overlays of geospatial species sample data (e.g. records with location information from museum collections or spatially extensive annual surveys; henceforth sample data) with environmental variables (e.g. elevation, vegetation types, land use) are often used to determine wildlife-habitat relationships and predict distributions. As pressures on our biological heritage increase the need for quick decisions, modelling procedures often rely mostly or exclusively on existing data and regularly ignore error estimates. Time and money considerations, increased accessibility to sample data and the development of various modelling approaches make distribution estimates with extant sample data appealing and easy. When the data used for modelling are less than optimal, which is often the case, inherent data errors or biases can manifest and negatively affect predictions. Many of these modelling approaches deduce wildlife habitat relationships from similar databases and are, therefore, each exposed to similar data quality issues. Use of existing data is often our best option for addressing urgent issues, yet we know little about the effects of common biases. For example, few studies have investigated model performance (e.g. the bias and precision of a prediction) as a function of sample data quantity, quality, or spatial configuration. Hirzel and Stockwell and Peterson have shown that larger sample sizes lead to greater accuracy. Also, Hirzel and Guisan reported differences in prediction accuracy resulting from various 'optimal' survey designs, but did not evaluate effects from 'biased' survey designs. Kadmon found that nearroad surveying biases had little effect on prediction accuracy when a bioclimatic model was used.

Accessibility has long been an important consideration when field surveys are conducted. For example, roads provide vantage points from which annual surveys such as the North American Breeding Bird Survey are conducted; however, this nonrandom design creates obvious difficulties with respect to inference about the population in question. Roadside habitat can differ from the habitat composition of surrounding areas and some analyses are limited when sample data are concentrated along roads or are near accessible sites.

Additional modelling factors could confound the performance of models that predict species distributions. First, the spatial contiguity of a species geographic range (a measure of whether the distribution is connected throughout or broken into disjunct units) could interact with a survey design to affect model performance. Second, surveys can incorrectly identify a species as present that is actually absent (false presence) or fail to detect a species that is actually present (false absence). Furthermore, because absence data (i.e. survey conducted but species not detected) are often not available, false absences could also occur when geographic areas without confirmation of species presence are treated as 'absences'. In either case, the prevalence of false absence or false presence records will affect empirical attempts to predict distributions based on environmental variables. Third, mapped environmental variables from which habitat affinities are derived contain errors. It is important to understand how and to what degree these factors affect model performance because as Dean found, overlaying predicted species distributions with as little as 5 per cent error could considerably alter the estimated distribution of species richness.

ENDEMISM

Endemism is the ecological state of being unique to a defined geographic location, such as an island, nation or other defined zone, or habitat type, and found only there; organisms that are indigenous to a place are not endemic to it if they are also found elsewhere. For example, all species of lemur are endemic to the island of Madagascar; none are native elsewhere. The extreme opposite of endemism is cosmopolitan distribution.

Physical, climatic, and biological factors can contribute to endemism. The Orange-breasted Sunbird is exclusively found in the Fynbos vegetation zone of southwestern South Africa. Political factors can play a part if a species is protected, or actively hunted, in one jurisdiction but not another.

There are two subcategories of endemism—paleoendemism and neoendemism. Paleoendemism refers to a species that was formerly widespread but is now restricted to a smaller area. Neoendemism refers to a species that has recently arisen such as a species that has diverged and become reproductively isolated, or one that has formed following hybridisation and is now classified as a separate species. This is a common process in plants, especially those which exhibit polyploidy.

Endemic types or species are especially likely to develop on biologically isolated areas such as islands because of their geographical isolation. This includes remote island groups, such as Hawaii, the Galápagos Islands, and Socotra, and biologically isolated but not island areas such as the highlands of Ethiopia or large bodies of water like Lake Baikal.

Endemics can easily become endangered or extinct if their restricted habitat changes, particularly but not only due to the actions of man, including the introduction of new organisms. There were millions of both Bermuda Petrels and 'Bermuda cedars' (actually *junipers*) in Bermuda when it was settled at the start of the seventeenth century. By the end of the century the petrels were thought to be extinct. Cedars, already ravaged by centuries of shipbuilding, were driven nearly to extinction in the twentieth century by the introduction of a parasite. Bermuda petrels and cedars, although not actually extinct, are very rare today, as are other species endemic to Bermuda.

Some of the principal threats to these special ecosystems are:

1. Large scale logging operations.
2. Slash-and-burn techniques sometimes a part of shifting cultivation.
3. Destruction of habital or vegetation leads to endangering of the endemic species.

ENDEMIC FLORA

The endemic flora of the Sinharaja occupies a unique status it accounts for 64–75 per cent of the total number of species recorded among the trees and lianes over 30 centimeters in girth. The contribution of the endemics to the forest stand ranges from 75–92 per cent. In a single site studied for the vegetation below 30 centimeters in girth, 60 per cent of the total number of species were endemic, and their contribution to the density was 85 per cent. It is also interesting to note that of the 25 genera endemic to Sri Lanka, 13 are represented in the Sinharaja. Several of these genera are monotypic, i.e. they are represented by one single species (Table 16.1).

Table 16.1. Endemic plant genera found in Sinharahja.

Schumacheria (Dilleniaceae)
*Trichadenia** (Flacourtiaceae)
Stemonoporus (Dipterocarpaceae)
Scutinanthe (Burseraceae)
*Pseadocarapa** (Meliaceae)
*Glenniea** (Sapindaceae)
*Leucocodon** (Rubiaceae)
*Schizostigma** (Rubiaceae)
*Championea** (Gesneriaceae)
Hortonia (Monimiaceae)
*Podadeniya** (Euporbiaceae)
*Cyphostigma** (Zingiberaceae)
*Loxococcus** (Palmae)

*Genus contains single species.

Endemic tree species considered to be rare have been identified in the Sinharaja as well as in other lowland rain forests. Studies show that 98 per cent of sub canopy species and 85 per cent of all understorey species fall into the categories of 'Rare', 'Vulnerable' and 'Endangered' as described by the International Union for Nature and Natural Resource Conservation (IUCN) Red Data Book. Of these endemic tree species, almost 25 species were restricted to a single forest site.

Similarly, of the 217 endemic trees and woody climbers of the rain forest region, 65 per cent, i.e. 140 species have so far been recorded in the Sinharaja. This percentage probably does not represent the actual value which could be greater. Although the smaller life-forms have been studied, their percentage values have not been computed as yet.

Flora of Australia

The flora of Australia comprises a vast assemblage of plant species estimated to over 20,000 vascular and 14,000 non-vascular plants, 2,50,000 species of fungi and over 3000 lichens. The flora has strong affinities with the flora of Gondwana, and below the family level has a highly endemic angiosperm flora whose diversity was shaped by the effects of continental drift and climate change since the Cretaceous. Prominent features of the Australian flora are adaptations to aridity and fire which include scleromorphy and serotiny. These adaptations are common in species from the large and well-known families Proteaceae (*Banksia*), Myrtaceae (*Eucalyptus*—gum trees), and Fabaceae (*Acacia*-wattle).

The settlement of Australia by Indigenous Australians, and by Europeans from 1788, has had a significant impact on the flora. The use of fire-stick farming by the Aborigines led to significant changes in the distribution of plant species over time, and the large-scale modification or destruction of vegetation for agriculture and urban development since 1788 has altered the composition of most terrestrial ecosystems, leading to the extinction of 61 plant species and endangering over 1000 more.

Vegetation Types

Australia's terrestrial flora can be collected into characteristic vegetation groups. The most important determinant is rainfall, followed by temperature which affects water availability. Several schemes of varying complexity have been created, the most recent scheme developed by the Natural Heritage Trust divides Australia's terrestrial flora into 30 major vegetation groups, and 67 major vegetation subgroups.

According to the scheme the most common vegetation types are those that are adapted to arid conditions where the area has not been significantly reduced by human activities such as land clearing for agriculture. The dominant vegetation type in Australia is the hummock grasslands that occur extensively in arid Western Australia, South Australia and the Northern Territory. It accounts for 23 per cent of the native vegetation, the predominant species of which are from the genus *Triodia. Zygochloa* also occurs in inland sandy areas like the Simpson Desert.

Vascular plants

Australia has over 30,000 described species of vascular plants, these include the angiosperms, seed-bearing non-angiosperms (like the conifers and cycads), and the spore-bearing ferns and fern allies. Of these about 11 per cent are naturalised species; the remainder are native or endemic. The vascular plant flora has been extensively catalogued, the work being published in the ongoing 'Flora of Australia' series. At the higher taxonomic levels the Australian flora is similar to that of the rest of the world; most vascular plant families are represented within the native flora, with the exception of the cacti, birch and a few others, while 9 families occur only in Australia. Australia's vascular flora is estimated to be 85 per cent endemic; this high level of vascular plant endemism is largely attributable to the radiation of some families like the Proteaceae, Myrtaceae, and Fabaceae.

Non-vascular plants

The algae are a large and diverse group of photosynthetic organisms. Many studies of algae include the cyanobacteria, in addition to micro and macro eukaryotic types that inhabit both fresh and saltwater. Currently, about 10,000 to 12,000 species of algae are known for Australia. The algal flora of Australia is unevenly documented: northern Australia remains largely uncollected for seaweeds and marine phytoplankton, descriptions of freshwater algae are patchy, and the collection of terrestrial algae has been almost completely neglected.

The bryophytes—mosses, liverworts and hornworts—are primitive, usually terrestrial, plants that inhabit the tropics, cool-temperate regions and montane areas; there are some specialised members that are adapted to semi-arid and arid Australia. There are slightly fewer that 1000 recognised species of moss in Australia. The five largest genera are the *Fissidens*, *Bryum*, *Campylopus*, *Macromitrium* and *Andreaea*. There are also over 800 species of liver— and horn-worts in 148 genera in Australia.

BIOME

Biomes are climatically and geographically defined as similar climatic conditions on the earth, such as communities of plants, animals, and soil organisms, and are often referred to as ecosystems. Some parts of the earth have more or less the same kind of abiotic and biotic factors spread over a large area creating a typical ecosystem over that area. Such major ecosystems are termed as biomes. Biomes are defined by factors such as plant structures (such as trees, shrubs, and grasses), leaf types (such as broadleaf and needleleaf), plant spacing (forest, woodland, savanna), and climate. Unlike ecozones, biomes are not defined by genetic, taxonomic, or historical similarities. Biomes are often identified with particular patterns of ecological succession and climax vegetation (quasi-equilibrium state of the local ecosystem). An ecosystem has many biotopes and a biome is a major habitat type. A major habitat type, however, is a compromise, as it has an intrinsic inhomogeneity. The biodiversity characteristic of each biome, especially the diversity of fauna and subdominant plant forms, is a function of abiotic factors and the biomass productivity of the dominant vegetation. In terrestrial biomes, species diversity tends to correlate positively with net primary productivity, moisture availability, and temperature. Ecoregions are grouped into both biomes and ecozones. A fundamental classification of biomes is:

1. Terrestrial (land) biomes.
2. Aquatic biomes (including Freshwater biomes and Marine biomes).

Biomes are often known in English by local names. For example, a temperate grassland or shrubland biome is known commonly as *steppe* in central Asia, *prairie* in North America, and *pampas* in South America. Tropical grasslands are known as *savanna* in Australia, whereas in Southern Africa it is known as certain kinds of *veld* (from Afrikaans). Sometimes an entire biome may be targeted for protection, especially under an individual nation's Biodiversity Action Plan.

Climate is a major factor determining the distribution of terrestrial biomes. Among the important climatic factors are:

1. Latitude: Arctic, boreal, temperate, subtropical, tropical.
2. Humidity: humid, semi-humid, semi-arid, and arid.
   (a) Seasonal variation: Rainfall may be distributed evenly throughout the year or be marked by seasonal variations.
   (b) Dry summer, wet winter: Most regions of the earth receive most of their rainfall during the summer months; Mediterranean climate regions receive their rainfall during the winter months.
3. Elevation: Increasing elevation causes a distribution of habitat types similar to that of increasing latitude.

The most widely used systems of classifying biomes correspond to latitude (or temperature zoning) and humidity. Biodiversity generally increases away from the poles towards the equator and increases with humidity.

Biome Classification Schemes

Biome classification schemes seek to define biomes using climatic measurements. Particularly in the 1970s and 1980s there was a significant push to understand the relationships between these measurements and properties of ecosystem energetics because such discoveries would enable the prediction of rates of energy capture and transfer among components within ecosystems. Such a study was conducted by Sims on North American grasslands. The study found a positive logistic correlation between evapotranspiration in mm/yr and above ground net primary production in g/m^2/yr. More general results from the study were that precipitation and water use lead to above ground primary production, solar radiation and temperature lead to belowground primary production (roots), and temperature and water lead to cool and warm season growth habit. These findings help explain the categories used in Holdridge's bioclassification scheme, which were then later simplified in Whittaker's. The number of classification schemes and the variety of determinants used in those schemes, however, should be taken as a strong indicator that biomes do not all fit perfectly into the classification schemes created.

Holdridge scheme

The Holdridge classification scheme was developed by L.R. Holdridge, a botanist. It maps climates based on four categories:
1. Average total precipitation (cm) on a logarithmic scale.
2. Potential evapotranspiration ratio: the potential evapotranspiration divided by the precipitation; the ratio increases from humid to arid regions.
3. Potential evapotranspiration.
4. Mean annual biotemperature (°C): calculated from monthly mean temperatures after converting any mean temperature to 0°C, based on the assumption that temperatures at or below freezing all have the same effect on plants, and delineating between –10°C and –30°C would yield unrealistic results.

In this scheme, climates are classified based on the biological effects of temperature and rainfall on vegetation under the assumption that these two abiotic factors are the largest determinants of the type of vegetation found in an area. Holdridge uses the 4 axis to define 30 so called 'humidity provinces', which are clearly visible in the Holdridge diagram. While the scheme largely ignores soil and sun exposure, Holdridge did acknowledge that these, too, were important factors in biome determination.

Whittaker's biome-type classification scheme

Whittaker appreciated biome-types as a representation of the great diversity of the living world, and saw the need to establish a simple way to classify these biome-types. Whittaker based his classification scheme on two abiotic factors: Precipitation and Temperature. His scheme can be seen as a simplification of Holdridge's, one more readily accessible, but perhaps missing the greater specificity that Holdrige's provides. Whittaker based his representation of global biomes on both previous theoretical assertions as well as an ever increasing empirical sampling of global ecosystems. Whittaker was in a unique position to make such a holistic assertion as he had previously compiled a review of biome classification. The Whittaker Classification Scheme can be viewed at the following address:

Key definitions for understanding Whittaker's scheme
1. Physiognomy: The apparent characteristics, outward features or appearance of ecological communities or species.

2. Biome: A grouping of terrestrial ecosystems on a given continent that are similar in vegetation structure, physiognomy, features of the environment and characteristics of their animal communities.

3. Formation: A major kind of community of plants on a given continent.

4. Biome-type: Grouping of convergent biomes or formations of different continents; defined by physiognomy.

5. Formation-type: Grouping of convergent formations.

Whittaker's distinction between biome and formation can be simplified: formation is used when applied to plant communities only, while biome is used when concerned with both plants and animals. Whittaker's convention of biome-type or formation-type is simply a broader method to categorise similar communities.

Whittaker's parameters for classifying biome-types

Whittaker, seeing the need for a simpler way to express the relationship of community structure to the environment, used what he called 'gradient analysis' of ecocline patterns to relate communities to climate on a worldwide scale. Whittaker considered four main ecoclines in the terrestrial realm.

1. Intertidal levels: The wetness gradient of areas that are exposed to alternating water and dryness with intensities that vary by location from high to low tide.

2. Climatic moisture gradient.

3. Temperature gradient by altitude.

4. Temperature gradient by latitude.

Along these gradients, Whittaker noted several trends that allow him to qualitatively establish biome-types:

1. The gradient runs from favourable to extreme with corresponding changes in productivity.

2. Changes in physiognomic complexity vary with the favourably of the environment (decreasing community structure and reduction of stratal differentiation as the environment becomes less favourable).

3. Trends in diversity of structure follow trends in species diversity; alpha and beta species diversities decrease from favourable to extreme environments.

4. Each growth-form (i.e. grasses, shrubs, etc.) has its characteristic place of maximum importance along the ecoclines.

5. The same growth forms may be dominant in similar environments in widely different parts of the world.

Whittaker summed the effects of gradients (3) and (4), to get an overall temperature gradient and combined this with gradient (2), the moisture gradient, to express the above conclusions in what is known as the Whittaker Classification Scheme. The scheme graphs average annual precipitation (x-axis) versus average annual temperature (y-axis) to classify biome-types.

Walter system

The Heinrich Walter classification scheme was developed by Heinrich Walter, a German ecologist. It differs from both the Whittaker and Holdridge schemes because it takes into account the seasonality of temperature and precipitation. The system, also based on precipitation and temperature, finds 9 major biomes, with the important climate traits and vegetation types summarised in the accompanying table.

The boundaries of each biome correlate to the conditions of moisture and cold stress that are strong determinants of plant form, and therefore the vegetation that defines the region.

1. Equatorial:
    (a) Always moist and lacking temperature seasonality.
    (b) Evergreen tropical rain forest.
2. Tropical:
    (a) Summer rainy season and cooler 'winter' dry season.
    (b) Seasonal forest, scrub, or savanna.
3. Subtropical:
    (a) Highly seasonal, arid climate.
    (b) Desert vegetation with considerable exposed surface.
4. Mediterranean:
    (a) Winter rainy season and summer drought.
    (b) Sclerophyllous (drought-adapted), frost-sensitive shrublands and woodlands.
5. Warm temperate:
    (a) Occasional frost, often with summer rainfall maximum.
    (b) Temperate evergreen forest, somewhat frost-sensitive.
6. Nemoral:
    (a) Moderate climate with winter freezing.
    (b) Frost-resistant, deciduous, temperate forest.
7. Continental:
    (a) Arid, with warm or hot summers and cold winters.
    (b) Grasslands and temperate deserts.
8. Boreal:
    (a) Cold temperate with cool summers and long winters.
    (b) Evergreen, frost-hardy needle-leaved forest (taiga).
9. Polar:
    (a) Very short, cool summers and long, very cold winters.
    (b) Low, evergreen vegetation, without trees, growing over permanently frozen soils.

Bailey system

Robert G. Bailey almost developed a biogeographical classification system for the United States in a map published in 1976. Bailey subsequently expanded the system to include the rest of South America in 1981 and the world in 1989. The Bailey system is based on climate and is divided into seven domains, with further divisions based on other climate characteristics (subarctic, warm temperate, hot temperate, and subtropical; marine and continental; lowland and mountain).

1. 100 Polar domain:
    (a) 120 Tundra Division.
    (b) M120 Tundra Division—Mountain Provinces.
    (c) 130 Subarctic Division.
    (d) M130 Subarctic Division—Mountain Provinces.
2. 200 Humid temperate Domain:
    (a) 210 Warm Continental Division.
    (b) M210 Warm Continental Division—Mountain Provinces.

   (c) 220 Hot Continental Division.
   (d) M220 Hot Continental Division—Mountain Provinces.
   (e) 230 Subtropical Division.
   (f) M230 Subtropical Division—Mountain Provinces.
   (g) 240 Marine Division.
   (h) M240 Marine Division—Mountain Provinces.
   (i) 250 Prairie Division.
   (j) 260 Mediterranean Division.
   (k) M260 Mediterranean Division—Mountain Provinces.
3. 300 dry Domain:
   (a) 310 Tropical/Subtropical Steppe Division
   (b) M310 Tropical/Subtropical Steppe Division—Mountain Provinces.

World Wide Fund (WWF) system

A team of biologists convened by the World Wide Fund for Nature (WWF) developed an ecological land classification system that identified fourteen biomes, called major habitat types, and further divided the world's land area into 867 terrestrial ecoregions. Each terrestrial Ecoregion has a specific EcoID, fomat XXnnNN (XX is the Ecozone, nn is the Biome number, NN is the individual number). This classification is used to define the Global 200 list of ecoregions identified by the WWF as priorities for conservation. The WWF major habitat types are:

1. Tropical and subtropical moist broadleaf forests (tropical and subtropical, humid).
2. Tropical and subtropical dry broadleaf forests (tropical and subtropical, semi-humid).
3. Tropical and subtropical coniferous forests (tropical and subtropical, semi-humid).
4. Temperate broadleaf and mixed forests (temperate, humid).
5. Temperate coniferous forests (temperate, humid to semi-humid).
6. Boreal forests/taiga (subarctic, humid).
7. Tropical and subtropical grasslands, savannas, and shrublands (tropical and subtropical, semi-arid).
8. Temperate grasslands, savannas, and shrublands (temperate, semi-arid).
9. Flooded grasslands and savannas (temperate to tropical, fresh or brackish water inundated).
10. Montane grasslands and shrublands (alpine or montane climate).
11. Tundra (Arctic).
12. Mediterranean forests, woodlands, and scrub or sclerophyll forests (temperate warm, semi-humid to semi-arid with winter rainfall).
13. Deserts and xeric shrublands (temperate to tropical, arid).
14. Mangrove (subtropical and tropical, salt water inundated).

Freshwater biomes

According to the World Wildlife Fund, the following are classified as freshwater biomes:

1. Large lakes.
2. Large river deltas.
3. Polar freshwaters.
4. Montane freshwaters.
5. Temperate coastal rivers.
6. Temperate floodplain rivers and wetlands.
7. Temperate upland rivers.

8. Tropical and subtropical coastal rivers.
9. Tropical and subtropical floodplain rivers and wetlands.
10. Tropical and subtropical upland rivers.
11. Xeric freshwaters and endorheic basins.
12. Oceanic islands.

Realms or Ecozones (terrestrial and freshwater, WWF)

1. NA Nearctic.
2. PA Palearctic.
3. AT Afrotropic.
4. IM Indomalaya.
5. AA Australasia.
6. NT Neotropic.
7. OC Oceania.
8. AN Antarctic.

Marine biomes

Marine biomes (H) (major habitat types), Global 200 (WWF). Biomes of the coastal and continental shelf areas (Neritic zone—list of ecoregions (WWF))

1. Polar.
2. Temperate shelves and sea.
3. Temperate upwelling.
4. Tropical upwelling.
5. Tropical coral.

Realms or Ecozones (marine, WWF):

1. North Temperate Atlantic.
2. Eastern Tropical Atlantic.
3. Western Tropical Atlantic.
4. South Temperate Atlantic.
5. North Temperate Indo-Pacific.
6. Central Indo-Pacific.
7. Eastern Indo-Pacific.
8. Western Indo-Pacific.
9. South Temperate Indo-Pacific.
10. Southern Ocean.
11. Antarctic.
12. Arctic.
13. Mediterranean.

Other marine habitat types:

1. Hydrothermal vents.
2. Cold seeps.
3. Benthic zone.
4. Pelagic zone (trades and westerlies).
5. Abyssal.
6. Hadal (ocean trench).

Major Habitats, Non Global 200 (WWF):
1. Littoral/Intertidal zone.
2. Kelp forest.
3. Pack ice.

Temperate Forest Biome

Temperate forests are often called deciduous forests. In a temperate forest, most of the trees lose their leaves in the winter. During the fall, when the weather gets cooler, the trees begin to shut down. Their leaves turn beautiful shades of colours. Come winter time, the leaves fall off of the trees. Why do you think this is so?

The deciduous trees must lose their leaves, because water is not available for the leaves to survive. Also, the trees are not strong enough to keep their leaves and hold up all the snow that comes in the winter. To learn some basic facts about temperate forests, follow the links below:
1. Climate.
2. Plants.
3. Animals.

Here are some interesting temperate forest facts:
1. The largest tree in the world is found right here in the United States. You can find it in California. This giant sequoia tree is 275 feet tall and 95 feet around.
2. The largest forest in the world covers parts of Scandinavia and northern Russia. It has over 3.5 million square miles of land covered in trees.
3. Most hardwood trees are used for firewood, construction or art. Lots of forests are being cut down for farm land.

To learn more about temperate forests, visit these wonderful web sites:
1. Explore the Fantastic Forest.
2. Temperate Deciduous Forest Biome.
3. Temperate Woodlands.

Is a temperate forest hot or cold?

In a temperate forest, there are four definate seasons. In some biomes, like the tundra, it is usually cold, but not in the temperate forest. Spring time brings new life to the trees and plants, warm temperatures, and rain which helps everything grow. Summer is hot, and everything is green for all the trees now have leaves. Fall is cool. The trees are beginning to change colours such as red, yellow, and bright orange. Winter months are extremely cold. All the trees loose their leaves. The average temperature for a year in the temperate forest is about 50°F. The forests usually gets about 30 to 60 inches of rainfall every year. The temperate forests are always changing due to the weather, animals, and people.

What kind of trees grow in the temperate forest?

A forest has many layers of plant life. There are the small plants such as shrubs, moss, ferns, and lichens which grow quietly beneath the shadows of the tall trees. The trees found in a temperate forest are called hardwoods. This means the trees loose their leaves in the winter. Their trunks are also made of a bark that is very hard. Most of the trees are maple, birch, beech, oak, hickory, and sweet gum. There are many other types of trees found in the forest too. A temperate forest also contains a few pine trees and other coniferous trees. The Taiga is usually found north of the temperate forests, but sometimes they overlap.

What kind of animals live in the temperate forest?

Animals in a forest come in all different shapes and sizes. Below are a few examples of the kinds of animals you might find in a forest. Have you ever seen any of these animals before?

Cardinal: A cardinal is a bird that is often found in the United States. You can tell it is a cardinal by its bright red feathers.

Squirrel: Here is an animal that you might have seen before. A squirrel is a furry animal with a big, bushy tail. It eats nuts and seeds and lives in the trees.

Bear: There are many types of bears around the world. There are grizzly bears, black bears, and polar bears just to name a few. Take a look at the site called The Cub Den to learn more about bears. There are so many other animals that live in the forests.

DESERT BIOME

Deserts cover about one fifth of the earth's surface and occur where rainfall is less than 50 cm/year. Although most deserts, such as the Sahara of North Africa and the deserts of the southwestern US, Mexico, and Australia, occur at low latitudes, another kind of desert, cold deserts, occur in the basin and range area of Utah and Nevada and in parts of western Asia. Most deserts have a considerable amount of specialised vegetation, as well as specialised vertebrate and invertebrate animals. Soils often have abundant nutrients because they need only water to become very productive and have little or no organic matter. Disturbances are common in the form of occasional fires or cold weather, and sudden, infrequent, but intense rains that cause flooding.

There are relatively few large mammals in deserts because most are not capable of storing sufficient water and withstanding the heat. Deserts often provide little shelter from the sun for large animals. The dominant animals of warm deserts are nonmammalian vertebrates, such as reptiles. Mammals are usually small, like the kangaroo mice of North American deserts.

Desert biomes can be classified according to several characteristics. There are four major types of deserts: (i) hot and dry, (ii) semiarid, (iii) coastal, and (iv) cold.

Hot and Dry Desert

The four major North American deserts of this type are the Chihuahuan, Sonoran, Mojave and Great Basin. Others outside the US include the Southern Asian realm, Neotropical (South and Central America), Ethiopian (Africa) and Australian. The seasons are generally warm throughout the year and very hot in the summer. The winters usually bring little rainfall.

Temperatures exhibit daily extremes because the atmosphere contains little humidity to block the Sun's rays. Desert surfaces receive a little more than twice the solar radiation received by humid regions and lose almost twice as much heat at night. Many mean annual temperatures range from 20°–25°C. The extreme maximum ranges from 43.5°–49°C. Minimum temperatures sometimes drop to –18°C.

Rainfall is usually very low and/or concentrated in short bursts between long rainless periods. Evaporation rates regularly exceed rainfall rates. Sometimes rain starts falling and evaporates before reaching the ground. Rainfall is lowest on the Atacama Desert of Chile, where it averages less than 1.5 cm. Some years are even rainless. Inland Sahara also receives less than 1.5 cm a year. Rainfall in American deserts is higher—almost 28 cm a year.

Soils are course-textured, shallow, rocky or gravely with good drainage and have no subsurface water. They are coarse because there is less chemical weathering. The finer dust and sand particles are blown elsewhere, leaving heavier pieces behind.

Canopy in most deserts is very rare. Plants are mainly ground-hugging shrubs and short woody trees. Leaves are 'replete' (fully supported with nutrients) with water-conserving characteristics. They tend to be small, thick and covered with a thick cuticle (outer layer). In the cacti, the leaves are much-reduced (to spines) and photosynthetic activity is restricted to the stems. Some plants open their stomata (microscopic openings in the epidermis of leaves that allow for gas exchange) only at night when evaporation rates are lowest. These plants include: yuccas, ocotillo, turpentine bush, prickly pears, false mesquite, sotol, ephedras, agaves and brittlebush. The animals include small nocturnal (active at night) carnivores. The dominant animals are burrowers and kangaroo rats. There are also insects, arachnids, reptiles and birds. The animals stay inactive in protected hideaways during the hot day and come out to forage at dusk, dawn or at night, when the desert is cooler.

Semiarid Desert

The major deserts of this type include the sagebrush of Utah, Montana and Great Basin. They also include the Nearctic realm (North America, Newfoundland, Greenland, Russia, Europe and northern Asia). The summers are moderately long and dry, and like hot deserts, the winters normally bring low concentrations of rainfall. Summer temperatures usually average between 21°–27°C. It normally does not go above 38°C and evening temperatures are cool, at around 10°C. Cool nights help both plants and animals by reducing moisture loss from transpiration, sweating and breathing. Furthermore, condensation of dew caused by night cooling may equal or exceed the rainfall received by some deserts. As in the hot desert, rainfall is often very low and/or concentrated. The average rainfall ranges from 2–4 cm annually.

The soil can range from sandy and fine-textured to loose rock fragments, gravel or sand. It has a fairly low salt concentration, compared to deserts which receive a lot of rain (acquiring higher salt concentrations as a result). In areas such as mountain slopes, the soil is shallow, rocky or gravely with good drainage. In the upper bajada (lower slopes) they are coarse-textured, rocky, well-drained and partly 'laid by rock bench'. In the lower bajada (bottom land) the soil is sandy and fine-textured, often with 'caliche hardpan'. In each case there is no subsurface water.

The spiny nature of many plants in semiarid deserts provides protection in a hazardous environment. The large numbers of spines shade the surface enough to significantly reduce transpiration. The same may be true of the hairs on the woolly desert plants. Many plants have silvery or glossy leaves, allowing them to reflect more radiant energy. These plants often have an unfavourable odour or taste. Semiarid plants include: Creosote bush, bur sage (*Franseria dumosa* or *F. deltoidea*), white thorn, cat claw, mesquite, brittle bushes (*Encelia farinosa*), lyciums, and jujube.

During the day, insects move around twigs to stay on the shady side; jack rabbits follow the moving shadow of a cactus or shrub. Naturally, many animals find protection in underground burrows where they are insulated from both heat and aridity. These animals include mammals such as the kangaroo rats, rabbits, and skunks; insects like grasshoppers and ants; reptiles are represented by lizards and snakes; and birds such as burrowing owls and the California thrasher.

Coastal Desert

These deserts occur in moderately cool to warm areas such as the Nearctic and Neotropical realm. A good example is the Atacama of Chile. The cool winters of coastal deserts are followed by moderately long, warm summers. The average summer temperature ranges from 13°–24°C; winter temperatures are 5°C or below. The maximum annual temperature is about 35°C and the minimum is about –4°C. In Chile, the temperature ranges from –2° to 5°C in July and 21°–25°C in January.

The average rainfall measures 8–13 cm in many areas. The maximum annual precipitation over a long period of years has been 37 cm with a minimum of 5 cm. The soil is fine-textured with a moderate salt content. It is fairly porous with good drainage. Some plants have extensive root systems close to the surface where they can take advantage of any rain showers. All of the plants with thick and fleshy leaves or stems can take in large quantities of water when it is available and store it for future use. In some plants, the surfaces are corrugated with longitudinal ridges and grooves. When water is available, the stem swells so that the grooves are shallow and the ridges far apart. As the water is used, the stem shrinks so that the grooves are deep and ridges close together. The plants living in this type of desert include the salt bush, buckwheat bush, black bush, rice grass, little leaf horsebrush, black sage, and chrysothamnus.

Some animals have specialised adaptations for dealing with the desert heat and lack of water. Some toads seal themselves in burrows with gelatinous secretions and remain inactive for eight or nine months until a heavy rain occurs. Amphibians that pass through larval stages have accelerated life cycles, which improves their chances of reaching maturity before the waters evaporate. Some insects lay eggs that remain dormant until the environmental conditions are suitable for hatching. The fairy shrimps also lay dormant eggs. Other animals include: insects, mammals (coyote and badger), amphibians (toads), birds (great horned owl, golden eagle and the bald eagle), and reptiles (lizards and snakes).

Cold Desert

These deserts are characterised by cold winters with snowfall and high overall rainfall throughout the winter and occasionally over the summer. They occur in the Antarctic, Greenland and the Nearctic realm. They have short, moist, and moderately warm summers with fairly long, cold winters. The mean winter temperature is between –2° to 4°C and the mean summer temperature is between 21°–26°C.

The winters receive quite a bit of snow. The mean annual precipitation ranges from 15–26 cm. Annual precipitation has reached a maximum of 46 cm and a minimum of 9 cm. The heaviest rainfall of the spring is usually in April or May. In some areas, rainfall can be heavy in autumn. The soil is heavy, silty, and salty. It contains alluvial fans where soil is relatively porous and drainage is good so that most of the salt has been leached out.

The plants are widely scattered. In areas of shadscale, about 10 per cent of the ground is covered, but in some areas of sagebush it approaches 85 per cent. Plant heights vary between 15 cm and 122 cm. The main plants are deciduous, most having spiny leaves. Widely distributed animals are jack rabbits, kangaroo rats, kangaroo mice, pocket mice, grasshopper mice, and antelope ground squirrels. In areas like Utah, population density of these animals can range from 14–41 individuals per hectare. All except the jack rabbits are burrowers. The burrowing habit also applies to carnivores like the badger, kit fox, and coyote. Several lizards do some burrowing and moving of soil. Deer are found only in the winter.

GRASSLAND

Grasslands are areas where the vegetation is dominated by grasses (Poaceae) and other herbaceous (non-woody) plants (forbs). However, sedge (Cyperaceae) and rush (Juncaceae) families can also be found. Grasslands occur naturally on all continents except Antarctica. In temperate latitudes, such as northwestern Europe and the Great Plains and California in North America, native grasslands are dominated by perennial bunch grass species, whereas in warmer climates annual species form a greater component of the vegetation. Grasslands are found in most ecological regions of the earth. For example there are five terrestrial ecoregion classifications (subdivisions) of the temperate grasslands, savannas, and shrublands biome ('ecosystem'), which is one of eight terrestrial ecozones of the earth's surface.

Grassland vegetation can vary in height from very short, as in chalk where the vegetation may be less than 30 cm (12 in) high, to quite tall, as in the case of North American tallgrass prairie, South American grasslands and African savanna. Woody plants, shrubs or trees, may occur on some grasslands — forming savannas, scrubby grassland or semi-wooded grassland, such as the African. While grasslands in general support diverse wildlife, given the lack of hiding places for predators, the African Savanna regions support a much greater diversity in wildlife than do temperate grasslands.

As flowering plants, grasses grow in great concentrations in climates where annual rainfall ranges between 500 and 900 mm (20 and 35 in). The root systems of perennial grasses and forbs form complex mats that hold the soil in place. Mites, insect larvae, nematodes and earthworms inhabit deep soil, which can reach 6 metres (20 ft) underground in undisturbed grasslands on the richest soils of the world. These invertebrates, along with symbiotic fungi, extend the root systems, break apart hard soil, enrich it with urea and other natural fertilisers, trap minerals and water and promote growth. Some types of fungi make the plants more resistant to insect and microbial attacks.

Climate

Natural grasslands primarily occur in regions that receive between 250 and 900 mm (9.8 and 35 in) of rain per year, as compared with deserts, which receive less than 250 mm (9.8 in) and tropical rainforests, which receive more than 2,000 mm (79 in). Anthropogenic grasslands often occur in much higher rainfall zones, as high as 200 cm (79 in) annual rainfall. Grassland can exist naturally in areas with higher rainfall when other factors prevent the growth of forests, such as in serpentine barrens, where minerals in the soil inhibit most plants from growing. Average daily temperatures range between –20° and 30°C. Temperate grasslands have warm summers and cold winters with rain or some snow.

Grassland Biodiversity and Conservation

Grasslands dominated by unsown wild-plant communities ('unimproved grasslands') can be called either natural or 'semi-natural' habitats. The majority of grasslands in temperate climates are 'semi-natural'. Although their plant communities are natural, their maintenance depends upon anthropogenic activities such as low-intensity farming, which maintains these grasslands through grazing and cutting regimes. These grasslands contain many species of wild plants — grasses, sedges, rushes and herbs — 25 or more speerican prairie grasslands or lowland wildflower meadows in the UK are now rare and their associated wild flora equally threatened. Associated with the wild-plant diversity of the 'unimproved' grasslands is usually a rich invertebrate fauna; also there are many species of birds that are grassland 'specialists', such as the snipe and the Great Bustard. Agriculturally improved grasslands, which dominate modern intensive agricultural landscapes, are usually poor in wild plant species due to the original diversity of plants having been destroyed by cultivation, the original wild-plant communities having been replaced by sown monocultures of cultivated varieties of grasses and clovers, such as Perennial ryegrass and White Clover. In many parts of the world 'unimproved' grasslands are one of the least threatened habitats, and a target for acquisition by wildlife conservation groups or for special grants to landowners who are encouraged to manage them appropriately.

Human Impact and Economic Importance

Grasslands are of vital importance for raising livestock for human consumption and for milk and other dairy products. Grassland vegetation remains dominant in a particular area usually due to grazing, cutting, or natural or manmade fires, all discouraging colonisation by and survival of tree and shrub seedlings. Some of the world's largest expanses of grassland are found in African savanna, and these are maintained by wild herbivores as well as by nomadic pastoralists and their cattle, sheep or goats.

Grasslands may occur naturally or as the result of human activity. Grasslands created and maintained by human activity are called anthropogenic grasslands. Hunting peoples around the world often set regular fires to maintain and extend grasslands, and prevent fire-intolerant trees and shrubs from taking hold. The tallgrass prairies in the American Midwest may have been extended eastward into Illinois, Indiana, and Ohio by human agency. Much grassland in northwest Europe developed after the Neolithic Period, when people gradually cleared the forest to create areas for raising their livestock.

Grassland Types (Biomes)

Tropical and subtropical grasslands

These grasslands are classified with tropical and subtropical savannas and shrublands as the tropical and subtropical grasslands, savannas, and shrublands biome. Notable tropical and subtropical grasslands include the Llanos grasslands of northern South America.

Temperate grasslands

Mid-latitude grasslands, including the Prairie and Pacific Grasslands of North America, the Pampas of Argentina, Brazil and Uruguay, calcareous downland, and the steppes of Europe. They are classified with temperate savannas and shrublands as the temperate grasslands, savannas, and shrublands biome. Temperate grasslands are the home to many large herbivores, such as bison, gazelles, zebras, rhinoceroses, and wild horses. Carnivores like lions, wolves and cheetahs and leopards are also found in temperate grasslands. Other animals of this region include: deer, prairie dogs, mice, jack rabbits, skunks, coyotes, snakes, fox, owls, badgers, blackbirds, grasshoppers, meadowlarks, sparrows, quails, hawks and hyenas.

Flooded grasslands

Grasslands that are flooded seasonally or year-round, like the Everglades of Florida, the Pantanal of Brazil, Bolivia and Paraguay or the Esteros del Ibera in Argentina. They are classified with flooded savannas as the flooded grasslands and savannas biome and occur mostly in the tropics and subtropics.

Montane grasslands

High-altitude grasslands located on high mountain ranges around the world, like the Páramo of the Andes Mountains. They are part of the montane grasslands and shrublands biome and also constitute tundra.

Tundra grasslands

Similar to montane grasslands, polar arctic tundra can have grasses, but high soil moisture means that few tundras are grass-dominated today. However, during the Pleistocene ice ages, a polar grassland known as steppe-tundra occupied large areas of the Northern hemisphere. These are in the tundra biome.

Desert and xeric grasslands

Also called desert grasslands, this is composed of sparse grassland ecoregions located in the deserts and xeric shrublands biome.

Fauna

Grassland in all its form supports a vast variety of mammals, reptiles, birds, and insects. Typical large mammals include the Blue Wildebeest, American Bison, Giant Anteater and Przewalski's Horse. There is evidence for grassland being much the product of animal behaviour and movement; some examples include migratory herds of antelope trampling vegetation and African Bush Elephants eating Acacia saplings before the plant has a chance to grow into a mature tree.

Tundra Biomes

In physical geography, tundra is a biome where the tree growth is hindered by low temperatures and short growing seasons. The term tundra comes through Russian from the Kildin Sami word tundâr 'uplands', 'treeless mountain tract'. There are three types of tundra: Arctic tundra, alpine tundra, and Antarctic tundra. In tundra, the vegetation is composed of dwarf shrubs, sedges and grasses, mosses, and lichens. Scattered trees grow in some tundra. The ecotone (or ecological boundary region) between the tundra and the forest is known as the tree line or timberline.

Arctic

Arctic tundra occurs in the far Northern Hemisphere, north of the taiga belt. The word 'tundra' usually refers only to the areas where the subsoil is permafrost, or permanently frozen soil. (It may also refer to the treeless plain in general, so that northern Sápmi would be included.) Permafrost tundra includes vast areas of northern Russia and Canada. The polar tundra is home to several peoples who are mostly nomadic reindeer herders, such as the Nganasan and Nenets in the permafrost area (and the Sami in Sápmi).

Arctic tundra contains areas of stark landscape and is frozen for much of the year. The soil there is frozen from 25–90 cm (9.8–35.4 inches) down, and it is impossible for trees to grow. Instead, bare and sometimes rocky land can only support low growing plants such as moss, heath, and lichen. There are two main seasons, winter and summer, in the polar tundra areas. During the winter it is very cold and dark, with the average temperature around –28°C (–18°F), sometimes dipping as low as –50°C (–58°F). However, extreme cold temperatures on the tundra do not drop as low as those experienced in taiga areas further south. During the summer, temperatures rise somewhat, and the top layer of the permafrost melts, leaving the ground very soggy. The tundra is covered in marshes, lakes, bogs and streams during the warm months. Generally daytime temperatures during the summer rise to about 12°C (54°F) but can often drop to 3°C (37°F) or even below freezing. Arctic tundras are sometimes the subject of habitat conservation programs. In Canada and Russia, many of these areas are protected through a national Biodiversity Action Plan.

The tundra is a very windy area, with winds often blowing upwards of 48–97 km/hr. (30–60 miles an hour). However, in terms of precipitation, it is desert-like, with only about 15–25 cm (6–10 inches) falling per year (the summer is typically the season of maximum precipitation). During the summer, the permafrost thaws just enough to let plants grow and reproduce, but because the ground below this is frozen, the water cannot sink any lower, and so the water forms the lakes and marshes found during the summer months. Although precipitation is light, evaporation is also relatively minimal.

The biodiversity of the tundras is low: 1,700 species of vascular plants and only 48 land mammals can be found, although millions of birds migrate there each year for the marshes. There are also a few fish species such as the flatfish. There are few species with large populations. Notable animals in the Arctic tundra include caribou (reindeer), musk ox, arctic hare, arctic fox, snowy owl, lemmings, and polar bears (only the extreme north).

Due to the harsh climate of the Arctic tundra, regions of this kind have seen little human activity, even though they are sometimes rich in natural resources such as oil and uranium. In recent times this has begun to change in Alaska, Russia, and some other parts of the world. A severe threat to the tundras, specifically to the permafrost, is global warming. The melting of the permafrost in a given area on human time scales (decades or centuries) could radically change which species can survive there.

Another concern is that about one-third of the world's soil-bound carbon is in taiga and tundra areas. When the permafrost melts, it releases carbon in the form of carbon dioxide and methane, both of which are greenhouse gases. The effect has been observed in Alaska. In the 1970s the tundra was a carbon sink, but today, it is a carbon source.

Antarctic

Antarctic tundra occurs on Antarctica and on several Antarctic and subantarctic islands, including South Georgia and the South Sandwich Islands and the Kerguelen Islands. Most of Antarctica is too cold and dry to support vegetation, and most of the continent is covered by ice fields. However, some portions of the continent, particularly the Antarctic Peninsula, have areas of rocky soil that support plant life. The flora presently consists of around 300–400 lichens, 100 mosses, 25 liverworts, and around 700 terrestrial and aquatic algae species, which live on the areas of exposed rock and soil around the shore of the continent. Antarctica's two flowering plant species, the Antarctic hair grass (*Deschampsia antarctica*) and Antarctic pearlwort (*Colobanthus quitensis*), are found on the northern and western parts of the Antarctic Peninsula.

In contrast with the Arctic tundra, the Antarctic tundra lacks a large mammal fauna, mostly due to its physical isolation from the other continents. Sea mammals and sea birds, including seals and penguins, inhabit areas near the shore, and some small mammals, like rabbits and cats, have been introduced by humans to some of the subantarctic islands. The Antipodes Subantarctic Islands tundra ecoregion includes the Bounty Islands, Auckland Islands, Antipodes Islands, the Campbell Island group, and Macquarie Island. Species endemic to this ecoregion include *Nematoceras dienemum* and *Nematoceras sulcatum*, the only Subantarctic orchids; the royal penguin; and the Antipodean albatross. The flora and fauna of Antarctica and the Antarctic Islands (south of 60° south latitude) are protected by the Antarctic Treaty.

Alpine

Alpine tundra does not contain trees because it has high altitude. Alpine tundra is distinguished from arctic tundra, because alpine tundra typically does not have permafrost, and alpine soils are generally better drained than arctic soils. Alpine tundra transitions to subalpine forests below the tree line; stunted forests occurring at the forest-tundra ecotone are known as *Krummholz.* Alpine tundra occurs in mountains worldwide. The flora of the alpine tundra is characterised by dwarf shrubs close to the ground. The cold climate of the alpine tundra is caused by the low air pressure, and is similar to polar climate.

Climatic Classification

Tundra climates ordinarily fit the Köppen climate classification ET, signifying a local climate in which at least one month has an average temperature high enough to melt snow (0°C or 32°F), but no month with an average temperature in excess of (10°C/50°F). The cold limit generally meets the EF climates of permanent ice and snows; the warm-summer limit generally corresponds with the poleward or altitudinal limit of trees, where they grade into the subarctic climates designated *Dfd* and *Dwd* (extreme winters as in parts of Siberia), *Dfc* typical in Alaska, Canada, European Russia, and Western Siberia (cold winters with months of freezing), or even *Cfc* (no month colder than –3°C as in parts of Iceland and southernmost South America). Tundra climates as a rule are hostile to woody vegetation even where the winters are comparatively mild by polar standards, as in Iceland.

Despite the potential diversity of climates in the ET category involving precipitation, extreme temperatures, and relative wet and dry seasons, this category is rarely subdivided. Rainfall and snowfall are generally slight due to the low vapour pressure of water in the chilly atmosphere, but as a rule potential evapotranspiration is extremely low, allowing soggy terrain of swamps and bogs even in places that get precipitation typical of deserts of lower and middle latitudes. The amount of native tundra biomass depends more on the local temperature than the amount of precipitation.

Taiga

Taiga also known as the boreal forest, is a biome characterised by coniferous forests. Taiga is the world's largest terrestrial biome and covers in North America most of inland Canada and Alaska as well as parts of the extreme northern continental United States; and in most of Sweden, Finland, inland and northern Norway, much of Russia (especially Siberia), northern Kazakhstan, northern Mongolia, and northern Japan (on the island of Hokkaido).

The term boreal forest is sometimes, particularly in Canada, used to refer to the more southerly part of the biome, while the term taiga is often used to describe the more barren areas of the northernmost part of the taiga approaching the tree line. The term taiga is of Russian origin.

Climate and geography

Taiga is the world's *largest* land biome, and makes up 29 per cent of the world's forest cover; the largest areas are located in Russia and Canada. The taiga is the terrestrial biome with the lowest annual average temperatures after the tundra and permanent ice caps. However, extreme minimums in the taiga are typically lower than those of the tundra. The lowest reliably recorded temperatures in the Northern Hemisphere were recorded in the taiga of northeastern Russia. The taiga or boreal forest has a subarctic climate with very large temperature range between seasons, but the long and cold winter is the dominant feature. This climate is classified as *Dfc, Dwc, Dsc, Dfd, Dwd* and *Dsd* in the Köppen climate classification scheme, meaning that the short summer (24 hrs average 10°C or more) lasts 1–3 months and always less than 4 months. There are also some much smaller areas grading towards the oceanic *Cfc* climate with milder winters. The mean annual temperature generally varies from –5°C to 5°C, but there are taiga areas in both eastern Siberia and interior Alaska-Yukon where the mean annual reaches down to –10°C.

Soils

Taiga soil tends to be young and poor in nutrients. It lacks the deep, organically-enriched profile present in temperate deciduous forests. The thinness of the soil is due largely to the cold, which hinders the development of soil and the ease with which plants can use its nutrients. Fallen leaves and moss can remain on the forest floor for a long time in the cool, moist climate, which limits their organic contribution to the soil; acids from evergreen needles further leach the soil, creating spodosol, also known as podzol. Since the soil is acidic due to the falling pine needles, the forest floor has only lichens and some mosses growing on it.

Flora

Since North America and Asia used to be connected by the Bering land bridge, a number of animal and plant species (more animals than plants) were able to colonise both continents and are distributed throughout the taiga biome. Others differ regionally, typically with each genus having several distinct species, each occupying different regions of the taiga. Taigas also have some small-leaved deciduous trees like birch, alder, willow, and poplar; mostly in areas escaping the most extreme winter cold.

However, the Dahurian Larch tolerates the coldest winters in the northern hemisphere in eastern Siberia. The very southernmost parts of the taiga may have trees such as oak, maple, elm, and tilia scattered among the conifers, and there is usually a gradual transition into a temperate mixed forest, such as the Eastern forest-boreal transition of eastern Canada. In the interior of the continents with the driest climate, the boreal forests might grade into temperate grassland.

There are two major types of taiga. The southern part is the closed canopy forest, consisting of many closely-spaced trees with mossy ground cover. In clearings in the forest, shrubs and wildflowers are common, such as the fireweed. The other type is the lichen woodland or sparse taiga, with trees that are farther-spaced and lichen ground cover; the latter is common in the northernmost taiga.

The forests of the taiga are largely coniferous, dominated by larch, spruce, fir, and pine. The woodland mix varies according to geography and climate so for example the Eastern Canadian forests ecoregion of the higher elevations of the Laurentian Mountains and the northern Appalachian Mountains in Canada is dominated by balsam fir *Abies balsamea*, while further north the Eastern Canadian Shield taiga of northern Quebec and Labrador is notably black spruce *Picea mariana* and tamarack larch *Larix laricina*.

Evergreen species in the taiga (spruce, fir, and pine) have a number of adaptations specifically for survival in harsh taiga winters, although larch, the most cold-tolerant of all trees, is deciduous. Taiga trees tend to have shallow roots to take advantage of the thin soils, while many of them seasonally alter their biochemistry to make them more resistant to freezing, called 'hardening'. The narrow conical shape of northern conifers, and their downward-drooping limbs.

Fauna

The boreal forest, or taiga, supports a large range of animals. Canada's boreal forest includes 85 species of mammals, 130 species of fish, and an estimated 32,000 species of insects. Insects play a critical role as pollinators, decomposers, and as a part of the food chain. Many nesting birds rely on them for food. The cold winters and short summers make the taiga a challenging biome for reptiles and amphibians, which depend on environmental conditions to regulate their body temperatures, and there are only a few species in the boreal forest. Some hibernate underground in winter. The taiga is home to a number of large herbivorous mammals, such as moose and reindeer/caribou. Some areas of the more southern closed boreal forest also have populations of other deer species such as the elk (wapiti) and roe deer. There is also a range of rodent species including beaver, squirrel, mountain hare, snowshoe hare, and vole. These species have evolved to survive the harsh winters in their native ranges. Some larger mammals, such as bears, eat heartily during the summer in order to gain weight, and then go into hibernation during the winter. Other animals have adapted layers of fur or feathers to insulate them from the cold.

Threats

Human activities

Large areas of Siberia's taiga have been harvested for lumber since the collapse of the Soviet Union. In Canada, eight per cent of the boreal forest is protected from development, the provincial government allows forest management to occur on Crown land under rigorous constraints. The main forestry practice in the boreal forest of Canada is clearcutting, which involves cutting down most of the trees in a given area, then replanting the forest as a monocrop (one species of tree) the following season. Industry officials claim that this process emulates the natural effects of a forest fire, which they claim clearcutting suppresses, protecting infrastructure, communities and roads. However, from an ecological perspective, this is a falsehood, for several reasons, including: (i) removing most of the trees in a given area is usually done using large machines which disrupt the soil greatly, and the dramatic diminution of ground cover permits large-scale erosion and avalanches, which further damage the habitat and sometimes endangers infrastructure, roads, and communities, (ii) clearcutting removes most of the biomass from an area, and the various macro and micro-nutrients it contains. This sudden decrease in nutrients in an area contrasts with a forest fire, which returns most of the nutrients to the soil, and (iii) forest fires leave standing snags, and leave patches of unburned trees. This helps preserve structure and micro-habitats within the area, whereas clearcutting destroys most of these habitats.

Climate change

The zone of latitude occupied by the boreal forest has experienced some of the greatest temperature increases on earth, especially during the last quarter of the twentieth century. Winter temperatures have increased more than summer temperatures. The number of days with extremely cold temperatures (e.g. −20° to −40°C) has decreased irregularly but systematically in nearly all the boreal region, allowing better survival for tree-damaging insects. In summer, the daily low temperature has increased more than the daily high temperature. In Fairbanks, Alaska, the length of the frost-free season has increased from 60–90 days in the early twentieth century to about 120 days a century later. Summer warming has been shown to increased water stress and reduce tree growth in dry areas of the southern boreal forest in central Alaska, western Canada and portions of far eastern Russia. Precipitation is relatively abundant in Scandinavia, Finland, northwest Russia and eastern Canada, where warmer summers accelerate tree growth. As a consequence of this warming trend, the warmer parts of the boreal forests are susceptible to replacement by grassland, parkland or temperate forest. In Siberia, the taiga is converting from predominantly needle-shedding larch trees to evergreen conifers in response to a warming climate. This is likely to further accelerate warming, as the evergreen trees will absorb more of the sun's rays. Given the vast size of the area, such a change has the potential to affect areas well outside of the region. In much of the boreal forest in Alaska, the growth of white spruce trees are stunted by unusually warm summers, while trees on some of the coldest fringes of the forest are experiencing faster growth than previously. Lack of moisture in the warmer summers are also stressing the birch trees of central Alaska.

Insects

Recent years have seen outbreaks of insect pests in forest-destroying plagues: the spruce-bark beetle (Dendroctonus rufipennis) in the Yukon Territory, Canada, and Alaska; the aspen-leaf miner; the larch sawfly; the spruce budworm (Choristoneura fumiferana); the spruce coneworm.

Protection

Many nations are taking direct steps to protect the ecology of the taiga by prohibiting logging, mining, oil and gas production, and other forms of development. In February 2010 the Canadian government established protection for 13,000 square kilometres of boreal forest by creating a new 10,700 square kilometre park reserve in the Mealy Mountains area of eastern Canada and a 3,000 square kilometre waterway provincial park that follows alongside the Eagle River from headwaters to sea. The taiga stores enormous quantities of carbon, possibly more than the temperate and tropical forests combined, much of it in peatland.

Natural Disturbance

One of the biggest areas of research and a topic still full of unsolved questions is the recurring disturbance of fire and the role it plays in propagating the lichen woodland. The phenomenon of wildfire by lightning strike is the primary determinant of understory vegetation and because of this, it is considered to be predominate driving force behind community and ecosystem properties in the lichen woodland. The significance of fire is clearly evident when one considers that understory vegetation influences tree seedling germination in the short-term and decomposition of biomass and nutrient availability in the long-term. The recurrent cycle of large, damaging fire occurs approximately every 70 to 100 years. Understanding the dynamics of this ecosystem is entangled with discovering the successional paths that the vegetation exhibits after a fire. Trees, shrubs and lichens all recover from fire induced damage through vegetative reproduction as well as invasion by propagules. Seeds that have fallen and become buried provide little help in re-establishment of a species. The reappearance of lichens is reasoned to occur because of varying conditions and light/nutrient availability in each different microstate. Several different studies have been done that have led to the formation of the theory that post-fire development can be propagated by any of four pathways: self replacement, species-dominance relay, species replacement, or gap-phase self replacement. Self replacement is simply the re-establishment of the pre-fire dominant species. Species-dominance relay is a sequential attempt of tree species to establish dominance in the canopy. Species replacement is when fires occur in sufficient frequency to interrupt species dominance relay. Gap-Phase Self-Replacement is the least common and so far has only been documented in Western Canada. It is a self replacement of the surviving species into the canopy gaps after a fire kills another species. The particular pathway taken after a fire disturbance depends on how the landscape is able to support trees as well as fire frequency. Fire frequency has a large role in shaping the original inception of the lower forest line of the lichen woodland taiga.

Centuries ago, the southern limits of lichen woodland taiga were only being formed. It has been hypothesised and subsequently proved by Serge Payette that the Spruce-Moss forest ecosystem was changed into the lichen woodland biome due to the initiation of two compounded strong disturbances. The two disturbances were large fire and the appearance and attack of the spruce budworm. The spruce budworm is a deadly insect to the spruce populations in the southern regions of the taiga. J.P. Jasinski confirmed this theory five years later stating 'Their (lichen woodlands) persistence, along with their previous moss forest histories and current occurrence adjacent to closed moss forests, indicate that they are an alternative stable state to the spruce–moss forests'.

MINOR BIOMES

Some of the minor biomes are discussed below.

Tropical Savannas

Tropical savannas or grasslands are associated with the tropical wet and dry climate type (Koeppen's Aw), but they are not generally considered to be a climatic climax. Instead, savannas develop in regions where the climax community should be some form of seasonal forest or woodland, but edaphic conditions or disturbances prevent the establishment of those species of trees associated with the climax community. Seasonal forests of the tropics are also widespread and vary along a latitudinal/moisture gradient between the tropical broadleaf evergreen forest of the equatorial zone and the deserts of the subtropics.

The word savanna stems from an Amerind term for plains which became Hispanicised after the Spanish Conquest.

The vegetation: Savannas are characterised by a continuous cover of perennial grasses, often 3 to 6 feet tall at maturity. They may or may not also have an open canopy of drought-resistant, fire-resistant, or browse-resistant trees, or they may have an open shrub layer. Distinction is made between tree or woodland savanna, park savanna, shrub savanna and grass savanna. Furthermore, savannas may be distinguished according to the dominant taxon in the tree layer: for example, palm savannas, pine savannas, and acacia savannas.

Climate: A tropical wet and dry climate predominates in areas covered by savanna growth. Mean monthly temperatures are at or above 64°F and annual precipitation averages between 30 and 50 inches. The dry season is associated with the low sun period.

Soils: Soils vary according to bedrock and edaphic conditions. In general, however, laterisation is the dominant soil-forming process and low fertility oxisols can be expected.

Temperate Rain Forest

Temperate rainforests are coniferous or broad leaf forest that occur in temperate zone and receive high rain fall. For temperate rain forests of North America, Alaback's definition is widely recognised:
1. Annual precipitation 200–400 cm.
2. Mean annual temperature between 4°C and 12°C. (39° and 54° Fahrenheit).

However, required annual precipitation depends on factors such as distribution of rainfall over the year, temperatures over the year and fog presence, and definitions in other countries differ considerably. For example, Australian definitions are ecological-structural rather than climatic:
1. Closed canopy of trees excludes at least 70 per cent of the sky.
2. Forest is composed mainly of tree species which do not require fire for regeneration, but with seedlings able to regenerate under shade and in natural openings.

The latter would, for example, exclude a part of the temperate rain forests of western North America, as Coast Douglas-fir, one of its dominant tree species, requires stand-destroying disturbance to initiate a new cohort of seedlings. The North American definition would in turn exclude a part of temperate rain forests in other countries.

Global distribution: Temperate forests cover a large part of the globe, but temperate rain forests only occur in few regions around the world.

Temperate rain forest regions

Pacific temperate rain forests of western North America

A portion of the temperate rain forest region of North America, the largest area of temperate zone rain forests on the planet, is the Pacific temperate rain forests ecoregion which occur on west-facing coastal mountains along the Pacific coast of North America, from Kodiak Island in Alaska to northern California, and are part of the Nearctic ecozone.

Chaparral

Chaparral is a shrubland or heathland plant community found primarily in the US state of California and in the northern portion of the Baja California peninsula, Mexico. It is shaped by a Mediterranean climate (mild, wet winters and hot dry summers) and wildfire, having summer drought tolerant plants with hard sclerophyllous evergreen leaves, as contrasted with the associated soft leaved, drought deciduous, scrub community of Coastal sage scrub, found below the chaparral biome. Chaparral covers 5 per cent of the state of California, and associated Mediterranean scrubland an additional 3.5 per cent.

Chaparral is characterised by frequent fires, with intervals ranging between a few years to a hundred years. Before fires, mature chaparral is characterised by nearly impenetrable dense thickets (except the more open chaparral of the desert). The plants are highly flammable. They grow as woody shrubs with hard and small leaves, are non-leaf dropping (non-deciduous), and are drought tolerant. After the first rains following a fire, the landscape is dominated by soft leaved non-woody annual plants, known as fire followers, which die back with the summer dry period.

Cismontane chaparral (this side of the mountain) refers to chaparral growing on the coastal side of large mountain ranges, such as west of the Sierra Nevada or San Jacinto Mountains east of the San Diego area, or south of the Transverse ranges north of the Los Angeles area, while transmontane (the other side of the mountain) chaparral refers to the shrubland community growing in the rainshadow of these ranges, which has a desert, not Mediterranean climate, and is referred to as Desert chaparral. The foothills west of the Sierra Nevada may be covered with cismontain chaparral, and have a hot and subhumid climate. Some classify cismontane chaparral further into upper and lower chaparral.

Similar plant communities are found in the four other Mediterranean climate regions around the world, including the Mediterranean Basin (where it is known as maquis), central Chile (where it is called matorral), South African Cape Region (known there as fynbos), and in Western and Southern Australia. According to the California Academy of Sciences, Mediterranean shrubland contains more than 20 per cent of the world's plant diversity. The word chaparral is a loan word from Spanish. The Spanish word comes from the word chaparro, which means both small and dwarf evergreen oak, which itself comes from the Basque word txapar, with the same meaning.

Conservation International and other conservation organisations consider the chaparral to be a biological community with a large number of different species that is under threat by human activity, know as a biodiversity hotspot.

SECTION V

Animal Ecology

Chapter 17

Animal Interrelationships

INTRODUCTION

Animal agriculture is closely interrelated to both the natural environment and human systems, including rural communities. Accordingly, changes in animal agriculture can have wide-ranging consequences across many areas. During the past 50 year, there has been tremendous change in animal agriculture, involving an increase in the size of production units, greater reliance on technology, a corresponding decrease in human labour, increased confinement of animals, and a general trend toward monoculture or specialised production systems. At least in part, these changes were brought about as a consequence of animal science research in nutrition, breeding, reproduction, growth, and so on. A long-term goal for animal scientists has been to increase the biological efficiency of animal-based food production, and the success in reaching this goal has been remarkable, with the time to market, growth rates, milk and egg production, etc. per animal increasing two- to threefold in some cases during the last 50 year. The increase in the efficiency of animal agriculture has brought about a parallel decrease in food prices. Nonetheless, whereas animal science in one sense has been very successful, new questions or issues have emerged. The scale of animal systems today sometimes concentrates large numbers of animals into smaller areas that cannot handle the resultant animal manure. Stream and ground water pollution is increasingly a concern in some regions. Odour is a nuisance problem that increasingly places neighbours and urban growth in conflict with confinement animal systems. Possibly one of the biggest issues can be stated in terms of sustainability: Can all current food animal production systems continue as they currently exist? Additionally, the decrease in the number of producers has affected rural communities, and in some cases has brought about the demise of small towns. Animal scientists typically contend that they serve the interests of producers and strive to promote practices that are environmentally sound. Bringing about a discussion among animal scientists as to whether these goals are always met or could be better met, is important if both the industry and our rural communities are to survive and thrive.

INTERSPECIFIC AND INTRASPECIFIC RELATIONSHIPS

Intraspecific Relationships

Intraspecific interactions are interactions among organisms of the same species. Members of a group must be able to communicate with each other to stay in touch with others in the group. Many animals use vocal communication of some sort while both plants and animals use various chemical signals to communicate with each other. In animals, these chemicals are generally called *pheromones* and are used for chemical communication over a distance (these are analogous to hormones within the body). There are various kinds of pheromones including things like sex pheromones in a variety of insects (used in gypsy moth traps), trail pheromones in ants, alarm pheromones in ants (and probably some other insects, too), and others in other animals, including humans. Plants often secrete inhibitory chemicals into the surrounding soil to prevent the growth of other plants nearby. This is called *allelopathy.* Amur Honeysuckle and Black Walnut are two species that are especially notorious for producing allelopathic chemicals.

Interspecific Relationships

Interspecific interactions are interactions among organisms of different species. Typically, these interactions are classified based on whether they are beneficial to one or both of the species involved or whether they are detrimental to one of the species involved. An organism's niche is its functional role within the community, including its activities and relationships, its 'address', its 'job' or function within the community, and how it relates to other organisms. The niche of each species is a little different to avoid competition. Different species, even closely-related ones, will have different food preferences, seasonality, daily feeding rhythms, and location within the habitat. For some species of katydids within the same genus, the difference may be as subtle as a preference for perching on the top vs the middle of a stem on a grass plant.

Types of relationships

Symbiosis

Any relationship that involves two (or more) species living together and interacting. This is a general term which includes predation, parasitism, commensalism, mutualism, etc. but often is used to mean mutualism.

Predation

When a larger animal eats other, smaller animals. Lions may eat antelope, and wolves may eat deer. Spiders, like this orbweaver, capture and eat insects such as the cricket she's eating.

Commensalism

A relationship between two species that is beneficial to one but of neutral benefit to the other. Cattle egrets follow cattle to feed on the insects stirred up by the grazing cattle.

Mutualism

A relationship between two species where both benefit. The yucca moth both pollinates and feeds on the yucca plant; acacia ants live in the thorns of, defend, and are fed by the acacia tree in which they live; and trees cannot get along without mycorrhizae living in/on their roots and absorbing food for them. Many plants and their pollinators have evolved mutualistic relationships. Butterfly-weed provides food for and is pollinated by butterflies like pipevine swallowtails.

Parasitism

When a smaller organism feeds on a larger, weakening or killing it. This is a relationship where one organism benefits and the other is harmed. Often the host is not killed outright. Because a parasite lives in/on the body of its host and needs the host to remain alive, it is usually advantageous for the parasite to not kill its host. Humans and domestic animals are occasionally infected with or bothered by tapeworms, roundworms, mosquitoes and/or leeches.

Parasitoid

A parasite that eventually causes the death of its host. By the time the parasitoid undergoes metamorphosis, all of the host's innards have been eaten. Often, insect larvae that are parasitoids of other insects eat the host's tissues, timing things such that just as they're ready to pupate, they have eaten up the whole insides of their host, and it dies.

Braconid wasps do this to tomato hornworms, and this hornworm, covered with cocoons of pupating braconids, probably has almost no body parts left inside. If you see a caterpillar like this on your tomato plants, leave it alone. The wasps will eventually hatch, mate, and lay eggs in any other tomato hornworms they can find—a good means of biological control.

SYMBIOSIS THEORY

Symbiosis means living together. There is some confusion over the use of word symbiosis. The term symbiosis has been used as a synonym of mutualism by some authors. Others have used it in a wider sense to include mutualism, commensalism and even parasitism.

Characteristics of symbiosis: There is probably no individual organism that does not play a host to at least one symbiont. A symbiont is an organism which takes part in a symbiotic association. Symbionts are associated to get mutual benefits like food, shelter, substratum for attachment or transportation. The relationship is more intimidate when compared to commensalism. Generally both the symbionts are benefitted. Symbiosis is mostly an obligatory association.

Types of Symbiosis

There are two types of symbiosis namely symbiosis with continuous contact and symbiosis without continuous contact.

Symbiosis with continuous contact

In this type of symbiosis, the association of the symbionts is continuous, permanent and obligatory. This type of association occurs between 'plants and plants', 'animals and animals' and 'animals and plants'.

Symbiosis between animal species

A good example for this type is the association of the protozoan flagellate *trichonympha termites*. The termites feed on wood but cannot digest the cellulose for the termites. The flagellates which live in the intestine digest the cellulose for the termite. While the trichonympha digests the cellulose for the termite, the termite provides food, shelter and constant internal environment to the with the trichonympha. They can, in fact, survive only in the intestines of the termites. The newly hatched termites instinctively lick the anus of another termite to get a supply of the flagellates.

Symbiosis between plant species

A more intimate form of mutualism is the association of two plants species as illustrated by Lichens which are seen to grow on the rocks and barks of trees. Lichens are associations of thread like fungi and photosynthesising algae. The algae produce food for themselves and for the fungus. The fungus in turn contributes water, nitrogen containing waste compounds and carbon dioxide for continued algae photosynthesis. Here, the intimacy of the association is such that neither the fungus nor the algae can grow independently in nature.

Symbiosis between animal and plant species

This kind of association exists between the flat worm *Convoluta* and the unicellular green algae *zoochlorella*. For sometime after hatching the flat worm does not possess zoochlorella and it is holozoic in nutrition. Later, when it acquires zoochlorella it appears green in colour and looses the digestive organs since zoochlorella is photosynthetic and provides food for its host. The flat worm provides shelter and other materials like carbon dioxide, etc. to zoochlorella.

Symbiosis without continuous contact

In this type of mutualism the symbionts the symbionts are not permanently associated. They come in contact with one another for a short period of time. Here the association of the symbionts is not continuous, but is temporary and facultative. The tickbird-rhinoceros association is very good example for this type of mutualism. The tickbirds sit on the backs of the rhinoceros and feed on their skin parasites. Thus the rhinoceros are relived of the parasites and they are also warned of any danger when the sharp eyed tickbird suddenly flies off temporarily. The tickbird, on other hand gets its food by way of feeding on the skin parasites. This is an example of facultative symbiosis because both the tickbirds and rhinoceros can get along without each other if necessary.

Dinoflagellate

The dinoflagellates are a large group of flagellate protists. Most are marine plankton, but they are common in fresh water habitats as well. Their populations are distributed depending on temperature, salinity or depth. About half of all dinoflagellates are photosynthetic, and these make up the largest group of marine eukaryotic aside from the diatoms. Being primary producers makes them an important part of the aquatic food chain. Some species, called zooxanthellae, are endosymbionts of marine animals and play an important part in the biology of coral reefs. Other dinoflagellates are colourless predators on other protozoa, and a few forms are parasitic. An algal bloom of dinoflagellates can result in a visible coloration of the water colloquially known as red tide. Part of the challenge in dinoflagellate taxonomy and nomenclature is that they have been independently classified by the rules of zoology and botany, and only recently have the disciplines converged.

Most dinoflagellates are unicellular forms with two *flagella*. One of these extends towards the posterior, called the *longitudinal* flagellum, while the other forms a lateral circle, called the *transverse* flagellum. In many forms these are set into grooves, called the *sulcus* and *cingulum*. The transverse flagellum, which is coiled, provides most of the force propelling the cell, and often imparts to it a distinctive whirling motion, which is what gives them their name. The longitudinal flagellum acts mainly as a rudder, but provides a small amount of propulsive force as well.

Most dinoflagellates have a peculiar form of nucleus, called a *dinokaryon*, in which the chromosomes are attached to the nuclear membrane. These lack histones and remain condensed throughout interphase rather than just during mitosis, which is closed and involves a unique external spindle. This sort of nucleus was once considered to be an intermediate between the nucleoid region of prokaryotes and the true nuclei of eukaryotes, and so were termed *mesokaryotic*, but now are considered advanced rather than primitive traits.

INSECT SYMBIOSIS

Insect symbiosis is a collection of microbes that can be called symbionts that are associated with insects and mites. Whether bacteria, fungi or spiroplasmids, and whether endosymbionts or casual gut symbionts.

Fungi and Insect Symbiosis

The most interesting of the fungus-insect symbiotic relationships are those involving colonial insects. One of the most important driving forces that result in symbiotic relationships between micro-organisms is the inability of animals to digest cellulose. When you think of herbivores, such as horses, sheep, cows, goats, etc. they do not actually have the ability to digest the cellulose from the plant material that they consume. Instead, they have symbiotic bacteria, in their stomach, that have the cellulolytic enzymes that digest the plant material for them. Other animals, such as detritivores, do not carry micro-organisms in their gut, but rather consume mycelium in well decomposed plant material as their food source. Thus, symbiotic relationships between animals and various micro-organisms are common.

Ants, termites and mushrooms

Social insects have always been of interest because of their seemingly, well ordered societies. In some of these social insects, the mound-building termites of Africa and Asia, and the leaf-cutting ants of Central and South America, there has evolved a rather unique strategy in the utilisation of cellulose-rich plant material. These insects cultivate cellulolytic fungi, in underground gardens. They establish pure cultures of their fungus. That is they grow only one fungus in their garden, which is not easily done, since there are so many sources of contamination that can occur and prevent their gardens from being successful. However, these insects are able to keep their gardens pure by constantly weeding out foreign fungi. They also care for their garden by providing suitable a food source, i.e. plant material, and moisture. So the fungi obviously benefit from this arrangement, but the ants and termites also benefit from this relationship. These insects are exclusively mycophagous, i.e. they eat fungi. The fungi that they cultivate decompose the wood and leaves brought in by the termites and ants, respectively, and provide them with digestible and nutritious mycelium.

Leaf-cutting ants, leucoagaricus and lepiota

These gardening ants are from the New World Tropics and are commonly referred to as the Attine ants. They represent hundreds of species of ants, from approximately fifty genera. Although many people have never heard of these ants, to the people of South America they are an all too familiar sight. In their search for food, these ants will devastate the natural vegetation and crops that are in their path, as they search for plant material to feed their fungus. At the base of their fruit trees could be seen the ant nests which were 'white as snow', presumably from the mycelium that they were growing. Of all the known species, *Atta sexdens* is the most economically important and the one which is most intensely studied, and the species that we will look at in detail as representative of this group of ants.

MUTUALISM RELATIONSHIPS

Like humans, other living species, such as animals, plants and other bacteria too, help each other for better survival. No living organism can survive in isolation. All the living organisms depend on each other for food or shelter. For example, snakes eat rats, eagles eat snakes, hunter birds eat eagles, so on and so forth. But, this food chain is common in normal weather conditions. In extreme weather conditions, some plant or animal species cannot afford to eliminate the other species but they need each other for survival. Mutualism relationships is all about two species living together for mutual benefits.

Mutualism relationships are characterised by positive reciprocal relationship between two species for survival. There can be many reasons for two species to get in to mutual relationships. The alliance may benefit them in the form of food, shelter, defense, transport, pollination, nutrition or any other mutual need, etc. There are two types of mutualism relationships. Symbiotic relationship, which is an obligate relationship, where the two species live in close proximity and at least one of the species need to contribute in order to survive. For example, parasitic fungus, that initially grows on the plant roots and depends on the plant for shelter. Later, when it grows significantly, it provides mineral nutrients to the plant, that helps the plant for better survival. The fungus in turn gets more carbohydrates from the plant necessary for its growth. In non symbiotic relationships, the two species may not live together or may not be dependent of each other, but they come together at times for certain mutual benefits. For example, birds and flower plants. Birds come to the plants for flower juices or to eat other organisms and in the process, pick up the pollens and spread it else where. This helps the plant for pollination and provides a greater opportunity for genetic diversity. Here, birds and flower plants are not directly dependent on each other, but their occasional alliance benefits both of them.

Examples of mutualism relationships: The animals get in to alliances for certain benefits. The best example for mutualism relationships between animals, can be of Egyptian plover and the crocodile. In the tropical African jungles, the crocodile lies keeping its mouth open. The plover flies in to the mouth of the crocodile and eats the decaying meat stuck in its teeth. The crocodile does not eat the plover, but appreciates the free dental care. This way, both of them are benefited from each other. Another example of mutualism relationship between animals can be of the clown fish and Ritteri sea anemones. The clown fish resides in the stinging tentacles of the sea anemone, which are otherwise very harmful. The sea anemone gets nutrition from the fecal matter of the clown fish and clown fish gets protection from its predators due to stingy tentacles.

Commensalism

In ecology, commensalism is a class of relationship between two organisms where one organism benefits but the other is neutral (there is no harm or benefit). There are three other types of association: mutualism (where both organisms benefit), competition (where both organisms are harmed) and parasitism (one organism benefits and the other one is harmed). Commensalism derives from the English word commensal, meaning 'sharing of food' in human social interaction, which in turn derives from the Latin cum mensa, meaning. Originally, the term was used to describe the use of waste food by second animals, like the carcass eaters that follow hunting animals, but wait until they have finished their meal.

Examples of commensal relationships

Commensalism is harder to demonstrate than parasitism and mutualism, for it is easier to show a single instance whereby the host is affected, than it is to prove or disprove that possibility. Often, a detailed investigation will show that the host indeed has become affected by the relationship.

Cattle egrets and livestock

An example of commensalism: Cattle egrets foraging in fields among cattle or other livestock. As cattle, horses and other livestock graze on the field, they cause movements that stir up various insects. As the insects are stirred up, the cattle egrets following the livestock catch and feed upon them. The egrets benefit from this relationship because the livestock have helped them find their meals, while the livestock are typically unaffected by it.

Tigers and golden jackals

In India, lone golden jackals expelled from their pack have been known to form commensal relationships with tigers. These solitary jackals, known as *kol-bahl*, will attach themselves to a particular tiger, trailing it at a safe distance in order to feed on the big cat's kills. A kol-bahl will even alert a tiger to a kill with a loud *pheal* (there by straying into mutualism). Tigers have been known to tolerate these jackals: one report describes how a jackal confidently walked in and out between three tigers walking together a few feet away from each other. Tigers will however kill jackals on occasion: the now extinct tigers of the Amu Darya region were known to eat jackals frequently.

Other examples

Another example of commensalism: birds following army ant raids on a forest floor. As the army ant colony travels on the forest floor, they stir up various flying insect species. As the insects flee from the army ants, the birds following the ants catch the fleeing insects. In this way, the army ants and the birds are in a commensal relationship because the birds benefit while the army ants are unaffected.

Orchids and mosses are plants that can have a commensal relationship with trees. The plants grow on the trunks or branches of trees. They get the light they need as well as nutrients that run down along the tree. As long as these plants do not grow too heavy, the tree is not affected.

Barnacles

Barnacles are highly sedentary crustaceans that must attach themselves permanently to a hard substrate, such as rocks, shells, whales or anything else on which they can gain a foothold. When they attach to the shell of a scallop, for instance, barnacles benefit by having a place to stay, leaving the scallop presumably unaffected.

Types of commensalism

Like all ecological interactions, commensalisms vary in strength and duration from intimate, long-lived symbioses to brief, weak interactions through intermediaries.

Phoresy

One animal attaching to another for transportation only. This concerns mainly arthropods, examples of which are mites on insects (such as beetles, flies or bees), *pseudoscorpions* on mammals or beetles, and millipedes on birds. Phoresy can be either obligate or facultative (induced by environmental conditions).

Inquilinism

Using a second organism for housing. Examples are epiphytic plants (such as many orchids) that grow on trees or birds that live in holes in trees.

Metabiosis

A more indirect dependency, in which one organism creates or prepares a suitable environment for a second. Examples include maggots, which feast and develop on corpses, and hermit crabs, which use gastropod shells to protect their bodies.

ANTAGONISM

Antagonism, in ecology, is an association between organisms in which one benefits at the expense of the other. As life has evolved, natural selection has favoured organisms that are able to efficiently extract energy and nutrients from their environment. Because organisms are concentrated packages of energy and nutrients in themselves, they can become the objects of antagonistic interactions. Although antagonism is commonly thought of as an association between different species, it may also occur between members of the same species through competition and cannibalism.

One way of understanding the diversity of antagonistic interactions is through the kinds of host or prey that species attack. Carnivores attack animals, herbivores attack plants, and fungivores attack fungi. Other species are omnivorous, attacking a wide range of plants, animals, and fungi. Regardless of the kinds of foods they eat, however, there are some general patterns in which species interact. Parasitism, grazing, and predation are the three major ways in which species feed on one another. The parasite lives on and feeds off its host, usually decreasing the host's ability to survive but not killing it outright. Grazing species are not as closely tied to their food source as parasites and often vary their diet between two or more species without directly killing them. Predators, however, capture and kill members of other species for food.

Antagonistic interactions may also involve defensive strategies that make use of chemical and physical deterrents. Many plant species may secrete chemicals into the soil to prevent other plants from taking root nearby or into their tissues to deter grazing. Some plants and animals may develop physical structures, such as hard coverings and spines, to discourage grazers and predators. In addition, some species possess adaptations that help them resemble others. Such adaptations may be used for both attack and defense.

PARASITISM

Parasitism is a type of symbiotic relationship between organisms of different species where one organism, the parasite, benefits at the expense of the other, the host. Traditionally parasite referred to organisms with lifestages that went beyond one host (e.g. *Taenia solium*), which are now called macroparasites (typically protozoa and helminths). Parasites can now also refer to microparasites, which are typically smaller, such as viruses and bacteria and can be directly transmitted between hosts of one species.

Unlike predators, parasites are generally much smaller than their host, although both are special cases of consumer-resource interactions. Parasite show a high degree of specialisation for their mode of life, and reproduce at a faster rate than their hosts. Classic examples of parasitism include interactions between vertebrate hosts and diverse animals such as tapeworms, flukes, the *Plasmodium* species, and fleas.

Parasitism is differentiated from the parasitoid relationship, though not sharply, by the fact that parasitoids generally kill or sterilise their hosts. Parasitoidy occurs in about as many classes of organism as parasitism does. The harm and benefit in parasitic interactions concern the biological fitness of the organisms involved. Parasites reduce host fitness in many ways, ranging from general or specialised pathology (such as parasitic castration), impairment of secondary sex characteristics, to the modification of host behaviour. Parasites increase their fitness by exploiting hosts for resources necessary for the parasite's survival: (i.e. food, water, heat, habitat, and dispersal).

Although the concept of parasitism applies unambiguously to many cases in nature, it is best considered part of a continuum of types of interactions between species, rather than an exclusive category. Particular interactions between species may satisfy some but not all parts of the definition. In many cases, it is difficult to demonstrate that the host is harmed. In others, there may be no apparent specialisation on the part of the parasite or the interaction between the organisms may be short-lived. In medicine, only eukaryotic organisms are considered parasites, with the exclusion of bacteria and viruses. Some branches of biology, however, regard members of these groups as parasitic.

Types of Parasitism

Parasites are classified based on their interactions with their hosts and on their life cycles. Parasites that live on the surface of the host are called ectoparasites (e.g. some mites) and those that live inside the host are called endoparasites (including all parasitic worms). Endoparasites can exist in one of two forms: intercellular (inhabiting spaces in the host's body) or intracellular (inhabiting cells in the host's body). Intracellular parasites, such as bacteria or viruses, tend to rely on a third organism which is generally known as the carrier or vector. The vector does the job of transmitting them to the host. An example of this interaction is the transmission of malaria, caused by a protozoan of the genus *Plasmodium*, to humans by the bite of an anopheline mosquito. An epiparasite is one that feeds on another parasite. This relationship is also sometimes referred to as hyperparasitism which may be exemplified by a protozoan (the hyperparasite) living in the digestive tract of a flea living on a dog.

Social parasites take advantage of interactions between members of social organisms such as ants or termites. In *kleptoparasitism*, parasites appropriate food gathered by the host. An example is the brood parasitism practiced by many species of cuckoo and cowbird, which do not build nests of their own but rather deposit their eggs in nests of other species and abandon them there. The host behaves as a 'babysitter' as they raise the young as their own. If the host removes the cuckoo's eggs, some cuckoos will return and attack the nest to compel host birds to remain subject to this parasitism. The cowbird's parasitism does not necessarily harm its host's brood; however, the cuckoo may remove one or more host eggs to avoid detection, and furthermore the young cuckoo may heave the host's eggs and nestlings from the nest.

Parasitism can take the form of isolated *cheating* or *exploitation* among more generalised mutualistic interactions. For example, broad classes of plants and fungi exchange carbon and nutrients in common mutualistic mycorrhizal relationships; however, some plant species known as myco-heterotrophs 'cheat' by taking carbon from a fungus rather than donating it.

Parasitoids are organisms whose larval development occurs inside or on the surface of another organism, resulting in the death of the host. This means that the interaction between the parasitoid and the host is fundamentally different from that of a true parasite and shares some of the characteristics of predation.

An adelpho-parasite is a parasite in which the host species is closely related to the parasite, often being a member of the same family or genus. An example of this is the citrus blackfly parasitoid, *Encarsia perplexa*, unmated females of which may lay haploid eggs in the fully developed larvae of their own species. These result in the production of male offspring. The marine worm *Bonellia viridis* has a similar reproductive strategy, although the larvae are planktonic.

Adaptations to Parasitism

Parasitic adaptations are responses to features in the parasite's environment and this environment is the body of another organism, the host. This seems to be a difficult environment to invade but those organisms that have done so have often been very successful both in terms of numbers of individuals and numbers of species. Blood and tissues seem to be harder to invade than the gut, as is shown by the smaller number of blood and tissue parasites. This is probably in part related to the difficulties of getting eggs to the outside from sites within the host. Almost all phyla have some parasitic members (at least 50 per cent of all species are parasites). None of the deuterostome phyla are truely parasitic (echinoderms, chordates, chaetognaths). Whilst amongst the protostomes, the only groups that have no known parasitic members are the ectoprocts, endoprocts, phoronids and brachiopods.

Morphological adaptations

1. Size: Many parasites are large compared with their free-living relatives. This could be related to increased egg production.

2. Shape: Most parasites are dorso-ventrally flattened and this is related to the need to cling on to the host. Fleas are laterally flattened and rely on escape through the hairs. Nematodes are the obvious exception to the trend of flattening in parasites and parasitic nematodes, as a whole, show little morphological specialisation.
3. In parasites and particularly in endoparasites there is loss of locomotory organs.
4. A characteristic feature of many parasites are organs of attachment. Despite the wide variety of parasites there are only two trends running through the evolution of attachment organs, the development of either hooks or suckers. Suckers occur in such widely divergent groups as protozoa, monogeneans, digeneans, cestodes, parasitic crustaceans and parasitic annelids. Spines and hooks are present in many parasitic groups and the elaboration of spines or suckers or both into an eversible proboscis has occurred in the cestodes, acanthocephalans, and the acarines (ticks). Other types of attachment organ include claws in parasitic insects and the ctenidia (comb organ) of fleas. Penetrative filaments occur in a number of groups of parasite (Oxyurid nematodes, Microspora protozoans).
5. In many parasites, particularly endoparasites, there is often a reduction in the CNS and sense organs.
6. In endoparasites, again there is a trend to reduce the gut and absorb nutrients through the whole body surface.
7. In those intestinal parasites, which do not absorb nutrients through the body surface, there is usually a thick cuticle. So helminths tend either to loose their gut and absorb nutrients through their teguments or else retain their gut and have a thick resistant cuticle.
8. In many parasites there is a tremendous elaboration of the reproductive organs, associated with increased gamete production. Cestodes, for example, basically consist of a small head and neck region and the rest is serially repeated gonads. Parasites can be described as being solely adapted for reproduction.
9. Parasitic protozoa are in an essentially isosmotic environment and so lack a contractile vacuole.
10. In parasitic insects there are often elaborate tracheal trunks, so the insect can remain air breathing even when it is in its host.

Life cycle adaptations

1. There is usually an increase in reproductive potential compared with free-living relatives. Parasites usually produce more eggs and sperm than their free-living relatives do and there may be a great elaboration of the reproductive organs. Other adaptations, which increase egg production, are hermaphroditism and parthenogenesis, where every individual produces eggs and loss of seasonal reproductive cycles, so eggs and sperm are produced all the year round. Rapid maturation and extended life span also increase total reproductive capacity. The reproductive potential of the parasite can again be increased by asexual reproduction at different stages of the life cycle. One of the best examples are the digeneans where a single sporocyst can give rise to daughter sporocysts each of which can give rise to several generations of redia, before the cercariae are produced. It has been estimated that the reproductive potential of a single liver fluke (*Fasciola hepatica*) is four hundred million offspring in its lifetime.
2. Infection of secondary and tertiary hosts. This has three advantages:
    (a) Increased reproductive potential, since asexual reproduction can take place in the alternative host.

(b) It increases the range of the parasite in space and time. That is infection of more than one host and can increase the geographical range of a parasite, particularly if one host is say terrestrial and the other aquatic. By infecting more than one host species the parasite can survive periods when one host is temporarily scarce.

(c) An intermediate host can channel the parasite towards its definitive host since the intermediate host is frequently part of the final host's food chain or else closely related ecologically.

3. There is a marked trend amongst the major parasitic groups to reduce the extent of the free-living phase of the life cycle (this avoids the variable external environment).

4. Many parasites have no provision for infecting new hosts beyond the provision of large numbers of eggs or larvae. However, the infective stages of many parasites show adaptations that help to increase their chances of infecting a host. These include:

(a) Behavioural responses to locate favourable environments.

(b) Responding to chemical stimuli from their host.

(c) Changing the behaviour of the infected intermediate host to increase the chances of them being eaten by the final host.

4. Integration of life cycles. There are many ways in which the life cycle of a parasite becomes integrated with that of its host, but they fall into two broad mechanisms:

(a) Regulation of infection by the host. Many parasites require a specific pattern of stimuli from their host before they are able to infect them. This is particularly clear in those parasites that infect their hosts passively via the gut in the form of cysts or eggs. Such stages may require pre-digestion with host enzymes and the presence of specific bile salts as well as the correct pH, temperature, redox potential, pO_2 and pCO_2 before they can hatch.

(b) Regulation of the adult parasite by the host. That is reproduction of the parasite is controlled by hormonal or physiological changes in the host (e.g. the periodicity of microfilariae, reproduction in *Opalina* and *Polystoma* in the frog).

Immunological adaptations

Vertebrates react to the presence of foreign material in their tissues by the production of a humoral and cell mediated response and this depends on the ability of the host to recognise the difference between self and non-self. In mammals it takes approximately 9 days for the immune response to become fully effective, so any parasite that persists for significantly longer than 9 days must have some mechanism for avoiding or mitigating the hosts immune response. These include:

1. Absorption of host antigen.
2. Antigenic variation.
3. Occupation of immunologically privileged sites.
4. Disruption of the host's immune response.
5. Molecular mimicry.
6. Loss or masking of surface antigens.

Biochemical adaptations

The major biochemical adaptation ones are:

1. Energy metabolism: The energy metabolism of adult helminth parasites is essentially anaerobic and helminths characteristically breakdown carbohydrate into a range of organic acids such as acetate, lactate, succinate, propionate and branched chain fatty acids such as 2-methylbutyrate and 2-methylvalerate. The TCA cycle is usually reduced or modified and many parasites fix

carbon dioxide and have a partial reversed cycle with phosphoenolpyruvate playing a central role. The cytochrome chain in helminths is often modified. The reduction in the TCA cycle and cytochrome chain results in a low ATP production/mole glucose catabolised. In *Ascaris*, for example, you get 6ATP/mole of glucose, compared with 36 ATP/mole in aerobic tissues. This low ATP/mole of glucose is often compensated for in parasites by high rates of glucose utilisation. In keeping with the anaerobic nature of adult helminth metabolism there is no beta-oxidation of fatty acids in adult helminths and only limited amino acid catabolism. So helminths would seem to be well adapted to live in anaerobic or microaerobic sites within the body such as the gut or lumen of the excretory system. What is not clear is why this essentially anaerobic metabolism is also found in helminths such as schistosomes that live in aerobic sites in the body (in this case the blood stream).

2. Synthetic reactions: In general the synthetic capacities of parasites are reduced when compared with their free-living relatives. This could be related to the low ATP/mole glucose produced in parasites or to the abundant source of nutrients in the parasite's environment.

3. Nutrient uptake: A number of parasites such as cestodes and acanthocephalans have no gut and produce no digestive enzymes of their own. Instead they rely on their host's digestive enzymes to breakdown food to low molecular weight compounds (amino acids, monosaccharides, fatty acids) which the parasites then absorb through their teguments. The absorption mechanisms of cestodes and acanthocephalans compete with the uptake mechanisms of their hosts intestine for the available nutrients. The amino acid, monosaccharide and fatty acid uptake mechanisms of cestodes and acanthocephalans (and digeneans) are kinetically very similar to those of the vertebrate intestine. So although there is a reduction of synthetic and catabolic pathways in parasites, there is an elaboration of transport mechanisms.

Chapter 18

Animal Adaptations

INTRODUCTION

All animals live in habitats. Habitats provide food, water, and shelter which animals need to survive, but there is more to survival than just the habitat. Animals also depend on their physical features to help them obtain food, keep safe, build homes, withstand weather, and attract mates. These physical features are called called physical adaptations. Physical adaptations do not develop during an animal's life but over many generations. The shape of a bird's beak, the number of fingers, colour of the fur, the thickness or thinness of the fur, the shape of the nose or ears are all examples of physical adaptations which help different animals to survive.

PREADAPTATION

In evolutionary biology, preadaptation describes a situation where a species evolves to use a preexisting structure or trait inherited from an ancestor for a potentially unrelated function. One example of preadaptation is dinosaurs having used feathers for insulation and display before using them to fly or sweat glands in mammals being transformed into mammary glands.

Another example is the hypothesis proposed by zoologist Jonathan Kingdon that before early humans became bipedal, they began engaging in squat feeding, i.e. turning over rocks and leaves to find insects, worms, snails and other food. Consequently, they adapted flatter feet than were necessary in their previous tree-dwelling ancestors, since that makes squatting much easier. Flatter feet are also extremely useful for bipedal animals, so they can be described as a preadaptation to bipedalism, even though (or rather because) the adaptation had nothing to do with bipedalism originally.

Arthropods provide the earliest identifiable fossils of land animals, from about 419 million years ago in the Late Silurian, and terrestrial tracks from about 450 million years ago appear to have been made by arthropods. Arthropods were well pre-adapted to colonise land, because their existing jointed exoskeletons provided protection against desiccation, support against gravity and a means of locomotion that was not dependent on water.

Some biologists dislike the term 'preadaptation' as it could imply an intentional plan, which is contrary to the nature of evolution. Some alternative terms that have been suggested include 'co-option' and exaptation, to avoid the implication of foresight.

Some phenomena could give the appearance of foresight, however, without actually involving it, being instead attributable to simple probability. For example, future environments (for example, hotter or drier ones), may resemble those already encountered by a population at one of its current spatial or temporal margins. This is not actual foresight, but rather the luck of having adapted to a climate which later becomes more prominent. Cryptic genetic variation may have the most strongly deleterious mutations purged from it, leaving an increased chance of useful adaptations, but this represents selection acting on current genomes with consequences for the future, rather than foresight.

Animal Adaptation to Desert

High temperatures and scarcity of water makes life very difficult in the desert. Adaptations in desert animals to acquire and retain water and to regulate body temperatures help them to survive in the harsh conditions of the desert.

Plant and animal bodies are made up of a number of complex biological processes. These processes can take place within a narrow range of temperatures. If the range is exceeded the organism dies. The problem with the desert regions is that temperatures reach extreme limits. Adding to the problem of extreme temperature is the scarcity of water in the desert. Water is the major constituents of the living bodies. To survive such harsh conditions, desert animals have developed certain features that have enabled them to survive in the desert.

Adaptations in desert animals

Adaptations that the desert animals have undergone are for attaining the following purposes.

To avoid heat

Most of the desert animals avoid being out in the sun during the hottest part of the day. Many desert mammals, reptiles and amphibians live in burrows to escape the intense desert heat. Rodents also plug the entrance of their burrows to keep the hot and dry desert winds out. Most of the animals of the desert either come out during the early morning or in the evening. Some of them like snakes, foxes and most rodents are nocturnal. They sleep during the daytime in their burrows or dens and hunt only during the night when the temperatures are low. Certain animals like the Round-tailed Ground Squirrel restore to estivation when they slow down their metabolism to conserve water and energy when the days become very hot.

To dissipate heat

Due to constant exposure to high temperatures, desert animals need to maintain their body temperatures at an optimum level so that the various processes that are important for their survival can be carried on. For this reason, some of them have developed long body parts that provide greater body surface to dissipate heat. For example, jackrabbits have large ears that are supplied with a large number of blood vessels from which excess heat can be easily lost. It is a known fact that light colours are better absorbers of heat than dark colours. Most desert animals are pale in colour. This prevents their bodies from absorbing more heat from the sun. However, turkeys and black vultures are dark in colour and hence they absorb considerable amount of heat during the day. To prevent their bodies from getting overheated, they have evolved the process of urohydrosis. In this process, they urinate on their legs that have numerous blood vessels. As the urine evaporates it absorbs the heat from the blood in the blood vessels of the legs.

To absorb water

In deserts where water is scarce, plants like cactus are a main source of water. These succulent plants have developed their own ways of storing water to help them tide through the dry days of the desert. Certain insects also depend upon nectar from flowers and sap from stems to get water. Kangaroo rats are known to be able to manufacture water by some metabolic process from the digestion of dry seeds. Many rodents of the desert have extra tubules in their kidneys that help them to extract most of the water from their urine and return it to the bloodstream. They also filter the moisture out of their exhaled breath through specialised organs in their nasal cavities.

To preserve water

Animals like the Gila Monster is known to store water in the fatty tissues in their tails and other parts of the body. Also, the hump of the camel has fatty tissue. When this fatty tissue is metabolised, it produces both energy as well as water. Desert animals like reptiles have minimised loss of water by excreting waste in the form of an insoluble white compound uric acid. This adaptation ensures very little wastage of water. Most of the scavengers and the predators have evolved ways of extracting water from the food that they eat. These are just a few examples of the amazing adaptations that the desert animals have evolved to survive the extreme conditions of the desert. Without these adaptations of the animals, the deserts would have been absolutely lifeless with no living creature or thing around.

Adaptation of Desert Animals to Their Environment — Some Examples

Desert animals live in an environment that can be very harsh. Major problems include an average 10 to 12 hours hot sunshine a day, a scarcity of water and often of food, and predators. Animals respond to such conditions, however, and have evolved mechanisms to deal with various extremes. Such evolved devices are not found in related species, such as deer, living in more temperate climates.

Obvious ways of coping with heat include seeking shade and burrowing. Red and fennec foxes, and most reptiles, burrow but local hares do not, preferring to rest up under a bush or in rock or sandstone crevice during the day. Some animals may also migrate seasonally.

Evolved mechanisms are largely physiological. To survive in a very arid environment animals must be able to regulate: (i) body temperature, (ii) water loss, and (iii) digestion.

The core of the problem is the ability to cope with very high ambient temperatures. Desert animals, particularly mammals, are able to reduce their metabolic rate to the minimum necessary for functioning efficiently. Given a choice, violent activity is rare, so as not to raise the basic metabolic rate. This is a problem with pregnant animals, where the metabolic rate is necessarily higher, but evolution has brought about adaptation in this area, too. The normal body temperature for larger desert animals, such as the camel and larger oryx, is around 97°F; for smaller animals, such as gazelle, it may be around 101°F.

External sources of heat are conduction (direct heat gained from actual contact with the ground, as when lying down), convection (heat brought by wind and air movements, which can often raise desert temperatures dramatically), and radiation (heat transmitted both from the sun downwards, and reflected from the ground surface upwards into the animals undersides).

Water balance depends on the relationship between loss and gain. Prime causes of water loss are sweating, evaporation via the lungs, and infection. Maintaining the correct water balance is a constant problem in desert areas for most of the year but animals have adapted to certain eating patterns to counter this. Such eating habits are also important in helping to regulate digestion.

Type samples

Desert wolf

This animal is a true hunter, constantly on the move tracking and killing prey as well as seeking out carrion. It has the stamina to cover long distances, but also has to cope with heat gain and water loss. It is brown in colour and generally a nocturnal animal, holing up during the day in a burrow. The true desert wolf looks very lean but in fact to survive it must carry no excess fat; it is honed down to optimum weight and physique by constant exercise and the ceaseless effort of seeking prey. It travels in small packs.

One unusual feature, in common with foxes, is that the wolf produces two or three cubs annually per litter. Most desert animals produce only single young, to increase the survival chances of both offspring and mother. Since the wolf, like dogs and foxes, possesses no sweat glands, it helps to control body temperature by evaporation from the lungs by rapid panting.

Eland

This animal is not strictly a desert type but has adapted to arid conditions. It regulates body temperature by increasing its respiration rate from a normal 10 to over 70 breaths per minute in a very short time (barely two minutes). It is important to control the body temperature because, in common with all mammals, the brain must receive blood at a constant temperature, without sudden fluctuations. In some animals an increase of 3°–5°F at the brain can easily bring about heatstroke. The eland permits its body temperature at the brain rarely exceeds 4°F above normal. This is because the carotid artery takes blood destined for the brain to the nasal area first. Rapid panting serves to cool the blood in the nose area. This is a major physiological adaptation.

The eland regulates water balance by lung evaporation. It only requires approximately four litres per 110 kg body weight per day compared to 25 litres at the same ratio for cows. Of course, water intake becomes more vital as the aridity increases. The major water source in deserts is not rainfall or oasis pools, but the vegetation. Succulent plants in particular are capable of retaining a lot of moisture, and there is the bonus for herbivores of high dewfall levels at dawn in desert environments. Experiments have shown that a plant's moisture content can be up to 30 per cent higher by night than by day, and this is the prime reason for nocturnal grazing. Animals such as the eland also possess an efficient nitrogen recycling mechanism that prevents excess moisture loss through urea.

The eland is capable of focussing on its food when eating while also being able to detect any movement on a horizontal plane ahead and on either side. Such a device enables it to be aware of the early approach of a would-be predator.

Arabian dorcas gazelle

Whereas large animals tend to congregate in small groups, smaller species such as gazelle are frequently found in herds of up to 40 or more. There is usually a dominant bull, with other males lower down the hierarchical scale straggling on the fringes of the main herd which consists typically of females and young. The Arabian Dorcas gazelle is all white underneath to reflect heat radiated upwards from the ground surface. It is brown on top for desert camouflage. Its hair is short but dense enough to act as a shield against direct heat.

Arabian gazelle

This animal may be distinguished from the previous gazelle by its black horizontal stripe along the body. Its hooves are small and hard, not splayed out; thus it is not well suited to an all-sand terrain.

This gazelle has a fatty layer beneath the skin to prevent not heat loss, but heat gain. It should be noted that animals like gazelle have a large surface area in proportion to body weight; hence the need for such a fatty layer as an extra shield besides the body coat. Arctic animals, by comparison, usually carry extra poundage to compensate.

Pregnancy places great demands on would-be mothers, and gazelles produce single offspring at two-year intervals, not annually. This helps the overall survival rate of the herd, which tends to frequent localities where good and abundant grazing is not easily come by. If necessary, an imminent birth can be delayed by up to 12 hours.

Fringe-eared oryx

The water intake requirement for this oryx is 80 per cent less than that for a normal cow. That is to say, the loss rate is much less. This oryx assimilates some seven or eight litres per day compared to the domestic cow's 25 or so. The coat is overall brown, indicating that the fringe-eared oryx is not a true desert animal. However, the hide is still light enough to reflect some 20 per cent of direct heat.

Scimitar-horned oryx

This animal is white in colour and well adapted. It is native to Chad and Central Africa. Its coat is capable of reflecting some 40 per cent of direct heat, and there also exists the fatty layer shield. The hair is short to keep the coat clean and reasonably free of parasites, in order to maintain the efficiency of the reflection mechanism.

This oryx can go for nine months without drinking partly due to its efficient nitrogen recycling system. The energy it can absorb from given plants is much higher than that absorbed by goats from the same vegetation. It is a ruminant, like all herbivores being unable to digest cellulose before it is sufficiently broken down by bacteria in the stomach. Body temperature may be raised by up to 6°F without physiological precautions being necessary. If the body temperature goes beyond this level, the sweating mechanism is activated. In contrast to most desert animals, the body temperature of the scimitar-horned oryx can go higher than the ambient temperature around it.

Addax

This North African gazelle is small with an all-white coat and is well suited to an arid environment. By day it spends much time lying or sitting in any available shade thus effectively reducing direct heat-load by as much as two-thirds. Radiated heat is also considerably reduced by keeping to shade. The hair of this animal is longer than that of some gazelles, but combined with its colour, it can reflect up to 60 per cent of direct heat.

Arabian oryx

The Arabian oryx is the specialist of the local desert environment. It has a white coat consisting of dense, longish hair. The hooves are wider than those of other similar animals, enabling it to take to the sands. Under the coat is a very thick fatty layer, and in between the skin is black which cuts down the absorption of ultraviolet rays. Brain temperature is well regulated and the kidneys are supremely efficient.

In the desert ecosystem, a group hierarchy system reduces stress in individuals. Eight or then is the usual number for a group of Arabian oryx in the wild, with a maximum of five or six groups in any given locality. Recently almost extinct, this oryx has been successfully bred in captivity and efforts are being made to reintroduce it to its original habitats. Apart from man, predators are not a problem.

The Arabian oryx is a shade seeker. It must regulate its body temperature to within 3°F of normal. It tends to feed at dusk because digestion is more effective during the cooler nocturnal hours. It can smell water miles away and make towards it over any kind of terrain. Unlike most herbivores, it can detect succulent root tubers up to half a meter deep in the ground and root them out. When disturbed it has no qualm about heading directly into areas like the empty quarter. The calves are brown and tend to crouch and hide while the adults flee from predators, to return later. However, within a fortnight of birth, the young are essentially prepared for long distance travel with the group.

DESERT ANIMAL SURVIVAL

Lack of water creates a survival problem for all desert organisms, animals and plants alike. But animals have an additional problem — they are more susceptible to extremes of temperature than are plants. Animals receive heat directly by radiation from the sun, and indirectly, by conduction from the substrate (rocks and soil) and convection from the air.

The biological processes of animal tissue can function only within a relatively narrow temperature range. When this range is exceeded, the animal dies. For four or five months of the year, the daily temperatures in the desert may actually exceed this range, called the range of thermoneutrality. Combined with the scarcity of life-sustaining water, survival for desert animals can become extremely tenuous.

Fortunately, most desert animals have evolved both behavioral and physiological mechanisms to solve the heat and water problems the desert environment creates. Among the thousands of desert animal species, there are almost as many remarkable behavioral and structural adaptations developed for avoiding excess heat. Equally ingenious are the diverse mechanisms various animal species have developed to acquire, conserve, recycle, and actually manufacture water.

Avoiding Heat

Behavioural techniques for avoiding excess heat are numerous among desert animals. Certain species of birds, such as the Phainopepla, a slim, glossy, black bird with a slender crest, breed during the relatively cool spring, then leave the desert for cooler areas at higher elevations or along the Pacific coast. The Costa's Hummingbird, a purple-crowned and purple-throated desert species, begins breeding in late winter, then leaves in late spring when temperatures become extreme. Many birds are active primarily at dawn and within a few hours of sunset, retiring to a cool, shady spot for the remainder of the day. Some birds, such as the kingbird, continue activity throughout the day, but always perch in the shade. Many animals (especially mammals and reptiles) are crepuscular, that is, they are active only at dusk and again at dawn. For this reason, humans seldom encounter rattlesnakes and Gila Monsters. Many animals are completely nocturnal, restricting all their activities to the cooler temperatures of the night. Bats, many snakes, most rodents and some larger mammals like foxes and skunks, are nocturnal, sleeping in a cool den, cave or burrow by day.

Some smaller desert animals burrow below the surface of the soil or sand to escape the hightemperatures at the desert surface. These include many mammals, reptiles, insects and all the desert amphibians. Rodents may plug the entrances to their burrows to keep out hot, desiccating air.

A few desert animals, such as the Round-tailed Ground Squirrel, a diurnal mammal, enter a state of estivation when the days become too hot and the vegetation too dry. They sleep away the hottest part of the summer. Some desert animals such as Desert Toads, remain dormant deep in the ground until the summer rains fill ponds. They then emerge, breed, lay eggs and replenish their body reserves of food and water for another long period. Some arthropods, such as the fairy shrimps and brine shrimps, survive as eggs, hatching in saline ponds and playas during summer or winter rains, and completing their life cycles.

Certain desert lizards are active during the hottest seasons, but move extremely rapidly over hot surfaces, stopping in cooler 'islands' of shade. Even their legs may be longer so they absorb less surface heat while running.

Dissipating Heat

Some animals dissipate heat absorbed from their surroundings by various mechanisms. Owls, Poorwills and nighthawks gape open-mouthed while rapidly fluttering their throat region to evaporate water from their mouth cavities. (Only animals with a good supply of water from prey can afford this type of cooling, however.) Many desert mammals have evolved long appendages to dissipate body heat into their environment. The enormous ears of jackrabbits, with their many blood vessels, release heat when the animal is resting in a cool, shady location. Their relatives in cooler regions have much shorter ears.

New World vultures, such as the Turkey and Black Vultures, are dark in colour and thus absorb considerable heat in the desert. But they excrete urine on their legs, cooling them by evaporation, and circulate the cooled blood back through the body. This behavior, called urohydrosis, is shared with their relatives the storks, successful birds of the African deserts. Both vultures and storks may escape the hot midday temperatures of the desert by soaring effortlessly, high on thermals of cooler air. Many desert animals are paler than their relatives elsewhere in more moderate environments. Pale colours may be seen in feathers, fur, scales or skin. Pale colours not only ensure that the animal takes in less heat from the environment, but help to make it less conspicuous to predators in the bright, pallid surroundings.

Retaining Water

The mechanisms some desert animals have evolved to retain water are even more elaborate. They range from simple to physiologically complex. Some retain water by burrowing into moist soil during the dry daylight hours (all desert toads). Some predatory and scavenging animals can obtain their entire moisture needs from the food they eat (e.g. Turkey Vulture) but still may drink when water is available. Reptiles and birds excrete metabolic wastes in the form of uric acid, an insoluble white compound, wasting very little water in the process. Mammals, however, excrete urea, a soluble compound that accounts for considerable water loss. Most mammals, therefore, need access to a good supply of fresh water, at least every few days, if not daily.

Acquiring Water

Desert creatures derive water directly from plants, particularly succulent ones, such as cactus. Many species of insects thrive in the deserts this way. Some insects tap plant fluids such as nectar or sap from stems, while others extract water from the plant parts they eat, such as leaves and fruit. The abundance of insect life permits insectivorous birds, bats and lizards to thrive in the desert.

Some desert creatures utilise all of these physical and behavioral mechanism to survive the extremes of heat and dryness. Certain desert mammals, such as Kangaroo Rats, live in underground dens which they seal off to block out midday heat and to recycle the moisture from their own breathing. These ingenious rodents (there are a number of species) also have specialised kidneys with extra microscopic tubules to extract most of the water from their urine and return it to the blood stream. And much of the moisture that would be exhaled in breathing is recaptured in the nasal cavities by specialised organs.

If that weren't enough, Kangaroo Rats, and some other desert rodents, actually manufacture their water metabolically from the digestion of dry seeds. These highly specialised desert mammals will not drink water even when it is given to them in captivity. These are just a few examples of the ingenious variety of adaptations animals use to survey in the desert, overcoming the extremes of heat and the paucity of water.

Aquatic Adaptations

When animals live in the water, they must have special adaptations to help them survive in an aquatic habitat. The more time an animal spends in the water, the more adaptations the animal will have for an aquatic life. Below are examples of some of these adaptations:

1. Streamlined body reduces friction when the animal moves through the water.
2. Smooth, almost furless body helps aquatic mammals move through the water with little friction.
3. Dense fur helps streamline the bodies of some aquatic mammals and keeps them warm.
4. Dense waterproof feathers keep cold water away from bird's skin and prevent wetting of the feathers.
5. Webbed feet, formed from thin skin between the toes, work like paddles.
6. Long legs and necks keep the bodies of wading birds out of the water and are thin, light, and easy to move, and the long neck helps the birds to reach the water or below it, for food.
7. Strainers in the mouth filter food particles from the water.
8. Flippers provide a large surface for pushing against water and act like paddles.
9. Eyes positioned on top of the head allow animals to hide almost fully submerged in water and still detect predators or prey above the water.
10. Nostrils positioned near the top of the head allow animals to come to the surface to breathe while only a small part of the body can be seen.
11. Nostrils close when the animal goes under the water.
12. Blubber is a thick layer of fat or oil stored between the skin and muscles of the body, provides insulation.
13. Transparent eyelids cover the eyes of animals swimming underwater.
14. Flattened tails serve as paddles.

Moving in water

Aquatic organisms move in and through the water in a number of ways. Begin by asking the class to list the pond animals they are familiar with and record suggestions on the board. Using the Table 18.1 below prompt students to think about how some organisms move in the water highlighting the links between habitat, diet and movement.

Table 18.1. Move in the water highlighting the links between habitat, diet and movement.

Organism	Where it lives	What it eats	How it moves	Adaptation
Duck	Above surface	Pondweed, insects, snails, larvae, tadpoles, small fish	Flies, dives, paddles	Wings, waterproof feathers, webbed feet
Adult dragonfly	Above surface	Insects	Flies, hovers	Two paired-wings, streamlined shape
Adult mayfly	Above surface	Insects	Flies, hovers	
Frogs	Pond edge	Insects, snails, slugs, worms	Hops, swims	Amphibious, moist skin, webbed feet, long, strong hind legs, sticky tongue
Newts	Pond edge	Water fleas, snails, worms	Walks, swims	Amphibious, moist skin, long muscular tail
Water vole	Pond edge	Insects, worms, grasses	Walks, swims	Oily coat, sharp teeth
Pond snail	Pond surface	Plants, algae, dead matter	Muscle contraction	Shell, muscular foot

(Contd...)

Organism	Where it lives	What it eats	How it moves	Adaptation
Pond skater	Pond surface	Dead plants and animals	Skates across water surface	Long splayed legs, water-repellent hairs
Mosquito larvae	Pond surface	Micro-organisms, detritus	Swims	Breathing tube
Fish	Mid-water	Water fleas, tadpoles, shrimp	Swims	Gills, fins, streamlined body
Tadpole	Mid-water	Insects, plants, dead matter	Swims	Streamlined body, tail
Great diving beetle	Mid-water	Water fleas, snails, water boatmen, larvae, leeches,	Walks, dives, swims	Streamlined body, fringed jointed legs
Water boatman	Mid-water	Shrimp, worms, tadpoles	Rows using legs	Paddling legs, hair-lined body traps air
Leech	Mid-water	Snails, larvae, tadpoles	Swims	Sucker, flattened body
Insect larval stages (Dragonfly/ Mayfly)	Mid-water	Fish, water fleas, shrimp, tadpoles, micro-organisms	Swims, crawls	Gills on abdomen
Freshwater worm	Pond bottom	Micro-organisms, dead and decaying matter 'detritus'	Swims	Haemoglobin, thin body wall, segmented body
Water flea	Pond bottom	Micro-organisms, dead and decaying matter 'detritus'	Rows using antennae	Antennae, flattened body
Freshwater shrimp	Pond bottom	Micro-organisms, dead and decaying matter 'detritus'	Swims	Side-flattened body, swimming legs, gills
Water hog louse	Pond bottom	Micro-organisms, dead and decaying matter 'detritus'	Walks	Gills, six paired legs

Adaptation to Environment in Animals

Adaptations for flight are called as volant adaptations. Bats, birds and insects are well adapted for an active flight. Some important flight adaptations are:
1. Boat or spindle shaped body that offers no resistance to air.
2. Forelimbs are modified into wings.
3. An exoskeleton of feathers which provide an insulation, thereby preventing the loss of heat and help maintain a constant body temperature.
4. The short tail with tail feathers which serves as a rudder in steering and counter balance during perching.
5. Presence of well developed flight muscles and a keeled sternum.
6. Presence of hollow or pneumatic bones to make their body light.
7. Presence of air sacs which serve as reservoirs of air and provide lightness and buoyancy to the body.
8. A large cerebellum which allows the bird to maintain equilibrium and co-ordination.

Cursorial adaptations

These are the adaptations for running and are found in flightless birds such as ostrich, kiwi and many other animals.

Cursorial adaptations in ostrich and kiwi respectively

1. More or less stream lined body so as to offer minimum resistances to air during running.
2. Hind limbs are usually long and have claws.

3. In the forelimbs ulna and in the hind limbs the fibula bones are greatly reduced as an adaptation for fast running.

Fossorial adaptations

Animals which dig into burrows for shelter and food are called fossorial animals. They have adaptations for burrowing:
1. Spindle shaped body to offer little resistance during going in and out of burrows.
2. Small tapering head with snout for burrowing.
3. Eyesight is reduced as they are of no use in the dark.
4. External ears tend to disappear as they might be an obstruction in burrowing.
5. Short and stout limbs are provided with strong claws for digging.

They hibernate during winter or unfavourable conditions.

Arboreal adaptations

Arboreal locomotion is the locomotion of animals in trees. In every habitat in which trees are present, animals have evolved to move in them. Some animals may only scale trees occasionally, while others are exclusively arboreal. These habitats pose numerous mechanical challenges to animals moving through them, leading to a variety of anatomical, behavioral and ecological consequences. Furthermore, many of these same principles may be applied to climbing without trees, such as on rock piles or mountains.

The earliest known tetrapod with specialisations that adapted it for climbing trees, was Suminia, a synapsid of the late Permian, about 260 million years ago. Some invertebrate animals are exclusively arboreal in habitat. These adaptations help the animals in climbing and are found in various animals like squirrels, lemurs, sloths and rodents.

Arboreal adaptations in squirrel:
1. The limb girdles are very stout giving support to the body in climbing.
2. Well developed claws help in grasping and climbing.
3. Syndactyly or union of digits is seen in some animals.

Chameleon Adaptation for Arboreal and Terrestrial Habitats and Insectivorous Mode of Feeding:
1. In some animals like the tree frog Hyla, fingers and toes bear adhesive pads that help in climbing and clinging to the trees.
2. The presence of a prehensile tail is a common character in arboreal forms.
3. Accessory organs for climbing like spines and tubercles are found in the forearm of some lemurs.

Parasitic adaptations

All parasitic have successfully adapted to live either on or inside their hosts. Some of their important adaptations are:
1. Presence of suckers or other structures for attachment to the host.
2. Loss of locomotory, sense and digestive organs, especially in endoparasites as they are fully dependent on the host and draw their nourishment from it.
3. Intermediate host and vector: To increase their chance of survival, many parasites have more than one host.

Malarial parasite plasmodium has 2 hosts — female anopheles mosquito and man:
1. Increased reproductive capacity — All successful parasites have very well developed reproductive organs and enormous productive capacity.

2. Complicated life cycle — The production of enormous number of eggs and several larval forms by some endoparasites like Fasciola is an adaptation to maintain the continuity of the species.

ADAPTIVE RADIATION

An adaptive radiation is the evolution of ecological and phenotypic diversity within a rapidly multiplying lineage. Starting with a recent single ancestor, this process results in the speciation and phenotypic adaptation of an array of species exhibiting different morphological and physiological traits with which they can exploit a range of divergent environments. Adaptive radiation, a characteristic example of cladogenesis, can be graphically illustrated as a bush or clade, of coexisting species (on the tree of life).

Identification

Four features can be used to identify an adaptive radiation:
1. A common ancestry of component species: Specifically a recent ancestry. Note that this is not the same as a monophyly in which all descendants of a common ancestor are included.
2. A phenotype-environment correlation: A significant association between environments and the morphological and physiological traits used to exploit those environments.
3. Trait utility: The performance or fitness advantages of trait values in their corresponding environments.
4. Rapid speciation: Presence of one or more bursts in the emergence of new species around the time that ecological and phenotypic divergence is underway.

Causes

Innovation

The evolution of a novel feature may permit a clade to diversify by making new areas of morphospace accessible. A classic example is the evolution of a fourth cusp in the mammalian tooth. This trait permits a vast increase in the range of foodstuffs which can be fed on. Evolution of this character has thus increased the number of ecological niches available to mammals. The trait arose a number of times in different groups during the Cenozoic, and in each instance was immediately followed by an adaptive radiation. Birds find other ways to provide for each other, i.e. the evolution of flight opened new avenues for evolution to explore, initiating an adaptive radiation. Other examples include placental gestation (for eutherian mammals) or bipedal locomotion (in hominins).

Opportunity

Adaptive radiations often occur as a result of an organism arising in an environment with unoccupied niches, such as a newly formed lake or isolated island chain. The colonising population may diversify rapidly to take advantage of all possible niches. In Lake Victoria, an isolated lake which formed recently in the African rift valley, over 300 species of cichlid fish adaptively radiated from one parent species in just 15,000 years. Adaptive radiations commonly follow mass extinctions: following an extinction, many niches are left vacant. A classic example of this is the replacement of the non-avian dinosaurs with mammals at the end of the Cretaceous, and of brachiopods by bivalves at the Permo-Triassic boundary.

Adaptive Radiation: Mammlian Forelimbs

The variety of forelimbs — the bat's wing, the sea lion's flipper, the elephant's supportive column, the human's arm and hand — further illustrates the similar anatomical plan of all mammals due to a shared ancestry. Despite the obvious differences in shape, mammalian forelimbs share a similar arrangement and arise from the same embryonic, homologous structures.

The mammalian forelimb includes the shoulder, elbow, and wrist joints. The scapula or shoulder blade connects the forelimb to the trunk and forms part of the shoulder joint. The humerus or upper arm bone forms part of the shoulder joint above, and elbow joint below. The radius and ulna comprise the lower arm bones or forearm, and contribute to the elbow and wrist joints. Finally, the carpal or wrist bones, the metacarpals, and phalanges form the bat wing, the sea lion flipper, the tree shrew, mole, and wolf paws, the elephant foot, and the human hand and fingers.

Using light colours, begin with the tree shrew scapula in the center of the plate. Next, colour the scapula on each of the other animals: the mole, bat, wolf, sea lion, elephant, and human. Continue colouring the other bones in this manner: humerus, radius, ulna, carpal bones, metacarpals, and phalanges.

After you have coloured all the structures in each animal, notice the variation in the overall shape of the forelimb. Notice, too, how the form of the bones contributes to the function of the forelimb in each species. The tree shrew skeleton closely resembles that of early mammals and represents the ancestral forelimb skeleton. The tree shrew is small bodied, moves easily on the ground or in the trees, and has a flexible forelimb for these functions.

The mole's forelimb is relatively short and lies close to the body, giving it a somewhat streamlined shape. The shovellike paw comprises almost half the length of the limb. The slender rodlike scapula and the short, peculiarly shaped humerus help anchor the forelimb against the trunk and draw the paws very near to the head. The elbow joint is rotated so that the paws face backward. Powerful muscles attach on the long bony olecranon process; they straighten the elbow joint and help the paws dig and push the soil out to the side. The robust metacarpals and phalanges give strength to the paw and an extra bone (the falciform) adds breadth.

Thus, the forelimb is well suited for the mole to dig its way through moist soil as it searches for insects. The bat's forelimb is adapted for flight. The humerus is short relative to the longer, slender radius; the ulna is reduced and not part of the wrist joint. The metacarpals and phalanges provide a light but strong frame over which the skin is stretched, much as the silk of a kite covers its frame.

The wolf is a swift runner, the better to pursue prey. Its humerus, radius, and ulna are relatively long and make possible a long stride. In marked contrast to those of the bat, the metacarpals are closely packed together for bearing weight. The wolf walks somewhat up on its toes, on bent phalanges.

For its life in the water and on land, the sea lion has a broad scapula, short and robust humerus, radius, and ulna. The scapula and humerus lie within the body cavity and so help streamline the animal. The relatively long metacarpals and phalanges form a broad paddle. The robust bones support the sea lion's weight on land; this gives the animal mobility when it hauls out on land to mate and give birth.

To support its five ton bulk, the elephant's shoulder, elbow and wrist joints are stacked one above the other, giving an arrangement like an architectural column; the scapula is oriented downward, so it too is in line with the robust humerus and ulna. The radius is reduced so that the ulna carries most of the weight. The metacarpals and phalanges are short and robust, and a pad of fat and skin cushions the foot.

The human forelimb is long, slender and mobile and, unlike that of other mammals, does not bear weight in locomotion. The ball and socket shoulder joint enables a 360° range of motion, and slender finger bones and a prominent thumb enable the hand to carry out fine manipulations. The similarity of a bat's wing, an elephant's leg, and a human's arm may not be readily apparent without a closer look at the underlying bony structures. The basic design of the mammalian forelimb demonstrates the evolutionary phenomenon of adaptive radiation. Through natural selection, the form of mammalian forelimbs has been modified during the last 65 million years into many shapes to perform a variety of functions. By adapting to forest, plains, air, water, and underground, mammals have been able to radiate (like the sun's rays) into a diversity of habitats. Studies of comparative anatomy and embryology well illustrate the 'descent with modification' of Darwin and the branching out of species from a common ancestor.

Evolutionary Radiation

An evolutionary radiation is an increase in taxonomic diversity or morphological disparity, due to adaptive change or the opening of ecospace. Radiations may affect one clade or many, and be rapid or gradual; where they are rapid, and driven by a single lineage's adaptation to their environment, they are termed adaptive radiations.

Examples of evolutionary radiation

Perhaps the most familiar example of an evolutionary radiation is that of placental mammals immediately after the extinction of the dinosaurs at the end of the Cretaceous, about 65 million years ago. At that time, the placental mammals were mostly small, insect-eating animals similar in size and shape to modern shrews. By the Eocene (58–37 million years ago), they had evolved into such diverse forms as bats, whales, and horses. Other familiar radiations include the Cambrian explosion, the radiation of land plants after their colonisation of land, the Cretaceous radiation of angiosperms, and the diversification of insects, a radiation that has continued almost unabated since the Devonian, 400 million years ago.

Radiations may be discordant, with either diversity or disparity increasing almost independently of the other or concordant, where both increase at a similar rate.

Chapter 19

Zoogeography

INTRODUCTION

Zoogeography is the branch of the science of biogeography that is concerned with the geographic distribution of animal species and their attributes. That makes zoogeography the study of how patterns of animal biodiversity vary over space and through time. Zoogeography is the study of the patterns of the past, present, and future distribution of animals (and their attributes) in nature and the processes that regulate these distributions. Zoogeography integrates information on the historical and current ecology, genetics, and physiology of organisms and their interaction with environmental processes (continental drift, climate) in regulating geographic distributions of animals. Scientists use descriptive and analytical approaches useful in hypothesis testing in zoogeography which illustrates the applied aspects of zoogeography (e.g. refuge design in conservation).

Zoogeography is often divided into two main branches: ecological zoogeography and historical zoogeography. The former investigates the role of present biotic and abiotic interactions in influencing animal distributions; the latter is concerned with historical reconstruction of the origin, dispersal, and extinction of taxa. Faunistics is a study of the fauna of some territory or area.

Based on the presence or absence of certain distinct groups of animals, especially mammals, the surface of the earth has been divided into six distinct areas or units. These distinct areas of the world, each with its own characteristic groups of animals, are known as Zoogeographical realms. The entire world has been divided into various regions which contain specific fauna. Scatter in 1857, who first of all divided to earth into following six regions:

Palearctic ecozone: The Palearctic or Palaearctic is one of the eight ecozones dividing the earth's surface. Physically, the Palearctic is the largest ecozone. It includes the terrestrial ecoregions of Europe, Asia north of the Himalaya foothills, northern Africa, and the northern and central parts of the Arabian Peninsula.

Regions of Ethiopia: Ethiopia is divided into 9 ethnically-based administrative regions (kililoch; singular — kilil) and two chartered cities (astedader akababiwach, singular — astedader akabibi). The word 'kilil' more specifically means 'reservation' or 'protected area' and the ethnic basis of the regions and choice of the word 'kilil' has drawn fierce criticism from the opposition, who have drawn comparisons to the bantustans of Apartheid South Africa. The 9 regions (and two chartered cities, marked by asterisks) are:

1. Addis Ababa.
2. Afar.
3. Amhara.
4. Benishangul-Gumuz.

5. Dire Dawa.
6. Gambela.
7. Harari.
8. Oromiya.
9. Somali.
10. Southern Nations, Nationalities, and Peoples Region.
11. Tigray.

These administrative regions replaced the older system of provinces (which are still sometimes used to indicate location within Ethiopia).

Australia region: Australia Region or Region in Australia is a district or province located in the island continent of Australia with distinct topographical characteristics. Located in the southern hemisphere, this smallest continent of the world is also accompanied by the islands of Tasmania and New Zealand. Australia neighbors Indonesia, East Timor and Papua New Guinea in the south, the Solomon Islands, Vanuatu and the French colony of New Caledonia in the southwest, and New Zealand in the northwest.

Australia Region or Region in Australia can be broadly divided according to their Inter Territorial regions, viz. Capital Country — it is one of those sixteen regions whose history longs back to ancient times, located in New South Wales and Australian Capital Territory districts. Since it overlaps the capital city of Canberra, it is aptly named so. Lake Eyre Basin—classified being one of the largest internal drainage systems in the world, the region covers one-sixth of the continent including, inland Queensland, portions of South Australia and the Northern Territory, and some part of western New South Wales.

Nearctic ecozone: The Nearctic is one of the eight terrestrial ecozones dividing the earth's land surface. The Nearctic ecozone covers most of North America, including Greenland and the highlands of Mexico. Southern Mexico, southern Florida, Central America, and the Caribbean islands are part of the Neotropic ecozone, together with South America.

Neotropic ecozone: In biogeography, the Neotropic or Neotropical zone is one of the eight terrestrial ecozones. This ecozone includes South and Central America, the Mexican lowlands, the Caribbean islands, and southern Florida, because these regions share a large number of plant and animal groups.

Indian Himalayan region: The Indian Himalayan Region (IHR) is a mountain range that spans ten states of India namely, Jammu and Kashmir, Himachal Pradesh, Uttaranchal, Sikkim, Arunachal Pradesh, Meghalaya, Nagaland, Manipur, Mizoram and Tripura as well as the hill regions of two states–Assam and West Bengal. The region is responsible for providing water to a large part of the Indian subcontinent and contains varied flora and fauna.

The Karakoram ranges are the northern-most ranges of India. To the south of the Karakoram range lie the Zangskar ranges. Parallel to the Zangskar ranges lie the Pir Panjal ranges. These three mountain ranges lie parallel to each other in the north-western part of India, most of its area lying in the state of Jammu and Kashmir.

ORIGIN AND EVOLUTION OF MAMMALS

The evolution of mammals within the synapsid lineage (sometimes called 'mammal-like reptiles') was a gradual process that took approximately 70 million years, beginning in the mid-Permian. By the mid-Triassic, there were many species that looked like mammals, and the first true mammals appeared in the early Jurassic. The earliest known marsupial, Sinodelphys, appeared 125 million years ago in the early Cretaceous, around the same time as Eomaia, the first known eutherian (member of placentals' 'parent' group); and the earliest known monotreme, Teinolophos, appeared two million years later. After the Cretaceous-Tertiary extinction wiped out the non-avian dinosaurs (birds are generally regarded as the surviving dinosaurs) and several other mammalian groups, placental and marsupial mammals diversified into many new forms and ecological niches throughout the tertiary, by the end of which all modern orders had appeared.

Ancestry of Mammals

Amniotes

The first fully terrestrial vertebrates were amniotes — their eggs had internal membranes that allowed the developing embryo to breathe but kept water in. This allowed amniotes to lay eggs on dry land, while amphibians generally need to lay their eggs in water (a few amphibians, such as the Surinam toad, have evolved other ways of getting around this limitation). The first amniotes apparently arose in the late Carboniferous from the ancestral reptiliomorphs (Fig. 19.1).

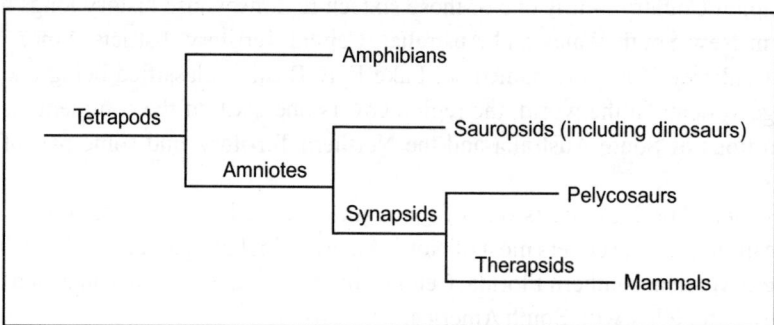

Fig. 19.1. The ancestry of mammals.

Synapsids

Synapsid skulls are identified by the distinctive pattern of the holes behind each eye, which served the following purposes:
1. Made the skull lighter without sacrificing strength.
2. Saved energy by using less bone.
3. Probably provided attachment points for jaw muscles. Having attachment points further away from the jaw made it possible for the muscles to be longer and therefore to exert a strong pull over a wide range of jaw movement without being stretched or contracted beyond their optimum range.

Therapsids

Therapsids descended from pelycosaurs in the middle Permian and took over their position as the dominant land vertebrates. They differ from pelycosaurs in several features of the skull and jaws, including larger temporal fenestrae and incisors that are equal in size. The therapsids went through a series of stages, beginning with animals that were very like their pelycosaur ancestors and ending with some that could easily be mistaken for mammals.

Therapsid family tree
Simplified from; only those that are most relevant to the evolution of mammals are described in Fig. 19.2.

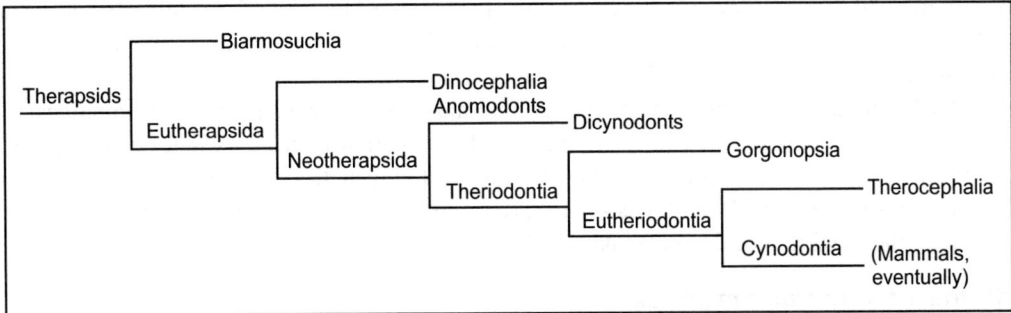

Fig. 19.2. Therapsid family tree.

Only the dicynodonts, therocephalians and cynodonts survived into the Triassic.

From Cynodonts to True Mammals

Many uncertainties: The Triassic takeover probably accelerated the evolution of mammals. The nearly-mammals are preserved in few good fossils, mainly because they were mostly smaller than rats.

Mammals or mammaliformes: One result of these uncertainties has been a change in the paleontologists' definition of 'mammal'. For a long time a fossil was considered a mammal if it met the jaw-ear criterion (the jaw joint consists only of the squamosal and dentary; and the articular and the quadrate bones have become the middle ear's malleus and incus).

Family tree—cynodonts to mammals
The details of family tree—cynodonts to mammals are shown in Fig. 19.3.

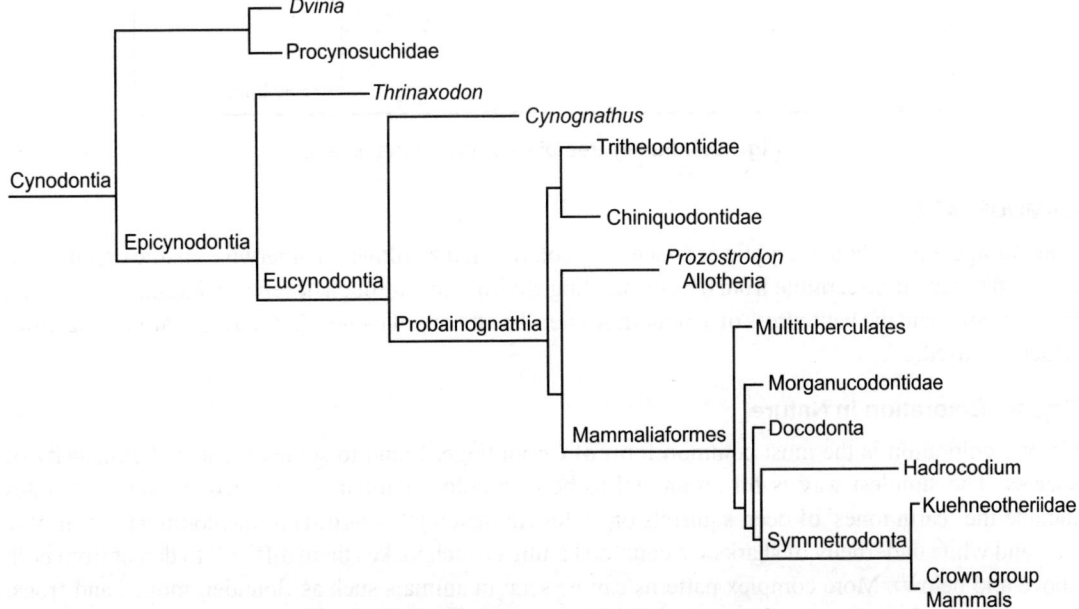

Fig. 19.3. Family tree—cynodonts to mammals.

Earliest Crown Mammals

The crown group mammals are the extant mammals and their relatives back to their last common ancestor, variously called 'crown mammals' or 'true mammals'. This part of the story introduces new complications, since the crown group are the only group that still has living members, enabling both anatomical and DNA analysis:

1. One has to distinguish between extinct groups and those that have living representatives.
2. One often feels compelled to try to explain the evolution of features that do not appear in fossils. This endeavour often involves molecular phylogenetics, a technique that has become popular since the mid-1980s but is still often controversial because of its assumptions, especially about the reliability of the molecular clock.

Family tree of early crown mammals

Family tree of early crown mammals are shown in Fig. 19.4. X marks extinct groups.

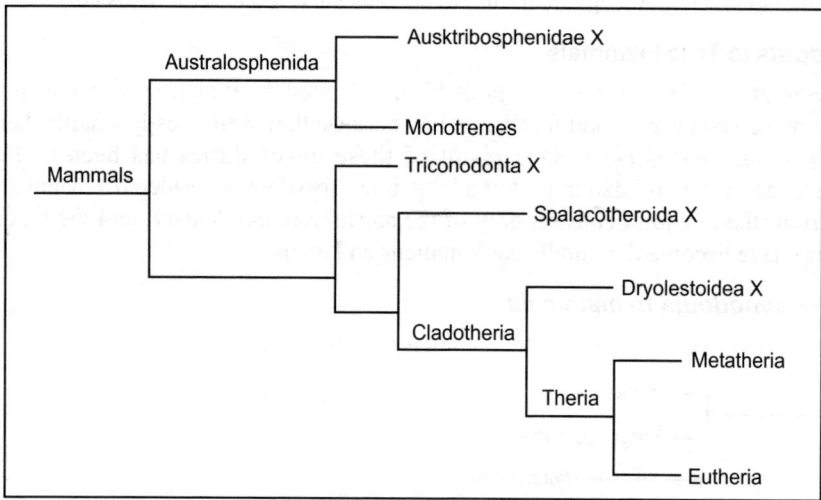

Fig. 19.4. Family tree of early crown mammals.

CAMOUFLAGE

Camouflage is a method of cryptic or concealing coloration that allows an otherwise visible organism or object to remain indiscernible from the surrounding environment through deception. Examples include a tiger's stripes and the battledress of a modern soldier. The theory of camouflage covers various strategies which are used.

Cryptic Coloration in Nature

Cryptic coloration is the most common form of camouflage, found to some extent in the majority of species. The simplest way is for an animal to be of a colour similar to its surroundings. Examples include the 'earth tones' of deer, squirrels or moles (to match trees or dirt) or the combination of blue skin and white underbelly of sharks via countershading (which makes them difficult to detect from both above and below). More complex patterns can be seen in animals such as flounder, moths, and frogs, among many others.

Type of camouflage a species

The type of camouflage a species will develop depends on several factors:

1. The environment in which it lives. This is usually the most important factor.
2. The physiology and behaviour of an animal. Animals with fur need different camouflage than those with feathers or scales. Likewise, animals who live in groups use different camouflage techniques than those that are solitary.
3. If the animal is preyed upon, then the behaviour or characteristics of its predator can influence how the camouflage develops. For example, if the predator has achromatic vision, then the animal will not need to match the colour of its surroundings.

Colour produced by animals

Animals produce colours in two ways:

1. Biochromes natural microscopic pigments that absorb certain wavelengths of light and reflect others, creating a visible colour that is targeted towards its primary predator.
2. Microscopic physical structures, which act like prisms to reflect and scatter light to produce a colour that is different from the skin, such as the translucent fur of the Polar Bear, which actually has black skin.

Cryptic coloration can change as well. This can be due to just a changing of the seasons or it can be in response to more rapid environmental changes. For example, the Arctic fox has a white coat in winter, and a brown coat in summer. Mammals and birds require a new fur coat and new set of feathers respectively, but some animals, such as cuttlefish, have deeper-level pigment cells, called chromatophores, that they can control. Other animals such as certain fish species or the nudibranch can actually change their skin coloration by changing their diet. However, the most well-known creature that changes colour, the chameleon, usually does not do so for camouflage purposes, but instead to express its mood.

Beyond colours, skin patterns are often helpful in cryptic coloration as well. The Craik-O' Brien-Cornsweet illusion describes visual perception as occurring through contrasts of outlines. One recognises a dog, for example, not by its colour as much as by its shape. Often what matters most for good cryptic coloration is to break up the outline of a creature's body. This can be seen in common domestic pets such as tabby cats, but striping overall in other animals such as tigers and zebras help them blend into their environment, the jungle and the grasslands respectively. The latter two provide an interesting example, as one's initial impression might be that their coloration does not match their surroundings at all, but tigers' prey are usually colour blind to a certain extent such that they cannot tell the difference between orange and green, and zebras' main predators, lions, are colour blind. In the case of zebras, the stripes also blend together so that a herd of zebras looks like one large mass, making it difficult for a lion to pick out any individual zebra. This same concept is used by many striped fish species as well. Among birds, the white 'chinstraps' of Canada geese make a flock in tall grass appear more like sticks and less like birds' heads.

In nature, there is a strong evolutionary pressure for animals to blend into their environment or conceal their shape; for prey animals to avoid predators and for predators to be able to sneak up on prey. Natural camouflage is one method that animals use to meet these. There are a number of methods of doing so. One is for the animal to blend in with its surroundings, while another is for the animal to disguise itself as something uninteresting or something dangerous.

There is a permanent co-evolution of the sensory abilities of animals for whom it is beneficial to be able to detect the camouflaged animal, and the cryptic characteristics of the concealing species. Different aspects of crypsis and sensory abilities may be more or less pronounced in given predator-prey pairs of species. Some cryptic animals also simulate natural movement, e.g. of a leaf in the wind. This is called procryptic behaviour or habit. Other animals attach or attract natural materials to their body for concealment.

A few animals have chromatic response, changing colour in changing environments, either seasonally (ermine, snowshoe hare) or far more rapidly with chromatophores in their integument (the cephalopod family). Some animals, notably in aquatic environments, also take steps to camouflage the odours they create that may attract predators. Some herd animals adopt a similar pattern to make it difficult to distinguish a single animal. Examples include stripes on zebras and the reflective scales on fish.

Animal Behaviour/Mimicry and Camouflage

Mimicry

A major concern of animals and other critters is to protect themselves from predators in order to survive and reproduce and pass their genes off to a new generation. Many animals have evolved adaptations known as antipredator devices such as camouflage and chemical toxins. Animals use camouflage to blend in with their environments in an attempt to be unrecognisable by predators. Other organisms such as the monarch butterfly contain chemical toxins that are secreted into the predator's mouth when it attempts to eat the butterfly. The monarch butterfly also has warning coloration that gives a warning sign to predators to remind them that the butterfly is toxic and should not be eaten.

These antipredator devices are so successful that other organisms have been known to mimic them. The organism that is mimicked is known as the model and the third party that is deceived by the model and its mimic is known as the receiver. The mimics have learned to take advantage of the colour patterns and markings that predators have learned by experience to avoid. The model is usually a species that has an abundant population and has successfully warded off predators with an antipredator device.

Organisms have learned to mimic their surroundings or environment in an attempt to 'hide' from predators. For example, lizards have learned to mimic tree trunk colour which proves to be very successful as predators will simply move past them as they believe that they are simply looking at a tree. Another example of this type of mimicry can be seen with the Katydid who will mimic a leaf in both colour and shape in an attempt to be hidden.

Some prey animals have evolved certain patterns on their bodies that mimic other animals in an attempt to startle their predators. The most common example of this type of mimicry can be found in some moths and butterflies who flash eye spots on their wings to predators. These eye spots startle the predator who believes that the eyes belong to a much larger animal that may be a threat to them.

In one form of mimicry known as aggressive mimicry, an organism will mimic a signal that is either deceptive or attractive to its prey. One example of this involves the praying mantis who will mimic flowers to attract insects that they can then capture and eat. Organisms can also imitate the behaviours of other organisms. Moth caterpillars, for example, will imitate the motion and body movements of a snake in order to scare off predators that are usually a prey item for snakes.

One of the most popular types of mimicry involves the warning coloration found on inedible or toxic organisms such as the monarch butterfly. Once these toxic organisms have adapted this warning coloration which warns predators to stay away, other organisms may start to mimic this warning coloration in an attempt to stay alive. Batesian mimics are those mimics that imitate unpalatable species even though they are palatable. Therefore, one species is harmful while the other is harmless. The wasp is a great example of Batesian mimicry. The wasp is the model species in this example as it possesses a sting which enables it to escape from predators. The bright warning coloration of the wasp has been mimicked by many other insects. Even though the mimics are harmless, the predator will avoid them due to bad experiences with wasps with the same coloration. With Müllerian mimicry, many unpalatable species share a similar colour pattern. Müllerian mimicry proves to be successful as the predator only has to be exposed to one of the species in order to learn to stay away from all the other species with the same warning colour patterns. The black and yellow striped bodies of social wasps, solitary digger wasps, and caterpillars of the cinnabar moths warn predators that the organism is inedible. This is a great example of Müllerian mimicry as all of these unpalatable, unrelated species have a shared colour pattern that keeps predators away.

Mimicry is a very successful antipredator device that species have evolved over many generations. As one can see organisms have come to mimic many different characteristics such as colour patterns and behaviours. However, selection only favours the mimics when they are less common than the model. Therefore, the fitness of mimics is 'negatively frequency-dependent'.

BIRDS MIGRATION AND NAVIGATION

Why Do Birds Migrate?

Food, water, protective cover, and a sheltered place to nest and breed are basic to a bird's survival. But the changing seasons can transform a comfortable environment into an unlivable one — the food and water supply can dwindle or disappear, plant cover can vanish, and competition with other animals can increase.

Most wild animals face the problem of occupying a habitat that is suitable for only a portion of the year. Fortunately, however, nature has provided methods for coping with the situation. One method, known as hibernation, involves entering a dormant state during the winter season. The other method, known as migration, involves escaping the area entirely. Because of the powers of flight, most birds adapt to seasonal changes in the environment by migrating; only a few birds species, such as the Common Poorwill, hibernate.

In North America, the ratio of migratory to nonmigratory birds varies greatly from region to region. In high arctic regions (northern Alaska, northern Canada, and Greenland), where many shorebirds and water fowl nest, the entire population often consists of migratory birds who are only there during the summers.

In North America, the ratio of migratory to nonmigratory birds varies greatly from region to region. In high arctic regions (northern Alaska, northern Canada, and Greenland), where many shorebirds and water fowl nest, the entire population often consists of migratory birds who are only there during the summers. In the forest and open country of eastern United States, over 80 per cent of the nesting land birds are migratory, spending the winter in more hospitable southern climates. There is a similar high percentage of south-migrating birds in inland areas of the West. However, in areas where the climate is more equable, like the Pacific Coast, more species are nonmigratory; in tropical regions at least 80 per cent of the birds are nonmigratory.

In the Rockies and Sierras of the West, migration often consists of moving from the high to low elevations. Rosy Finches, Townsend's Solitaires, and Mountain Quail perform these movements quite regularly whereas others, such as Clark's Nutcracker, are much more erratic.

Some migration schedules do not always closely follow seasonal changes in the weather. For example, since the vegetative food supply of nomadic species such as the crossbills, redpolls, and Pine Grosbeaks fluctuates in abundance from year to year, these birds migrate in some winters and not in others. In contrast, insect-eating birds such as warblers, vireos, and flycatchers that live in the far north have no choice but to migrate from their summer habitats, since their food supply always disappears from sight in winter; their migration therefore tends to involve long distances and regular timing.

Evolution of Bird Migration

Bird migration is a behaviour that has evolved over many thousands of years. Scientists believe that migration began to evolve when individuals that moved from one area to another ultimately produced more young than those that remained in one area. Migratory behaviour continues to evolve because of the changing environment in which the birds live. If environmental conditions favour migration, the number of birds that migrate increases; if conditions permit the birds to stay in one place, the sedentary type predominates. A good example of such adaptive behaviour is a migratory North American bird called the dark-eyed Junco. It was undoubtedly some of these migrants gone astray that colonised Guadeloupe Island, some 150 miles off the coast of Baja California, where the junco is now established as a sedentary population.

Waterfowl Migration

Geese have long life spans and, like many other large water birds, they use regular stopover places along their flyways and return year after year to the same nesting and wintering areas. The migration pattern of the Snow Geese is typical of many of our ducks, geese, and swans. Snow Geese nest in high arctic regions from the North Slope of Alaska, eastward along the coast of northwestern Greenland, and southward along the western and southern shores of Hudson Bay. They migrate southward during the fall in large flocks, flying both day and night at high altitudes. The time of their flight is dependent upon weather — they prefer to fly in clear skies and with a good tail wind. When conditions are right they can cover many hundreds of miles during a single high-altitude flight. Snow Geese spend the winter on the mid-Atlantic coast, the Louisiana-Texas Gulf coast, and in California and the Southwest.

Shorebird Migration

Shorebirds, like most of our waterfowl, nest on the arctic tundra and migrate to southern wintering grounds. Yet unlike waterfowl, many shorebirds — sandpipers, plovers, godwits, curlews — migrate beyond the confines of the North American continent.

A classic example is the Lesser Golden Plover, which spends the northern winter on the vast Argentinian grasslands called the pampas. There, along with its fellow migrants — yellowlegs, Hudsonian Godwits, stilts, and Baird's, Pectoral, Upland, and Buff-breasted Sandpipers — the Lesser Golden-Plover spends the southern summer with resident South American shorebirds. In spring the golden-plovers migrate northward in flocks, crossing the Caribbean and Gulf of Mexico. They enter the United States mainly along the Texas and Louisiana coasts, and head up through the interior of North America, stopping to feed on insects in pastures and plowed fields of the agricultural Midwest. These long-distance migrants arrive in their breeding grounds in June, and nest during the long days of the brief northern summer.

Young shorebirds are precocial (capable at birth). Little parental care seems to be required in many of these species. In fact, the adult birds depart on their southward migration before summer is over, weeks before their youngsters begin their trip.

Though golden-plover migrate a bit later, the height of their passage period of many adult shorebirds in the eastern United States occurs from late July through August. In midsummer they reappear, well on their way to the wintering ground. The precocial young birds appear some weeks later, guided southwards by their instincts.

Lesser Golden-Plovers are one of a number of species of birds that follow different migration routes in spring and fall. In fall golden-plovers fly southeastward from their nesting areas to the coast of Labrador and Maritime Canada; from there many initiate a nonstop over-water flight all the way to South America. During the crossing, the golden-plovers ascend to great heights (over 20,000 feet), but it still takes several days and nights of continuous flying before they reach the continent.

Seabird Migration

Seabirds are marvelously adapted for covering great distances over seemingly trackless oceans and, as migrants go, they hold the records. The fabled Arctic Tern nests as far north as open land exists and travels the length of the oceans to winter at the other end of the world. It is a round trip of some 25,000 miles performed every year of the birds life.

More truly oceanic are the shearwaters, albatrosses, and storm-petrels, who spend most of their lives out of sight of land, coming ashore only to nest. One of the most numerous species in North American waters is the Sooty Shearwater. Sooty Shearwaters nest in islands deep in the Southern Hemisphere, mostly around New Zealand and the southern tip of South America. Breeding in burrows that they excavate on these islands during the southern summer, they depart northward after nesting to spend the southern winter at temperate North American latitudes. In late spring and throughout the summer, Sooty Shearwaters appear in large numbers off both American coasts, but they are especially prevalent on the West Coast, where thousands are often viewed from shore. By late summer the shearwaters are on their way back to the other end of the earth, literally circling its oceans in the process.

Land Bird Migration

It is a common misconception that many of our North American summer land birds go to South America in the winter. A few do so — Swainson's and Grey-cheeked Thrushes, Bobolinks, Northern Waterthrushes, and Blackpoll Warblers — but most travel only as far as Mexico, Central America, and the Caribbean. Birds that can eat seeds do not need to migrate even that far because their food is generally more abundant.

Many species of sparrow, junco, towhee, and longspur stay within the United States, migrating only as far as it is necessary to find winter weather that is less severe. Most individuals migrate only a few hundred to a thousand miles, and in some cases the northern edge of the winter range overlaps the southern part of their nesting area. When raising young, most of these birds augment their diets by harvesting insects. In winter, however, they feed almost entirely on small seeds.

Unlike water birds, most small land birds do not migrate in compact flocks, but in summer they nest in dispersed territories, each occupied by a single pair. Most summer land birds migrate at night. The Northern Oriole, a typical summer land bird, is found across North America wherever deciduous trees predominate. Like most songbirds, the Northern Oriole migrates at night and spends the winter primarily from central Mexico southward to northern South America and the West Indies. Eastern populations fly across the Gulf of Mexico in spring and fall.

The Blackpoll Warbler is a very remarkable traveler. Weighing in at about 20 grams, this bird performs an over-water flight from New England to the coast of South America, flying at only 25 miles per hour. The bird takes three to four days to make the trip.

Choosing the Route of Bird Migration

Each migratory species generally has a route of travel between its nesting and winter range, but for the majority of species these migration routes are quite broad. Waterfowl tend to be confined to some what narrower corridors determined by the availability of suitable habitat. In fact, it was once thought that there were distinct, narrow flyways (the Atlantic flyway, Central flyway, and so forth) used exclusively by various populations of water fowl. The discovery that birds from one nesting area could be found migrating in several different flyways put that concept to rest.

Although entire species might not be confined to narrow corridors during migration, individual birds often exhibit amazing loyalty to places occupied during previous breeding and non breeding seasons, as well as stopover points between the two, a phenomenon known as site fidelity.

Birds Migrating in Flocks or Alone

Some species of birds are highly social during migration, moving in flocks that may stay together for the whole journey. Flocked migrants are the most conspicuous of migrating birds and are the most familiar to us. They are also easiest to watch because they are generally daytime migrants. In at least some cases, flocks of migrating birds consist of family groups. It is thought that young birds learn details of the routes of travel and layover sites from the more experienced adults, though this is surely not the sole reason that these diverse types of birds travel so often in groups.

Flocked migrants include a wide variety of birds. Most of the large water birds travel in flocks, usually of impressive size. Among these are auks and puffins, cormorants, pelicans, ducks and geese, cranes, gulls, terns, sandpipers, and plovers. Often they congregate during migration at a few major stopover or staging areas where food is particularly abundant. Flocking is also common among many land birds, including doves, swifts, swallows, larks, pipits, crows, jays, waxwings, blackbirds, and starlings.

An equally diverse array of species seems to migrate in a more solitary fashion, perhaps occasionally forming more or less aggregations with others of their kind, but basically winging it alone. These birds include grebes, most herons, rails, some hawks, owls, nightjars, cuckoos, hummingbirds, kingfishers, woodpeckers, most flycatchers, creepers, wrens, kinglets, thrushes, vireos, wood warblers, and orioles.

Getting the Migration Timing Right

An experienced bird-watcher can mark the seasons in a given locale by the highly predictable times of arrival and departure of familiar migrating birds. Across much of the United States, for example, the first sign of spring is the arrival of flocks of Re-winged Blackbirds. The seasonal timing of migration is closely related to the likelihood that the necessities of life will be available at the time of arrival. Within each family of birds, however, there are great differences in timing among species. For example, among American wood warblers, the Louisiana Waterthrush may arrive at its breeding ground a full two months before the Connecticut Warbler passes through.

Variation also exists in the pace of migratory travel, even among related species. Shorebirds, waterfowl, cranes, and other species that use traditional stopover areas tend to make long flights interrupted by days of layover. Many land birds that are able to enjoy an abundance of resting and feeding habitats, such as warblers, flycatchers, and sparrows, typically fly one night, then rest for two or three nights, depending on the weather. These more leisurely travelers might cover as many as 2000 miles in a nights flight, especially if they have a tail wind. Some of these same kinds of birds are capable of making extended flights if, for example, they have to cross a large body of water.

There seems to be a premium on getting to nesting areas as soon as possible. In fact, many kinds of waterfowl begin move northward as soon as the lakes and ponds are released from the grip of ice. But at the same time there may be a heavy price to pay if the birds arrive too early — weather is less dependable in early spring.

Though it is possible to delineate the migration period of every species, the situation is actually quite complex. In may species, birds of different age and sexes tend to migrate at somewhat different times. In spring, the males of many species of songbirds precede the females, presumably because it is to a male's advantage to arrive in its breeding are early and stake out its territory before its rivals do.

Birds Fuelling for the Journey

Flying is a strenuous activity, and even though birds are marvelously adapted to their aerial life, with super-efficient haemoglobin, a lung and air-sac system that allows for maximum oxygen intake, and hollow bones, a calorie is still only a calorie, and it takes a lot of them to propel a body through the air over hundreds of miles. As the season of migration approaches, signaled by the changing length of the day, as well as by built-n biological clocks, a birds metabolism changes and it begins to deposit stores of fat under its skin. In species that make long, nonstop fligths, the amount of fat deposited can be quite impressive equaling half their body weight or more.

A single flight exhausts a large proportion of these 'fuel tanks', which must then be replenished before the next leg of the journey. For this reason, it is vitally important for migrating birds to find good, stable food supplies during their journey as well as before it. Stopover sites where fat stores can be replenished quickly can therefore be every bit important to long-distance migrants as suitable nesting and breeding grounds.

Birds Ascending the Heights

Because of reports of planes colliding with birds at very high altitudes and the stories about the bar-headed Geese that migrate over the Himalayas, cresting above Mount Everest, one could get the impression that bird migration normally occurs at immense heights. With the help of radar, however, we know which altitudes are common and which are unusual.

Birds behave somewhat differently from one species to the next, but songbirds fly at altitudes of less than 5,000 feet, and the majority travel no higher than 2500 feet. Waterfowl and shore birds tend to fly higher; it is not unusual to detect them above 10,000 feet or even as high as 20,000 feet, especially when they are making long, over-water flights.

Birds Checking the Weather

Just as long-term changes in climate have moulded the evolution of bird migration, seasonal and day-to-day changes in weather dramatically influence the timing and course of migration. When the weather conditions are right, the number of birds in flight can reach millions. Not only are the immediate flight conditions important, but the weather at the destination or starting point of the flight may also be critical to a bird's survival. For example, water birds must not arrive at northern latitudes before the ice has melted, and many tend to follow the spring thaw northward; in late fall many linger in the north until freezing temperatures force them to move. For most birds, however, it pays to anticipate seasonal changes in climate and to be gone well before conditions deteriorate. Many warblers, flycatchers, and other insect eaters begin their fall migration in late summer while the days are still warm and the insect life abundant.

During migration, the most critical weather factors are wind direction and changes in temperature. In spring, northbound birds select the warming temperatures and southerly winds that characterise the western side of high-pressure systems; in fall, they favour the lower temperatures and north winds that occur following the passage of a cold front. Birds also tend to avoid rainy, overcast weather, fog, and high winds, and even stop in the middle of their journey if they encounter deteriorating weather while over land. Recent research has shown that pigeons are quite sensitive to small changes in air pressure, birds are able to anticipate weather before any overt signs are evident. Insect-eaters like the Barn Swallow begin to migrate in late summer, well before their food supply is threatened by early frosts. Spring migrants, driven to reach the breeding grounds early, are often caught by late spring storms that cause high mortality.

Wind Factor for Birds

At the altitudes at which birds fly, wind speeds often exceed 20 miles per hour. A head wind can halt a bird's forward progress or even blow it backward, whereas a tail wind can easily double its speed. High winds can prevent small birds from migrating. Strong crosswinds can cause birds to drift far off course and may be disastrous for land birds carried over the ocean; such winds are often the reason that birds are sometimes found far outside their normal range.

As in most aspects of bird behaviour, there is variation in the speed at which different species fly. Most small songbirds, for example, fly at air speeds (speeds without any influence of wind) of only 20 miles per hour. Waterfowl, and especially the larger shorebirds, maintain speeds of about 40 miles per hour or even more. Migrating birds, however, generally propel themselves at quite moderate speeds, which means that the wind has an enormous impact on their progress. It is therefore no surprise that birds are supreme interpreters of weather and wind.

Long flights over water or other inhospitable terrain provide the ultimate tests of migratory strategy. In many cases, selecting a departure day on which there are tail winds may make the difference between life and death. In spring, millions of small birds fly northward across the Gulf of Mexico. Most of the time at that latitude, moderate southerly winds blow across the gulf, aiding the migrants. Occasionally, however, a cold front penetrates the gulf from the north. With the northerly winds behind them, these fronts often do not reach Yucatan, a major departure point for the Gulf crossing. Thus, birds can embark on the trip in fine weather with southerly winds only to run into problems out over the water. Trouble will likely come in the form of rain showers and, if the birds penetrate the front, potent head winds.

Under these conditions, what would have been a relatively easy trip can turn into a disaster, and birds that would have arrived on the northern Gulf Coast with fat to spare, arrive exhausted at the beaches, if they can make it at all. Thousands can die of starvation or dehydration while cresting on offshore oil rigs or in the water itself, as evidenced by the corpses that sometimes wash up on beaches. Clearly, it pays to be able to judge the weather.

Trends in Migration for Birds

One of the most critical weather factors affecting migration is wind direction. Winds blow clockwise around high — and counter-clockwise around low air-pressure systems. Migration tend to be heaviest in areas where the winds blow in the direction the birds are going and lightest where headwinds impede migration.

Birds Finding the Way Home

The ability of birds to return to a familiar place from any distance is a remarkable feat of nature. For centuries people have taken advantage of this ability in homing pigeons by using them to take messages from distant points back to familiar sites. Homing pigeons are domesticated nonmigratory birds with an instinct to return to their lofts (nesting sites) that is improved with training and by selective breeding. Training is started at short distances from the nesting site; over time, this distance is gradually increased to hundreds of miles from its loft at a completely unfamiliar location flies in the direction of home within a minute or two of its release.

Understanding homing behaviour is one of the greatest challenges to ornithologists. Fortunately, because they are able to carefully control the conditions under which the pigeons are released, researchers have been able to learn a great deal about how the birds navigate their way home. A Manx Shearwater, when released thousands of miles from its nest, will return within days. In one study, a Manx Shearwater averages 250 miles per day during a homing flight that lasted 12 days and covered 3200 miles.

Although homing ability has been fostered in pigeons by careful breeding and selecting of stock, it appears that training is not always necessary: Many species of wild birds perform similarly remarkable feats. One such bird is the migratory Manx Shearwater. Built like tiny albatrosses, these seabirds spend most of their lives skimming over the ocean surface far from the sight of land. They come ashore only to nest in burrows, which they dig in the ground on offshore islands in order to be safe from predators. The ease of locating and observing their nests make shearwaters ideal subjects for homing experiments.

In one such experiment, adult shearwaters taken from nesting burrows off the coast of Wales were flown thousands of miles from their nests to places in Europe and North America that were completely unfamiliar to them. Most of the birds returned to their burrows at astonishing speeds, speeds that would not have been possible had they wandered randomly or searched for the way home over a wide area. Even traveling over what to our eyes appears to be trackless ocean, these birds demonstrated that they knew the precise direction in which thy needed to go. Ornithologists believe that similar abilities are responsible for the return of small land birds to previous nesting and wintering places at the ends of long migrations.

Great Bird Navigators

Many migratory birds are remarkably faithful to previous nesting and overwintering places. Though a bird might be able to come close to these site merely by flying in a general direction during the course of migration, at some point more sophisticated navigating techniques must take over to guide the bird to its precise destination. Many animals are able to find their way home. One way of doing this is to directly sense the goal—to see, hear or smell it. Another way is to memorise the details of the outward journey and then reverse the route based on an integration of that information. Birds, however, apparently rely on a completely different process to find their way.

To understand the nature of the problem, imagine yourself in the following situation: You have been blindfolded and taken by a circuitous and unfamiliar route to a place you have never been before. There, in a forest without any view if distant landmarks, the blindfold is removed. You are left alone with a compass and a map, and you need to find your way back home. Unfortunately, before you can use the compass for information about direction, you must determine where you are in relation to your goal— you need to find your location on the map so you will know where you are in relation to home. A bird in an unfamiliar setting is quickly able to gain the information it needs to orient itself and navigate its way back home. To explain bird navigation, we have what is known as the 'map-and-compass' theory.

The compass component of this theory gives direction—north, south, east, west; the map component tells the bird where it is or gives locality. Scientists have learned a great deal more about the compass component than they have about mapping. They know that birds have several means of determining compass directions, but unfortunately, they still have no satisfactory explanation for how birds use biological 'maps' to guide then to a precise location from an unfamiliar starting point.

Bird Sun Navigators

Some observations indicate that birds might use the sun as a visual cue to determine compass directions. Starling, for example, seem able to negotiate the proper direction only if they have a view of the clear sky and sun; cloud cover seems to induce confusion. In an experiment in which the sun's apparent position was changed with mirrors attached to an orientation cage containing starlings, observers noted that the direction of the starlings' hopping, which earlier had been correlated to the direction they chose to migrate, was shifted accordingly.

Even birds that migrate exclusively at night pay considerable attention to the sun. At first this may seem odd because, after all, the sun is not visible to the nocturnal birds when they are flying. On the other hand, it is a predominate feature in the sky at a time of day (dusk) when birds may well be making decisions about whether to fly that night and in what direction. Radar studies have shown that most night migrants take off during this twilight period. Like many other animals (including insects, fish, reptiles, and mammals), birds are endowed with a built-in clock that tells them the time of day. Using this internal clock, young pigeons, at least, learn the sun's path of movement across the sky. Many birds are known to have an internal clock, and many are known to have a sun compass, but it is only in pigeons that ornithologists can watch the learning process develop.

Bird Star Navigators

Most birds migrate not by day, but at night. Employing conical cages, researchers quickly observed that night migrants exhibited hopping behaviour similar to that of day migrants and that they oriented themselves in the proper direction under clear, starry skies but became disoriented when it was cloudy. The decisive tests were performed in a planetarium, where star patterns can be manipulated at will. The experiments indicated that night-migrating birds learn and orient by spatial relationships among the constellations, rather than using information supplied by any single star. More recently, scientists have discovered that birds begin to develop star compass capability when they are quite young, and as experienced adults they can use many parts of the sky to decipher compass directions.

Magnetic Bird Navigators

A sun compass for migration during the day, a star compass for nocturnal migration — life would be much simpler had this been the end of the story. For a long time there has been a popular theory that birds have a magnetic compass guiding their navigatory behaviour. There is a good reason for that speculation — it would be a convenient system to use during overcast days or nights. But many respected biologists assured ornithologists that for birds to sense such force was almost impossible because the earth's magnetic field is such a weak force. A group of German ornithologists conducted research that provided evidence to the contrary. In the study, night migrants that were placed in orientation cages indoors in closed rooms showed weak but consistent and seasonally appropriate hopping directions, which suggested that they did not need the sun or stars to determine direction. By placing the cages within sets of wire coils through which a weak electric current was passed, it was possible to change the configuration of the magnetic field surrounding the birds to determine if they respond in a predicable way.

Although their responses continued to be weak, the birds did respond, and their orientation could be shifted by changing the magnetic directions. It therefore appears that is addition to sun and star compasses, at least several kinds of birds possess a magnetic compass, though as yet none of the field studies has found a clear indication of its influence on migratory birds in the wild.

Biological Bird Navigators

After about three decades of experiments on homing pigeons, scientists currently have two viable hypotheses concerning the 'mapping' ability of birds. Although only homing pigeons have been studied, there is good reason to believe that migratory birds also rely on some sort of biological map to find their way back to traditional nesting or wintering sites.

The first hypothesis, conceived and tested primarily by a group of Italian scientists, involves an 'odour map'. The scientists propose that young pigeons learn this map by smelling different odours that reach their home loft on winds from varying directions. They would, for example, learn that a certain odour arrives on winds blowing from the east. If a pigeon is transported eastward from its loft it should smell that odour more strongly either on the way to or at the release site. This should tell the pigeon that it needs to fly westward to return home. Although it may sound preposterous to some, there is a large amount of evidence supporting this hypothesis. However, even its strongest proponents do not extend the idea to include long-distance migrants.

The second hypothesis proposes that birds may be able to extract latitude and longitude from the earth's magnetic field. Unlike the compasses that are thought to help birds determine direction, this map is believed to help birds determine location. The main support for this hypothesis comes from observations of pigeons released in areas of 'magnetic anomalies' (places where the earth's magnetic field is distributed due to large iron deposits near the surface). When pigeons are first released in these areas, they depart in random directions, but after their initial confusion, most birds are able to correct their course and return home once they escape the influence of the anomaly. Because, in theory, magnetic anomalies are not a strong enough force to affect any of the birds' compasses — magnetic, sun or star — proponents argue that the fact that the pigeons are initially affected indicates the existence of a different aspect of navigation that is being affected — hence, the map. Neither hypothesis has been proven to the satisfaction of all the experts, so new and different experiments continue to be performed. It may turn out that neither of these alternatives is correct or a synthesis of the two may emerge.

Bird Navigational Techniques

Sun compass: Some birds seem able to negotiate the proper direction only if they have a clear view of the sun. Even night migrants appear touse the sun as a cue, as most take off during twilight.

Star compass: Night migrant orient themselves in the proper direction under clear, starry skies but become disoriented when it is cloudy. Night-migrating birds learn and orient by the spatial relationships among the constellations, rather than by using information supplied by any single star.

Odour map: Some short-distance migrants use an 'odour map' to return to nesting and wintering sites. Studies show that young pigeons learn the odours—carried by the wind—which reach their home sites.

Magnetic map: Migratory birds may rely on an instinctual map to find their way back to nesting or wintering sites. Magnetic disturbances may interrupt these abilities.

Magnetic compass: Sever kinds of birds appear to possess a built-in magnetic compass to use on cloudy days. Birds tested in a controlled environment showed that they knew which direction to migrate even without the sun or stars.

Bird Migration Altitudes

Birds choose altitudes based on a number of factors. Some of those factors can be the terrain traversed, distance to travel, time of year, and the strength of the bird. The weather also determines the altitude chosen on an individual trip.

Chapter 20

Animal Communities

INTRODUCTION

Animal communities focuses on the relationship between living and nonliving things in a particular environment, stressing how these are intertwined. In doing so, it provides the specialised vocabulary necessary to any consideration and discussion of these topics.

Plants and animals are not distributed randomly over the landscape. Each organism lives in an environment which best provides the needs of that organism: food, water, air, temperature, etc. The climate, geology, and topography of any spot on earth are the chief factors which determine these environments. The species of animals and plants that live in a particular area also are greatly affected by the intensity, frequency and recency of human activity.

ANIMAL POPULATION DENSITY

The calculations involved in estimating animal population density are relatively straightforward. However, reasonable estimates require accurate raw data. A random sample must be taken in which each animal has an equal chance of capture, and the sample population must randomly redistribute itself once re-released. While marked animals must stay marked, the second sample must be taken quickly to minimise the effects of births, deaths and migration on the population size. For sedentary populations, the small samples taken, known as quadrats, must be reasonably representative of the whole population. Additional quadrat sampling may be required if the data appears too varied.

Calculation of Mobile Animal Populations

1. Use the Lincoln-Peterson index to calculate density. This simple, capture-recapture method has been used effectively since the 1930s.
2. Assign variables. N (number) is the number that you are looking for, the total number of animals. Use m (marked) to represent the number of animals taken in the first capture and marked. Use r (remarked) to represent the number of recaptured animals. Utilise n to represent the number of animals captured the second time.
3. Apply the Lincoln-Peterson index which states that the percentage of marked individuals to the total population should equal the percentage of remarked individuals to the recaptured population. This statement is represented by the formula $m/N = r/n$.
4. Rearrange the equation to solve for the total population number. The formula becomes $N = mn/r$.

5. Consider using a standard deviation equation to check for accuracy. This will yield an error margin that enables you to state the population with scientifically accepted confidence. Use this formula — $S =$ the square root of $[(m + 1)(n + 1)(m - r)(n - r)/(r + 1)(r + 1)(r + 2)]$ — to calculate standard deviation.
6. Interpret the standard deviation calculation. Remember that the larger the deviation, the less accurate the estimate of actual population size is. For example, with an estimated population size of 500, a confidence interval range of ±25 (475–525) is more precise than one of ±100 (400–600).

Calculation of Sedentary Animal Populations

1. Use a quadrat technique to generalise population density estimates from several small areas into a larger one. A quadrat is a small area in which the actual animal population is counted.
2. Check the data to ensure that the number, size and arrangement of quadrats are reasonably likely to be representative of the population as a whole. For example, if you sample four quadrats and one has two animals, one has 800 animals and two others have 57 animals, you might need to question the sampling methodology.
3. Average the number of individuals found in each quadrat. Using the example cited above results in an average of 229 individuals per quadrat — $(2 + 800 + 57 + 57)/4 = 229$.
4. Multiply the average number obtained by the ratio of the larger area to the quadrat size (they should all be the same). For example, if your sample area is 200 m² and each quadrat is 2 m², the ratio calculation is 200 m²/2 m² = 100.
5. Estimate population density by multiplying the average number of animals per quadrat by the area ratio obtained. For example, the population density in this sample is calculated by multiplying 229 by 100 in order to come up with 22,900 individuals.

PHYTOPLANKTON

Phytoplankton are the autotrophic component of the plankton community. Most phytoplankton are too small to be individually seen with the unaided eye. However, when present in high enough numbers, they may appear as a green discoloration of the water due to the presence of chlorophyll within their cells (although the actual colour may vary with the species of phytoplankton present due to varying levels of chlorophyll or the presence of accessory pigments such as phycobiliproteins, xanthophylls, etc.).

Ecology of Phytoplankton

Phytoplankton obtain energy through the process of photosynthesis and must therefore live in the well-lit surface layer (termed the euphotic zone) of an ocean, sea, lake or other body of water. Phytoplankton account for half of all photosynthetic activity on earth. Thus phytoplankton are responsible for much of the oxygen present in the earth's atmosphere — half of the total amount produced by all plant life. Their cumulative energy fixation in carbon compounds (primary production) is the basis for the vast majority of oceanic and also many freshwater food webs (chemosynthesis is a notable exception). The effects of anthropogenic warming on the global population of phytoplankton is an area of active research. Changes in the vertical stratification of the water column, the rate of temperature-dependent biological reactions, and the atmospheric supply of nutrients are expected to have important effects on future phytoplankton productivity. Additionally, changes in the mortality of phytoplankton due to rates of zooplankton grazing may be significant. As a side note, one of the more remarkable food chains in the ocean — remarkable because of the small number of links — is that of phytoplankton-feeding krill (a type of shrimp) feeding baleen whales.

Phytoplankton are also crucially dependent on minerals. These are primarily macronutrients such as nitrate, phosphate or silicic acid, whose availability is governed by the balance between the so-called biological pump and upwelling of deep, nutrient-rich waters. However, across large regions of the World Ocean such as the Southern Ocean, phytoplankton are also limited by the lack of the micronutrient iron. This has led to some scientists advocating iron fertilisation as a means to counteract the accumulation of human-produced carbon dioxide (CO_2) in the atmosphere. Large-scale experiments have added iron (usually as salts such as iron sulphate) to the oceans to promote phytoplankton growth and draw atmospheric CO_2 into the ocean. However, controversy about manipulating the ecosystem and the efficiency of iron fertilisation has slowed such experiments.

While almost all phytoplankton species are obligate photoautotrophs, there are some that are mixotrophic and other, non-pigmented species that are actually heterotrophic (the latter are often viewed as zooplankton). Of these, the best known are dinoflagellate genera such as *Noctiluca* and *Dinophysis*, that obtain organic carbon by ingesting other organisms or detrital material.

The term phytoplankton encompasses all photoautotrophic micro-organisms in aquatic food webs. Phytoplankton serve as the base of the aquatic food web, providing an essential ecological function for all aquatic life. However, unlike terrestrial communities, where most autotrophs are plants, phytoplankton are a diverse group, incorporating protistan eukaryotes and both eubacterial and archaebacterial prokaryotes. There are about 5000 known species of marine phytoplankton. There is uncertainty in how such diversity has evolved in an environment where competition for only a few resources would suggest limited potential for niche differentiation.

In terms of numbers, the most important groups of phytoplankton include the diatoms, cyanobacteria and dinoflagellates, although many other groups of algae are represented. One group, the coccolithophorids, is responsible (in part) for the release of significant amounts of dimethyl sulphide (DMS) into the atmosphere. DMS is converted to sulphate and these sulphate molecules act as cloud condensation nuclei, increasing general cloud cover. In oligotrophic oceanic regions such as the Sargasso Sea or the South Pacific Gyre, phytoplankton is dominated by the small sized cells, called picoplankton, mostly composed of cyanobacteria (*Prochlorococcus*, *Synechococcus*) and picoeucaryotes such as Micromonas.

Environmental threats

A controversial 2010 has found that marine phytoplankton have declined substantially in the world's oceans over the past century. Since 1950 alone, algal biomass was reported to have decreased by about 40 per cent, probably in response to ocean warming.

Aquaculture

Phytoplankton are a key food item in both aquaculture and mariculture. Both utilise phytoplankton as food for the animals being farmed. In mariculture, the phytoplankton is naturally occurring and is introduced into enclosures with the normal circulation of seawater. In aquaculture, phytoplankton must be obtained and introduced directly. The plankton can either be collected from a body of water or cultured, though the former method is seldom used. Phytoplankton is used as a foodstock for the production of rotifers, which are in turn used to feed other organisms. Phytoplankton is also used to feed many varieties of aquacultured molluscs, including pearl oysters and giant clams.

The production of phytoplankton under artificial conditions is itself a form of aquaculture. Phytoplankton is cultured for a variety of purposes, including foodstock for other aquacultured organisms, a nutritional supplement for captive invertebrates in aquaria. Culture sizes range from small-scale laboratory cultures of less than 1L to several tens of thousands of litres for commercial aquaculture. Regardless of the size of the culture, certain conditions must be provided for efficient growth of plankton. The majority of cultured plankton is marine, and seawater of a specific gravity of 1.010 to 1.026 may be used as a culture medium. This water must be sterilised, usually by either high temperatures in an autoclave or by exposure to ultraviolet radiation, to prevent biological contamination of the culture. Various fertilisers are added to the culture medium to facilitate the growth of plankton. A culture must be aerated or agitated in some way to keep plankton suspended, as well as to provide dissolved carbon dioxide for photosynthesis. In addition to constant aeration, most cultures are manually mixed or stirred on a regular basis. Light must be provided for the growth of phytoplankton. The colour temperature of illumination should be approximately 6500 K, but values from 4000 K to upwards of 20,000 K have been used successfully. The duration of light exposure should be approximately 16 hours daily; this is the most efficient artificial day length.

ZOOPLANKTON

Zooplankton are heterotrophic (sometimes detritivorous) plankton. Plankton are organisms drifting in oceans, seas, and bodies of fresh water.

Ecology

Zooplankton is a categorisation spanning a range of organism sizes including small protozoans and large metazoans. It includes holoplanktonic organisms whose complete life cycle lies within the plankton, as well as meroplanktonic organisms that spend part of their lives in the plankton before graduating to either the nekton or a sessile, benthic existence. Although zooplankton are primarily transported by ambient water currents, many have locomotion, used to avoid predators (as in diel vertical migration) or to increase prey encounter rate.

Ecologically important protozoan zooplankton groups include the foraminiferans, radiolarians and dinoflagellates (the latter are often mixotrophic). Important metazoan zooplankton include cnidarians such as jellyfish; crustaceans such as copepods and krill; chaetognaths (arrow worms); molluscs such as pteropods; and chordates such as salps and juvenile fish. This wide phylogenetic range includes a similarly wide range in feeding behaviour: filter feeding, predation and symbiosis with autotrophic phytoplankton as seen in corals. Zooplankton feed on bacterioplankton, phytoplankton, other zooplankton (sometimes cannibalistically), detritus (or marine snow) and even nektonic organisms. As a result, zooplankton are primarily found in surface waters where food resources (phytoplankton or other zooplankton) are abundant.

Through their consumption and processing of phytoplankton and other food sources, zooplankton play a role in aquatic food webs, as a resource for consumers on higher trophic levels (including fish), and as a conduit for packaging the organic material in the biological pump. Since they are typically small, zooplankton can respond rapidly to increases in phytoplankton abundance, for instance, during the spring bloom.

Zooplankton can also act as an disease reservoir. They have been found to house the bacterium *Vibrio cholerae*, which causes cholera, by allowing the cholera vibrios to attach to their chitinous exoskeletons. This symbiotic relationship enhances the bacterium's ability to survive in an aquatic environment, as the exoskeleton provides the bacterium with carbon and nitrogen.

FRESHWATER MACRO-INVERTEBRATES

Freshwater benthic macro-invertebrates or more simply 'benthos', are animals without backbones that are larger than ½ millimeter (the size of a pencil dot). These animals live on rocks, logs, sediment, debris and aquatic plants during some period in their life. The benthos include crustaceans such as crayfish, mollusks such as clams and snails, aquatic worms and the immature forms of aquatic insects such as stonefly and mayfly nymphs.

These animals are widespread in their distribution and can live on all bottom types, even on manmade objects. They can be found in hot springs, small ponds and large lakes. Some are even found in the soil beneath puddles. Many species of benthos are able to move around and expand their distribution by drifting with currents to a new location during the aquatic phase of their life or by flying to a new stream during their terrestrial phase. Most benthic species can be found throughout the year, but the largest numbers occur in the spring just before the reproductive period. In colder months, many species burrow deep within the mud or remain inactive on rock surfaces. Many aquatic insects undergo a complete metamorphosis — the transition from egg to larva to pupa and finally to adult. They remain in the water for most of their lives (typically one month to four years). After becoming adults, the majority of insects live for only a brief time, usually a few hours to a few days, while they locate mates and reproduce.

Ecological Importance of Benthic Macroinvertebrates

Benthos is an important part of the food chain, especially for fish. Many invertebrates feed on algae and bacteria, which are on the lower end of the food chain. Some shred and eat leaves and other organic matter that enters the water. Because of their abundance and position as 'middlemen' in the aquatic food chain, benthos plays a critical role in the natural flow of energy and nutrients. As benthic invertebrates die, they decay, leaving behind nutrients that are reused by aquatic plants and other animals in the food chain.

Advantages of using the benthos to monitor water quality

Unlike fish, benthos cannot move around much so they are less able to escape the effects of sediment and other pollutants that diminish water quality. Therefore, benthos can give us reliable information on stream and lake water quality. Their long life cycles allow studies conducted by aquatic ecologists to determine any decline in environmental quality. Benthos represents an extremely diverse group of aquatic animals, and the large number of species possess a wide range of responses to stressors such as organic pollutants, sediments, and toxicants. Many benthic macroinvertebrates are long-lived, allowing detection of past pollution events such as pesticide spills and illegal dumping.

Analysis of Benthos Samples

Sampling is based on the type of aquatic habitat under study. In turbulent riffles (shallow areas with fast flows), the most commonly sampled stream habitat, various nets are used to capture benthos. The nets are secured and the water is stirred up causing the benthos to float down stream into the net. In slow moving or still water, a dip net is often used to sample shore areas under bank overhangs or tree roots. In ponds or lakes with soft mud bottoms, grab samplers may be used to collect benthos.

Benthic communities can be used to monitor stream quality conditions over a broad area or they can be used to determine the effects of point source discharges from sources such as sewage treatment plants and factories. Ecologists who evaluate environmental quality using the benthos often consider the following characteristics of a benthic sample to be important indicators of stream, river or lake quality:

1. Taxa richness: A measure of the number of different types of animals; greater taxa richness generally indicates better water quality.
2. Pollution tolerance: Many types of benthos are sensitive to pollutants such as metals and organic wastes. Mayflies, stoneflies, and caddisflies are generally intolerant of pollution. If a large number of these insect types are collected in a sample, the water quality in the stream is likely to be good. If only pollution-tolerant organisms such as non-biting midges and worms are found, the water is likely to be polluted.
3. Functional groups: The presence or absences of certain feeding groups (such as scrapers and filterers) may indicate a disturbance in the food supply of the benthic animals in the stream and the possible effects of toxic chemicals.

When water quality, habitat condition, land use and fish are evaluated along with benthos, the result is a comprehensive picture of environmental quality. This information helps resource managers at do not resuscitate (DNR) to determine which rivers and streams in Maryland have good habitat quality and should be protected. The benthos can also help identify those rivers and stream showing signs of stress. Resource managers can then apply management actions that will improve environmental quality, such as storm water control in urban areas and best management practices on farmlands to control nutrient runoff.

INVERTEBRATES AS INDICATORS

Aquatic invertebrates live in the bottom parts of our waters. They are also called benthic macro-invertebrates or benthos, (benthic = bottom, macro = large, invertebrate = animal without a backbone) and make good indicators of watershed health because they:

1. Live in the water for all or most of their life.
2. Stay in areas suitable for their survival.
3. Are easy to collect.
4. Differ in their tolerance to amount and types of pollution.
5. Are easy to identify in a laboratory.
6. Often live for more than one year.
7. Have limited mobility.
8. Are integrators of environmental condition.

Some benthos are found more often, and in larger amounts, in waters that are generally clean or unpolluted by organic wastes. Without too much organic matter, the waters usually have lots of oxygen for the benthos. This use as an 'indicator' of water quality has been occurring for many years. For example, stoneflies are often considered to be clean water benthos. But when thinking about worms and midges, water quality professionals often view these as indicators of dirty water, especially in rivers and streams.

Unfortunately, it is not always a clear decision to make. Oxygen is only one factor affecting the benthos. Others include toxic chemicals, nutrients, and habitat quality. Some types of stoneflies may actually be found in waters that are not so clean, and likewise some types of worms and midges can be found in cleaner waters. So it is important to understand that there are some more complex methods to make these types of decisions, and to determine whether waters are healthy or polluted for aquatic life. Depending upon the type of aquatic environment, such as standing waters like lakes and wetlands, the categories of clean, somewhat pollution tolerant, and pollution tolerant don't necessarily apply.

Secondary Production

Production which occurs in animals is known as secondary production. The net quantity of energy transferred and stored in the somatic and reproductive tissues of heterotrophs over a period of time is called secondary productivity.

Gross secondary production in animals equals the amount of biomass assimilated or biomass eaten less feces. As in the case of plants herbivores and carnivores ingest the food material out of which a part gets assimilated and a part is egested. A large part of the assimilated energy is consumed during metabolic process like respiration, growth, reproduction, etc. and the rest is available to be laid down as new biomass. The net secondary production (NSP) = Energy assimilated from the food eaten – faeces – energy consumed for respiration. Secondary productivity indicates about food resources available to the heterotrophic populations, including man, in the food chain.

Spatial Pattern

A succession of spatially explicit ecological models in the early 1990s indicated that large-scale regular spatial patterns could arise within homogeneous landscapes from local biotic interactions alone, with potentially profound implications for the maintenance of biodiversity and ecological stability.

Over the past decade, however, multiple studies have shown that regular patterns are both common and persistent across a range of ecosystems. But the crucial questions of whether and how these patterns influence ecosystem functioning remain unanswered. Here, we show that the even spacing of subterranean termite mounds in an apparently homogeneous African savanna provides a template for parallel spatial patterning in tree-dwelling animal communities. We further show that the uniformity of this pattern at small spatial scales elevates the productivity of the entire landscape, providing support for models linking spatial pattern with ecosystem functioning.

Taylor's power law for dependence of variance on mean in animal populations: Ecologists studying the behaviour of population of animals of a certain species are often concerned with the spatial aggregation of the population, and how this changes with time. Suppose such a populations is sampled simultaneously at geographically different sites of a common size, repeatedly on several occasions. On each occasion the site-counts may be regarded as a random sample from some probability distribution that relates to the aggregation displayed by the population. On different occasions the population density may differ over several orders of magnitude. Realistic theoretical models of aggregation are difficult to build because the probability distribution of site-counts is unknown, and may be different in kind (not merely differing in parameter values) on different occasions, especially when the mean density differs markedly between occasions. On occasion i let the distribution of site-counts have variance V_i and mean μ_i. Clapham noted that counts from a population of plants dispersed randomly over an area follow the Poisson distribution for which $V_i = \mu_i$, but generally populations are more aggregated and $V_i > \mu_i$. Studies of animals have usually involved indices of aggregation based exclusively on some proposed parametric relationship between V_i and μ_i.

Taylor suggested that variance and mean are related in the form of a power function:

$$V_i = \alpha \mu_i^\beta \qquad \qquad \dots (20.1)$$

and this relationship is independent of the form of distribution pertaining on any occasion i. Taylor called the parameter β and index of aggregation, and considered its value ecologically important, species-specific, and indicative of the rate of change of spatial behaviour of an animal in response to a change in the population density. He considered α to be a parameter of less immediate ecological interest, partly dependent on sampling procedure.

Mark and recapture method: Mark and recapture is a method commonly used in ecology to estimate population size. This method is most valuable when a researcher fails to detect all individuals present within a population of interest every time that researcher visits the study area. Other names for this method or closely related methods, include capture-recapture, capture-mark-recapture, mark-recapture, sight-resight, mark-release-recapture, multiple systems estimation and band recovery.

Another major application for these methods is in epidemiology, where they are used to estimate the completeness of ascertainment of disease registers. Typical applications include estimating the number of people needing particular services (i.e. services for children with learning disabilities, services for medically frail elderly living in the community) or with particular conditions (i.e. illegal drug addicts, people infected with HIV, etc.).

Typically a researcher visits a study area and uses traps to capture a group of individuals alive. Each of these individuals is marked with a unique identifier (e.g. a numbered tag or band), and then is released unharmed back into the environment. A mark recapture method was first used for ecological study in 1896 by Petersen to estimate plaice, *Platichthys platessa*, populations.

Sufficient time is allowed to pass for the marked individuals to redistribute themselves among the unmarked population. Next, the researcher returns and captures another sample of individuals. Some of the individuals in this second sample will have been marked during the initial visit and are now known as recaptures. Other animals captured during the second visit will not have been captured during the first visit to the study area. These unmarked animals usually are given a tag or band during the second visit and then are released.

Population size can be estimated from as few as two visits to the study area. Commonly, more than two visits are made, particularly if estimates of survival or movement are desired. Regardless of the total number of visits, the researcher simply records the date of each capture of each individual. The 'capture histories' generated are analysed mathematically to estimate population size, survival or movement.

Lincoln–Petersen method of analysis: The Lincoln–Petersen method can be used to estimate population size if only two visits are made to the study area. This method assumes that the study population is 'closed'. In other words, the two visits to the study area are close enough in time so that no individuals die, are born, move into the study area (immigrate) or move out of the study area (emigrate) between visits. The model also assumes that no marks fall off animals between visits to the field site by the researcher, and that the researcher correctly records all marks.

Given those conditions, estimated population size is:

$$N = \frac{MC}{R},$$

where,

N = Estimate of total population size.

M = Total number of animals captured and marked on the first visit.

C = Total number of animals captured on the second visit.

R = Number of animals captured on the first visit that were then recaptured on the second visit.

Adaptation of the Lincoln–Petersen method: It is assumed that all individuals have the same probability of being captured in the second sample, regardless of whether they were previously captured in the first sample (with only two samples, this assumption cannot be tested directly).

This implies that, in the second sample, the proportion of marked individuals that are caught (R/M) should equal the proportion of the total population that is caught (C/N). For example, if half of the marked individuals were recaptured, it would be assumed that half of the total population was included in the second sample.

In symbols:

$$\frac{R}{M} = \frac{C}{N}.$$

A rearrangement of this gives:

$$N = \frac{MC}{R},$$

the formula used for the Lincoln–Petersen method.

Sample calculation

A biologist wants to estimate the size of a population of turtles in a lake. She captures 10 turtles on her first visit to the lake, and marks their backs with paint. A week later she returns to the lake and captures 15 turtles. Five of these 15 turtles have paint on their backs, indicating that they are recaptured animals.

$$N = \frac{MC}{R} = \frac{10 \times 15}{5} = 30$$

In this example, the Lincoln–Petersen method estimates that there are 30 turtles in the lake.

A refined form. The schnabel method of M–R

Known as the Schnabel Method of M–R. A less biased estimator of population size can be obtained with a modified version of the first formula above:

$$N = \frac{(M+1)(C+1)}{R+1} - 1,$$

where, as before,

N = Estimate of total population size.

M = Total number of animals captured and marked on the first visit.

C = Total number of animals captured on the second visit.

R = Number of animals captured on the first visit that were then recaptured on the second visit.

An approximately unbiased variance of N or Var(N), can be estimated as:

$$\mathrm{Var}(N) = \frac{(M+1)(C+1)(M-R)(C-R)}{(R+1)(R+1)(R+2)}.$$

Mammal: Mammals are members of a class of air-breathing vertebrate animals characterised by the possession of hair, three middle ear bones, and mammary glands functional in mothers with young. Most mammals also possess sweat glands and specialised teeth, and the largest group of mammals, the placentals, have a placenta which feeds the offspring during gestation. The mammalian brain, with its characteristic neocortex, regulates endothermic and circulatory systems, including a four-chambered heart. Mammals range in size from the 30–40 millimetre (1 to 1.5 inch) Bumblebee Bat to the 33 metre (108 foot) Blue Whale.

The number of species accepted by the zoological community on any given calendar date depends on the classification scheme, and the figure changes continually as new species are discovered and the classification of others is revised. According to a major reference work, Mammal Species of the World, which is updated through periodic editions, 5676 species were known in 2007, distributed in 1229 genera, 153 families and 29 orders. In 2008 the IUCN completed a 5 year, 17,000 scientist Global Mammal Assessment for its IUCN red list, which counted 5488 accepted species at the end of that period. The class is divided into two subclasses (not counting fossils): the Prototheria (order of Monotremata) and the Theria, the latter containing the infraclasses Metatheria (including marsupials) and Eutheria (the placentals). The classification of mammals between the relatively stable class and family levels having changed often, different treatments of subclass, infraclass and order appear in contemporaneous literature, especially for Marsupialia.

Except for the five species of monotremes (which lay eggs), all living mammal species give birth to live young. Most mammals, including the six most species-rich orders, belong to the placental group. The three largest orders, in descending order, are Rodentia (mice, rats, porcupines, beavers, capybaras, and other gnawing mammals), Chiroptera (bats), and Soricomorpha (shrews, moles and solenodons). The next three largest orders include the Primates, to which the human species belongs, the Cetartiodactyla (including the even-toed hoofed mammals and the whales) and the Carnivora (dogs, cats, weasels, bears, seals, and their relatives).

The early synapsid mammalian ancestors, a group which included pelycosaurs such as Dimetrodon, diverged from the amniote line that would lead to reptiles at the end of the Carboniferous period. Although they were preceded by many diverse groups of non-mammalian synapsids (sometimes referred to as mammal-like reptiles), the first true mammals appeared 220 million years ago in the Triassic period. Modern mammalian orders appeared in the Palaeocene and Eocene epochs of the Palaeogene period. Phylogenetically, the clade Mammalia is defined as all descendants of the most recent common ancestor of monotremes (e.g. echidnas and platypuses) and therian mammals (marsupials and placentals). This means that some extinct groups of 'mammals' are not members of the crown group Mammalia, even though most of them have all the characteristics that traditionally would have classified them as mammals. These 'mammals' are now usually placed in the unranked clade Mammaliaformes.

Arthropod: An arthropod is an invertebrate animal having an exoskeleton (external skeleton), a segmented body, and jointed appendages. Arthropods are members of the phylum Arthropoda, and include the insects, arachnids, crustaceans, and others. Arthropods are characterised by their jointed limbs and cuticles, which are mainly made of α-chitin; the cuticles of crustaceans are also biomineralised with calcium carbonate. The rigid cuticle inhibits growth, so arthropods replace it periodically by molting. The arthropod body plan consists of repeated segments, each with a pair of appendages. It is so versatile that they have been compared to Swiss Army knives, and it has enabled them to become the most species-rich members of all ecological guilds in most environments. They have over a million described species, making up more than 80 per cent of all described living animal species, and are one of only two animal groups that are very successful in dry environments—the other being the amniotes.

Soil animals: Soil animals are an extremely diverse group of organisms. Soil animals are grouped roughly according to their size, into three groups. The first group is the microfauna. These are the smallest of the soil animals ranging from 20–200 µm. The main soil animals in this group are protozoa. The mesofauna is the next largest group and range in size from 200 µm – 10 mm. The most important animals in this group are mites, collembola (or spring tails) and nematodes. The macrofauna contain the largest soil animals such as earthworms, beetles and termites. Generally, the most common soil animals are protozoa, nematodes, mites and collembola.

Microfauna: Microfauna are small animals and unicellular organisms visible only under a microscope. Usually microfauna is defined as creatures smaller than 0.1 mm (100 microns) in size, with mesofauna as organisms between 0.1 mm and 2 mm in size, though definitions may vary.

In the soil, microfauna can be found in large numbers—generally several thousand per gram. Anyone can take a bit of wet soil, put it under a microscope, and find microfauna. Some of the most common and important microfauna are protozoa (unicellular eukaryotes), mites (among the most diverse and successful of all animals), springtails (related to insects), nematodes (transparent wormlike creatures), rotifers (named for their wheel-like ciliated mouthparts), and tardigrades, also known as 'water bears', one of the hardiest organisms in nature. Microfauna can be found worldwide, wherever there is wet soil, and some other places as well. Springtails have been found in the McMurdo Dry Valleys of Antarctica, one of the coldest and driest places on earth. Microfauna are accompanied by microflora, which includes algae, bacteria, fungi and yeasts, capable of digesting just about any organic substance, and some inorganic substances, such as TNT and synthetic rubber.

Nematode: The nematodes or roundworms (phylum nematoda) are the most diverse phylum of pseudocoelomates, and one of the most diverse of all animals. Nematode species are very difficult to distinguish; over 28,000 have been described, of which over 16,000 are parasitic. It has been estimated that the total number of nematode species might be approximately 10,00,000. Unlike cnidarians or flatworms, roundworms have a digestive system that is like a tube with openings at both ends.

Nematodes have successfully adapted to nearly every ecosystem from marine to fresh water, from the polar regions to the tropics, as well as the highest to the lowest of elevations. They are ubiquitous in freshwater, marine, and terrestrial environments, where they often outnumber other animals in both individual and species counts, and are found in locations as diverse as mountains, deserts, oceanic trenches, and within the earth's lithosphere. They represent, for example, 90 per cent of all life forms on the ocean floor. Their many parasitic forms include pathogens in most plants and animals (including humans). Some nematodes can undergo cryptobiosis.

One group of carnivorous fungi, the nematophagous fungi, are predators of soil nematodes. They set enticements for the nematodes in the form of lassos or adhesive structures. Nematodes have even been found at great depth (0.9–3.6 km) below the surface of the earth.

Plankton: Plankton (singular plankter) are any drifting organisms (animals, plants, archaea or bacteria) that inhabit the pelagic zone of oceans, seas or bodies of fresh water. Plankton are defined by their ecological niche rather than phylogenetic or taxonomic classification. They provide a crucial source of food to larger, more familiar aquatic organisms such as fish and cetacea.

Chapter 21

Marine Ecology

INTRODUCTION

Marine ecology is the branch of ecology dealing with the interdependence of all organisms living in the ocean, in shallow coastal waters, and on the seashore. The marine environment for all organisms consists of non-living, abiotic factors and living, biotic factors.

Abiotic: The abiotic factors include all the physical, chemical and geological variables that have a bearing on the type of life that can exist in an area. Included are: (i) water, (ii) light, (iii) temperature, (iv) pH, (v) salinity, (vi) substratum, (vii) nutrient supply, (viii) dissolved gases, (ix) pressure, (x) tides, (xi) currents, (xii) waves, and (xiii) exposure to air.

Biotic: The biotic factors are the interactions among living organisms. Zonation two major divisions in the marine world: (i) pelagic zone—waters of the world, and (ii) benthic zone—the ocean bottom.

The pelagic zone include the productive coastal waters—neritic zone and deep waters of the open ocean—oceanic zone. Another division in the pelagic zone is related to light penetration—the photic and aphotic zones.

The benthic zone extends from the seashore to the deepest parts of the sea. The material that makes up the bottom is the substratum and the organisms living there are the benthos.

Tides uncover parts of this zone and the area uncovered is the intertidal zone, above is the supratidal zone, affected by salt spray but not covered by sea water and below the intertidal zone is the subtidal zone—submerged and extending seaward. The elevation and slope determines the length of time it is exposed. This affects organisms living there because some are restricted to zones according to their adaptations to this type of zone (intertidal, etc.).

Organisms living in pelagic waters also put up with changes in salinity, temperature, etc. and inhabit the coastal areas, etc. which fit their adaptations [can withstand large changes (eury—prefix) and narrow tolerance (steno)].

Other zones include the surface waters of the coastal areas called the neritic zone and the waters of the ocean called the epipelagic zone. The open ocean is less productive than the neritic zone which contains plant plankton, fish larva, invertebrate larva that will eventually end up near the coast.

The open ocean is divided into zones depending on the amount of light it receives—from the epipelagic layer to the mesopelagic zone 200–1000 m in which daytime inhabitants migrate upwards during the night, bringing back nutrients and some exhibit bioluminescence (light producing organs called photophores). The deep sea layers bathypelagic 1000–4000 m and the abyssopelagic zone (below 4000 m) have limited food supplies although bacteria have been found that can make their own food.

Trophic (feeding) relationships: Energy transfer is accomplished in a series of steps by groups of organisms known as autotrophs, heterotrophs, and decomposers. Each level on the pyramid represents a trophic level.

Autotrophs absorb sunlight energy and transfer inorganic mineral nutrients into organic molecules. The autotrophs of the marine environment include algae and flowering plants and in the deep sea are chemosynthetic bacteria that harness inorganic chemical energy to build organic matter — autotrophic nutrition — supply food molecules to organisms that can't absorb sunlight.

Heterotrophs: Consumers that must rely on primary producers as a source of energy — heterotrophic nutrition. The energy stored in the organic molecules is passed to consumers in a series of steps of eating and being eaten and is known as a food chain. Each step represents a trophic level and the complex food chains within a community interconnect and is known as a food web.

Decomposers: The final trophic level that connects consumer to producer is that of the decomposers. They live on dead plant and animal material and the waste products excreted by living things. The nutritional activity of these replenish nutrients that are essential ingredients for primary production. The dead and partially decayed plant and animal tissue and organic wastes from the food chain are DETRITUS. This contains an enormous amount of energy and nutrients. Many filter/deposit feeding animals use detritus as food. Saprophytes decompose detritus completing the cycle.

Energy transfers in marine environments: Primary producers usually outnumber consumers and at each succeeding step of the food chain the numbers decrease. The numerical relationship is called the pyramid of numbers. (base as opposed to each step.) The energy pyramid is the energy distribution at each trophic level as it passes from producers through the consumers. Some energy is lost as it passes to the next level because:

1. Consumers don't usually consume the entire organism.
2. Energy is used to capture food.
3. Organisms used energy during their metabolism.
4. Energy is lost as heat.

Generally only 10 per cent will pass on to the next level.

Food relationships: Predator-prey — predator kills and eats another organism — the prey. Antipredator defenses have evolved like:

1. Poisonous secretions.
2. Sharp spines and thorns.
3. The hard shell-like construction of coralline algae also deters grazing sea urchins.

Scavengers: Feed on dead plants and animals that they have not killed — crabs ripping chunks of flesh from fish on the beach are scavengers. Most scavengers consume detritus rather than flesh and deep sea animals can feed on both.

Symbiotic — refers to close nutritional relationship between two different species:

1. Commensalism — one benefits.
2. Mutualism both benefit and parasitism.
3. One benefits at hosts expense.

Population cycles: density or numbers of individuals depends on

1. Natality or rate of production of new organisms.
2. Mortality rate of death in a population.

Now to be stable the two must be in equilibrium but under favourable conditions, populations can increase numbers (can be seasonal and geographical) but this also increases mortality because of decreased food supply and living space and increased predation. If mortality is greater, then the population decreases. These favorable conditions depend on:

1. High concentration of nutrient rich water.
2. Rapid cycling of materials by decomposers.
3. High numbers or rapid turnover of producer organisms.
4. Light.
5. Nutrients including nitrate, phosphate, silicon, potassium, magnesium, copper, iron.

Silicon dioxide needed for outer glass covering of diatoms and forms internal structural parts of sponges, K and NO_4 and PO_4 needed in plant proteins, lipids and carbohydrates during photosynthesis, and the nutrients can be considered a limiting factor as well as pH temperature, light, depth salinity nesting sites and predation.

MARINE ECOSYSTEM

Marine ecosystems are among the largest of earth's aquatic ecosystems. They include oceans, salt marsh and intertidal ecology, estuaries and lagoons, mangroves and coral reefs, the deep sea and the sea floor. They can be contrasted with freshwater ecosystems, which have a lower salt content. Marine waters cover two-thirds of the surface of the earth. Such places are considered ecosystems because the plant life supports the animal life and vice-versa.

Marine ecosystems are very important for the overall health of both marine and terrestrial environments. According to the World Resource Center, coastal habitats alone account for approximately 1/3 of all marine biological productivity, and estuarine ecosystems (i.e. salt marshes, seagrasses, mangrove forests) are among the most productive regions on the planet. In addition, other marine ecosystems such as coral reefs, provide food and shelter to the highest levels of marine diversity in the world.

Like all ecosystems, marine ecosystems are mostly self-sustaining systems of life forms and the physical environment. In these ecosystems materials are cycled and recycled. All ecosystems have certain things in common and marine ecosystems have a few unique twists to these cycles. We can begin this discussion with describing the make-up of life forms (critters).

Almost all life forms are made of a cell (if they are unicellular) or cells (if they are multicellular) and the products of the cell(s). To sustain life, these cells must carry on a chemical reaction called cellular respiration.

Cellular Respiration

Cellular respiration equation is given below:

$$\text{Cell food} + \text{Oxygen} \rightarrow \text{Energy} + \text{Carbon dioxide}$$

Cellular respiration is the life sustaining process for all life forms including both plant and animal types. It is where a cell uses what we call 'cell food' (a simple carbohydrate molecule - usually glucose made of six carbons, twelve hydrogens, and six oxygens) in the presence of oxygen to make the 'cell energy' (in the form of a molecule we call ATP) and with a waste product of carbon dioxide. The ATP is used to repair the cell and keeps it alive. A few organisms can cellularly respire without oxygen but it is not as efficient and we call these critters 'anaerobes.' Most critters need oxygen to stay alive and 'cell food.' The cell food is obtained by eating in animals (also called consumers) and by photosynthesis in critters we call plants (also called producers).

Photosynthesis

Photosynthesis equation is given below:

$$\text{Carbon dioxide} + \text{Water} \xrightarrow{\text{Light}} \text{Cell food} + \text{Oxygen}$$

Photosynthesis is the food-making process for producers which we often call plants. In the marine environment many of the producers are microscopic and classified in a kingdom called Protoctista, along with the algae. But, they are still like plants in that they photosynthesise no matter what group they are classified as. Photosynthesis is where carbon dioxide and water are combined within a specialised part of a cell (called a chloroplast), in the presence of light, to form the 'cell food' molecule called glucose and with a waste product of oxygen. In a sense it is cellular respiration in reverse but it will only happen in cells with chloroplasts (in which there is the molecule called chlorophyll that is involved). It is a complicated reaction but the end product provides the recycling of carbon dioxide back to oxygen for the cycle of these two gases. Without photosynthesis all life forms (plants and animals) would use up the oxygen on earth and it would only be found as carbon dioxide (not usable in cellular respiration).

Both unicellular (one celled) and multicellular (more than one cell) individuals are found in the marine ecosystem taking on various levels of interaction. There are unicellular plants and animals. There are multicellular plants and animals as well. The multicellular species may be very simple or more complex with cells specialised as tissues, organs, and organ systems depending on the species. Whether or not a species is simple or complex they each have their place in the ecosystem and interesting stories about their life histories.

Populations are groups of organisms of one species living in an area and interacting. Interactions between individuals of the same species are termed intraspecific interactions and can be positive (like cooperation), negative (like competition) or anything in between.

Communities are composed of populations of many species living in an area and interacting. Interactions between species are termed interspecific interactions and, just like with intraspecific interactions, can be positive, negative or anything in between.

Ecosystems are areas where the community, or communities, are rather self-sustaining this involves what we call food chains and webs — that is describing who eats who and following the energy flow (in the form of food) all the way to what we call the bottom of the food chain. The bottom is where the producers are found. The producers are the photosynthesisers and depend on carbon dioxide and light so that they can make their own food. Gases are cycled and recycled in ecosystems through cellular respiration and photosynthesis. Nutrients are cycled by decomposers (bacteria and fungi) in ecosystems that are involved with rotting (decomposing) any dead materials and releasing the nutrients from these cells. Food is cycled through what we call trophic pyramids — always with the producers at the base. Ecosystems can be interpreted as very large areas (like the entire planet earth) or smaller subunits (like a tidepool habitat) depending on how strict one uses the term 'self-sustaining' (Fig. 21.1).

Trophic pyramids in the ocean are similar to those on land: With the base always larger than the upper trophic levels. The base is the producers. The second level is the level of the critters that eat the producers (we call these herbivores, or plant eating animals). The third level is composed of critters that eat the herbivores (we would call these carnivores, or animal eating animals). As you move between the levels there is always a loss of biomass because the transfer of energy is usually only ten per cent efficient. This means that if there were 50 kg of producers in an area there would be enough food for 5 kg of herbivores but the 5 kg of herbivores could only support ½ kg of a carnivore. So you see, as you move up the food chain (or trophic pyramid) there is a smaller and smaller biomass that can be supported.

Sometimes we refer to the second and third levels of the trophic pyramid as primary consumers (the herbivores) and secondary consumers (the first order carnivores). There can be many more levels (each with a loss of 90 per cent biomass average) with tertiary consumers (carnivores eating the secondary consumers Fig. 21.2) .

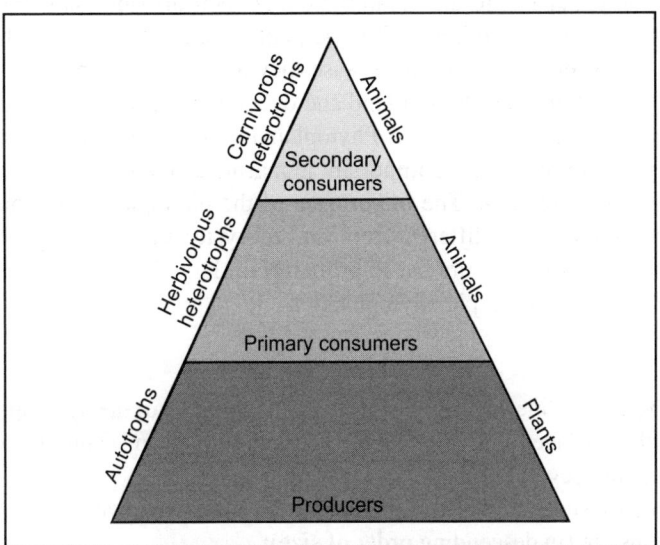

Fig. 21.1. The trophic pyramid of an ecosystem. Kelp could be placed on the bottom as a producer, the kelp crab on the middle layer (as a primary consumer or herbivore), and the two spot octopus on the top (as a secondary consumer or carnivore).

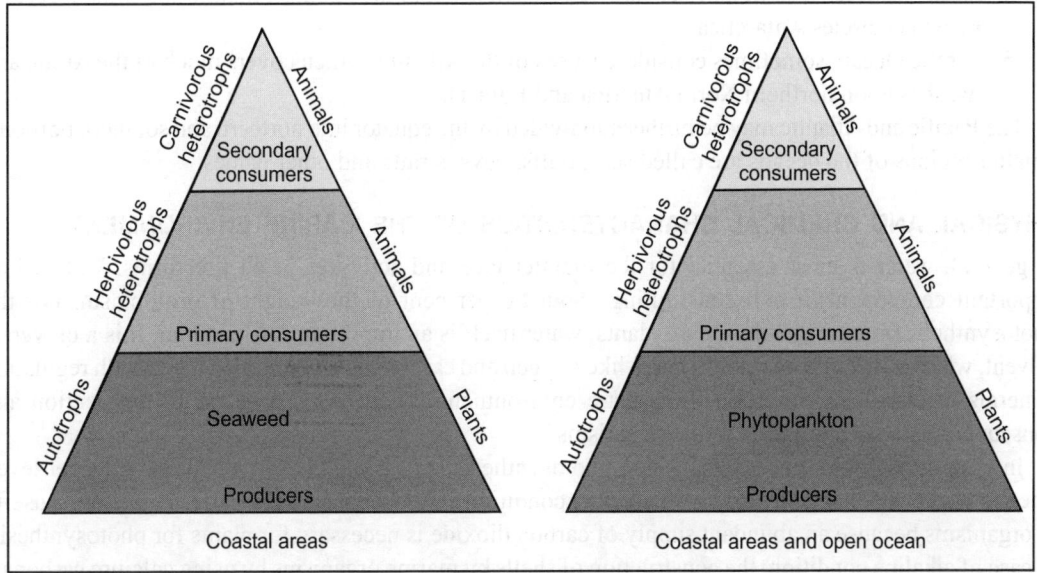

Fig. 21.2. Marine trophic pyramids are either seaweed based (coastal areas) or phytoplankton based (coastal areas and open ocean).

Marine producers are either seaweeds and/or phytoplankton: Marine producers are either seaweeds and/or phytoplankton and must be found where there is enough light for photosynthesis. Therefore the marine producers are always in the upper layers of the ocean, the area we call the photic zone. Animals can be found everywhere, in both the photic and aphotic zones, because they are not limited by light. Most seaweeds grow only attached to the ocean bottom so their distribution is limited to the edges of continents and islands where the depth is within the photic zone. These seaweeds account for only a small portion of the producers in the ocean because they are so geographically limited. Most of the ocean is open ocean, away from the edges of land and over deep water. It is here, near the surface, that the phytoplankton (plant plankton) dominates. Phytoplankton is also found in the same coastal areas as the seaweeds. Both types of producers are important in marine ecosystems however the phytoplankton based ecosystems are more common. The importance of the phytoplankton is one of the things that make marine ecosystems unique and different from land ecosystems because phytoplankton is generally microscopic. Land based ecosystems are based primarily on large land plants but marine ecosystems are based primarily on tiny, microscopic phytoplankters.

OCEAN

An ocean is a major body of saline water, and a principal component of the hydrosphere. Approximately 71 per cent of the earth's surface ($\sim 3.6 \times 10^8$ km^2) is covered by ocean, a continuous body of water that is customarily divided into several principal oceans and smaller seas.

The major oceanic divisions are defined in part by the continents, various archipelagos, and other criteria. These divisions are (in descending order of size):

1. Pacific Ocean, which separates Asia and Australia from the Americas.
2. Atlantic Ocean, which separates the Americas from Eurasia and Africa.
3. Indian Ocean, which washes upon southern Asia and separates Africa and Australia
4. Southern Ocean, sometimes considered an extension of the Pacific, Atlantic and Indian Oceans, which encircles Antarctica.
5. Arctic Ocean, sometimes considered a sea of the Atlantic, which covers much of the Arctic and washes upon northern North America and Eurasia.

The Pacific and Atlantic may be further subdivided by the equator into northern and southern portions. Smaller regions of the oceans are called seas, gulfs, bays, straits and other names.

PHYSICAL AND CHEMICAL CHARACTERISTICS OF THE MARINE ENVIRONMENT

In general, water is most essential for the maintenance and activities in all life forms. Water is an important component of cell constituting about 80 per cent of the weight of protoplasm. For the photosynthetic process in autotrophic plants, water itself is an important raw material. It is a universal solvent, which carries the necessary gases like oxygen and carbon dioxide and also the growth regulating minerals in dissolved condition. In aquatic environments, there is no problem of dessication and consequently no specialisation in the organisms.

In view of high transparency of water, photosynthesis is possible even at a relatively deeper level. The sea water can change from acid to alkaline condition and vice versa. This buffering nature is useful to organisms because an abundant supply of carbon dioxide is necessary for plants for photosynthesis. In case of alkaline condition, the construction of shells by marine organisms by using calcium carbonate is enhanced. The lower specific gravity of sea water is most beneficial to marine organisms. As the sea water contains large number of salts it is a most suitable, environment for living cells. Further it has been found that the ratio of total salt content of seawater is almost same as that of body fluids of many invertebrates.

In general, the marine environment offers a wide range of living conditions. The salinity ranges from very dilute estuarine condition to as high as 37 per cent. Varying light conditions exist with brilliant sunlight at the surface waters to no light in the deep waters. Likewise, the pressure varies from 1 atmosphere at the surface to 1000 atmospheres at greater depths. These gradients of environmental parameters are favourable to a number of sensitive animals. The more fluctuations of environmental features are especially encountered in coastal areas due to their peculiar physiographic characters. The water movement/circulation is useful in the oxygenation of subsurface water, for the dispersal of metabolic wastes and plant and animal (growth) nutrients, and also for the disposal of spores, eggs, larvae and even adults.

CORAL REEFS

Coral reefs are underwater structures made from calcium carbonate secreted by corals. Corals are colonies of tiny living animals found in marine waters that contain few nutrients. Most coral reefs are built from stony corals, which in turn consist of polyps that cluster in groups. The polyps are like tiny sea anemones, to which they are closely related. But unlike sea anemones, coral polyps secrete hard carbonate exoskeletons which support and protect their bodies. Reefs grow best in warm, shallow, clear, sunny and agitated waters.

Often called 'rainforests of the sea', coral reefs form some of the most diverse ecosystems on earth. They occupy less than one tenth of one per cent of the world's ocean surface, about half the area of France, yet they provide a home for twenty-five per cent of all marine species, including fish, molluscs, worms, crustaceans, echinoderms, sponges, tunicates and other cnidarians. Paradoxically, coral reefs flourish even though they are surrounded by ocean waters that provide few nutrients. They are most commonly found at shallow depths in tropical waters, but deep water and cold water corals also exist on smaller scales in other areas.

Coral reefs deliver ecosystem services to tourism, fisheries and shoreline protection. The annual global economic value of coral reefs has been estimated at $ US375 billion as on in year 2008. However, coral reefs are fragile ecosystems, partly because they are very sensitive to water temperature. They are under threat from climate change, ocean acidification, blast fishing, cyanide fishing for aquarium fish, overuse of reef resources, and harmful land-use practices, including urban and agricultural runoff and water pollution, which can harm reefs by encouraging excess algae growth.

Formation of Coral Reef

Most coral reefs were formed after the last glacial period when melting ice caused the sea level to rise and flood the continental shelves. This means that most coral reefs are less than 10,000 years old. As communities established themselves on the shelves, the reefs grew upwards, pacing rising sea levels. Reefs that rose too slowly could become drowned reefs, covered by so much water that there was insufficient light. Coral reefs are also found in the deep sea away from the continental shelves, around oceanic islands and as atolls. The vast majority of these islands are volcanic in origin. The few exceptions have tectonic origins where plate movements have lifted the deep ocean floor on the surface.

In 1842 in his first monograph, the structure and distribution of coral reefs Charles Darwin set out his theory of the formation of atoll reefs, an idea he conceived during the voyage of the Beagle. He theorised uplift and subsidence of the earth's crust under the oceans formed the atolls. Darwin's theory sets out a sequence of three stages in atoll formation. It starts with a fringing reef forming around an extinct volcanic island as the island and ocean floor subsides. As the subsidence continues, the fringing reef becomes a barrier reef, and ultimately an atoll reef.

Darwin's theory starts with a volcanic island which becomes extinct. As the island and ocean floor subside, coral growth builds a fringing reef, often including a shallow lagoon between the land and the main reef. As the subsidence continues the fringing reef becomes a larger barrier reef further from the shore with a bigger and deeper lagoon inside. Ultimately the island sinks below the sea, and the barrier reef becomes an atoll enclosing an open lagoon.

Darwin predicted that underneath each lagoon would be a bed rock base, the remains of the original volcano. Subsequent drilling proved this correct. Darwin's theory followed from his understanding that coral polyps thrive in the clean seas of the tropics where the water is agitated, but can only live within a limited depth range, starting just below low tide. Where the level of the underlying earth allows, the corals grow around the coast to form what he called fringing reefs, and can eventually grow out from the shore to become a barrier reef.

Where the bottom is rising, fringing reefs can grow around the coast, but coral raised above sea level dies and becomes white limestone. If the land subsides slowly, the fringing reefs keep pace by growing upwards on a base of older, dead coral, forming a barrier reef enclosing a lagoon between the reef and the land. A barrier reef can encircle an island, and once the island sinks below sea level a roughly circular atoll of growing coral continues to keep up with the sea level, forming a central lagoon. Barrier reefs and atolls don't usually form complete circles, but are broken in places by storms. Like sea level rise, a rapidly subsiding bottom subside can overwhelm coral growth, killing the animals and the reef.

The two main variables determining the geomorphology, or shape, of coral reefs are the nature of the underlying substrate on which they rest, and the history of the change in sea level relative to that substrate. Healthy tropical coral reefs grow horizontally from 1 to 3 centimetres (0.39 to 1.2 in) per year, and grow vertically anywhere from 1 to 25 centimetres (0.39 to 9.8 in) per year; however, they grow only at depths shallower than 150 metres (490 ft) due to their need for sunlight, and cannot grow above sea level.

Types of Coral Reef

The three principal reef types are:
1. Fringing reef: A reef that is directly attached to a shore or borders it with an intervening shallow channel or lagoon.
2. Barrier reef: A reef separated from a mainland or island shore by a deep channel or lagoon.
3. Atoll reef: A more or less circular or continuous barrier reef extending all the way around a lagoon without a central island.

Other reef types or variants are:
1. Patch reef: An isolated, comparatively small reef outcrop, usually within a lagoon or embayment, often circular and surrounded by sand or seagrass.
2. Apron reef: A short reef resembling a fringing reef, but more sloped; extending out and downward from a point or peninsular shore.
3. Bank reef: A linear or semi-circular shaped-outline, larger than a patch reef.
4. Ribbon reef: A long, narrow, possibly winding reef, usually associated with an atoll lagoon
5. Table reef: An isolated reef, approaching an atoll type, but without a lagoon.
6. Habili: Reef in the red sea that does not reach the surface near enough to cause visible surf, although it may a hazard to ships (from the Arabic for 'unborn').

7. Microatolls: Certain species of corals form communities called microatolls. The vertical growth of microatolls is limited by average tidal height. By analysing growth morphologies, microatolls offer a low resolution record of patterns of sea level change. Fossilised microatolls can also be dated using radioactive carbon dating. Such methods have been used to reconstruct Holocene sea levels.

8. Cays: Small, low-elevation, sandy islands formed on the surface of a coral reef. Material eroded from the reef piles up on parts of the reef or lagoon, forming an area above sea level. Plants can stabilise cays enough to become habitable by humans. Cays occur in tropical environments throughout the Pacific, Atlantic and Indian Oceans (including the Caribbean and on the Great Barrier Reef and Belize Barrier Reef), where they provide habitable and agricultural land for hundreds of thousands of people.

9. When a coral reef cannot keep up with the sinking of a volcanic island, a seamount or guyot is formed. The tops of seamounts and guyots are below the surface. Seamounts are rounded at the top and guyots are flat. The flat top of the guyot, also called a tablemount, is due to erosion by waves, winds, and atmospheric processes.

Zones of Coral Reef

Coral reef ecosystems contain distinct zones that represent different kinds of habitats. Usually three major zones are recognised: the fore reef, reef crest, and the back reef (frequently referred to as the reef lagoon). All three zones are physically and ecologically interconnected. Reef life and oceanic processes create opportunities for exchange of seawater, sediments, nutrients, and marine life among one another.

Thus, they are integrated components of the coral reef ecosystem, each playing a role in the support of the reefs' abundant and diverse fish assemblages.

Most coral reefs exist in shallow waters less than fifty metres deep. Some inhabit tropical continental shelves where cool, nutrient rich upwelling does not occur, such as great barrier reef. Others are found in the deep ocean surrounding islands or as atolls, such as in the Maldives. The reefs surrounding islands form when islands subside into the ocean, and atolls form when an island subsides below the surface of the sea. Alternatively, Moyle and Cech distinguish six zones, though most reefs possess only some of the zones:

1. The reef surface is the shallowest part of the reef. It is subject to the surge and the rise and fall of tides. When waves pass over shallow areas, they shoal, as shown in Fig. 21.3. This means that the water is often agitated. These are the precise condition under which coral flourish. Shallowness means there is plenty of light for photosynthesis by the symbiotic zooxanthellae, and agitated water promotes the ability of coral to feed on plankton. However other organisms must be able to withstand the robust conditions to flourish in this zone.

2. The off-reef floor is the shallow sea floor surrounding a reef. This zone occurs by reefs on continental shelves. Reefs around tropical islands and atolls drop abruptly to great depths, and don't have a floor. Usually sandy, the floor often supports seagrass meadows which are important foraging areas for reef fish.

3. The reef drop-off is, for its first 50 metres, habitat for many reef fish who find shelter on the cliff face and plankton in the water nearby. The drop-off zone applies mainly to the reefs surrounding oceanic islands and atolls.

4. The reef face is the zone above the reef floor or the reef drop-off. 'It is usually the richest habitat. Its complex growths of coral and calcareous algae provide cracks and crevices for protection, and the abundant invertebrates and epiphytic algae provide an ample source of food.'

5. The reef flat sandy bottomed flat can be behind the main reef, containing chunks of coral. 'The reef flat may be a protective area bordering a lagoon, or it may be a flat, rocky area between the reef and the shore. In the former case, the number of fish species living in the area often is the highest of any reef zone.'

6. The reef lagoon 'many coral reefs completely enclose an area, thereby creating a quiet-water lagoon that usually contains small patches of reef.'

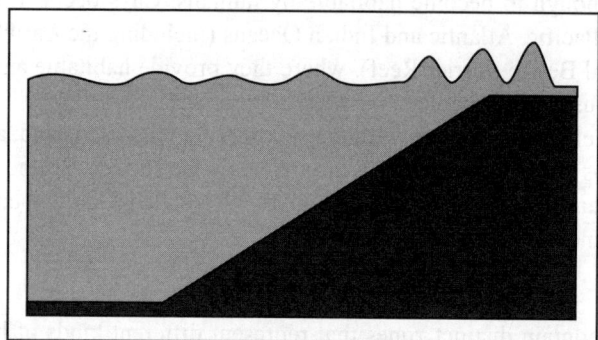

Fig. 21.3. Water in the reef surface zone is often agitated. This diagram represents a reef on a continental shelf. The water waves at the left travel over the *off-reef floor* until they encounter the *reef slope* or *fore reef*. Then the waves pass over the shallow *reef crest*. When a wave enters shallow water it shoals, that is, it slows down and the wave height increases.

However, the 'topography of coral reefs is constantly changing. Each reef is made up of irregular patches of algae, sessile invertebrates, and bare rock and sand. The size, shape and relative abundance of these patches changes from year to year in response to the various factors that favour one type of patch over another. Growing coral, for example, produces constant change in the fine structure of reefs. On a larger scale, tropical storms may knock out large sections of reef and cause boulders on sandy areas to move.'

Biodiversity

Coral reefs form some of the world's most productive ecosystems, providing complex and varied marine habitats that support a wide range of other organisms. Fringing reefs just below low tide level also have a mutually beneficial relationship with mangrove forests at high tide level and sea grass meadows in between: the reefs protect the mangroves and seagrass from strong currents and waves that would damage them or erode the sediments in which they are rooted, while the mangroves and seagrass protect the coral from large influxes of silt, fresh water and pollutants. This additional level of variety in the environment is beneficial to many types of coral reef animals, which for example may feed in the sea grass and use the reefs for protection or breeding.

Reefs are home to a large variety of organisms, including fish, seabirds, sponges, Cnidarians (which includes some types of corals and jellyfish), worms, crustaceans (including shrimp, cleaner shrimp, spiny lobsters and crabs), molluscs (including cephalopods), echinoderms (including starfish, sea urchins and sea cucumbers), sea squirts, sea turtles and sea snakes. Aside from humans, mammals are rare on coral reefs, with visiting cetaceans such as dolphins being the main exception. A few of these varied species feed directly on corals, while others graze on algae on the reef. Reef biomass is positively related to species diversity.

Fish

Over 4000 species of fish inhabit coral reefs. The reasons for this diversity remain controversial. Hypotheses include the 'lottery', in which the first (lucky winner) recruit to a territory is typically able to defend it against latecomers, 'competition', in which adults compete for territory, and less-competitive species must be able to survive in poorer habitat, and 'predation', in which population size is a function of post-settlement piscivoremortality. Healthy reefs can produce up to 35 tons of fish per square kilometer each year, but damaged reefs produce much less. Reef species include:

1. Fish that influence the coral. These feed either on small animals living near the coral, seaweed/ algae, or on the coral itself. Fish that feed on small animals include *Labridae* (cleaner fish) who notably feed on organisms that inhabit larger fish, bullet fish and sea-urchin-eating *Balistidae* (triggerfish) while seaweed-eating fish include the *Pomacentridae* (damselfishes). *Serranidae* (groupers) cultivate the seaweed by removing creatures feeding on it (such as sea urchins), and they remove inedible seaweeds. Fish that eat coral itself include *Scaridae* (parrotfish) and *Chaetodontidae* (butterflyfish).

2. Fish that cruise the boundaries of the reef or nearby seagrass meadows. These include predators such as *Trachinotus* (pompanos), groupers, horse mackerels, certain types of shark, barracudas and *Lutjanidae* (snappers). Herbivorous and plankton-eating fish also populate reefs. Seagrass-eating fish include horse mackerel, snapper, *Pagellus* (porgies) and *Conodon* (grunts). Plankton-eating fish include *Caesio* (fusilier), ray, chromis, and the nocturnal *Holocentridae* (squirrelfish), *Apogonidae* (cardinalfish) and *Myctophidae* (lanternfish).

Fish that swim in coral reefs can be as colourful as the reef. Examples are the parrotfish, *Pomacanthidae* (angelfish), damselfish, *Clinidae* (blennies) and butterflyfish. At night, some change to a less vivid colour.

Invertebrates

Sea urchins, Dotidae and sea slugs eat seaweed. Some species of sea urchins, such as *Diadema antillarum*, can play a pivotal part in preventing algae overrunning reefs. Nudibranchia and sea anemones eat sponges.

A number of invertebrates, collectively called cryptofauna, inhabit the coral skeletal substrate itself, either boring into the skeletons (through the process of bioerosion) or living in pre-existing voids and crevices. Those animals boring into the rock include sponges, bivalve mollusks, and sipunculans. Those settling on the reef include many other species, particularly crustaceans and *polychaete* worms.

Algae

Reefs are chronically at risk of algal encroachment. Overfishing and excess nutrient supply from onshore can enable algae to outcompete and kill the coral. In surveys done around largely uninhabited US Pacific islands, algae inhabit a large percentage of surveyed coral locations. The algae population consists of turf algae, coralline algae, and macroalgae.

Seabirds

Coral reef systems provide important habitats for seabird species, some endangered. For example, Midway Atoll in Hawaii supports nearly three million seabirds, including two-thirds (1.5 million) of the global population of Laysan Albatross, and one-third of the global population of black-footed albatross. Each seabird species has specific sites on the atoll where they nest. Altogether, 17 species of seabirds live on Midway. The short-tailed albatross is the rarest, with fewer than 2200 surviving after excessive feather hunting in the late nineteenth century.

Threats

Coral reefs are dying around the world. In particular, coral mining, agricultural and urban runoff, pollution (organic and non-organic), overfishing, blast fishing, disease, and the digging of canals and access into islands and bays are localised threats to coral ecosystems. Broader threats are sea temperature rise, sea level rise and pH changes from ocean acidification, all associated with greenhouse gas emissions.

In El Nino-year 2010, preliminary reports show global coral bleaching reached its worst level since another El Nino year, 1998, when 16 per cent of the world's reefs died as a result of increased water temperature. In Indonesia's Aceh province, surveys showed some 80 per cent of bleached corals died. In July, Malaysia closed several dive sites where virtually all the corals were damaged by bleaching.

In order to find answers for these problems, researchers study the various factors that impact reefs. The list includes the ocean's role as a carbon dioxide sink, atmospheric changes, ultraviolet light, ocean acidification, viruses, impacts of dust storms carrying agents to far flung reefs, pollutants, algal blooms and others. Reefs are threatened well beyond coastal areas.

General estimates show approximately 10 per cent of the world's coral reefs are dead. About 60 per cent of the world's reefs are at risk due to destructive, human-related activities. The threat to the health of reefs is particularly strong in Southeast Asia, where 80 per cent of reefs are endangered.

Protection

Marine protected areas (MPA) have become increasingly prominent for reef management. MPAs promote responsible fishery management and habitat protection. Much like national parks and wildlife refuges, and to varying degrees, MPAs restrict potentially damaging activities. MPA encompass both social and biological objectives, including reef restoration, aesthetics, biodiversity, and economic benefits. Conflicts surrounding MPAs involve lack of participation, clashing views, effectiveness, and funding. In some situations, as in the Phoenix Islands Protected Area, MPA's can also provide revenue, potentially equal to the income that they would have generated without controls as Kiribati did for its Phoenix Islands.

Biosphere reserve, marine park, national monument and world heritage status can protect reefs. For example Belize's Barrier reef, Chagos archipelago, Sian Ka'an, the Galapagos islands, Great Barrier Reef, Henderson Island, Palau and Papahanaumokuakea Marine National Monument are world heritage sites. In Australia, the Great Barrier Reef is protected by the Great Barrier Reef Marine Park Authority, and is the subject of much legislation, including a Biodiversity Action Plan. Inhabitants of Ahus Island, Manus Province, Papua New Guinea, have followed a generations-old practice of restricting fishing in six areas of their reef lagoon. Their cultural traditions allow line fishing but not net or spear fishing. The result is that both the biomass and individual fish sizes are significantly larger than in places where fishing is unrestricted.

Artificial reefs

Efforts to expand the size and number of coral reefs generally involve supplying substrate to allow more corals to find a home. Substrate materials include discarded vehicle tyres, scuttled ships, subway cars, and formed concrete such as reef balls. Reefs also grow unaided on marine structures such as oil rigs. In large restoration projects, propagated hermatypic coral on substrate can be secured with metal pins, superglue or milliput. Needle and thread can also attach A-hermatype coral to substrate.

Low voltage electrical currents applied through seawater crystallise dissolved minerals onto steel structures. The resultant white carbonate (aragonite) is the same mineral that makes up natural coral reefs. Corals rapidly colonise and grow at accelerated rates on these coated structures. The electrical currents also accelerate formation and growth of both chemical limestone rock and the skeletons of corals and other shell-bearing organisms. The vicinity of the anode and cathode provides a high pH environment which inhibits the growth of competitive filamentous and fleshy algae. The increased growth rates fully depend on the accretion activity.

ESTUARY

An estuary is a partly enclosed coastal body of water with one or more rivers or streams flowing into it, and with a free connection to the open sea. Estuaries form a transition zone between river environments and ocean environments and are subject to both marine influences, such as tides, waves, and the influx of saline water; and riverine influences, such as flows of fresh water and sediment. The inflow of both seawater and freshwater provide high levels of nutrients in both the water column and sediment, making estuaries among the most productive natural habitats in the world.

Estuaries are typically classified by their geomorphological features or by water circulation patterns and can be referred to by many different names, such as bays, harbours, lagoons, inlets, or sounds, although sometimes these water bodies do not necessarily meet the above criteria of an estuary and may be fully saline. Estuaries are amongst the most heavily populated areas throughout the world, with about 60 per cent of the world's population living along estuaries and the coast. As a result, estuaries are suffering degradation by many factors, including sedimentation from soil erosion from deforestation; overgrasing and other poor farming practices; overfishing; drainage and filling of wetlands; eutrophication due to excessive nutrients from sewage and animal wastes; pollutants including heavy metals, PCBs, radionuclides and hydrocarbons from sewage inputs; and diking or damming for flood control or water diversion.

Definition of estuary. The most widely accepted definition is: 'a semi-enclosed coastal body of water, which has a free connection with the open sea, and within which sea water is measurably diluted with freshwater derived from land drainage.' However, this definition excludes a number of coastal water bodies such as coastal lagoons and brackish seas. A more thorough definition of an estuary would be 'a semi-enclosed body of water connected to the sea as far as the tidal limit or the salt intrusion limit and receiving freshwater runoff; however the freshwater inflow may not be perennial, the connection to the sea may be closed for part of the year and tidal influence may be negligible.' This definition includes classical estuaries as well as fjords, lagoons, river mouths, and tidal creeks. Estuaries are a dynamic ecosystem with a connection with the open sea through which the seawater enters accordingly to the rhythm of the tides. The seawater entering the estuary is diluted by the freshwater flowing from rivers and streams. The pattern of dilution varies in different estuaries and is dependent on the volume of freshwater, tidal amplitude range, and the extent of evaporation from the water within the estuary.

Classification Based on Geomorphology

Drowned river valleys

Many drowned river valley estuaries were formed between about 15,000 and 6000 years ago following the end of the Wisconsin (or Devensian) glaciation when a eustatic rise in sea level of 100 to 130 m (330 to 430 ft) flooded river valleys that were cut into the landscape when sea level was lower, creating the estuarine systems.

Lagoon-type or bar-built

These estuaries are semi-isolated from ocean waters by barrier beaches (barrier islands and barrier spits). Formation of barrier beaches partially encloses the estuary with only narrow inlets allowing contact with the ocean waters. Bar-built estuaries typically develop on gently sloping plains located along tectonically stable edges of continents and marginal sea coasts. They are extensive along the Atlantic and Gulf coasts of the US in areas with active coastal deposition of sediments and where tidal ranges are less than 4 m (13 ft).

Fjord-type

Fjord type estuaries are formed in deeply eroded valleys formed by glaciers. These U-shaped estuaries typically have steep sides, rock bottoms, and underwater sills contoured by glacial movement. The shallowest area of the estuary occurs at the mouth, where terminal glacial deposits or rock bars form sills that restrict water flow. In the upper reaches of the estuary, the depth can exceed 300 m (980 ft). The width-to-depth ratio is generally small. When estuaries contain very shallow sills, tidal oscillations only affect near surface waters to sill depth, and waters below sill depth may remain stagnant for very long periods of time, resulting in only an occasional exchange of the deep water of the estuary with the ocean.

Tectonically produced

These estuaries are formed by subsidence or land cut off from the ocean by land movement associated with faulting, volcanoes, and landslides. Inundation from eustatic sea level rise during the Holocene Epoch has also contributed to the formation of these estuaries. There are only a small number of tectonically produced estuaries; one example is the San Francisco Bay, which was formed by the crustal movements of the San Andreas fault system causing the inundation of the lower reaches of the Sacramento and San Joaquin rivers.

Classification Based on Water Circulation

Salt wedge

In this type of estuary, river output greatly exceeds marine input and tidal effects have a minor importance. Fresh water floats on top of the seawater in a layer that gradually thins as it moves seaward. The denser seawater moves landward along the bottom of the estuary, forming a wedge-shaped layer that is thinner as it approaches land. As a velocity difference develops between the two layers, shear forces generate internal waves at the interface, mixing the seawater upward with the freshwater. An example of a salt wedge estuary is the Mississippi River.

Partially mixed

As tidal forcing increases, river output becomes less than the marine input. Here, current induced turbulence causes mixing of the whole water column such that salinity varies more longitudinally rather than vertically, leading to a moderately stratified condition. Examples include the Chesapeake Bay and Narragansett Bay.

Vertically homogenous

Tidal mixing forces exceed river output, resulting in a well mixed water column and the disappearance of the vertical salinity gradient. The freshwater-seawater boundary is eliminated due to the intense turbulent mixing and eddy effects. The lower reaches of the Delaware Bay and the Raritan River in New Jersey are examples of vertically homogenous estuaries.

Inverse

Inverse estuaries occur in dry climates where evaporation greatly exceeds the inflow of fresh water. A salinity maximum zone is formed, and both riverine and oceanic water flow close to the surface towards this zone. This water is pushed downward and spreads along the bottom in both the seaward and landward direction. An example of an inverse estuary is Spencer Gulf, South Australia.

Intermittent

Estuary type varies dramatically depending on freshwater input, and is capable of changing from a wholly marine embayment to any of the other estuary types.

Implications for Marine Life

Estuaries provide habitats for a large number of organisms and support very high productivity. Estuaries provide habitats for many fish nurseries, depending upon their locations in the world, such as salmon and sea trout. Also, migratory bird populations, such as the black-tailed godwit, Limosa limosa islandica make essential use of estuaries.

Two of the main challenges of estuarine life are the variability in salinity and sedimentation. Many species of fish and invertebrates have various methods to control or conform to the shifts in salt concentrations and are termed osmoconformers and osmoregulators. Many animals also burrow to avoid predation and to live in the more stable sedimental environment. However, large numbers of bacteria are found within the sediment which have a very high oxygen demand. This reduces the levels of oxygen within the sediment often resulting in partially anoxic conditions, which can be further exacerbated by limited water flux.

Phytoplankton are key primary producers in estuaries. They move with the water bodies and can be flushed in and out with the tides. Their productivity is largely dependant upon the turbidity of the water. The main phytoplankton present are diatoms and dinoflagellates which are abundant in the sediment. It is important to remember that a primary source of food for many organisms on estuaries, including bacteria, is detritus from the settlement of the sedimentation.

Human Impacts

Of the 32 largest cities in the world, 22 are located on estuaries. For example, New York City is located at the orifice of the Hudson River estuary.

As ecosystems, estuaries are under threat from human activities such as pollution and overfishing. They are also threatened by sewage, coastal settlement, land clearance and much more. Estuaries are affected by events far upstream, and concentrate materials such as pollutants and sediments. Land run-off and industrial, agricultural, and domestic waste enter rivers and are discharged into estuaries. Contaminants can be introduced which do not disintegrate rapidly in the marine environment, such as plastics, pesticides, furans, dioxins, phenols and heavy metals.

Such toxins can accumulate in the tissues of many species of aquatic life in a process called bioaccumulation. They also accumulate in benthic environments, such as estuaries and bay muds: a geological record of human activities of the last century.

For example, Chinese and Russian industrial pollution, such as phenols and heavy metals, in the Amur River have devastated fish stocks and damaged its estuary soil. Estuaries tend to be naturally eutrophic because land runoff discharges nutrients into estuaries. With human activities, land run-off also now includes the many chemicals used as fertilisers in agriculture as well as waste from livestock and humans. Excess oxygen depleting chemicals in the water can lead to hypoxia and the creation of dead zones. It can result in reductions in water quality, fish, and other animal populations.

Overfishing also occurs: Chesapeake Bay once had a flourishing oyster population which has been almost wiped out by overfishing. Historically the oysters filtered the estuary's entire water volume of excess nutrients every three or four days. Today that process takes almost a year, and sediment, nutrients, and algae can cause problems in local waters. Oysters filter these pollutants, and either eat them or shape them into small packets that are deposited on the bottom where they are harmless.

MARINE DEBRIS

Marine debris, also known as marine litter, is human-created waste that has deliberately or accidentally become afloat in a lake, sea, ocean or waterway. Oceanic debris tends to accumulate at the centre of gyres and on coastlines, frequently washing aground, when it is known as beach litter or tidewrack. Deliberate disposal of wastes at sea is called ocean dumping.

Some seeming forms of marine debris, such as driftwood, occur naturally, and human activities have been discharging similar material into the oceans for thousands of years. Recently however, with the increasing use of plastic, human influence has become an issue as many types of plastics do not biodegrade. Waterborne plastic poses a serious threat to fish, seabirds, marine reptiles, and marine mammals, as well as to boats and coastal habitations. Ocean dumping, accidental container spillages, litter washed into storm drains, and wind-blown landfill waste are all contributing to this problem.

Environmental Impact

Many animals that live on or in the sea consume flotsam by mistake, as it often looks similar to their natural prey. Plastic debris, when bulky or tangled, is difficult to pass, and may become permanently lodged in the digestive tracts of these animals, blocking the passage of food and causing death through starvation or infection. Tiny floating particles also resemble zooplankton, which can lead filter feeders to consume them and cause them to enter the ocean food chain.

Toxic additives used in the manufacture of plastic materials can leach out into their surroundings when exposed to water. Waterborne hydrophobic pollutants collect and magnify on the surface of plastic debris, thus making plastic far more deadly in the ocean than it would be on land. Hydrophobic contaminants are also known to bioaccumulate in fatty tissues, biomagnifying up the food chain and putting great pressure on apex predators. Some plastic additives are known to disrupt the endocrine system when consumed; others can suppress the immune system or decrease reproductive rates.

MARINE POLLUTION

Marine pollution occurs when harmful effects, or potentially harmful effects, can result from the entry into the ocean of chemicals, particles, industrial, agricultural and residential waste, noise, or the spread of invasive organisms. Most sources of marine pollution are land based. The pollution often comes from nonpoint sources such as agricultural runoff and wind blown debris.

Many potentially toxic chemicals adhere to tiny particles which are then taken up by plankton and benthos animals, most of which are either deposit or filter feeders. In this way, the toxins are concentrated upward within ocean food chains. Many particles combine chemically in a manner highly depletive of oxygen, causing estuaries to become anoxic.

When pesticides are incorporated into the marine ecosystem, they quickly become absorbed into marine food webs. Once in the food webs, these pesticides can cause mutations, as well as diseases, which can be harmful to humans as well as the entire food web. Toxic metals can also be introduced into marine food webs. These can cause a change to tissue matter, biochemistry, behaviour, reproduction, and suppress growth in marine life. Also, many animal feeds have a high fish meal or fish hydrolysate content. In this way, marine toxins can be transferred to land animals, and appear later in meat and dairy products.

Pathways of Pollution

There are many different ways to categorise, and examine the inputs of pollution into our marine ecosystems. There are three main types of inputs of pollution into the ocean: direct discharge of waste into the oceans, runoff into the waters due to rain, and pollutants that are released from the atmosphere. One common path of entry by contaminants to the sea are rivers. The evaporation of water from oceans exceeds precipitation. The balance is restored by rain over the continents entering rivers and then being returned to the sea. Pollution is often classed as point source or nonpoint source pollution. Point source pollution occurs when there is a single, identifiable, and localised source of the pollution. An example is directly discharging sewage and industrial waste into the ocean. Pollution such as this occurs particularly in developing nations. Nonpoint source pollution occurs when the pollution comes from ill-defined and diffuse sources. These can be difficult to regulate. Agricultural runoff and wind blown debris are prime examples.

Direct discharge

Pollutants enter rivers and the sea directly from urban sewerage and industrial waste discharges, sometimes in the form of hazardous and toxic wastes. Inland mining for copper, gold, etc. is another source of marine pollution. Most of the pollution is simply soil, which ends up in rivers flowing to the sea. However, some minerals discharged in the course of the mining can cause problems, such as copper, a common industrial pollutant, which can interfere with the life history and development of coral polyps.

Land runoff

Surface runoff from farming, as well as urban runoff and runoff from the construction of roads, buildings, ports, channels, and harbours, can carry soil and particles laden with carbon, nitrogen, phosphorus, and minerals. This nutrient-rich water can cause fleshy algae and phytoplankton to thrive in coastal areas; known as algal blooms, which have the potential to create hypoxic conditions by using all available oxygen. Polluted runoff from roads and highways can be a significant source of water pollution in coastal areas. About 75 per cent of the toxic chemicals that flow into Puget Sound are carried by stormwater that runs off paved roads and driveways, rooftops, yards and other developed land.

Ship pollution

Ships can pollute waterways and oceans in many ways. Oil spills can have devastating effects. While being toxic to marine life, polycyclic aromatic hydrocarbons (PAHs), the components in crude oil, are very difficult to clean up, and last for years in the sediment and marine environment.

Discharge of cargo residues from bulk carriers can pollute ports, waterways and oceans. In many instances vessels intentionally discharge illegal wastes despite foreign and domestic regulation prohibiting such actions. It has been estimated that container ships lose over 10,000 containers at sea each year (usually during storms). Ships also create noise pollution that disturbs natural wildlife, and water from ballast tanks can spread harmful algae and other invasive species. Ballast water taken up at sea and released in port is a major source of unwanted exotic marine life.

Atmospheric pollution

Another pathway of pollution occurs through the atmosphere. Wind blown dust and debris, including plastic bags, are blown seaward from landfills and other areas. Climate change is raising ocean temperatures and raising levels of carbon dioxide in the atmosphere. These rising levels of carbon dioxide are acidifying the oceans. This, in turn, is altering aquatic ecosystems and modifying fish distributions, with impacts on the sustainability of fisheries and the livelihoods of the communities that depend on them. Healthy ocean ecosystems are also important for the mitigation of climate change.

Deep sea mining

Deep sea mining is a relatively new mineral retrieval process that takes place on the ocean floor. Ocean mining sites are usually around large areas of polymetallic nodules or active and extinct hydrothermal vents at about 1400–3700 metres below the ocean's surface. The vents create sulphide deposits, which contain precious metals such as silver, gold, copper, manganese, cobalt, and zinc. The deposits are mined using either hydraulic pumps or bucket systems that take ore to the surface to be processed. As with all mining operations, deep sea mining raises questions about environmental damages to the surrounding areas

Because deep sea mining is a relatively new field, the complete consequences of full scale mining operations are unknown. However, experts are certain that removal of parts of the sea floor will result in disturbances to the benthic layer, increased toxicity of the water column and sediment plumes from tailings. Removing parts of the sea floor disturbs the habitat of benthic organisms, possibly, depending on the type of mining and location, causing permanent disturbances. Aside from direct impact of mining the area, leakage, spills and corrosion would alter the mining area's chemical makeup.

Among the impacts of deep sea mining, sediment plumes could have the greatest impact. Plumes are caused when the tailings from mining (usually fine particles) are dumped back into the ocean, creating a cloud of particles floating in the water. Two types of plumes occur: near bottom plumes and surface plumes. Near bottom plumes occur when the tailings are pumped back down to the mining site. The floating particles increase the turbidity, or cloudiness, of the water, clogging filter-feeding apparatuses used by benthic organisms. Surface plumes cause a more serious problem. Depending on the size of the particles and water currents the plumes could spread over vast areas. The plumes could impact zooplankton and light penetration, in turn affecting the food web of the area.

SECTION VI

Soil and Water Ecology

Chapter 22

Water Ecology

INTRODUCTION

Water is the lifeblood of our planet. It is fundamental to the biochemistry of all living organisms. The earth's ecosystems are linked and maintained by water, it drives plant growth and provides a permanent habitat for many species, including some 8500 species of fish, and a breeding ground or temporary home for others, such as most of the world's 4200 species of amphibians and reptiles described so far. These ecosystems offer environmental security to humankind by providing goods, such as fish, plants for food and medicines and timber products, services, such as flood protection and water quality improvement, and biodiversity.

Part of the success of the human species has been our ability to control the hydrological cycle, storing water for drinking, growing food and driving industrial processes, harnessing its power for generating energy, and reducing vulnerability to natural hazards, such as floods and droughts. However, this drive to overcome nature is now seen to have disadvantages. It has destroyed much of the earth's natural beauty and degraded many of the vital ecological support systems that keep the planet fit for life. Degradation of the Aral Sea and destruction of the Amazon rainforest are high profile examples. Whilst no one wants to give up reliable water supplies and flood prevention infrastructure, a balance needs to be struck between allocating water directly for people for industry, agriculture and public supply and indirectly for people through the good, services and attributes provided functioning ecosystems. However, this requires a suitable way to allocate water to various uses in an objective and equitable manner. Economic value of water is often used to support such decisions. A stumbling block is that although environmental economists have developed techniques for valuing some ecosystem functions, there is no satisfactory method to measure the ethical values of, on one hand to conserve biodiversity, and on the other to provide food and water to starving thirsty people.

This chapter describes some of the linkages between the ecology and water and explores the hydrological functions of ecosystems and describes the benefits of combining ecological and water management. It also highlights how an optimum water allocation can be achieved through a trade-off between natural and highly managed systems.

WATER CRISIS

Whilst water is created and destroyed in biochemical processes such as respiration and photosynthesis, the total amount of water on earth is stable at around 1.4 billion km^3. Of this, about 41,000 km^3 circulates through the hydrological cycle, the remaining being stored for long periods in the oceans, ice caps and aquifers. It only moves from place to place and changes in quality. Furthermore, the renewal rate provided by rainfall varies around the world. In the Atacama dessert in southern Peru it almost never rains, whilst 6000 mm of rain per year is not uncommon in parts of New Zealand. In any one place rainfall also varies from year to year. In the early 1980s the world witnessed tragic scenes of drought and starvation in the Sahel, but by August 1988 floods ravaged the same region. Water availability also varies over a longer time scale. Some 10,000–20,000 years ago, during glacial phases in high latitudes, rainfall over the current Sahara desert and Middle East was much higher and percolation of water to underlying rocks led to the build up of substantial groundwater resources. However, the recent drier climate in these regions means that recharge is much reduced and groundwater exploited is not being replaced at the same rate. Superimposed upon natural climate cycles are man-induced global changes. The consensus is that by 2050 global temperatures will rise by about 0.2°C per decade, with some areas exceeding this rate and some areas cooling. The implications for water resources are not clear. However, for the Mediterranean region, Estrela have estimated that a 1.5°–2.0°C rise in temperature could result in 10 per cent reduction in rainfall and a 40 to 70 per cent reduction in renewable water resources.

In addition, the burden of insufficient water for domestic use is increasingly being borne disproportionately by women and children. Because they are the primary water collectors, longer collection times mean that women have less time for agricultural production and less time for child care. Water is vital to women for many small-scale food processing or craft activities, which are important sources of income. Women are also the main care providers, thus sickness in the family due to contaminated water impacts on women more severely than on men. In some households, children are involved in water collection and have insufficient time for school.

With such a water crisis facing many countries, it seems an immense task just to manage water so that there is enough for people to drink let alone for agricultural and industrial uses. Thus, providing water to other users, such as 'the environment' surely ought to be given a low priority. Indeed the situation is often presented as a conflict of competing demand, as though it was a matter of choice between water for people and water for wildlife. This ignores the indirect benefits to mankind of functioning ecosystems. Research by Sullivan to develop a Water Poverty Index has shown that the link between water and poverty involves five elements: water resource (surface and groundwater availability), water use (demand for domestic use, agriculture or industry), access (distance to source and legal rights), institutional capacity (human and financial capacity to manage the system) and environment (hydrological functions of ecosystems including protection from floods, water quality improvement and provision of water resources). Some of the poorest people in the world are the most dependent on the environment because they rely on natural resources for food and building material and are most vulnerable to natural disasters, such as floods. Conserving the world's ecosystems is an essential component of addressing world poverty and thus an important ethical consideration for those concerned with overcoming the inequalities between rich and poor people.

CONTROLLING WATER

In an attempt to overcome the vagaries of the hydrological cycle, throughout human history water has been increasingly controlled to benefit mankind. As early as 6000 BC, the Egyptians manipulated water to irrigate crops and built the Sadd el-Kafar ('dam of the Pagans') around 2800 BC. Dams can store water during the wet season and release it during the dry season or when needed for irrigation or hydropower generation. The Romans are noted for the huge aqueducts that they constructed (between 50 BC and 100 AD) to distribute water, such as the Pont du Gard, that supplies Nîmes (in southern

France). These projects, including the Alicante dam (41 metres high), built in 1594, were considered as a miracle of modern engineering, reducing floods, improving agriculture and securing water supplies. More recent mega-hydrological projects include the pumping of water from wells 450 metres deep in the remote southern Libyan dessert and piping it 1000 km to the coast to grow crops. During the eighteenth century, the River Guadalquivir in Southwest Spain was straightened, reducing its lengths by 50 km (40 per cent) to reduce flood risk. In the nineteenth century widespread floodplain reclamation began to improve agriculture, involving the construction of embankments along major rivers, such as the Rhone, to prevent flooding of this land. Intensification of agriculture and industrial and urban development continued during the twentieth century as the population increased and engineering techniques improved, culminating in 1950s and 1960s with the construction of major dams, such as the Hoover dam in the United States of America, led to the belief that man could control the environment totally.

During the past few decades, there has been an increasing realisation that the 'hard' engineering approach to water management has had its costs as well as its benefits. For example, installation of powerful pumps, whilst producing short-term economic benefits, has led to over-exploitation of groundwater in may parts of the Mediterranean, such as in the La Mancha region of Spain. In addition, the Aswam dam in Egypt, at the same time as generating power and providing irrigation water, has led to the loss of fisheries, and led to coastal erosion and salt-water intrusion. These problems have led to calls for more environmently-based water management, which works with nature rather than against it. For this to be widely accepted, the benefits for people of maintaining ecosystem processes must be demonstrated.

In 1997, following many years of increasingly antagonistic debate between pro and anti dam lobbies, the World Bank and IUCN-The World Conservation Union held a meeting in Gland, Switzerland. Participants agreed unanimously that insufficient data were available to conclude unambiguously whether dams were achieving their development objectives. They recommended the establishment of an international independent commission with a clear and achievable mandate. The World Commission on Dams started work in August 1998 to produce a global review if the development effectiveness of dams, a framework for options assessment and decision-making processes and internationally acceptable criteria and guidelines for planning, construction, operation, monitoring and decommissioning of dams. The Commission's report concluded that dams have made an important and significant contribution to human development, but the social and environmental costs have, in too many cases, been unacceptable and often unnecessary. A key principle of the Commission was equity, i.e. that decisions made concerning dams should not be biased towards any particular group, and all key stakeholders should perceive the process and outcomes to be fair and legitimate, which requires transparency in the procedures and decision-making criteria. The report makes many recommendations including increased participation from stakeholders and provision of environmental flows downstream of dams. A detailed assessment of dams, ecosystem functions and environmental flow restoration was undertaken for the Commission.

Whilst it was widely recognised that low flows downstream of dams need to be maintained to support instream ecology, less consideration had been given to the release of high flows for short periods to inundate downstream floodplain and deltaic ecosystems. When flooded periodically, these wetland ecosystems supply important products (e.g. arable land, fisheries, livestock grazing), functions (e.g. groundwater recharge, nutrient cycling) and attributes (e.g. biodiversity), which have contributed to the economic, social and environmental security of rural communities world-wide for many centuries. Floods are also very important for fish migration and sediment transport. Reduction in the frequency and magnitude of flooding by dams, whilst it will be beneficial in many locations to protect vulnerable urban areas, it alters the conditions to which ecosystems have adapted and may degrade the natural services that provide benefits to people. In some cases, the release of managed floods has been proposed, and in a few places implemented, as a mitigation measure to restore and conserve wetland ecosystems in order that traditional livelihood strategies may be maintained. There is clearly a trade-off associated with this action, as less water will remain in the reservoir for its primary design purposes, such as hydropower, irrigation, domestic or industrial supply (Fig. 22.1). The World Bank has adopted the idea of managed flood releases as best practice for dam operation. Managed floods are not a panacea for downstream environmental problems of dams. Nevertheless, they may be appropriate in many cases where downstream wetland ecosystems support dependent livelihoods (particularly where alternative livelihood strategies are limited) or important biodiversity and bio-productivity.

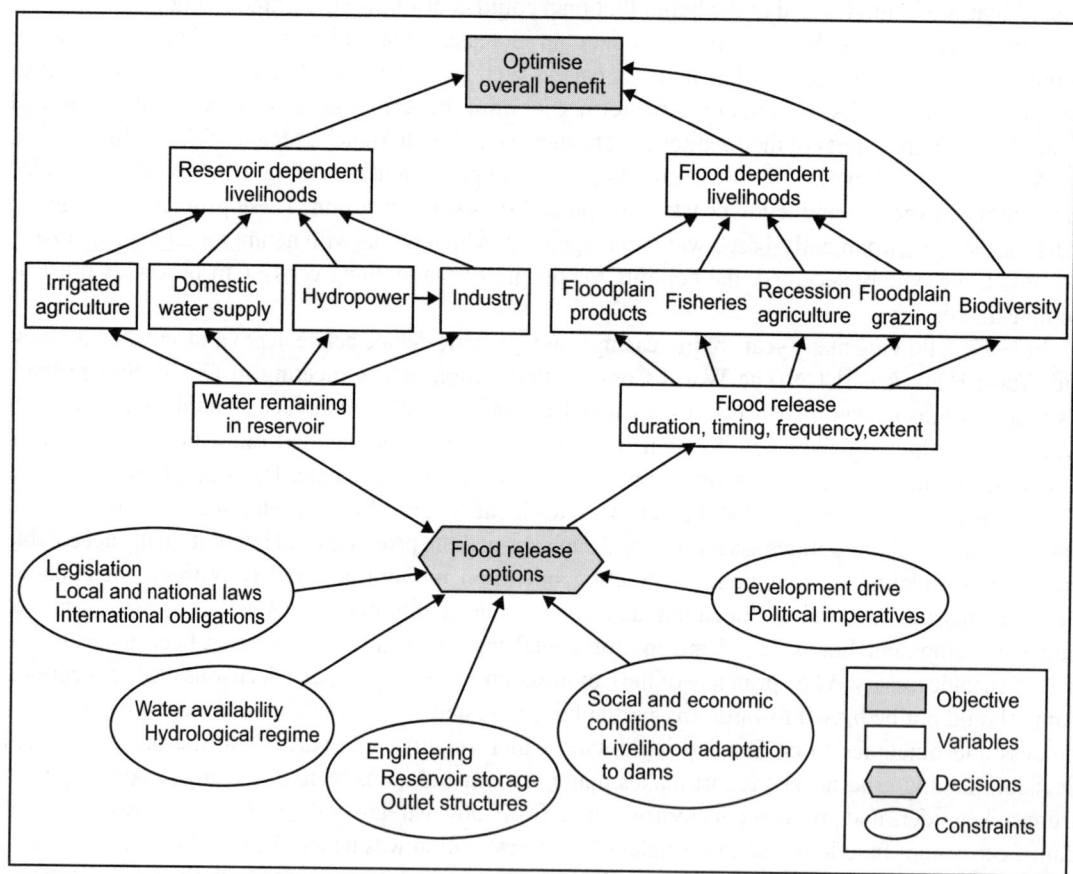

Fig. 22.1. The trade-off between using water for managed flood releases and for reservoir based activities.

WATER AND THE ENVIRONMENT

The Bruntland Report, Our Common Future, Caring for the earth and Agenda 21 from the UNCED Conference in Rio in 1992 marked a turning point in our thinking about water and ecosystems. A central principle that emerged was that the lives of people and the environment are profoundly inter-linked and that ecological processes keep the planet fit for life providing our food, air to breathe, medicines and much of what we call 'quality of life'. The immense biological, chemical and physical diversity of the earth forms the essential building blocks of the ecosystem. The sustainable development of water was the focus of the Dublin Conference in 1991 (a preparatory meeting for UNCED). It concluded that 'since water sustains all life, effective management of water resources demands a holistic approach, linking social and economic development with protection of natural ecosystems'. For example, upstream ecosystems need to be conserved if their vital role in regulating the hydrological cycle is to be maintained. Well-managed headwater grasslands and forests reduce runoff during wet periods, increase infiltration to the soil and aquifers and reducing soil erosion. Downstream ecosystems provide valuable resources, such as fish nurseries, floodplain forests or pasture, but these must be provided with freshwater and seen as a legitimate water user. At the UNCED Conference itself, it was agreed that 'in developing and using water resources priority has to be given to the satisfaction of basic needs and the safeguarding of ecosystems'. Thus whilst people need access to water directly to drink, irrigate crops or run industrial processes, providing water to the environment means using water indirectly for people. The declaration from the Second World Water Forum in The Hague 2000 highlighted the need to ensure the integrity of ecosystems through sustainable water resources management. The World Summit on Sustainable Development held in August 2002 in Johannesburg, reinforced the role of environmental protection as a key pillar of sustainable development. South Africa has taken a lead in implementing the concept. Principle 9 of the new water law of South Africa states that: 'the quantity, quality and reliability of water required to maintain the ecological functions on which humans depend shall be reserved so that the human use of water does not individually or cumulatively compromise the long term sustainability of aquatic and associated ecosystems'. Likewise, Tanzania is currently developing a new water policy that gives high priority to water allocation for ecosystems.

Many organisations have recognised the importance of understanding the links between water and ecosystems. The 1996–2001 Fifth International Hydrology Program of UNESCO included an Ecohydrology theme that focused on two projects: (i) interactions between river systems, floodplains and wetlands, and (ii) a comprehensive assessment of surficial ecohydrological processes. A particular focus was the resilience and resistance of ecosystems and their role in management of water quality, such as the use of buffer strips to ameliorate the impacts of agricultural pollution on river systems. UNESCO has strong links with the scientific committee on water research (SCOWAR), which is part of the International council for scientific unions (ICSU). SCOWAR has focused on hydrological impacts on ecological systems from different regions around the world and on forecasting the ecological consequences of changing water regimes. Similarly, the International Association of Hydrological Sciences has supported research on the links between hydrology and aquatic ecology.

HYDROLOGICAL FUNCTIONS OF NATURAL ECOSYSTEMS

Natural ecosystems, such as forests and wetlands can play a valuable role in managing the hydrological cycle. Vegetation encourages infiltration of water into the soil, aiding the recharge of underground aquifers, lowering flood risk and anchoring the soil, thus reducing erosion. In Honduras the La Tigra National Park, 7,500 hectares of cloud forest, sustains a high quality, well-regulated water flow throughout the year, yielding over 40 per cent of the water supply of Tegucigalpa, the capital city. Because of its value for watershed protection, La Tigra is today the focus of an investment program involving a series of economic incentives for villagers living in the buffer zones. The value of these services is considerable. Rather than build water treatment facilities at a cost of US$7 billion, the New York City water department has spent a tenth of this sum to ensure the protection of the biological and hydrological processes of the highlands of the catchment. Evidence for the beneficial role of forests in reducing floods at a local scale has been extrapolated to the regional scale. Various reports suggest that deforestation of the Himalayas has increased flood risk downstream in India and Bangladesh. However, Kaimowitz has raised questions concerning the evidence for large-scale impacts and contests that deforestation had limited influence on the impact of floods in central America caused by hurricane Mitch in October 1988.

Forests also take up water and release it into the atmosphere. A rain forest tree can pump 2.5 million gallons of water into the atmosphere during its lifetime but much of this is recycled and not lost from the forest. In the Amazon rainforest, 50 per cent of rainfall is derived from local evaporation. After forest cover is removed an area can become hotter and drier because water is no longer cycled between plants and the atmosphere. This can lead to a positive feedback cycle of desertification, with increasing loss of water resources in that area. Results of simulations using a global circulation model, in which the Amazon tropical forest and savannah was replaced by pasture land, predicted a weakened hydrological cycle with less precipitation and evaporation and an increase in surface temperate due to changes in albedo and roughness. Rainfall was reduced by 26 per cent for the year as a whole. Similarly, modelling the removal of natural vegetation in the Sahelian region of Africa suggests that rainfall has been reduced by 22 per cent between June and August and the rainy season has been delayed by half a month. Therefore these ecosystems function as water cycling systems between the earth and the atmosphere and in return for the water they use provide the service of regulating both global and local climate and maintaining local water resources.

Acreman advocated ecosystem management as one of the main principles of water management for people and the environment. The ecosystem management approach aims to integrate all the important physical, chemical and biological components and processes which interact with social, economic and institutional factors. This requires integrated management of mountains, drylands, forests, agriculture, housing, industry, transport, waste disposal, aquifers, rivers, lakes, wetlands and anything which has an effect on the environment. The appropriate management scale depends upon the relative importance of the components in the system. The fundamental unit for water issues is normally the drainage basin, as this demarcates a hydrological system, in which components and processes are linked by water movement. Deforestation of headwater catchments can, for example, affect water yield and frequency of flooding downstream. Hence the term integrated river basin management has developed as a broad concept which takes a holistic approach. However, frequently the underlying aquifer does not coincide with the surface river basin. Thus, where groundwater plays a significant role, a group of basins overlying the aquifer may constitute the appropriate unit of water resource management. For issues where air quality is influential, such as acid rain, the 'airshed' (as apposed to the watershed) will be more appropriate implying the integrated management of source areas, which may be industries in the UK, with affected areas in Scandinavia.

Ecosystem conservation can be a cost-effective solution to water management. For example, Mackinson has shown that the cost of establishment of protected areas, reforestation where necessary and other measures to protect the catchments of 11 irrigation projects in Indonesia, ranging from less than 1 to 5 per cent of the development costs of the individual irrigation projects. This compares very favourably with the estimated 30–40 per cent loss in efficiency of the irrigation systems if catchments were not properly safeguarded.

Wetlands, such as floodplains, marshes and reed beds, can also perform important hydrological functions within a catchment including storage of water during floods, nutrient cycling and recharging groundwater. The value of utilising the natural functions of aquatic ecosystems, as an alternative to major engineering investment, was recognised as early as 1972 by the US Corps of Army Engineers. They recommended that the most cost effective approach to flood control in the Charles River of Massachusetts lay in conserving the 3800 hectares of mainstream wetlands which provide natural valley storage of flood waters. Serious flooding of cities in Germany and the Netherlands along the River Rhine during 1994 was made worse by the presence of embankments upstream. These had separated the river from the floodplain wetland, protecting agricultural land, but preventing access by the river to natural floodwater storage. In 1995 two large flood storage wetlands were created on the German bank of the Rhine as part of a program to reduce flood damage downstream and restore degraded floodplain ecosystems. Hollis have demonstrated that recharge to the aquifer which supplies well-water to some 1,00,000 people in the Komodugu-Yobe basin, Nigeria, occurs during flooding of the Hadejia-Nguru wetlands. However, dams constructed upstream, which stored water for intensive irrigation, have degraded the wetlands by starving them of water. Following presentation of research on the natural functions of the wetlands, the Nigerian authorities realised the benefits of conserving the wetlands and have been exploring the potential for releasing water from the reservoirs to augment flooding of the floodplain. This is consistent with the ideas of Scudder and Acreman who have promoted more widely the benefits of making managed flood releases from dams to conserve important ecosystems downstream as a cornerstone of integrated catchment and water resources management.

Wetlands also perform important water quality functions. The Nakivubo papyrus swamp in Uganda receives semi-treated effluent from the Kampala sewage works and highly polluted storm water from the city and its suburbs. During the passage of the effluent through the wetland, sewage is absorbed and the concentrations of pollutants are considerably reduced. Water flowing out of the wetland enters Murchison Bay about 2 km from the intakes of the two Kampala water supply works. Consequently, the National Sewerage and Water Corporation is supporting conservation of swamps and other wetlands near Kampala because they purify the water, serving as a low cost alternative to industrial sewage treatment. Likewise, Smith described the important functions of the 75,000 hectare North Selangor Peat Swamp forest, which borders one of the largest rice schemes in Malaysia. These wetlands mitigate floods and maintain high water quality. In recent years the forests have been cleared for agriculture and tin mining, reducing the buffering effect on pollution and releasing sediment. It is forecast that further clearance would result in significant water quality problems in rice scheme. Because of this valuable water purifying function, in many parts of the Europe and North America artificial wetlands have been created to treat polluted water, including sewage effluent and mine waste.

Meynell and Qureshi reported on the vital functions of the mangrove ecosystem in the delta at the mouth of the River Indus delta in Pakistan. By breaking the force of wind and waves, they protect the coast and Port Qasim from damage. Wave height can reach six metres in the open sea beyond the mangroves, but in the sheltered creeks the maximum recorded has been 0.5 metres. Mangroves also stabilise the creek banks which maintains channel width. This focuses the currents, reducing sedimentation by encouraging scouring of the channels bed. The creeks are thus selfcleaning and able to maintain their geometry naturally. Without mangroves Port Qasim would need expensive engineering works such as sea walls and constant dredging costing around US$ 1 per cubic metre and thus would not be economical. It is clear from the above examples that natural ecosystems can perform valuable hydrological functions. Clearly, not all ecosystems perform all functions, for example, not all wetlands reduce floods, recharge groundwater and improve water quality. Nevertheless, each has its own role to play in the natural processes of the catchment. Thus conservation of ecosystems should be a key element in sustainable water resource management.

ECOSYSTEMS AND BIODIVERSITY

Many ecosystems support a wide range of species and large numbers of individuals. Water availability is often a key controlling factor in biodiversity. For example, in central Africa, Tchamba, Drijver and Njiforti describe how flood water from the River Logone inundates annually a large floodplain, originally around 6000 km². This wetland has a high biodiversity with large herds of giraffe, elephant, lions and various ungulates (including topi, antelope, reedbuck, gazelle, kob). Part of the floodplain has been designated as the Waza National Park which attracts around 6000 tourists per year, which bring direct financial benefit to the region. In the flood season, the entire floodplain becomes a vast fish nursery. Up to the 1960s, fishing was the primary economic activity amongst the local Kotoko people who could earn US$ 2000 in four months. The floodplain is also an important dry season pasture for 3,00,000 cattle and 10,000 sheep and goats. Since 1979, the area inundated has reduced, partly by climatic factors, but primarily due to construction of embankments and a barrage across the floodplain which created Lake Maga to supply water to an irrigation project. Flooding became insufficient in large areas to grow any floating rice and fish yields fall by 90 per cent. The irrigation schemes which cover around 5000 hectares were not making full use of water stored in Lake Maga and the potential to release water to rehabilitate the floodplain was identified. To implement this, the embankments along the river were modified in 1994 to allow flood waters to reach the floodplain, which has revitalised the wetland ecosystem and extensive fishing and grazing have been rejuvenated.

Coastal ecosystems in particular have been neglected in water management. For most water engineers, any freshwater that reaches the sea is a waste. This ignores the fact that valuable ecosystems, along the river and in the coastal zone, rely on inputs of freshwater. For instance, freshwater from the Zambezi supports extensive inshore fisheries on the Sofala bank at the mouth of the river. This provides Mozambique with an important source of foreign income worth some US$ 50–60 million per year. Gamelsrød has shown that shrimp abundance is directly related to wet season freshwater runoff (Fig. 22.2) and earnings could be increased by US$ 10 million per year by correctly releasing flood waters from the Cahora Bassa dam which are not currently utilised. Likewise, a positive relationship between freshwater runoff and shrimp production was found for the Tortugas grounds off the Florida peninsula of United States of America by Lugo and Snedaker. These estuarine wetlands receive water from the Everglades National Parks and further demonstrate the close link between ecosystems through the hydrological cycle.

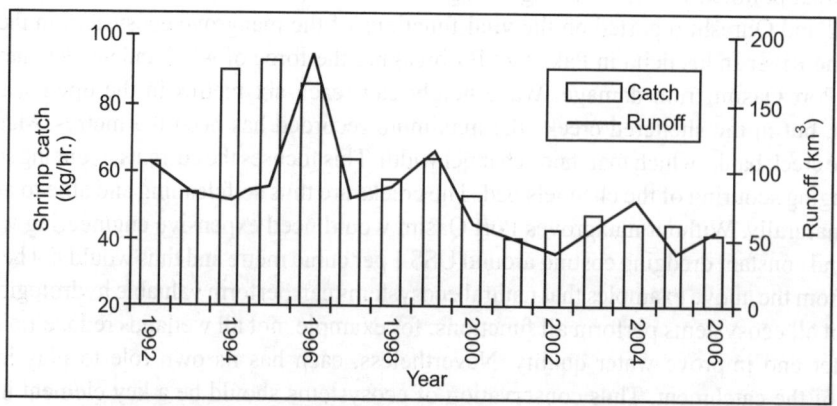

Fig. 22.2. Relationship between shrimp abundance and wet season freshwater runoff from the Zambezi.

A similar situation occurred in the Nile delta following completion of the high dam at Aswam in Egypt in 1968. Nutrients brought to the sea by the river supported a rich sardine fishery, but fish catches declined from 22,618 million tons in 1968 to only 13,450 million tons in 1980 and rates are still falling. In lower Nile, fish populations have also declined. Of the 47 commercial species present in 1948, only 17 now exist. The reservoir behind Aswam has created new fishing opportunities and produced some 34,000 tons in 1987, but much is due to the increased fishing effort and it is unclear if this rate is sustainable.

Water control may not always be environmentally detrimental. Masundire reported that Lake Kariba, created by the construction of Kariba dam on the Zambezi River, supports an important inland fishery and the whole shoreline has been declared a 'recreational park' as the availability of water during the dry season attracts large herds of buffalo, eland and other species, enhancing the eco-tourism value of the area, which brings in foreign currency. However, the dam has had negative effects on the ecology downstream and on the health of local people as disease vectors such as snails have proliferated.

The above examples demonstrate that the products from many ecosystems can have a direct benefit to mankind. To maintain the valuable products and services they provide, they must be treated as legitimate water users and allocated sufficient water to remain healthy.

WATER ALLOCATION, SUSTAINABILITY AND ETHICS

In the water debate, it is useful to divide human needs into three areas:
1. Economic security, e.g. drinking water, shelter, food and other consumable goods.
2. Social security, e.g. protection from natural hazards, such as floods.
3. Ethical security, e.g. upholding the rights of people and other species to water.

Figure 22.3 summarises the implications of allocating water to indirect human use, by supporting ecosystem processes and direct use. The upper part of Fig. 22.3 shows the impact of allocating water to natural ecosystems, which in turn provide valuable goods (e.g. fish), services (e.g. water regulation) and amenity/touristic value (landscape and species). In this case the impact on the hydrological cycle is frequently positive, as, for example, ecosystems improve water quality. Additionally, it satisfies the growing belief amongst many people that humans have a moral duty to protect wildlife, through providing sufficient water to maintain flora and fauna. The idea that the natural environment has a right to water *per se* was taken up at UNCED in 1982, where the governments of the United Nations made an ethical commitment to the environment in the form of the World Charter for Nature. This expresses absolute support of the governments for the principle of conserving biodiversity. It recognises that every form of life is unique and warrants respect, regardless of its direct worth to humankind and that the lasting benefits of nature depend on maintenance of essential ecological processes and life-support systems and upon the diversity of life forms. It promotes conservation of ecosystems as a public good independent of their utility as a resource and hence water rights to species and ecosystems. Whilst we may all support this at a superficial level, especially when thinking of flagship species such the giant panda, few would want to extend this to include the small-pox virus. Indeed, human choice influences our objectives for the environment significantly. Many of the cherished landscapes of the world, from the terraced paddy fields of Bali to the rolling fields and hedgerows of rural England, are highly managed systems. Indeed, very little of the earth is natural or completely unaffected by human influences; almost all is managed, intentionally or unintentionally, to a greater or lesser extent. It is unrealistic to expect to return much of the earth to a natural ecosystem, and it is undesirable as many are classified as World Heritage Sites. The status of the ecosystems is as much a result of societal choice.

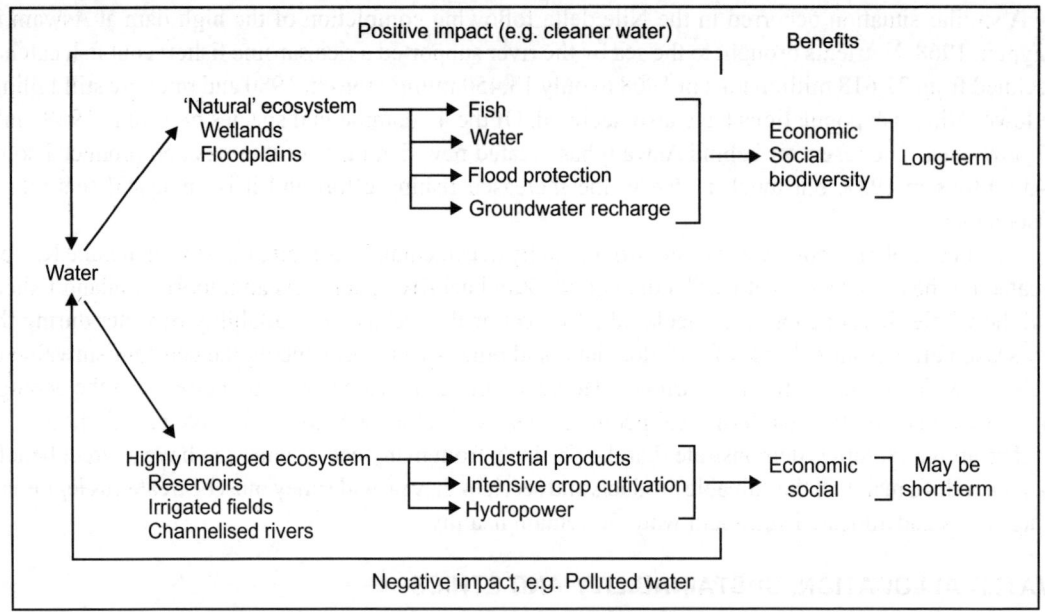

Fig. 22.3. Natural and highly managed ecosystem benefits.

The lower part of the Fig. 22.3 shows the direct use of water through the development of highly managed systems, including reservoirs, intensive irrigation schemes, dams, river embankments and water purification plants. This has led to production of crops, industrial products, electricity, protection from floods and provision of clean water, thus improving economic and social security. However, this has often caused negative impacts in the form of pollution. To some extent economic and social security have been improved. In addition, through the provision of food and water to starving and thirsty people in drought stricken countries, technology has contributed to the ethical security of those who do not face this problem.

The important question is 'at what level to maintain the earth's ecosystems?' The concept of sustainability suggest that we need to maintain the earth's ecosystems so that they yield the greatest benefit to present generations, whilst maintaining the potential to meet the needs and aspirations of future generations. The problem is to decide how much water should be utilised directly for people for domestic use, agriculture and industry and how much water should be used indirectly by people to maintain ecosystems that provide environmental goods and elemental services. Figure 22.4 shows the problem conceptually as a trade-off between natural and highly managed systems. As natural systems are modified into highly managed systems, the benefits of the natural system obviously decline (solid line); e.g. hydrological functions, products and biodiversity are lost. At the same time, benefits from the highly managed system increase (dotted line); e.g. food production rises. It is suggested that the benefits from highly managed systems reaches a plateau, whilst the benefits of the natural system will decline to zero at some point. The total long-term benefits (dashed line) can be calculated by adding the benefits of the natural and highly managed systems. The total rises to a maximum before declining. It is at this point that the balance between naturalness and high management is optimised. Obviously, the value that society places on goods and services and ethical considerations will determine the exact form of these curves. Indeed the perceived benefits will vary between different groups and individuals. It is essential therefore that the costs and benefits to society of allocating water alternatively to maintain ecosystems and to support direct use in the form of agricultural, industrial and domestic uses are quantified.

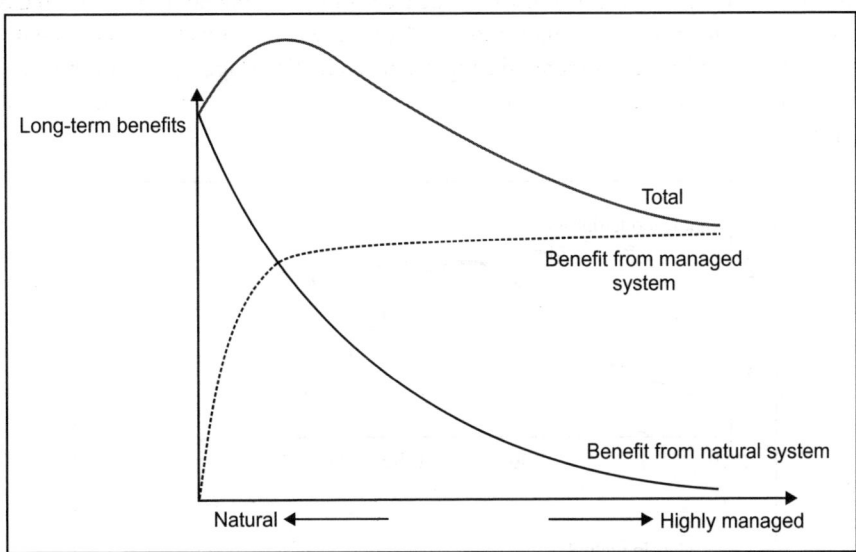

Fig. 22.4. Maximising benefits from freshwater ecosystems.

A major question is whether highly managed systems are sustainable. Developed countries have faced many problems over the past 200 years. However, humankind has been ingenious enough continually to find technological fixes to problems as they arise, such that, as a whole, the highly managed system is sustainable. The upper graph in Fig. 22.5 shows the increase and subsequent decrease of different forms of pollution facing developed countries since 1800. For example, human waste increasingly polluted rivers up to the 1850s, when sewage treatment plants were developed to deal with the problem. Currently pesticide and fertiliser pollution in surface and groundwaters is causing considerable problems. Future issues, such as endocrine disrupters, which have created hermaphrodite fish in European rivers, may not be as easy to remedy as sewage. A further counter argument is that more reliance on technology makes us more vulnerable when the technology fails. People along the River Mississippi in the United States of America lived and farmed behind the embankments, which protected them from small-scale floods. However, when a large flood came in 1993, the impacts were worse because of intensive cultivation and embankments preventing the water from returning to the river. The central graph in Fig. 22.5 shows that similar water quality problems are facing rapidly developing countries, but within a much more condensed period, which began in the early 1900s. In recent years the same problems are being faced by developing countries. The developed countries' technological fixes are available from developed countries (Fig. 22.5, lower graph), but it not clear who would finance their implementation.

QUANTIFYING WATER NEEDS OF ECOSYSTEMS

The water needs of ecosystems and their various component species is a complex issue that has been the subject of considerable study world-wide. Much of the focus has been on minimum flow requirements, but many species rely on high flows and flooding. For example, the acacia trees in the riverine forests of the Indus river valley require inundation from flood water for their moisture, which also brings important nutrients. At least in their early stages of growth, the trees must be flooded for at least 10 days per year. Once acacias are about 8–10 years old their roots are normally able to reach the permanent water table. Some species have specific requirements during a particular life stage. Common reeds (Phragmites australis) on the other hand require permanent inundation of around 200 mm, but can tolerate short periods of drying.

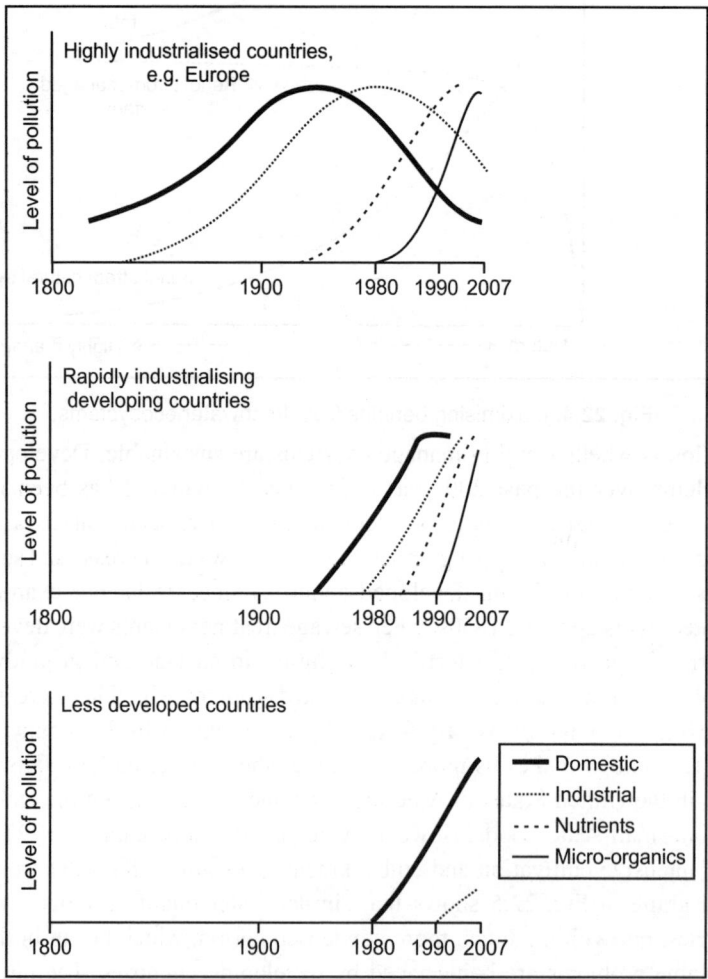

Fig. 22.5. Water quality problems in developed, rapidly industrialising and developing countries.

The Palla fish of Asia, for example, requires a minimum depth of 1.8 metres for breeding. Welcomme studied the fish catches in various floodplains in Africa as a surrogate for fish productivity. However, his data show a linear relationship between flooded area and fish catch (Fig. 22.6). There is no clear threshold point below which the flooded area is insufficient to maintain the fish population. The water needs of such an ecosystem depend on how many fish people wish to catch.

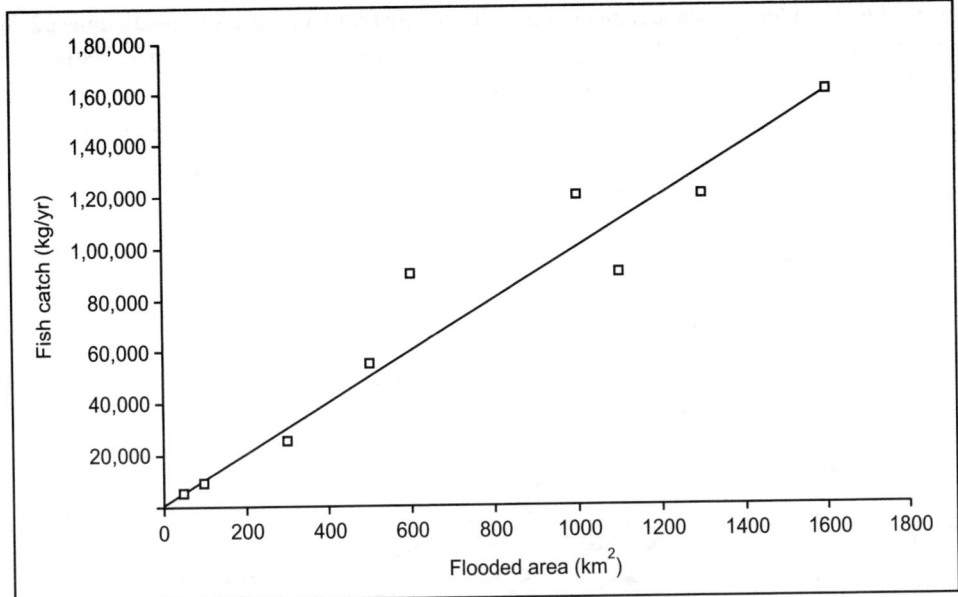

Fig. 22.6. Relationship between flooded area and fish catch.

The environment agency of England and Wales has developed a procedure called the Resource Assessment and Management (RAM) framework for defining the water needs of rivers in terms of a proportion of the flow duration curve (where the flow duration curve defines the relationship between flow and the percentage of the time this flow is equalled or exceeded, Fig. 22.7). This is a default procedure when no other detailed method is available. The exact proportion is based on sensitivity of the river to removal of water and is determined though consideration of four elements: (i) physical character, (ii) fisheries, (iii) macrophytes, and (iv) macro-invertebrates. Each element is given a score from 1–5 based on its sensitivity to abstraction. For physical characterisation, rivers with steep gradients and /or wide shallow cross-sections score 5 (most sensitive). At the other extreme, deep lowland river reaches score 1 (least sensitive). Photographs of typical river reaches in each class are provided to aid scoring of physical character. Scoring for fisheries is determined by using expert opinion to interpret the available monitoring data and classify the river using common indicator species. In the case of macrophytic vegetation, various approaches are possible depending upon the availability and quality of the data. Once a score for each of the four elements has been defined, the scores are combined to categorise the river into one of five environmental weighting categories, ranging from very high to very low. A look-up table is used to determine the percentage of the natural flow that can be taken at different flow percentiles (e.g. Table 22.1 gives figures for Q_{95}). In this way the ecological need can be derived based on the proportion of the flow percentiles not abstracted. The figures in the table are based largely on professional judgement of specialists, since critical levels have not been defined directly by scientific studies. Any such figures are open to revision, but with no clear alternative, this provides a pragmatic way forward.

As discussed above, the water allocation to a river ecosystem will depend upon the expectations (or objectives) that have been set for it. As part of the derivation of the environmental flows agreed for the South African water law, the Department of Water Affairs and Forestry produced a classification of rivers according to ecological management targets. There are four target classes, A-D (Table 22.2). Two additional classes, E and F may describe present ecological status but not a target. Water resources currently in category E or F must have a target class of D or above.

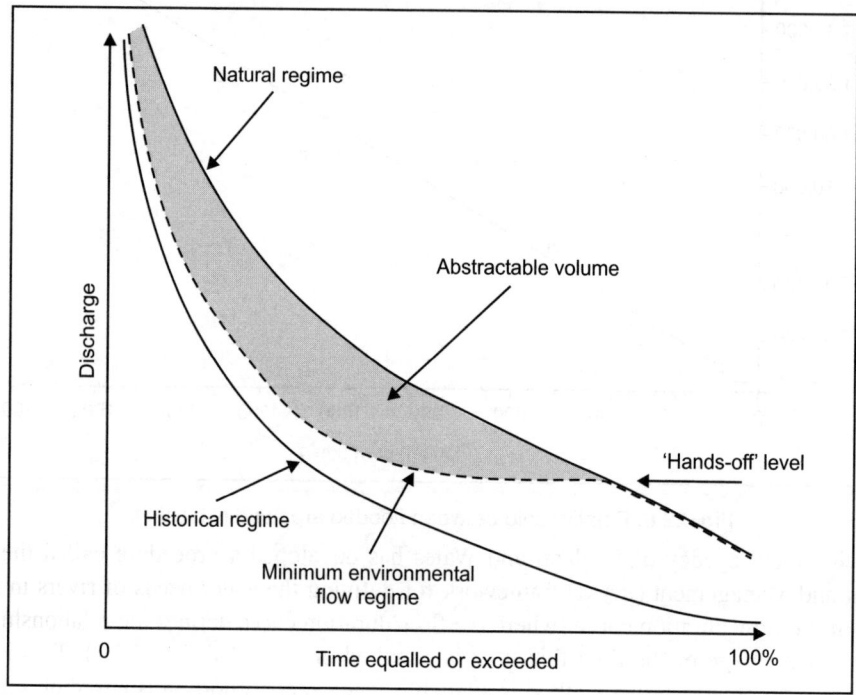

Fig. 22.7. Environmental and natural flow duration curves.

Table 22.1. Percentages of Q_{95} flow that can be abstracted for different environmental weighting bands.

Environmental weighting band	% of Q_{95} that can be abstracted
A	0–5
B	5–10
C	10–15
D	15–25
E	25–30
Others	Special treatment

Table 22.2. Ecological management classes.

Class	Description
A	Negligible modification from natural conditions. Negligible risk to sensitive species
B	Slight modification from natural conditions. Slight risk to intolerant biota
C	Moderate modification from natural conditions. Especially intolerant biota may be reduced in number and extent
D	High degree of modification from natural conditions. Intolerant biota unlikely to be present

Once the river has been assigned to a class, the ecological Reserve flow regime is defined using the Building Block approach. This involves an expert panel, including a geomorphologist, hydrologist and ecologists, undertake a field visit to the river and study available data to determine the various elements (building blocks) of the Reserve, such as low flows, average flows and small and large floods. As a rapid first approximation to setting the Reserve, which does not require the expert panel, Hughes and Münster have a devised a hydrological assessment method. This assumes that if the flow regime is highly variable, biota will be adjusted to a relative scarcity of water and will hence require a lower proportion of natural flow. Biota in less variable rivers are more sensitive to reductions in flow and a larger proportion of the mean will be required.

Research by the US fish and wildlife service on the flow requirements of riverine species, including fish, invertebrates and plants, led to the development of a system called PHABSIM (Physical Habitat Simulation) that relates river flow to in-stream ecology. PHABSIM assumes that a given species has preferences for certain habitat characteristics, such as water depth or flow velocity. The graph in Fig. 22.8 shows changes in in-stream physical habitat (indexed by weighted useable area–WUA) for the fry/juvenile life stage of brown trout in two contrasting UK rivers, the Piddle, a lowland groundwater-fed river, and the Wye, a upland river in an impermeable catchment. PHABSIM has been used to determine the impacts of changing river flows on brown trout. This species is used because of it's sensitivity to river flows and it's acceptance by many stakeholders as a good indicator of river health. PHABSIM has also been used to estimate the ecological effects (in terms of available physical habitat) for historical or future anticipated changes in flow caused by abstraction or dam construction. The method has been adapted for use in many countries including UK, Canada, Austria and New Zealand. However, PHABSIM has been applied primarily to the physical habitat needs for species and has not normally considered indirect impacts, for example reduced river flows may increase concentrations of pollutants or may reduce dissolved oxygen.

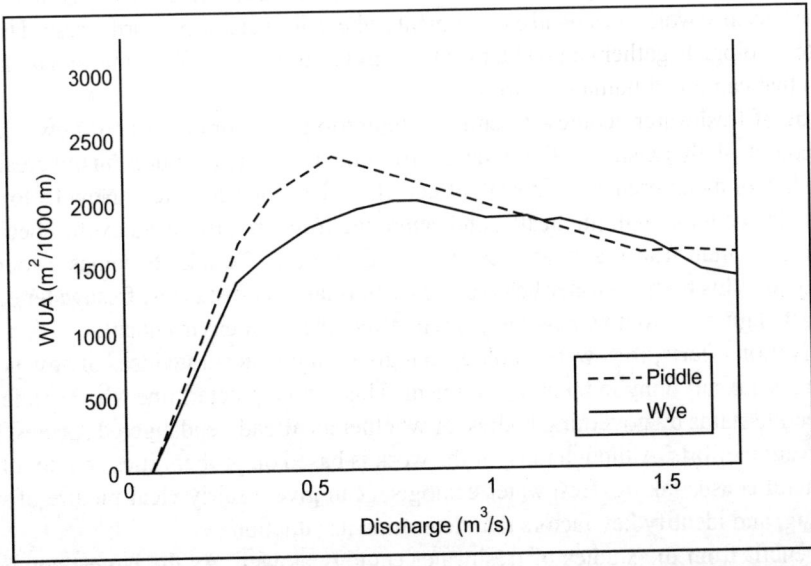

Fig. 22.8. Relationship between discharge and habitat availability (WUA-weighted usable area) for brown trout in two UK rivers.

Although the quantity of water flowing in a river is a key control over the health of the ecosystem, it is the interaction of the flow and the channel structure that defines the conditions that make it suitable for different species. One of the strengths of habitat approaches like PHABSIM is that they combine river flow and channel morphology. Acreman and Elliott used PHABSIM to show how straightening and deepening of the channel of the River Wey in the UK reduced its suitability for fish, even though the flow was unchanged. Subsequent restoration of the channel by narrowing and reinstatement of an irregular path improved conditions for fish. Booker showed how both different levels of channel and flow modification in urban areas influenced the habitat for fish.

In some cases detailed local studies have revealed critical timing of water requirements for specific ecosystems. For example, the ecology of the Diawling National Park, in the delta of the River Senegal in Mauritania, is controlled by seasonal variations in water availability and salinity, which has generated particular vegetation types, such as mangroves with associated species, including penaeid shrimp and mullet. In the late 1980s, the delta was separated hydrologically from the river by construction of the Diama dam and right-bank embankment. These maintain water levels in the Senegal River for gravity irrigation and navigation, but caused degradation of biodiversity of Park and loss of natural resources, including grazing and fisheries, in the buffer zone, increasing poverty in local communities.

FRESHWATER ECOLOGY

Freshwater ecology is a specialised subcategory of the overall study of organisms and the environment. Unlike biology, ecology refers to the study of not just organisms but how they react, and are affected by the natural surrounding environment or ecosystem. By studying the plants and animals in a body of water as well as the components of the water itself, a scientist specialising in freshwater ecology can discover vital information about the health and needs of a freshwater system.

Rather than study the vast world of saltwater like marine ecologists, scientists that work in freshwater ecology concentrate on the ecosystems of bodies of non-brackish water, such as lakes, ponds, and streams. Some may also work in wetland environments where the water is primarily fresh. The information that freshwater ecologists gather can be helpful to conservation efforts for plants and animals, but also provides data that can effect humans as well.

In the study of freshwater ecology, scientists attempt to get accurate ideas of how a body of fresh water goes about its daily existence. Every detail, from the microbic creatures busily creating algae, to the large reptilian or avian predators present, effects the life of the ecosystem. New factors can disrupt and reorganise the ecosystem dramatically, and can range from an introduced exotic species, chemical runoff from a new industrial plant, or even increased usage if the lake becomes a tourist spot. By understanding how this body of water behaves under normal circumstances, freshwater ecologists can make an educated guess as to how new factors will effect the local environment.

For conservation efforts, freshwater ecology can give roughly accurate ideas of how populations of plants or animals are surviving in their environment. This can help determine whether a fading species is given protective status by governing bodies, or whether an already endangered species is recovering due to conservation efforts. Although most of the work is based on probabilities and population graphs rather than literal census taking, freshwater ecologists can give a fairly clear picture of which way a species is going, and identify key factors that determine its situation.

Humans benefit from the studies of freshwater ecology as well. As the largest component of the ecosystem, the water is constantly tested and analysed for important data such as chemical composition and possible hazards. The work of freshwater ecologists can be used to determine the viability of a new drinking water source, or test a current water source for possible contamination. By protecting drinking water sources, freshwater ecologists are contributing not only to the good of the environment, but the good of their own species as well.

AQUATIC ANIMAL

An aquatic animal is an animal, either vertebrate or invertebrate, which lives in water for most or all of its life. It may breathe air or extract its oxygen from that dissolved in water through specialised organs called gills, or directly through its skin. Natural environments and the animals that live in them can be categorised as aquatic (water) or terrestrial (land). Animals that move readily from water to land and vice versa are often referred to as amphibious.

The term aquatic can in theory be applied to animals that live in either freshwater (freshwater animals) or saltwater (seawater animals). However, the adjective marine is most commonly used for animals that live in saltwater, i.e. in oceans, seas, etc. Aquatic animals (especially freshwater animals) are often of special concern to conservationists because of the fragility of their environments. Aquatic animals are subject to pressure from overfishing, destructive fishing, marine pollution and climate change.

Air Breathing Aquatic Animals

In addition to water breathing animals, e.g. fishes, mollusks, etc. the term 'aquatic animal' can be applied to air-breathing aquatic or sea mammals such as those in the order Cetacea (whales), which cannot survive on land, as well as four-footed mammals like the river otter (*Lontra canadensis*) and beavers (family Castoridae).

Aquatic animals include for example the seabirds, such as gulls (family Laridae), pelicans (family Pelecanidae), and albatrosses (family Diomedeidae), and most of the Anseriformes (ducks, swans and geese). Amphibious and amphibiotic animals, like frogs (the order Anura), while they do require water, are separated into their own environmental classification. The majority of amphibians (class Amphibia) have an aquatic larval stage, like a tadpole, but then live as terrestrial adults, and may return to the water to mate.

HYDROLOGICAL CYCLE/WATER CYCLE

The water cycle, also known as the hydrologic cycle or H_2O cycle, describes the continuous movement of water on, above and below the surface of the earth. Water can change states among liquid, vapour, and ice at various places in the water cycle. Although the balance of water on earth remains fairly constant over time, individual water molecules can come and go. The water moves from one reservoir to another, such as from river to ocean, or from the ocean to the atmosphere, by the physical processes of evaporation, condensation, precipitation, infiltration, runoff, and subsurface flow. In so doing, the water goes through different phases: liquid, solid, and gas.

The hydrologic cycle also involves the exchange of heat energy, which leads to temperature changes. For instance, in the process of evaporation, water takes up energy from the surroundings and cools the environment. Conversely, in the process of condensation, water releases energy to its surroundings, warming the environment.

The water cycle figures significantly in the maintenance of life and ecosystems on earth. Even as water in each reservoir plays an important role, the water cycle brings added significance to the presence of water on our planet. By transferring water from one reservoir to another, the water cycle purifies water, replenishes the land with freshwater, and transports minerals to different parts of the globe. It is also involved in reshaping the geological features of the earth, through such processes as erosion and sedimentation. In addition, as the water cycle involves heat exchange, it exerts an influence on climate as well. Water cycle is shown in Fig. 22.9.

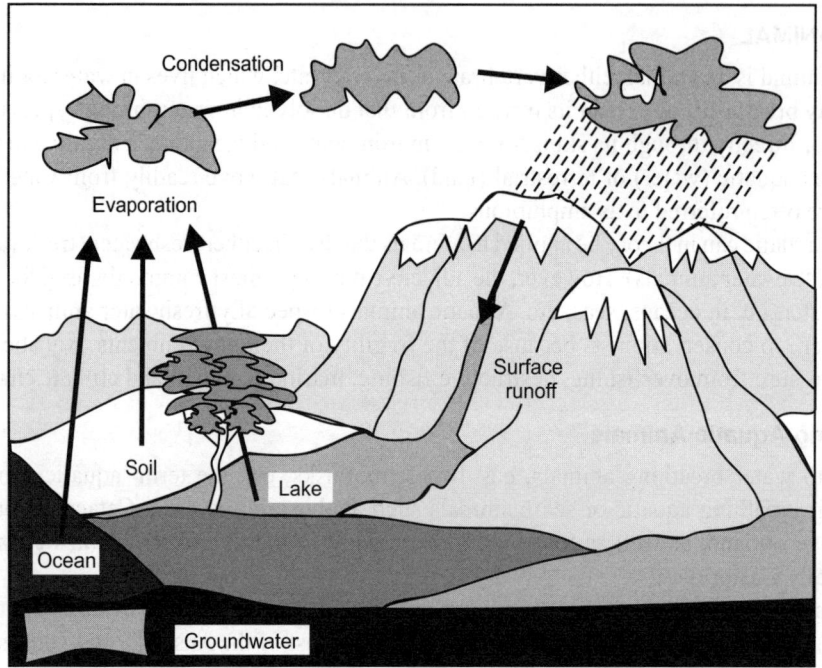

Fig. 22.9. Water cycle.

The sun, which drives the water cycle, heats water in oceans and seas. Water evaporates as water vapour into the air. Ice and snow can sublimate directly into water vapour. Evapotranspiration is water transpired from plants and evaporated from the soil. Rising air currents take the vapour up into the atmosphere where cooler temperatures cause it to condense into clouds. Air currents move water vapour around the globe, cloud particles collide, grow, and fall out of the sky as precipitation. Some precipitation falls as snow or hail, and can accumulate as ice caps and glaciers, which can store frozen water for thousands of years. Snowpacks can thaw and melt, and the melted water flows over land as snowmelt. Most water falls back into the oceans or onto land as rain, where the water flows over the ground as surface runoff. A portion of runoff enters rivers in valleys in the landscape, with streamflow moving water towards the oceans. Runoff and groundwater are stored as freshwater in lakes. Not all runoff flows into rivers, much of it soaks into the ground as infiltration. Some water infiltrates deep into the ground and replenishes aquifers, which store freshwater for long periods of time. Some infiltration stays close to the land surface and can seep back into surface-water bodies (and the ocean) as groundwater discharge. Some groundwater finds openings in the land surface and comes out as freshwater springs. Over time, the water returns to the ocean, where our water cycle started.

Different Processes

Precipitation: Condensed water vapour that falls to the earth's surface. Most precipitation occurs as rain, but also includes snow, hail, fog drip, graupel, and sleet. Approximately 5,05,000 km³ (1,21,000 cu mi) of water fall as precipitation each year, 3,98,000 km³ (95,000 cu mi) of it over the oceans.

Canopy interception: The precipitation that is intercepted by plant foliage and eventually evaporates back to the atmosphere rather than falling to the ground.

Snowmelt: The runoff produced by melting snow.

Runoff: The variety of ways by which water moves across the land. This includes both surface runoff and channel runoff. As it flows, the water may seep into the ground, evaporate into the air, become stored in lakes or reservoirs, or be extracted for agricultural or other human uses.

Infiltration: The flow of water from the ground surface into the ground. Once infiltrated, the water becomes soil moisture or groundwater.

Subsurface flow: The flow of water underground, in the vadose zone and aquifers. Subsurface water may return to the surface (e.g. as a spring or by being pumped) or eventually seep into the oceans. Water returns to the land surface at lower elevation than where it infiltrated, under the force of gravity or gravity induced pressures. Groundwater tends to move slowly, and is replenished slowly, so it can remain in aquifers for thousands of years.

Evaporation: The transformation of water from liquid to gas phases as it moves from the ground or bodies of water into the overlying atmosphere. The source of energy for evaporation is primarily solar radiation. Evaporation often implicitly includes transpiration from plants, though together they are specifically referred to as evapotranspiration. Total annual evapotranspiration amounts to approximately 5,05,000 km^3 (1,21,000 cu mi) of water, 4,34,000 km^3 (104,000 cu mi) of which evaporates from the oceans.

Sublimation: The state change directly from solid water (snow or ice) to water vapour.

Advection: The movement of water—in solid, liquid, or vapour states—through the atmosphere. Without advection, water that evaporated over the oceans could not precipitate over land.

Condensation: The transformation of water vapour to liquid water droplets in the air, creating clouds and fog.

Transpiration: The release of water vapour from plants and soil into the air. Water vapour is a gas that cannot be seen.

Residence Times

The residence time of a reservoir within the hydrologic cycle is the average time a water molecule will spend in that reservoir. It is a measure of the average age of the water in that reservoir.

Groundwater can spend over 10,000 years beneath earth's surface before leaving. Particularly old groundwater is called fossil water. Water stored in the soil remains there very briefly, because it is spread thinly across the earth, and is readily lost by evaporation, transpiration, stream flow, or groundwater recharge. After evaporating, the residence time in the atmosphere is about 9 days before condensing and falling to the earth as precipitation.

The major ice sheets—Antarctica and Greenland—store ice for very long periods. Ice from Antarctica has been reliably dated to 8,00,000 years before present, though the average residence time is shorter.

In hydrology, residence times can be estimated in two ways. The more common method relies on the principle of conservation of mass and assumes the amount of water in a given reservoir is roughly constant. With this method, residence times are estimated by dividing the volume of the reservoir by the rate by which water either enters or exits the reservoir. Conceptually, this is equivalent to timing how long it would take the reservoir to become filled from empty if no water were to leave (or how long it would take the reservoir to empty from full if no water were to enter).

An alternative method to estimate residence times, which is gaining in popularity for dating groundwater, is the use of isotopic techniques. This is done in the subfield of isotope hydrology.

Changes Over Time

The water cycle describes the processes that drive the movement of water throughout the hydrosphere. However, much more water is 'in storage' for long periods of time than is actually moving through the cycle. The storehouses for the vast majority of all water on earth are the oceans. It is estimated that of the 332,500,000 mi^3 (1,386,000,000 km^3) of the world's water supply, about 321,000,000 mi^3 (1,338,000,000 km^3) is stored in oceans, or about 95 per cent. It is also estimated that the oceans supply about 90 per cent of the evaporated water that goes into the water cycle.

During colder climatic periods more ice caps and glaciers form, and enough of the global water supply accumulates as ice to lessen the amounts in other parts of the water cycle. The reverse is true during warm periods. During the last ice age glaciers covered almost one-third of earth's land mass, with the result being that the oceans were about 400 ft (122 m) lower than today. During the last global 'warm spell', about 1,25,000 years ago, the seas were about 18 ft (5.5 m) higher than they are now. About three million years ago the oceans could have been up to 165 ft (50 m) higher.

The scientific consensus expressed in the 2007 Intergovernmental panel on climate change (IPCC) Summary for Policymakers is for the water cycle to continue to intensify throughout the 21st century, though this does not mean that precipitation will increase in all regions. In subtropical land areas — places that are already relatively dry — precipitation is projected to decrease during the 21st century, increasing the probability of drought. The drying is projected to be strongest near the poleward margins of the subtropics. Annual precipitation amounts are expected to increase in near-equatorial regions that tend to be wet in the present climate, and also at high latitudes. These large-scale patterns are present in nearly all of the climate model simulations conducted at several international research centers as part of the 4th Assessment of the IPCC.

Glacial retreat is also an example of a changing water cycle, where the supply of water to glaciers from precipitation cannot keep up with the loss of water from melting and sublimation. Glacial retreat since 1850 has been extensive.

Human activities that alter the water cycle include:

1. Agriculture.
2. Industry.
3. Alteration of the chemical composition of the atmosphere.
4. Construction of dams.
5. Deforestation and afforestation.
6. Removal of groundwater from wells.
7. Water abstraction from rivers.
8. Urbanisation.

Effects on Climate

The water cycle is powered from solar energy and 86 per cent of the global evaporation occurs from the oceans, reducing their temperature by evaporative cooling. Without the cooling, the effect of evaporation on the greenhouse effect would lead to a much higher surface temperature of 67°C (153°F), and a warmer planet.

Effects on Biogeochemical Cycling

While the water cycle is itself a biogeochemical cycle, flow of water over and beneath the earth is a key component of the cycling of other biogeochemicals. Runoff is responsible for almost all of the transport of eroded sediment and phosphorus from land to waterbodies. The salinity of the oceans is derived from erosion and transport of dissolved salts from the land. Cultural eutrophication of lakes is primarily due to phosphorus, applied in excess to agricultural fields in fertilisers, and then transported overland and down rivers. Both runoff and groundwater flow play significant roles in transporting nitrogen from the land to waterbodies. The dead zone at the outlet of the Mississippi River is a consequence of nitrates from fertiliser being carried off agricultural fields and funnelled down the river system to the Gulf of Mexico. Runoff also plays a part in the carbon cycle, again through the transport of eroded rock and soil.

Slow Loss Over Geological Time

The hydrodynamic wind within the upper portion of a planet's atmosphere allows light chemical elements such as hydrogen to move up to the exobase, the lower limit of the exosphere, where the gases can then reach escape velocity, entering outer space without impacting other particles of gas. This type of gas loss from a planet into space is known as planetary wind. Planets with hot lower atmospheres could result in humid upper atmospheres that accelerate the loss of hydrogen.

FISHES

Fishes are aquatic animals, cold-blooded animal. Their body is covered by scales and has two sets of paired or unpaired fins, one or two dorsal fins, an anal fin, and a tail fin; has jaws. Fish has a streamlined body that allows it to swim rapidly; extracts oxygen from the water using gills. They are found abundantly in sea or fresh water. Fishes are oviparous, they shed their eggs and the eggs are fertilised outside of the female's body by the male squirting milt onto or around them. Fish range in size from the 16 m (51 ft) whale shark to an 8 mm (just over ¼ of an inch) long stout infant fish.

Facts about fishes and water animals:

1. Sea turtles absorb a lot of salt from the sea water in which they live. They excrete excess salt from their eyes, so it often looks as though they're crying.
2. Only one out of a thousand baby sea turtles survives after hatching.
3. Prehistoric turtles may have weighed as much as 5000 pounds.
4. Turtles have no teeth.
5. As it gets older the cockle just adds another layer to its outer shell. To work out their age you can count the rings on them just like a tree.
6. A goldfish has a memory span of three seconds.
7. A male Angler fish attaches itself to a female and never lets go. Their vascular systems unite and the male becomes entirely dependent on the female's blood for nutrition.
8. Beavers can hold their breathe for 45 minutes under water.
9. The sailfish is the fastest swimmer, reaching 109 km/hr (68 mph).
10. The slowest fish is the Sea Horse, which moves along at about 0.016 km/hr (0.01 mph).
11. The heart of a blue whale is the size of a small car.
12. The tongue of a blue whale is as long as an elephant.
13. The largest jellyfish ever caught measured 2.3 m (7'6") across the bell with a tentacle of 36 m (120 ft) long.

14. Fish and insects do not have eyelids—their eyes are protected by a hardened lens.
15. The stickleback is one of the few fishes that builds a nest. The male, in his red breeding colours, makes a nest of weeds where the female lays her eggs. Then the male stays by the nest to guard the eggs until they hatch.

RAIN MEASUREMENT

Instrument: Rain Gauge (or Rain Gage)

Rainfall impacts all of us, from the lack of rain during times of drought to the dangers of flash floods when we receive too much rain too fast. For some interests—such as gardeners, farmers, and meteorologists—rain gauge data is critical information. For the rest of us, obtaining rain measurement information with a rain gauge is interesting and fun!

A rain gauge collects falling precipitation and funnels it to a rain measurement device. In a standard manual rain gage (one that has to be manually emptied), the rain measurement device is a graduated cylinder. In home weather stations, a self-emptying tipping bucket rain gauge is the most common type of rain gauge used for rain measurement. The reading is then transmitted to the inside console for recording/display of the information. This is done either wirelessly using radio waves (wireless rain gauge) or via electronic cabling. Stand alone wireless rain gauges are also available.

Rainfall rate is generally described as light, moderate or heavy. Light rainfall is considered less than 0.10 inches of rain per hour. Moderate rainfall measures 0.10 to 0.30 inches of rain per hour. Heavy rainfall is more than 0.30 inches of rain per hour. Rainfall amount is described as the depth of water reaching the ground, typically in inches or millimeters (25 mm equals one inch). An inch of rain is exactly that, water that is one inch deep. One inch of rainfall equals 4.7 gallons of water per square yard or 22,650 gallons of water per acre. There are other terms used to describe precipitation: type (rain, snow, etc.) intensity (light, moderate, or heavy), and character (showery, intermittent, or continuous). Meteorologists use these terms in their forecasts or an actual weather event, such as 'today's light rain showers resulted in six tenths of an inch of precipitation'.

HUMIDITY

Humidity is a term for the amount of water vapour in air, and can refer to any one of several measurements of humidity. Formally, humid air is not 'moist air' but a mixture of air and water vapour, and humidity is defined in terms of the water content of this mixture, called the Absolute humidity. In everyday usage, it commonly refers to relative humidity, expressed as a percent in weather forecasts and on household humidistats; it is so called because it measures the current absolute humidity relative to the maximum. Specific humidity is a ratio of the water vapour content of the mixture to the dry air content. The water vapour content of the mixture can be measured either as mass per volume or as a partial pressure, depending on the usage. In meteorology, humidity indicates the likelihood of precipitation, dew, or fog. High relative humidity reduces the effectiveness of sweating in cooling the body by reducing the rate of evaporation of moisture from the skin. This effect is calculated in a heat index table, used during summer weather.

Measurement of Humidity

There are various devices used to measure and regulate humidity. A device used to measure humidity is called a psychrometer or hygrometer. A humidistat is used to regulate the humidity of a building with a dehumidifier. These can be analogous to a thermometer and thermostat for temperature control.

Humidity is also measured on a global scale using remotely placed satellites. These satellites are able to detect the concentration of water in the troposphere at altitudes between 4 and 12 km. Satellites that can measure water vapour have sensors that are sensitive to infrared radiation. Water vapour specifically absorbs and re-radiates radiation in this spectral band. Satellite water vapour imagery plays an important role in monitoring climate conditions (like the formation of thunderstorms) and in the development of future weather forecasts.

Climate

While humidity itself is a climate variable, it also interacts strongly with other climate variables. The humidity is affected by winds and by rainfall. At the same time, humidity affects the energy budget and thereby influences temperatures in two major ways. First, water vapour in the atmosphere contains 'latent' energy. During transpiration or evaporation, this latent heat is removed from surface liquid, cooling the earth's surface. This is the biggest non-radiative cooling effect at the surface. It compensates for roughly 70 per cent of the average net radiative warming at the surface. Second, water vapour is the most important of all greenhouse gases. Water vapour, like a green lens that allows green light to pass through it but absorbs red light, is a 'selective absorber'. Along with other greenhouse gases, water vapour is transparent to most solar energy, as you can literally see. But it absorbs the infrared energy emitted (radiated) upward by the earth's surface, which is the reason that humid areas experience very little nocturnal cooling but dry desert regions cool considerably at night. This selective absorption causes the greenhouse effect. It raises the surface temperature substantially above its theoretical radiative equilibrium temperature with the sun, and water vapour is the cause of more of this warming than any other greenhouse gas.

The most humid cities on earth are generally located closer to the equator, near coastal regions. Cities in South and Southeast Asia are among the most humid, such as Kolkata and Chennai in India, the cities of Manila in the Philippines and Bangkok in Thailand: these places experience extreme humidity during their rainy seasons combined with warmth giving the feel of a lukewarm sauna. Darwin, Australia experiences an extremely humid wet season from December to April. Shanghai and Hong Kong in China also have an extreme humid period in their summer months. Kuala Lumpur and Singapore have very high humidity all year round because of their proximity to water bodies and the equator and overcast weather. Perfectly clear days are dependent largely upon the season in which one decides to travel. During the South-west and North-east Monsoon seasons (respectively, late May to September and November to March), expect heavy rains and a relatively high humidity post-rainfall. Outside the monsoon seasons, humidity is high (in comparison to countries North of the Equator), but completely sunny days abound. In cooler places such as Northern Tasmania, Australia, high humidity is experienced all year due to the ocean between mainland Australia and Tasmania. In the summer the hot dry air is absorbed by this ocean and the temperature rarely climbs above 35°C.

Cloud, Mist and Fog

A **cloud** is a visible mass of water droplets or frozen ice crystals suspended in the atmosphere above the surface of the earth or another planetary body. Clouds in the earth's atmosphere are studied in the nephology or cloud physics branch of meteorology. Two processes, possibly acting together, can lead to air becoming saturated: cooling the air or adding water vapour to the air. Generally, precipitation will fall to the surface; an exception is virga which evaporates before reaching the surface. Clouds can show convective development like cumulus, be in the form of layered sheets such as stratus, or appear in thin fibrous wisps as with cirrus. Prefixes are used in connection with clouds: strato for low cumulus-category clouds that show some stratiform characteristics, nimbo for low to middle stratiform clouds that can produce moderate to heavy precipitation, alto for middle clouds, and cirro for high clouds. Whether or not a cloud is low, middle, or high level depends on how far above the ground its base forms. Some cloud types, especially those with significant vertical extent, can form in the low or middle ranges depending on the moisture content of the air. Clouds have Latin names due to the popular adaptation of Luke Howard's cloud categorisation system, which began to spread in popularity during December 1802. Synoptic surface weather observations use code numbers for the types of tropospheric cloud visible at each scheduled observation time based on the height and physical appearance of the clouds. While a majority of clouds form in the earth's troposphere, there are occasions where clouds in the stratosphere and mesosphere are observed. Clouds have been observed on other planets and moons within the Solar system, but due to their different temperature characteristics, they are composed of other substances such as methane, ammonia, or sulphuric acid.

Mist: Mist is a phenomenon of small droplets suspended in air. It can occur as part of natural weather or volcanic activity, and is common in cold air above warmer water, in exhaled air in the cold, and in a steam room of a sauna. It can also be created artificially with aerosol canisters if the humidity conditions are right. Mist makes a beam of light visible from the side via refraction and reflection on the suspended water droplets. Mist usually occurs near the shores, and is often associated with fog. Mist can be as high as mountain tops when extreme temperatures are low.

Freezing mist: Freezing mist is similar to freezing fog, only the density is less and the visibility greater. When mist falls below 0°C in temperature it becomes known as freezing mist.

Fog: Fog is a collection of water droplets or ice crystals suspended in the air at or near the earth's surface. While fog is a type of a cloud, the term 'fog' is typically distinguished from the more generic term 'cloud' in that fog is low-lying, and the moisture in the fog is often generated locally (such as from a nearby body of water, like a lake or the ocean, or from nearby moist ground or marshes).

Fog is distinguished from mist only by its density, as expressed in the resulting decrease in visibility: Fog reduces visibility to less than 1 km (5/8 statute mile), whereas mist reduces visibility to no less than 1 km (5/8 statute mile). For aviation purposes in the UK, a visibility of less than 2 km but greater than 999 m is considered to be mist if the relative humidity is 95 per cent or greater—below 95 per cent haze is reported.

FROST

Frost is the solid deposition of water vapour from saturated air. It is formed when solid surfaces are cooled to below the dew point of the adjacent air as well as below the freezing point of water. Frost crystals' size differ depending on time and water vapour available. Frost is also usually translucent in appearance. There are many types of frost, such as radiation and window frost. Frost causes economic damage when it destroys plants or hanging fruits.

If a solid surface is chilled below the dew point of the surrounding air and the surface itself is colder than freezing, frost will form on the surface. Frost consists of spicules of ice which grow out from the solid surface. The size of the crystals depends on time, temperature, and the amount of water vapour available. Based on wind direction, 'Frost arrows' might form.

In general, for frost to form the deposition surface must be colder than the surrounding air. For instance frost may be observed around cracks in cold wooden sidewalks when moist air escapes from the ground below. Other objects on which frost tends to form are those with low specific heat or high thermal emissivity, such as blackened metals; hence the accumulation of frost on the heads of rusty nails. The apparently erratic occurrence of frost in adjacent localities is due partly to differences of elevation, the lower areas becoming colder on calm nights. It is also affected by differences in absorptivity and specific heat of the ground which in the absence of wind greatly influences the temperature attained by the superincumbent air.

DEW

Dew is water in the form of droplets that appears on thin, exposed objects in the morning or evening. As the exposed surface cools by radiating its heat, atmospheric moisture condenses at a rate greater than that at which it can evaporate, resulting in the formation of water droplets. When temperatures are low enough, dew takes the form of ice; this form is called frost (frost is, however, not frozen dew). Because dew is related to the temperature of surfaces, in late summer it is formed most easily on surfaces which are not warmed by conducted heat from deep ground, such as grass, leaves, railings, car roofs, and bridges. Dew should not be confused with guttation, which is the process by which plants release excess water from the tips of their leaves.

Formation of Dew

Water vapour will condense into droplets depending on the temperature. The temperature at which droplets can form is called the Dew Point. When surface temperature drops, eventually reaching the dew point, atmospheric water vapour condenses to form small droplets on the surface. This process distinguishes dew from those hydrometeors (meteorological occurrences of water) which are formed directly in air cooling to its dew point (typically around condensation nuclei) such as fog or clouds. The thermodynamic principles of formation, however, are virtually the same.

Measurement of Dew

A classical device for dew measurement is the drosometer. A small, artificial condenser surface is suspended from an arm attached to a pointer or a pen that records the weight changes of the condenser on a drum. Besides being very wind sensitive, however, this, like all artificial surface devices, only provides a measure of the meteorological potential for dew formation. The actual amount of dew in a specific place is strongly dependent on surface properties. For its measurement, plants, leaves, or whole soil columns are placed on a balance with their surface at the same height and in the same surroundings as would occur naturally, thus providing a small lysimeter. Further methods include estimation by means of comparing the droplets to standardised photographs, or volumetric measurement of the amount of water wiped from the surface. It has to be kept in mind that some of these methods include guttation, while others only measure dewfall and/or distillation.

Significance of Dew

Due to its dependence on radiation balance, dew amounts can reach a theoretical maximum of about 0.8 mm per night, measured values, however, rarely exceeding 0.5 mm. In most climates of the world, the annual average is too small to compete with rain. In regions with considerable dry seasons, adapted plants like lichen or pine seedlings benefit from dew. Large-scale, natural irrigation without rainfall, such as in the Atacama Desert and Namib desert, however, is mostly attributed to fog water. Another effect of dew on plants is its role as a habitat for pathogens such as the fungus Phytophthora infestans which infects potato plants.

Chapter 23

Soil Ecology

INTRODUCTION

Soil is a natural body consisting of layers (soil horizons) of mineral constituents of variable thicknesses, which differ from the parent materials in their morphological, physical, chemical, and mineralogical characteristics. It is composed of particles of broken rock that have been altered by chemical and environmental processes that include weathering and erosion. Soil differs from its parent rock due to interactions between the lithosphere, hydrosphere, atmosphere, and the biosphere. It is a mixture of mineral and organic constituents that are in solid, gaseous and aqueous states.

Soil particles pack loosely, forming a soil structure filled with pore spaces. These pores contain soil solution (liquid) and air (gas). Accordingly, soils are often treated as a three state system. Most soils have a density between 1 and 2 g/cm^3. Soil is also known as earth: it is the substance from which our planet takes its name. Little of the soil composition of planet earth is older than the Tertiary and most no older than the Pleistocene. In engineering, soil is referred to as regolith or loose rock material.

SOIL FORMING FACTORS

Soil formation, or pedogenesis, is the combined effect of physical, chemical, biological, and anthropogenic processes on soil parent material. Soil genesis involves processes that develop layers or horizons in the soil profile. These processes involve additions, losses, transformations and translocations of material that compose the soil. Minerals derived from weathered rocks undergo changes that cause the formation of secondary minerals and other compounds that are variably soluble in water, these constituents are moved (translocated) from one area of the soil to other areas by water and animal activity. The alteration and movement of materials within soil causes the formation of distinctive soil horizons.

The weathering of bedrock produces the parent material from which soils form. An example of soil development from bare rock occurs on recent lava flows in warm regions under heavy and very frequent rainfall. In such climates, plants become established very quickly on basaltic lava, even though there is very little organic material. The plants are supported by the porous rock as it is filled with nutrient-bearing water which carries, for example, dissolved minerals and guano. The developing plant roots, themselves or associated with mycorrhizal fungi, gradually break up the porous lava and organic matter soon accumulates.

But even before it does, the predominantly porous broken lava in which the plant roots grow can be considered a soil. How the soil 'life' cycle proceeds is influenced by at least five classic soil forming factors that are dynamically intertwined in shaping the way soil is developed, they include: parent material, regional climate, topography, biotic potential and the passage of time.

Parent Material

The material from which soils form is called parent material. It includes: weathered primary bedrock; secondary material transported from other locations, e.g. colluvium and alluvium; deposits that are already present but mixed or altered in other ways—old soil formations, organic material including peat or alpine humus; and anthropogenic materials, like landfill or mine waste. Few soils form directly from the breakdown of the underlying rocks they develop on. These soils are often called 'residual soils', and have the same general chemistry as their parent rocks. Most soils derive from materials that have been transported from other locations by wind, water and gravity.

Weathering is the first stage in the transforming of parent material into soil material. In soils forming from bedrock, a thick layer of weathered material called saprolite may form. Saprolite is the result of weathering processes that include: hydrolysis (the replacement of a mineral's cations with hydrogen ions), chelation from organic compounds, hydration (the absorption of water by minerals), solution of minerals by water, and physical processes that include freezing and thawing or wetting and drying. The mineralogical and chemical composition of the primary bedrock material, plus physical features, including grain size and degree of consolidation, plus the rate and type of weathering, transforms it into different soil materials.

Climate

Soil formation greatly depends on the climate, and soils from different climate zones show distinctive characteristics. Temperature and moisture affect weathering and leaching. Wind moves sand and other particles, especially in arid regions where there is little plant cover. The type and amount of precipitation influence soil formation by affecting the movement of ions and particles through the soil, aiding in the development of different soil profiles. Seasonal and daily temperature fluctuations affect the effectiveness of water in weathering parent rock material and affect soil dynamics. The cycle of freezing and thawing is an effective mechanism to break up rocks and other consolidated materials. Temperature and precipitation rates affect biological activity, rates of chemical reactions and types of vegetation cover.

Biological Factors

Plants, animals, fungi, bacteria and humans affect soil formation. Animals and micro-organisms mix soils to form burrows and pores allowing moisture and gases to seep into deeper layers. In the same way, plant roots open channels in the soils, especially plants with deep taproots which can penetrate many meters through the different soil layers to bring up nutrients from deeper in the soil. Plants with fibrous roots that spread out near the soil surface, have roots that are easily decomposed, adding organic matter. Micro-organisms, including fungi and bacteria, affect chemical exchanges between roots and soil and act as a reserve of nutrients. Humans can impact soil formation by removing vegetation cover; this removal promotes erosion.

Vegetation impacts soils in numerous ways. It can prevent erosion from rain or surface runoff. It shades soils, keeping them cooler and slowing evaporation of soil moisture, or it can cause soils to dry out by transpiration. Plants can form new chemicals that break down or build up soil particles. Vegetation depends on climate, land form topography and biological factors. Soil factors such as soil density, depth, chemistry, pH, temperature and moisture greatly affect the type of plants that can grow in a given location. Dead plants, dropped leaves and stems of plants fall to the surface of the soil and decompose. The organisms feed on them and mix the organic material with the upper soil layers; these organic compounds become part of the soil formation process, ultimately shaping the type of soil formed.

Time

Time is a factor in the interactions of all the above factors as they develop soil. Over time, soils evolve features dependent on the other forming factors, and soil formation is a time-responsive process dependent on how the other factors interplay with each other. Soil is always changing. For example, recently-deposited material from a flood exhibits no soil development because there has not been enough time for soil-forming activities. With additions, removals and alterations, soils are always subject to new conditions. Whether these are slow or rapid changes depend on climate, landscape position and biological activity.

Characteristics

Soil colour is often the first impression one has when viewing soil. Striking colours and contrasting patterns are especially memorable. Soil colour is primarily influenced by soil mineralogy. Many soil colours are due to the extensive and various iron minerals. The development and distribution of colour in a soil profile result from chemical and biological weathering, especially redox reactions. Aerobic conditions produce uniform or gradual colour changes, while reducing environments result in disrupted colour flow with complex, mottled patterns and points of colour concentration.

Soil structure is the arrangement of soil particles into aggregates. These may have various shapes, sizes and degrees of development or expression. Soil structure affects aeration, water movement, resistance to erosion and plant root growth. Structure often gives clues to texture, organic matter content, biological activity, past soil evolution, human use, and chemical and mineralogical conditions under which the soil formed.

Soil texture refers to sand, silt and clay composition. Soil content affects soil behaviour, including the retention capacity for nutrients and water. Sand and silt are the products of physical weathering, while clay is the product of chemical weathering. Clay content has retention capacity for nutrients and water. Clay soils resist wind and water erosion better than silty and sandy soils, because the particles are more tightly joined to each other. In medium-textured soils, clay is often translocated downward through the soil profile and accumulates in the subsoil.

SOIL HORIZONS

A soil horizon is a specific layer in the land area that is parallel to the soil surface and possesses physical characteristics which differ from the layers above and beneath. Horizon formation (horizonation) is a function of a range of geological, chemical, and biological processes and occurs over long time periods. Soils vary in the degree to which horizons are expressed. Relatively new deposits of soil parent material, such as alluvium, sand dunes, or volcanic ash, may have no horizon formation, or only the distinct layers of deposition. As age increases, horizons generally are more easily observed. The exception occurs in some older soils, with few horizons expressed in deeply weathered soils, such as the oxisols in tropical areas with high annual precipitation.

The term 'horizon' describes each of the distinctive layers that occur in a soil. Each soil type has at least one, usually three or four horizons and these are described by soil scientists when seeking to classify soils (Soil-Net). Horizons are defined in most cases by obvious physical features, colour and texture being chief among them. These may be described both in absolute terms (particle size distribution for texture, for instance) and in terms relative to the surrounding material, i.e. 'coarser' or 'sandier' than the horizons above and below. Soil generally consists of visually and texturally distinct layers, which are shown in Fig. 23.1.

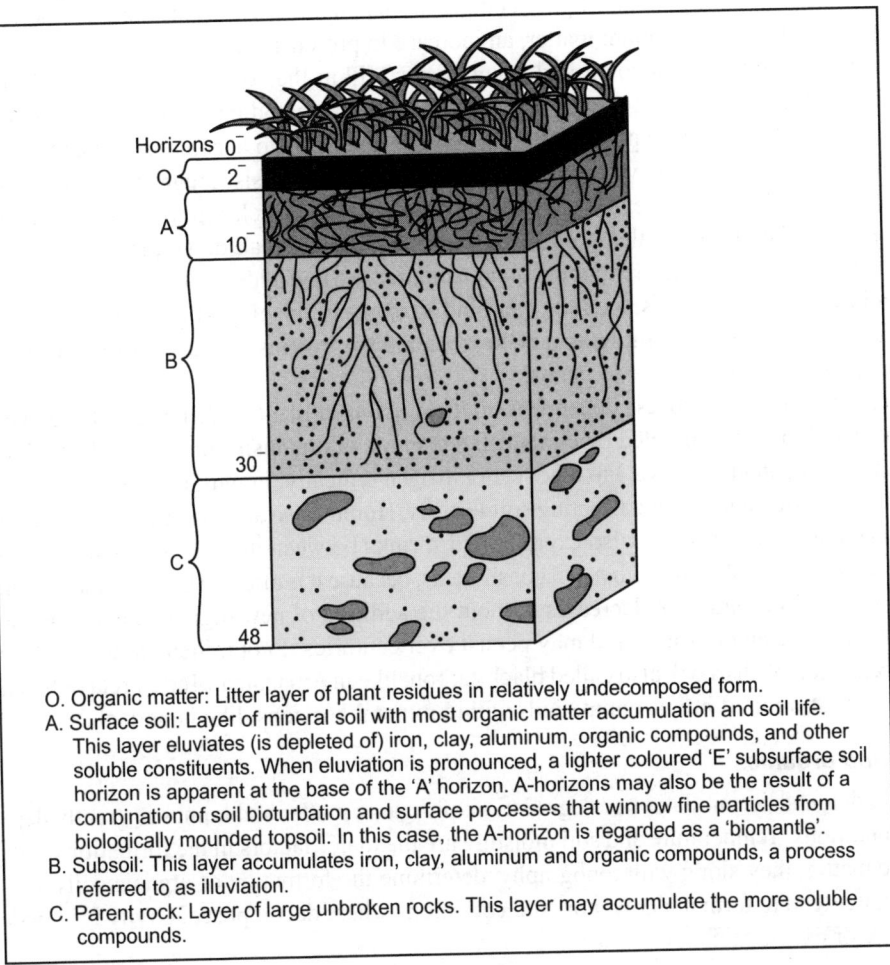

O. Organic matter: Litter layer of plant residues in relatively undecomposed form.
A. Surface soil: Layer of mineral soil with most organic matter accumulation and soil life. This layer eluviates (is depleted of) iron, clay, aluminum, organic compounds, and other soluble constituents. When eluviation is pronounced, a lighter coloured 'E' subsurface soil horizon is apparent at the base of the 'A' horizon. A-horizons may also be the result of a combination of soil bioturbation and surface processes that winnow fine particles from biologically mounded topsoil. In this case, the A-horizon is regarded as a 'biomantle'.
B. Subsoil: This layer accumulates iron, clay, aluminum and organic compounds, a process referred to as illuviation.
C. Parent rock: Layer of large unbroken rocks. This layer may accumulate the more soluble compounds.

Fig. 23.1. Various layers of soil.

Organic Matter

Most living things in soils, including plants, insects, bacteria and fungi, are dependent on organic matter for nutrients and energy. Soils often have varying degrees of organic compounds in different states of decomposition. Many soils, including desert and rocky-gravel soils, have no or little organic matter. Soils that are all organic matter, such as peat (histosols), are infertile.

Humus

Humus refers to organic matter that has decomposed to a point where it is resistant to further breakdown or alteration. Humic acids and fulvic acids are important constituents of humus and typically form from plant residues like foliage, stems and roots. After death, these plant residues begin to decay, starting the formation of humus. Humus formation involves changes within the soil and plant residue, there is a reduction of water soluble constituents including cellulose and hemicellulose; as the residues are deposited and break down, humin, lignin and lignin complexes accumulate within the soil; as micro-organisms live and feed on the decaying plant matter, an increase in proteins occurs.

Lignin is resistant to breakdown and accumulates within the soil; it also chemically reacts with amino acids which add to its resistance to decomposition, including enzymatic decomposition by microbes. Fats and waxes from plant matter have some resistance to decomposition and persist in soils for a while. Clay soils often have higher organic contents that persist longer than soils without clay. Proteins normally decompose readily, but when bound to clay particles they become more resistant to decomposition. Clay particles also absorb enzymes that would break down proteins. The addition of organic matter to clay soils, can render the organic matter and any added nutrients inaccessible to plants and microbes for many years, since they can bind strongly to the clay. High soil tannin (polyphenol) content from plants can cause nitrogen to be sequestered by proteins or cause nitrogen immobilisation, also making nitrogen unavailable to plants.

Humus formation is a process dependent on the amount of plant material added each year and the type of base soil; both are affected by climate and the type of organisms present. Soils with humus can vary in nitrogen content but have 3 to 6 per cent nitrogen typically; humus, as a reserve of nitrogen and phosphorus, is a vital component affecting soil fertility. Humus also absorbs water, acting as a moisture reserve, that plants can utilise; it also expands and shrinks between dry and wet states, providing pore spaces. Humus is less stable than other soil constituents, because it is affected by microbial decomposition, and over time its concentration decreases without the addition of new organic matter. However, some forms of humus are highly stable and may persist over centuries if not millennia: they are issued from the slow oxidation of charcoal, also called black carbon, like in Amazonian Terra preta or black earths,or from the sequestration of humic compounds within mineral horizons, like in podzols.

Climate and organics

The production and accumulation or degradation of organic matter and humus is greatly dependent on climate conditions. Temperature and soil moisture are the major factors in the formation or degradation of organic matter, they along with topography, determine the formation of organic soils. Soils high in organic matter tend to form under wet or cold conditions where decomposer activity is impeded by low temperature or excess moisture.

Soil Solutions

Soils retain water that can dissolve a range of molecules and ions. These solutions exchange gases with the soil atmosphere, contain dissolved sugars, fulvic acids and other organic acids, plant nutrients such as nitrate, ammonium, potassium, phosphate, sulphate and calcium, and micronutrients such as zinc, iron and copper. Some arid soils have sodium solutions that greatly impact plant growth. Soil pH can affect the type and amount of anions and cations that soil solutions contain and that exchange with the soil atmosphere and biological organisms.

Geologists also have a particular interest in the patterns of soil on the surface of the earth. Soil texture, colour and chemistry often reflect the underlying geologic parent material, and soil types often change at geologic unit boundaries. Buried paleosols mark previous land surfaces and record climatic conditions from previous eras. Geologists use this paleopedological record to understand the ecological relationships in past ecosystems. According to the theory of biorhexistasy, prolonged conditions conducive to forming deep, weathered soils result in increasing ocean salinity and the formation of limestone.

Geologists use soil profile features to establish the duration of surface stability in the context of geologic faults or slope stability. An offset subsoil horizon indicates rupture during soil formation and the degree of subsequent subsoil formation is relied upon to establish time since rupture.

Soil examined in shovel test pits is used by archaeologists for relative dating based on stratigraphy. What is considered most typical is to use soil profile features to determine the maximum reasonable pit depth than needs to be examined for archaeological evidence in the interest of cultural resources management. Soils altered or formed by man (anthropic and anthropogenic soils) are also of interest to archaeologists, such as terra preta soils.

Uses of Soil

Soil is used in agriculture, where it serves as the primary nutrient base for plants; however, as demonstrated by hydroponics, it is not essential to plant growth if the soil-contained nutrients could be dissolved in a solution. The types of soil used in agriculture (among other things, such as the purported level of moisture in the soil) vary with respect to the species of plants that are cultivated.

Soil material is a critical component in the mining and construction industries. Soil serves as a foundation for most construction projects. Massive volumes of soil can be involved in surface mining, road building and dam construction. Earth sheltering is the architectural practice of using soil for external thermal mass against building walls.

Soil resources are critical to the environment, as well as to food and fibre production. Soil provides minerals and water to plants. Soil absorbs rainwater and releases it later, thus preventing floods and drought. Soil cleans the water as it percolates. Soil is the habitat for many organisms: the major part of known and unknown biodiversity is in the soil, in the form of invertebrates (earthworms, woodlice, millipedes, centipedes, snails, slugs, mites, springtails, enchytraeids, nematodes, protists), bacteria, archaea, fungi and algae; and most organisms living above ground have part of them (plants) or spend part of their life cycle (insects) below ground. Above-ground and below-ground biodiversities are tightly interconnected, making soil protection of paramount importance for any restoration or conservation plan.

Degradation

Land degradation is a human-induced or natural process which impairs the capacity of land to function. Soils are the critical component in land degradation when it involves acidification, contamination, desertification, erosion or salination.

While soil acidification of alkaline soils is beneficial, it degrades land when soil acidity lowers crop productivity and increases soil vulnerability to contamination and erosion. Soils are often initially acid because their parent materials were acid and initially low in the basic cations (calcium, magnesium, potassium and sodium). Acidification occurs when these elements are removed from the soil profile by normal rainfall, or the harvesting of forest or agricultural crops. Soil acidification is accelerated by the use of acid-forming nitrogenous fertilisers and by the effects of acid precipitation.

Soil contamination at low levels is often within soil capacity to treat and assimilate. Many waste treatment processes rely on this treatment capacity. Exceeding treatment capacity can damage soil biota and limit soil function. Derelict soils occur where industrial contamination or other development activity damages the soil to such a degree that the land cannot be used safely or productively. Remediation of derelict soil uses principles of geology, physics, chemistry and biology to degrade, attenuate, isolate or remove soil contaminants to restore soil functions and values. Techniques include leaching, air sparging, chemical amendments, phytoremediation, bioremediation and natural attenuation.

Desertification is an environmental process of ecosystem degradation in arid and semi-arid regions, often caused by human activity. It is a common misconception that droughts cause desertification. Droughts are common in arid and semiarid lands. Well-managed lands can recover from drought when the rains return. Soil management tools include maintaining soil nutrient and organic matter levels, reduced tillage and increased cover. These practices help to control erosion and maintain productivity during periods when moisture is available. Continued land abuse during droughts, however, increases land degradation. Increased population and livestock pressure on marginal lands accelerates desertification.

Soil erosional loss is caused by wind, water, ice and movement in response to gravity. Although the processes may be simultaneous, erosion is distinguished from weathering. Erosion is an intrinsic natural process, but in many places it is increased by human land use. Poor land use practices including deforestation, overgrazing and improper construction activity. Improved management can limit erosion by using techniques like limiting disturbance during construction, avoiding construction during erosion prone periods, intercepting runoff, terrace-building, use of erosion-suppressing cover materials, and planting trees or other soil binding plants.

Soil piping is a particular form of soil erosion that occurs below the soil surface. It is associated with levee and dam failure, as well as sink hole formation. Turbulent flow removes soil starting from the mouth of the seep flow and subsoil erosion advances upgradient. The term sand boil is used to describe the appearance of the discharging end of an active soil pipe.

Soil salination is the accumulation of free salts to such an extent that it leads to degradation of soils and vegetation. Consequences include corrosion damage, reduced plant growth, erosion due to loss of plant cover and soil structure, and water quality problems due to sedimentation. Salination occurs due to a combination of natural and human caused processes. Arid conditions favour salt accumulation. This is especially apparent when soil parent material is saline. Irrigation of arid lands is especially problematic. All irrigation water has some level of salinity. Irrigation, especially when it involves leakage from canals and overirrigation in the field, often raises the underlying water table. Rapid salination occurs when the land surface is within the capillary fringe of saline groundwater. Soil salinity control involves watertable control and flushing with higher levels of applied water in combination with tile drainage or another form of subsurface drainage. Soil salinity models like SWAP, DrainMod-S, UnSatChem, SaltMod and SahysMod are used to assess the cause of soil salination and to optimise the reclamation of irrigated saline soils.

IMPORTANT COMPONENTS OF SOIL

pH

One of the most important components of soil is the pH. The pH of soil can be modified by adding different chemicals. Soil pH indicates how acid or alkaline the soil is. The pH scale ranges from 0 to 14. Any substance with a pH near the lower end of the scale is very acidic. Substances in the upper range of the scale have a high alkalinity or are very basic. The pH of a soil is crucial because crops grow best in a narrow pH range which can vary among crops. For example, blueberries and a few types of flowers grow best when the pH is 5.5 or less. Potatoes, a more familiar crop, grow best with a soil pH range of

5.5 to 6.0. Most garden vegetables, shrubs, trees and lawns grow best when the soil pH is over 6.0 or 6.5. The range between 5.5 and 7.5 is favourable for two reasons. It allows sufficient micro-organisms to break down organic matter. It is also the best range for nutrient availability.

Liming

When farmers originally cleared the lands on the Island, the soil quality was adequate for growing potatoes. However, potatoes were not the only crop at that time. Therefore, farmers needed a way to increase the pH of the soil to make it suitable for other crops. The pH of soil can be increased by liming. This is why people sometimes spread white powder on their lawns or gardens. This white powder is lime. Calcitic limestone ($CaCO_3$) provides a good three source of Calcium (Ca) and helps neutralise soil acidity.

SOIL TEXTURE

Soil texture is an important soil characteristic that drives crop production and field management. The textural class of a soil is determined by the percentage of sand, silt, and clay. Soils can be classified as one of four major textural classes: (i) sands, (ii) silts, (iii) loams, and (iv) clays.

Soil texture determines the rate at which water drains through a saturated soil; water moves more freely through sandy soils than it does through clayey soils. Once field capacity is reached, soil texture also influences how much water is available to the plant; clay soils have a greater water holding capacity than sandy soils. In addition, well drained soils typically have good soil aeration meaning that the soil contains air that is similar to atmospheric air, which is conducive to healthy root growth, and thus a healthy crop. Soils also differ in their susceptibility to erosion (erodibility) based on texture; a soil with a high percentage of silt and clay particles has a greater erodibility than a sandy soil under the same conditions. Differences in soil texture also impacts organic matter levels; organic matter breaks down faster in sandy soils than in fine-textured soils, given similar environmental conditions, tillage and fertility management, because of a higher amount of oxygen available for decomposition in the light-textured sandy soils. The cation exchange capacity of the soil increases with percent clay and organic matter and the pH buffering capacity of a soil (its ability to resist pH change upon lime addition), is also largely based on clay and organic matter content. Soil tilth (how easily or difficult a field is tilled) is influenced by texture, soil moisture, aeration, and organic matter as well.

SOIL STRUCTURE

Soil structure is determined by how individual soil granules clump or bind together and aggregate, and therefore, the arrangement of soil pores between them. Soil structure has a major influence on water and air movement, biological activity, root growth and seedling emergence.

Soil structure describes the arrangement of the solid parts of the soil and of the pore space located between them. It is dependent on: what the soil developed from; the environmental conditions under which the soil formed; the clay present, the organic materials present; and the recent history of management.

Aggregation of primary soil particles is a critical determinant of soil structure. Clay colloids - minute particles (diameters smaller than 2 micrometres) play a significant role in aggregation between the full range of soil particles. Adhesion between particles is via electrostatic force (flocculation) or cementing substances, such as organic matter and minerals. Other factors important in considering soil structure are: the stability of aggregates under wetting and drying conditions; the stability of aggregates to physical disturbance; the fabric and nature of the aggregates.

There are five major classes of structure seen in soils: platy, prismatic, columnar, granular, and blocky. There are also structureless conditions. Some soils have simple structure, each unit being an entity without component smaller units. Others have compound structure, in which large units are composed of smaller units separated by persistent planes of weakness.

CLASSIFICATION OF SOIL WATER

Soil water has been classified from a physical and biological point of view.

Physical Classification of Soil Water

Gravitational water: Gravitational water occupies the larger soil pores (macro pores) and moves down readily under the force of gravity. Water in excess of the field capacity is termed gravitational water. Gravitational water is of no use to plants because it occupies the larger pores. It reduces aeration in the soil. Thus, its removal from soil is a requisite for optimum plant growth. Soil moisture tension at gravitational state is zero or less than 1/3 atmosphere.

Factors Affecting Gravitational Water

Texture

Plays a great role in controlling the rate of movement of gravitational water. The flow of water is proportional to the size of particles. The bigger the particle, the more rapid is the flow or movement. Because of the larger size of pore, water percolates more easily and rapidly in sandy soils than in clay soils.

Structure

It also affects gravitational water. In platy structure movement of gravitational water is slow and water stagnates in the soil. Granular and crumby structure helps to improve gravitational water movement. In clay soils having single grain structure, the gravitational water, percolates more slowly. If clay soils form aggregates (granular structure), the movement of gravitational water improves.

Capillary water

Capillary water is held in the capillary pores (micro pores). Capillary water is retained on the soil particles by surface forces. It is held so strongly that gravity cannot remove it from the soil particles. The molecules of capillary water are free and mobile and are present in a liquid state. Due to this reason, it evaporates easily at ordinary temperature though it is held firmly by the soil particle; plant roots are able to absorb it. Capillary water is, therefore, known as available water. The capillary water is held between 1/3 and 31 atmosphere pressure.

Hygroscopic Water

The water that held tightly on the surface of soil colloidal particle is known as hygroscopic water. It is essentially non-liquid and moves primarily in the vapour form.

Hygroscopic water held so tenaciously (31 to 10000 atmospheres) by soil particles that plants cannot absorb it. Some micro-organism may utilise hygroscopic water. As hygroscopic water is held tenaciously by surface forces its removal from the soil requires a certain amount of energy. Unlike capillary water which evaporates easily at atmospheric temperature, hygroscopic water cannot be separated from the soil unless it is heated.

SOIL ORGANISMS

Heterotroph

A heterotroph is an organism that cannot fix carbon and uses organic carbon for growth. This contrasts with autotrophs, such as plants and algae, which can use energy from sunlight (photoautotrophs) or inorganic compounds (lithoautotrophs) to produce organic compounds such as carbohydrates, fats, and proteins from inorganic carbon dioxide. These reduced carbon compounds can be used as an energy source by the autotroph and provide the energy in food consumed by heterotrophs. Ninety-five per cent or more of all types of living organisms are heterotrophic. Heterotrophs, by consuming reduced carbon compounds, are able to use all the energy that they obtain from food for growth and reproduction, unlike autotrophs, which must use some of their energy for carbon fixation. Heterotrophs are unable to make their own food, however, and whether using organic or inorganic energy sources, they can die from a lack of food. This applies not only to animals and fungi but also to bacteria.

NUTRIENTS OF SOILS

A nutrient is a chemical that an organism needs to live and grow or a substance used in an organism's metabolism which must be taken in from its environment. Nutrients are the substances that enrich the body. They are used to build and repair tissues, regulate body processes. Methods for nutrient intake vary, with animals and protists consuming foods that are digested by an internal digestive system, but most plants ingest nutrients directly from the soil through their roots or from the atmosphere. An exception are the carnivorous plants, which externally digest nutrients from animals, before ingesting them. The effects of nutrients are dose-dependent. Organic nutrients include carbohydrates, fats, proteins, and vitamins. Inorganic chemical compounds such as dietary minerals, water, and oxygen may also be considered nutrients. A nutrient is essential if it must be obtained from an external source, either because the organism cannot synthesised it or produces insufficient quantities. Nutrients needed in very small amounts are micronutrients and those that are needed in larger quantities are called macronutrients. The effects of nutrients are dose-dependent and shortages are called deficiencies.

TYPES OF NUTRIENT

Micronutrients

Micronutrients are required only in very small quantities. Their concentrations in plant tissues are only a small proportion of the concentrations of macronutrients.

Micronutrients play many complex roles in plant nutrition, but most of them are used in the functioning of a number of enzyme systems. However, there is considerable variation in the specific functions of the various micronutrients in plants and in microbial growth processes. For example, copper, iron, and molybdenum are an essential part of the complex reactions which make up photosynthesis and many other metabolic processes. Zinc and manganese function in many plant enzyme systems as bridges. They connect the enzyme with the substrate upon which it is meant to act.

Macronutrients

Macronutrients is defined in several different ways:
1. The chemical elements humans consume in the largest quantities are carbon, hydrogen, nitrogen, oxygen, phosphorus, and sulphur.

2. The classes of chemical compounds humans consume in the largest quantities and which provide bulk energy is carbohydrates, proteins, and fats. Water and atmospheric oxygen also must be consumed in large quantities, but are not always considered 'food' or 'nutrients'.

3. Calcium, salt (sodium and chloride), magnesium, and potassium (along with phosphorus and sulphur) are sometimes added to the list of macronutrients because they are required in large quantities compared to other vitamins and minerals. They are sometimes referred to as the macrominerals.

The remaining vitamins, minerals, fats or elements, are called micronutrients because they are required in relatively small quantities.

ECOLOGY OF SOIL

Most heterotrophs are chemoorganoheterotrophs (or simply organotrophs) and utilise organic compounds both as a carbon source and an energy source. The term 'heterotroph' very often refers to chemoorgano-heterotrophs. Heterotrophs function as consumers in food chains: they obtain organic carbon by eating other heterotrophs or autotrophs. They break down complex organic compounds (e.g. carbohydrates, fats, and proteins) produced by autotrophs into simpler compounds (e.g. carbohydrates into glucose, fats into fatty acids and glycerol, and proteins into amino acids). They release energy by oxidising carbon and hydrogen atoms present in carbohydrates, lipids, and proteins to carbon dioxide and water, respectively.

All animals and fungi are heterotrophic, as well as most protists and prokaryotes. Some animals, such as corals, form symbiotic relationships with autotrophs and obtain organic carbon in this way. Furthermore, some parasitic plants have also turned fully or partially heterotrophic, while carnivorous plants consume animals to augment their nitrogen supply while remaining autotrophic.

Ecosystem Effects

Not unexpectedly, soil contaminants can have significant deleterious consequences for ecosystems. There are radical soil chemistry changes which can arise from the presence of many hazardous chemicals even at low concentration of the contaminant species. These changes can manifest in the alteration of metabolism of endemic micro-organisms and arthropods resident in a given soil environment. The result can be virtual eradication of some of the primary food chain, which in turn have major consequences for predator or consumer species. Even if the chemical effect on lower life forms is small, the lower pyramid levels of the food chain may ingest alien chemicals, which normally become more concentrated for each consuming rung of the food chain. Many of these effects are now well known, such as the concentration of persistent DDT materials for avian consumers, leading to weakening of egg shells, increased chick mortality and potential extinction of species.

Effects occur to agricultural lands which have certain types of soil contamination. Contaminants typically alter plant metabolism, most commonly to reduce crop yields. This has a secondary effect upon soil conservation, since the languishing crops cannot shield the earth's soil mantle from erosion phenomena. Some of these chemical contaminants have long half-lives and in other cases derivative chemicals are formed from decay of primary soil contaminants.

Chapter 24

Soil and Water Conservation

INTRODUCTION

Soil erosion is one form of soil degradation along with soil compaction, low organic matter, loss of soil structure, poor internal drainage, salinisation, and soil acidity problems. These other forms of soil degradation, serious in themselves, usually contribute to accelerated soil erosion. Soil erosion is a naturally occurring process on all land. The agents of soil erosion are water and wind, each contributing a significant amount of soil loss each year in Ontario. Soil erosion may be a slow process that continues relatively unnoticed or it may occur at an alarming rate causing serious loss of topsoil. The loss of soil from farmland may be reflected in reduced crop production potential, lower surface water quality and damaged drainage networks.

EROSION BY WATER

The rate and magnitude of soil erosion by water is controlled by the following factors.

Rainfall Intensity and Runoff

Both rainfall and runoff factors must be considered in assessing a water erosion problem. The impact of raindrops on the soil surface can break down soil aggregates and disperse the aggregate material. Lighter aggregate materials such as very fine sand, silt, clay and organic matter can be easily removed by the raindrop splash and runoff water; greater raindrop energy or runoff amounts might be required to move the larger sand and gravel particles.

Soil movement by rainfall (raindrop splash) is usually greatest and most noticeable during short-duration, high-intensity thunderstorms. Although the erosion caused by long-lasting and less-intense storms is not as spectacular or noticeable as that produced during thunderstorms, the amount of soil loss can be significant, especially when compounded over time. Runoff can occur whenever there is excess water on a slope that cannot be absorbed into the soil or trapped on the surface. The amount of runoff can be increased if infiltration is reduced due to soil compaction, crusting or freezing. Runoff from the agricultural land may be greatest during spring months when the soils are usually saturated, snow is melting and vegetative cover is minimal.

Soil Erodibility

Soil erodibility is an estimate of the ability of soils to resist erosion, based on the physical characteristics of each soil. Generally, soils with faster infiltration rates, higher levels of organic matter and improved soil structure have a greater resistance to erosion. Sand, sandy loam and loam textured soils tend to be less erodible than silt, very fine sand, and certain clay textured soils.

Tillage and cropping practices which lower soil organic matter levels, cause poor soil structure, and result of compacted contribute to increases in soil erodibility. Decreased infiltration and increased runoff can be a result of compacted subsurface soil layers. A decrease in infiltration can also be caused by a formation of a soil crust, which tends to 'seal' the surface. On some sites, a soil crust might decrease the amount of soil loss from sheet or rain splash erosion, however, a corresponding increase in the amount of runoff water can contribute to greater rill erosion problems.

Past erosion has an effect on a soils' erodibility for a number of reasons. Many exposed subsurface soils on eroded sites tend to be more erodible than the original soils were, because of their poorer structure and lower organic matter. The lower nutrient levels often associated with subsoils contribute to lower crop yields and generally poorer crop cover, which in turn provides less crop protection for the soil.

Slope Gradient and Length

Naturally, the steeper the slope of a field, the greater the amount of soil loss from erosion by water. Soil erosion by water also increases as the slope length increases due to the greater accumulation of runoff. Consolidation of small fields into larger ones often results in longer slope lengths with increased erosion potential, due to increased velocity of water which permits a greater degree of scouring (carrying capacity for sediment).

Vegetation

Soil erosion potential is increased if the soil has no or very little vegetative cover of plants and/or crop residues. Plant and residue cover protects the soil from raindrop impact and splash, tends to slow down the movement of surface runoff and allows excess surface water to infiltrate.

The erosion-reducing effectiveness of plant and/or residue covers depends on the type, extent and quantity of cover. Vegetation and residue combinations that completely cover the soil, and which intercept all falling raindrops at and close to the surface and the most efficient in controlling soil (e.g. forests, permanent grasses). Partially incorporated residues and residual roots are also important as these provide channels that allow surface water to move into the soil.

The effectiveness of any crop, management system or protective cover also depends on how much protection is available at various periods during the year, relative to the amount of erosive rainfall that falls during these periods. In this respect, crops which provide a food, protective cover for a major portion of the year (for example, alfalfa or winter cover crops) can reduce erosion much more than can crops which leave the soil bare for a longer period of time (e.g. row crops) and particularly during periods of high erosive rainfall (spring and summer). However, most of the erosion on annual row crop land can be reduced by leaving a residue cover greater than 30 per cent after harvest and over the winter months or by inter-seeding a forage crop (e.g. red clover). Soil erosion potential is affected by tillage operations, depending on the depth, direction and timing of plowing, the type of tillage equipment and the number of passes. Generally, the less the disturbance of vegetation or residue cover at or near the surface, the more effective the tillage practice in reducing erosion.

CONSERVATION MEASURES

Certain conservation measures can reduce soil erosion by both water and wind. Tillage and cropping practices, as well a land management practices, directly affect the overall soil erosion problem and solutions on a farm. When crop rotations or changing tillage practices are not enough to control erosion on a field, a combination of approaches or more extreme measures might be necessary. For example, contour plowing, strip cropping or terracing may be considered.

EFFECTS

Sheet and Rill Erosion

Sheet erosion is soil movement from raindrop splash resulting in the breakdown of soil surface structure and surface runoff; it occurs rather uniformly over the slope and may go unnoticed until most of the productive topsoil has been lost. Rill erosion results when surface runoff concentrates forming small yet well-defined channels. These channels are called rills when they are small enough to not interfere with field machinery operations. The same eroded channels are known as gullies when they become a nuisance factor in normal tillage.

Gully Erosion

There are farms in Ontario that are losing large quantities of topsoil and subsoil each year due to fully erosion. Surface runoff, causing gull formation or the enlarging of existing gullies, is usually the result of improper outlet design for local surface and subsurface drainage systems. The soil instability of fully banks, usually associated with seepage of ground water, leads to sloughing and slumping (caving-in) of bank slopes. Such failures usually occur during spring months when the soil water conditions are most conducive to the problem.

Gully formations can be difficult to control if remedial measures are not designed and properly constructed. Control measures have to consider the cause of the increased flow of water across the landscape. This where the multitude of conservation measures come into play. Operations with farm machinery adjacent to gullies can be quite hazardous when cropping or attempting to reclaim lost land.

Stream and Ditch Bank Erosion

Poor construction or inadequate maintenance, of surface drainage systems, uncontrolled livestock access, and cropping too close to both stream banks has led to bank erosion problems.

The direct damages from bank erosion include:
1. The loss of productive farmland.
2. The undermining of structures such as bridges.
3. The washing out of lanes, roads and fence rows.

Poorly constructed tile outlets may also contribute to stream and ditch bank erosion. Some do not function properly because they have no rigid outlet pipe or have outlet pipes that have been damaged by erosion, machinery, inadequate or no splash pads, and bank cave-ins.

On-site effects: The implications of soil erosion extend beyond the removal of valuable topsoil. Crop emergence, growth and yield are directly affected through the loss of natural nutrients and applied fertilisers with the soil. Seeds and plants can be disturbed or completely removed from the eroded site. Organic matter from the soil, residues and any applied manure, is relatively light-weight and can be readily transported off the field, particularly during spring thaw conditions. Pesticides may also be carried off the site with the eroded soil.

Soil quality, structure, stability and texture can be affected by the loss of soil. The breakdown of aggregates and the removal of smaller particles or entire layers of soil or organic matter can weaken the structure and even change the texture. Textural changes can in turn affect the water-holding capacity of the soil, making it more susceptible to extreme condition such a drought.

Off-site effects: Off-site impacts of soil erosion are not always as apparent as the on-site effects. Eroded soil, deposited down slope can inhibit or delay the emergence of seeds, bury small seedling and necessitate replanting in the affected areas. Sediment can be deposited on down slope properties and can contribute to road damage.

Sediment which reaches streams or watercourses can accelerate ban erosion, clog drainage ditches and stream channels, silt in reservoirs, cover fish spawning grounds and reduce downstream water quality. Pesticides and fertilisers, frequently transported along with the eroding soil can contaminate or pollute downstream water sources and recreational areas. Because of the potential seriousness of some of the off-site impacts, the control of 'non-point' pollution from agricultural land has become of increasing importance in Ontario.

EROSION BY WIND

The rate and magnitude of soil erosion by wind is controlled by the following factors.

Erodibility of Soil

Very fine particles can be suspended by the wind and then transported great distances. Fine and medium size particles can be lifted and deposited, while coarse particles can be blown along the surface (commonly known as the saltation effect). The abrasion that results can reduce soil particle size and further increase the soil erodibility.

Soil Surface Roughness

Soil surfaces that are not rough or ridged offer little resistance to the wind. However, over time, ridges can be filled in and the roughness broken down by abrasion to produce a smoother surface susceptible to the wind. Excess tillage can contribute to soil structure breakdown and increased erosion.

Climate

The speed and duration of the wind have a direct relationship to the extent of soil erosion. Soil moisture levels can be very low at the surface of excessively drained soils or during periods of drought, thus releasing the particles for transport by wind. This effect also occurs in freeze drying of the surface during winter months.

Unsheltered Distance

The lack of windbreaks (trees, shrubs, residue, etc.) allows the wind to put soil particles into motion for greater distances thus increasing the abrasion and soil erosion. Knolls are usually exposed and suffer the most.

Vegetative Cover

The lack of permanent vegetation cover in certain locations has resulted in extensive erosion by wind. Loose, dry, bare soil is the most susceptible, however, crops that produce low levels of residue also may not provide enough resistance. As well, crops that produce a lot of residue also may not protect the soil in severe cases. The most effective vegetative cover for protection should include an adequate network of living windbreaks combined with good tillage, residue management, and crop selection.

Resulting Effect

Wind erosion may create adverse operating conditions in the field. Crops can be totally ruined so that costly delay and reseeding is necessary — or the plants may be sandblasted and set back with a resulting decrease in yield, loss of quality, and market value.

Soil drifting is a fertility-depleting process that can lead to poor crop growth and yield reductions in areas of fields where wind erosion is a recurring problem. Continual drifting of an area gradually causes a textural change in the soil. Loss of fine sand, silt, clay and organic particles from sandy soils serves to lower the moisture holding capacity of the soil. This, in turn, increases the erodibility of the soil and compounds the problem. The removal of wind blown soils from fence rows, ditches, roads and from around buildings is a costly process.

Water Erosion Control

Soil loss due to water erosion reduces crop yields. Managing your soil and water resources is the best way to prevent soil from being washed away. This section describes cost-effective ways to maintain successful crop production while protecting soil and water quality.

Snowmelt and rainfall are the driving forces for water erosion on the prairies. Bare soils are very vulnerable to erosion. Steep slopes and long, uninterrupted slopes are especially prone to water erosion. Silty soils, low in organic matter, and soils with an impermeable subsoil layer are also more susceptible to water erosion.

Plant cover — either growing plants or crop residue — protects soil from the erosive power of flowing water and rain drop impact. Conservation farming methods maintain a protective cover on the soil. These land use and management practices can be adapted to fit the needs of any farm operation. Some areas of Alberta suffer from severe water erosion. In these areas, special measures may be needed to control erosion.

Land use and management practices

Select appropriate land use

Farm management decisions should consider the potential for erosion under different practices, especially on land that is marginal for annual crop production. Areas at high risk for erosion due to steep slopes or erodible soils may be better suited for forage production or grazing. Steeply sloped lands under cultivation can be converted to permanent cover to minimise erosion. Wooded areas with poor soils and steep slopes can be left in their natural state and managed profitably as woodlots. Alternative land uses can conserve the soil and have environmental benefits, while remaining profitable to the farm operation.

Maintain organic matter

Soil organic matter is very important for good crop production and for reducing soil erosion. Organic matter is made up of dead plant material. During decomposition, this material releases nutrients for plants. Organic matter also improves soil structure and tilth. Organic matter and micro-organisms cement individual soil particles into larger aggregates. Soils high in organic matter have large, stable aggregates which resist erosion. A soil with stable aggregates also has more large pore spaces to hold water. With this increased moisture-holding ability, there is less ponding in fields, and less runoff and erosion.

To maintain soil quality and fertility, new additions of plant material must equal the rate of organic matter decomposition and nutrient use by plants. Conventional tillage and fallowing practices increase soil temperature and also mix and aerate the soil, causing faster organic matter decomposition. The result has been a long-term decline in soil organic matter on the prairies.

Returning crop residue to the soil helps to replace organic matter and plant nutrients. Rotations which include forages return more residues to the soil and increase fertility. Manure applications and legume plowdown are also good sources of organic matter and nutrients.

Maintain crop residue cover

One of the best ways to reduce erosion is to protect the soil surface with a cover of growing plants or crop residue. Surface cover cushions the impact of rain drops so soil particles are not as easily dislodged and moved. It also slows the flow of water, giving the soil time to absorb more water and thereby reducing runoff and erosion. An Alberta research study has shown that any increase in infiltration is directly related to a decrease in runoff. As well, crop residue traps snow and reduces evaporation for higher soil moisture which can improve crop yields, especially in a dry year.

Standing stubble and evenly spread straw and chaff protect the soil during spring runoff. Tillage should be kept to a minimum because it reduces the crop residue cover. Conservation tillage systems that leave most of the crop residue on the surface will reduce erosion and may have other benefits, such as lower equipment operating costs and labour inputs.

Reduce tillage

Conventional tillage buries the protective crop residue cover and disturbs the soil. The loose soil particles are easily detached by rain drops and running water. These factors lead to increased runoff and erosion.

Alberta research shows that switching to reduced or zero tillage systems is needed to protect soils on steeper and longer slopes from erosion. Reduced and minimum tillage systems leave a good crop residue cover to prevent erosion and conserve soil moisture. These systems also save time and energy, and costs are usually similar to or lower than those for conventional tillage systems. Tillage is reduced by replacing some tillage operations for weed control with herbicide applications or by using alternative tillage equipment that helps maintain a good residue cover.

Residue management is important in all conservation tillage systems. Straw and chaff must be spread evenly at harvest to avoid or reduce such problems as: plugging during subsequent operations; poor seed germination; disease, weed and insect infestations; and nutrient tie-up.

Use zero tillage or direct seeding

Zero tillage systems minimise soil disturbance to maintain as much crop residue cover as possible to conserve soil moisture and prevent erosion. Long-term zero tillage also increases soil organic matter and improves soil quality and fertility.

Direct seeding also aims to conserve both soil moisture and soil. It differs from zero tillage in that some tillage options remain open. Minimal disturbance tillage operations (which leave the stubble standing) are sometimes used to apply fertiliser or incorporate herbicides or manure. A high percentage of crop residue remains on the surface to protect against erosion. In both zero tillage and direct seeding systems, straw and chaff should be spread evenly across the entire width of the cut during combining. Harrowing may be needed to achieve uniform distribution, especially for heavy crop residues. The crop is seeded into the previous crop's stubble with minimal soil disturbance. Fertiliser is usually banded in a row near the seed. Herbicide applications replace tillage to control weeds. Management practices such as crop and herbicide rotations can be used to reduce weed problems.

Direct seeding and zero tillage systems save time and may have lower operating costs than conventional tillage systems. Although herbicide costs may increase, tillage-related costs decrease. Improvements in herbicides and sprayers, and the availability of seed drills able to operate in crop residues have made it easier to switch to zero tillage and direct seeding.

Use conservation fallow

The long-term use of conventional fallow has caused serious soil degradation problems. In conventional fallow, weed control and seedbed preparation are done by tillage. Tillage buries crop residues, leaving the soil at risk from erosion for a long period. While the soil is fallow, organic matter decomposes. This releases nitrogen and other nutrients for the fallowing crop. However, with no residue input from crops during the fallow period, the amount of organic matter declines. The resulting poorer soil structure lowers the soil's ability to absorb water and increases runoff and erosion.

If summer fallow is necessary, maintain a crop residue cover by minimising surface disturbance. Herbicides can be used to control weeds, and one spray operation can replace two tillage operations. Wide blade cultivators or rod weeders minimise residue disturbance. Reducing tillage will protect the soil, conserve soil moisture and may also lower equipment operating costs and labour needs.

Grow forages and use crop rotations

Forage crops are a component in many conservation farming systems. Forages can be grown on poorer soils or steep slopes not suitable for other crops or used in rotations to build organic matter or break disease cycles. Forage cover protects the soil from erosion, and the fibrous roots hold the soil in place. As a perennial crop or plowdown, forages add organic matter and improve soil quality and structure. Improved soil structure allows the soil to absorb more water which reduces runoff and erosion.

Crop rotations for erosion control alternate forages with cereals and oilseeds or legumes. A well-planned rotation will improve soil quality and reduce erosion. Legumes in the rotation also add nitrogen and improve soil fertility. In drier areas, forages are harder to establish and may deplete moisture in a short-term rotation. An alternative annual crop such as a legume can be grown in these areas or the forages can be maintained as a longer term crop.

Use direct seeding for pasture conversion

Direct seeding is a good option for converting hay or pasture land to annual crop production. It produces crop yields similar to those from conventionally plowed systems, and also prevents soil erosion and moisture loss. In conventional systems, intense operations such as plowing, heavy discing and cultivations are used. They are costly and time consuming, and expose soil to erosion.

Annual crops such as barley, oats and peas can be direct seeded into pasture sod after the pasture vegetation has been killed by a herbicide, usually glyphosate. Fall spraying is usually preferred over spring spraying for better annual crop yields, weed/pasture plant control, and moisture conservation.

Controlling Severe Erosion

The following measures control gullies and other severe erosion problems. For severe erosion, it is a good idea to get technical advice to find the best solution for your situation.

Grassed waterways

Gully erosion can often be controlled with a grassed waterway. A grassed waterway is a wide, shallow grassed channel that can carry a large volume of water quickly down a steep slope. It can be crossed by lifting tillage equipment.

Grassed waterways on agricultural land need to be able to carry peak runoff events from snowmelt and rainstorms (up to one cubic metre of water per second). The size of the waterway depends on the size of the area to be drained. A typical grassed waterway cross-section is saucer-shaped with a nearly flat-bottomed channel, a bottom width of 3 m and channel depth of at least 30 cm. Side slopes usually rise about 1 m for every 10 m horizontal distance. The waterway should follow the natural drainage path if possible.

Cross-section of a typical grassed waterway

The grass cover must be well established to handle high flows without erosion. A fast-growing cover crop, such as oats, provides initial, temporary protection for the waterway until the grass cover is established. Steeper portions of the waterway which are very susceptible to erosion can be protected by bio-degradable erosion control mats until the grass is established. Commercially available mats are made from straw, jute or aspen wood shavings. A well-built and maintained grassed waterway is very durable and erosion-resistant. The waterway should be mowed regularly, and weeds and brush must be controlled for the waterway to remain effective.

Lined channels

Lined channels are a means of dropping water to lower elevations along steep parts of a waterway. Those portions of the waterway are precisely shaped and carefully lined with heavy-duty erosion control matting, a type of geotextile product. The lining is covered with a layer of soil and seeded to grass. The resulting channel is highly resistant to erosion. Lined channels are appropriate for waterways that only carry water occasionally and have slopes up to 10 per cent. Companies that sell geotextile products provide detailed information on installation of their products.

Drop structures

Drop structures are constructed along waterways to drop water to lower elevations without causing erosion. They are constructed of concrete, wood, metal or rock. Drop structures are the most costly but occasionally the most appropriate form of erosion control at specific locations along a waterway.

Small, intermittent waterways entering Alberta's deep river valleys are capable of causing very large gullies. Pipe drop structures are effective and economic for controlling this kind of erosion. Concrete sewer manholes or vertical corrugated steel pipes used with smaller diameter corrugated plastic or metal pipe, can transport water safely down long, steep slopes. A crawler tractor with a blade is used to form a firm bed down the length of the gully beginning at the top. A small track hoe is used to dig and install any buried sections of pipe. Above-ground portions of installed pipe can be secured with posts made of angle iron.

Cross section of a pipe drop structure

A slow release drop structure is an inexpensive and effective measure to control gully erosion. An earth berm is constructed upstream from the gully. Runoff water is held back temporarily by the berm. The water drains slowly through a small diameter plastic pipe (75 mm to 200 mm diameter) which runs under the berm and down the slope, and outlets at the bottom. A durable, high density polyethylene pipe is recommended. The small pipe can be held in place on the slope where needed with steel pins. This structure can only be used where there is an area with enough storage capacity upstream of the gully. The flooded area is fully drained within two days to prevent crop damage. In fact, the temporary backflooding benefits the crop by increasing soil moisture.

Terracing

Water erosion over long, wide slopes without well-developed channels can be controlled with terracing. A channel and berm with up to 1 m difference in elevation are constructed across the slope to intercept runoff and carry it safely off the field. The material excavated to create the channel is used to build the berm. A survey is essential to find the best terrace location on the slope and to maintain proper grade for drainage. The project should be staked before construction to guide the equipment operators. Heavy-duty road construction equipment, such as a motor scraper, is needed to construct terraces.

Cross-section of one type of terrace

Terraces are practical only when crop returns from the land are high enough to justify construction costs. Tillage and residue management options should be evaluated before considering terraces.

Many on-farm water erosion problems can be solved by the farm operator with minimal expense or inconvenience. Modifying tillage practices to keep crop residue on the surface can greatly reduce erosion. A crop residue cover also conserves soil moisture and improves soil tilth and fertility for better crop production. Costs for conservation tillage systems are usually similar to or lower than costs for conventional tillage systems over the long-term. Preventing soil erosion helps to ensure the sustainability of the farm operation. Grassed waterways, drop structures, lined channels or terraces are used to control more severe water erosion problems. Technical advice may be needed to implement some of these special measures.

WIND EROSION CONTROL

Wind erosion can occur when soils dry and small soil particles are moved by the wind. Wind erosion is usually more of a problem on sandy and peat soils. The drying out of the finely divided surface layers of these soils leaves them susceptible to wind erosion. Small soil particles are first detached and then moved by the wind. Once particles begin to move they have an abrasive action and dislodge other soil particles and intensify erosion. Dislodged soil particles bounce along the surface of the ground and may reach heights of a foot during salutation. Increased wind speed and smaller particle size can result in soil particles being suspended and carried by the wind.

Erosion Process

Factors affecting wind erosion are soil moisture content, wind velocity, soil surface roughness, soil characteristics, and the nature and orientation of vegetation or crop residues. Wind speeds of 12 miles per hour are sufficient to initiate soil erosion. As wind speeds increase above 12 miles per hour, the quantity of soil carried by the wind increases rapidly. A rough soil surface with large clods or ridges will reduce wind erosion. The presence of corn or small grain residues on the soil surface also will reduce soil movement. As the clay content of the soil increases, the stability of soil aggregates increases and wind erosion decreases. In contrast, as the sand content of the soil increases, aggregate stability decreases and soil movement by wind increases.

The sugarbeet crop is most susceptible to damage from wind erosion after planting and until the crop is big enough to begin shading the row. After emergence sugarbeet seedlings can easily be injured or killed by blowing soil particles. The problem is intensified if rain or irrigation reduces surface roughness, and as the soil dries it becomes susceptible to wind erosion. During the spring, wind speeds increase as weather fronts move from west to east across the intermountain sugarbeet growing region. Therefore the presence of sandy soils, frequent high intensity thunderstorms, and the absence of crop residues on the soil surface make wind erosion a serious threat to establishing the sugarbeet crop.

Reducing Wind Erosion

Several cultural methods can help reduce wind erosion. In the absence of crop residues, soil roughness and soil moisture content can reduce wind erosion. Also the planter can be equipped with tillage tools to roughen the soil surface adjacent to the crop row. This will generally reduce wind erosion until rainfall or irrigation reduce aggregate stability and clod size. As the soil dries, surface roughness must be re-established by rotary hoeing, cultivating or ditching the area between sugarbeet rows. Irrigation will temporarily stop wind erosion until the soil surface dries.

Crop Residues

Crop residues from the previous crop can be successfully utilised to provide wind and water erosion protection for the sugarbeet plant. Small grain stubble can be sprayed with a glyphosate product in the spring before sugarbeet planting to control emerged weeds. Sugarbeet can be planted directly into the stubble by equipping the planter with residue moving devices which remove the small grain residue directly over the crop row (Fig. 24.1). The residue remaining between the sugarbeet rows will protect sugarbeet seedlings. When the crop is established this residue can be buried with cultivation. Sugarbeet also can be planted into corn residue that has been disked before planting. Again, the planter needs to be equipped with some type of residue moving devices to minimise corn residue directly over the row.

Fig. 24.1. Roughening the soil between sugarbeet rows.

Corn and small grains produce sufficient residue after harvest to provide erosion protection during the winter and spring and for the following sugarbeet crop; however, other crops, particularly dry edible beans, do not provide enough residue after harvest to protect the soil or the following sugarbeet crop from wind erosion. Dry edible beans are harvested in early to mid-September. Cover crops of winter wheat or winter rye can be seeded immediately after bean harvest with a grain drill or seed can be spread with a fertiliser spreader and incorporated into the soil with a shallow tillage operation. A disk drill with narrow row spacing will provide a level planting surface in the spring for the following sugarbeet crop. The seeding rate for either wheat or rye is usually 1 to 1.5 bu/acre. Rye will provide more top growth and better wind erosion protection than wheat early in the spring. The cover crop should be planted by September 15 to assure adequate soil protection over winter. If soil moisture is lacking at the time of seeding, sprinkler or furrow irrigation can be beneficial in improving cover crop density and growth.

Cover Crops

A cover crop with sufficient growth will provide soil erosion protection during the fall, winter and spring. The fall seeded cover crop also can provide protection to a spring planted sugarbeet crop. Allow the cover crop to grow to a 3 to 5-inch height in the spring before killing with a glyphosate product like roundup. Sugarbeet can then be planted directly into the standing cover crop residue or strips can be tilled through the cover crop to provide a planting area for spring planted sugarbeet. An appropriate planter must be used for sugarbeet to obtain proper seed depth and to ensure that the cover crop residue is not punched into the seed furrow with the seed, creating inadequate seed-soil contact. A conventionally equipped, dedicated sugarbeet planter, such as a Milton or Deere 71 Flexi-Planter, will have difficulty placing sugarbeet seed at the proper depth and achieving good seed-soil contact in this cover crop situation.

Sugarbeet growers have devised an alternative practice for controlling fall planted broadcast or narrow-row cover crops while accommodating satisfactory performance of sugarbeet planters. When the cover crop reaches a height of 3–4 inches in the spring, narrow strips, approximately 12 inches wide, are sprayed and killed with herbicide. This spraying operation requires a band sprayer, straight rows and accurate 'guess' rows. By sugarbeet planting time, the cover crop in these rows has died and sugarbeet can be planted without interference from the residue. The remaining cover crop in the interrow area must be sprayed with an appropriate herbicide immediately prior to sugarbeet planting or at least before any sugarbeet begin to emerge unless the sugarbeet is tolerant to the herbicide. This system provides both excellent wind erosion protection and good planter performance with traditional sugarbeet planters. Growers should be cautious because cutworms can be attracted to fall planted cereal cover crops and feed on sugarbeet seedlings as they emerge. Sugarbeet fields should be scouted early in the growing season for cutworms and treated with an insecticide if crop damage is observed.

An alternative to planting the cover crop with a grain drill or broadcasting the seed, is to plant the cover crop in defined rows to match the row spacing of the sugarbeet crop. The cover crop can be planted with a row crop planter or with a grain drill which has appropriate openers shut off or raised. The cover crop rows must be planted straight using a marker to obtain accurate 'guess row' width. The cover crop rows should be perpendicular to the prevailing wind. The row units for the cover crop planter or drill should be positioned so the tractor tyres do not run over the soil where sugarbeet rows will be planted.

Seeding the cover crop in distinct rows provides a residue-free area for planting the spring row crop. Conventional sugarbeet planters can be used to plant the spring crop if the area between rows of cover crop is relatively level. An example of this technique would be to use a row crop planter to plant winter rye in 22 inch rows at the rate of 1 bu/acre in the fall after bean harvest. The following spring the cover crop should reach a height of 3 to 5 inches before being treated with Roundup at 1.5 to 2 pt/acre. Plant sugarbeet between the cover crop rows with a conventional sugarbeet planter. The cover crop provides early season protection for the developing seedlings until the sugarbeet are large enough to protect themselves. The remaining cover crop could then be removed with cultivation.

The timing of herbicide application to kill the cover crop is critical. The cover crop must be allowed to grow tall enough to provide adequate protection for both the soil and the crop to be planted. If allowed to grow too large, the cover crop will compete with the spring planted crop for soil moisture and may be more difficult to control. Rain or wind can delay herbicide application beyond the planned date. If a nonselective herbicide is used to kill the cover crop, it must be applied before any of the spring planted crop begins to emerge.

Spring planted sugarbeet on coarse textured soils can be injured by blowing soil particles. A spring planted cover crop can provide early season protection for sugarbeet until the crop is established. The seedbed can be prepared conventionally and barley or oats seeded at the rate of one bushel per acre with a row crop planter in March or early April. Most row crop planters can be used to seed the cover crop in rows speed far enough apart to facilitate the planting of sugarbeet in mid to late April. To prevent compaction over the sugarbeet row, the hitch attachment on the cover crop planter should be moved one-half row width on the planter frame so the tractor tyres are in line with the cover crop rows rather than where the sugarbeet rows will be. This spring seeded cover crop also could be planted in narrow rows with a disk drill. The resulting surface must be relatively level to allow planting directly into the growing cover crop without further tillage.

The cover crop should have emerged and begun to grow before sugarbeet are planted. Most conventional sugarbeet planters will perform satisfactorily in either wide or narrow rows if the surface between cover crop rows was left relatively level after cover crop planting. When the cover crop reaches a height of 6 to 8 inches if planted in wide rows or 3 to 5 inches if drilled in narrow rows, it should be treated with an approved graminicide, appropriate for the crop being grown. The cover crop will provide early season protection for the establishing crop and can be killed before it becomes too large and begins to compete with the crop. When the sugarbeet is sufficiently large, the cover crop can removed with cultivation.

Cover crop systems are very effective for sugarbeet production. Properly managed, these systems minimise spring tillage, eliminate the need for emergency soil roughening, and will help assure a good sugarbeet stand. Producers who use cover crop systems offer the following five keys for success:

1. The cover crop must attain sufficient growth in the fall. This means early planting and irrigation as needed.
2. Careful attention should be paid to timing of herbicide application to kill the cover crop in the spring. If it's applied too early, there will not be enough cover; if it's applied too late, there will be too much competition with the sugarbeet crop for soil moisture, fertility, and sunlight.
3. Correct seed depth control and complete seed to soil contact should be ensured. Residue or an irregular soil surface must not interfere with planting sugarbeet.
4. Irrigation to establish the cover crop in the fall and to establish the sugarbeet crop in the spring will be essential. This is easiest with a well-supplied center pivot.
5. Be prepared to deal with an increased risk of early season cutworm problems.

MULCH

In agriculture and gardening, mulch is a protective cover placed over the soil to retain moisture, reduce erosion, provide nutrients, and suppress weed growth and seed germination. Mulching in gardens and landscaping mimics the leaf cover that is found on forest floors.

Materials

Materials used as mulches vary and depend on a number of factors. Use takes into consideration availability, cost, appearance, the effect it has on the soil — including chemical reactions and pH, durability, combustibility, rate of decomposition, how clean it is — some can contain weed seeds or plant pathogens.

A variety of materials are used as mulch:

1. Organic residues: Grass clippings, leaves, hay, straw, kitchen scraps comfrey, shredded bark, whole bark nuggets, sawdust, shells, woodchips, shredded newspaper, cardboard, wool, but also animal manure, etc. Many of these materials also act as a direct composting system, such as the mulched clippings of a mulching lawn mower or other organics applied as sheet composting.
2. Compost: This should be fully composted material to avoid possible phytotoxicity problems, and the weed seed must have been eliminated, otherwise the mulch will actually produce weed cover.
3. Rubber mulch: Made from recycled tyre rubber.
4. Plastic mulch: Crops grow through slits or holes in thin plastic sheeting. This method is predominant in large-scale vegetable growing, with millions of acres cultivated under plastic mulch worldwide each year (disposal of plastic mulch is cited as an environmental problem).
5. Rock and gravel can also be used as a mulch. In cooler climates the heat retained by rocks may extend the growing season.

Organic Mulches

Organic mulches decay over time and are temporary. The way a particular organic mulch decomposes and reacts to wetting by rain and dew affects its usefulness.

Organic mulches can negatively affect plant growth when they are decomposed rapidly by bacteria and fungi, which require nitrogen that they remove from the surrounding soil. Organic mulches can mat down, forming a barrier that blocks water and air flow between the soil and the atmosphere. Some organic mulches can wick water from the soil to the surface, which can dry out the soil.

Commonly available organic mulches include:

1. Leaves from deciduous trees, which drop their foliage in the fall. They tend to be dry and blow around in the wind, so are often chopped or shredded before application. As they decompose they adhere to each other but also allow water and moisture to seep down to the soil surface. Thick layers of entire leaves, especially of Maples and Oaks, can form a soggy mat in winter and spring which can impede the new growth lawn grass and other plants. Dry leaves are used as winter mulches to protect plants from freezing and thawing in areas with cold winters, they are normally removed during spring.
2. Grass clippings, from mowed lawns are sometimes collected and used elsewhere as mulch. Grass clippings are dense and tend to mat down, so are mixed with tree leaves or rough compost to provide aeration and to facilitate their decomposition without smelly putrefaction. Rotting fresh grass clippings can damage plants; their rotting often produces a damaging buildup of trapped heat. Grass clippings are often dried thoroughly before application, which mediates against rapid decomposition and excessive heat generation. Fresh green grass clippings are relatively high in nitrate content, and when used as a mulch, much of the nitrate is returned to the soil, but the routine removal of grass clippings from the lawn results in nitrogen deficiency for the lawn.
3. Peat moss or sphagnum peat, is long lasting and packaged, making it convenient and popular as a mulch. When wetted and dried, it can form a dense crust that does not allow water to soak in. When dry it can also burn, producing a smoldering fire. It is sometimes mixed with pine needles

to produce a mulch that is friable. It can also lower the pH of the soil surface, making it useful as a mulch under acid loving plants.

4. Wood chips are a by-product of the pruning of trees by arborists, utilities and parks; they are used to dispose of bulky waste. Tree branches and large stems are rather coarse after chipping and tend to be used as a mulch at least three inches thick. The chips are used to conserve soil moisture, moderate soil temperature and suppress weed growth. The decay of freshly produced chips from recently living woody plants, consumes nitrate; this is often off set with a light application of a high-nitrate fertiliser. Wood chips are most often used under trees and shrubs. When used around soft stemmed plants, an unmulched zone is left around the plant stems to prevent stem rot or other possible diseases. They are often used to mulch trails, because they are readily produced with little additional cost outside of the normal disposal cost of tree maintenance.

5. Bark chips, of various grades are produced from the outer corky bark layer of timber trees. Sizes vary from thin shredded strands to large coarse blocks. The finer types are very attractive but have a large exposed surface area that leads to quicker decay. Layers two or three inches deep are usually used, bark is relativity inert and its decay does not demand soil nitrates.

6. Straw mulch or field hay or salt hay are lightweight and normally sold in compressed bales. They have an unkempt look and are used in vegetable gardens and as a winter covering. They are biodegradable and neutral in pH. They have good moisture retention and weed controlling properties but also are more likely to be contaminated with weed seeds. Salt hay is less likely to have weed seeds than field hay.

7. Cardboard or newspaper can be used as mulches. These are best used as a base layer upon which a heavier mulch such as compost is placed to prevent the lighter cardboard/newspaper layer from blowing away. By incorporating a layer of cardboard/newspaper into a mulch, the quantity of heavier mulch can be reduced, whilst improving the weed suppressant and moisture retaining properties of the mulch. However, additional labour is expended when planting through a mulch containing a cardboard/newspaper layer, as holes must be cut for each plant. Sowing seed through mulches containing a cardboard/newspaper layer is impractical. Application of newspaper mulch in windy weather can be facilitated by briefly pre-soaking the newspaper in water to increase its weight.

Application

In temperate climates, the effect of mulch is dependent upon the time of year at which it is applied as it tends to slow changes in soil temperature and moisture content. Mulch, when applied to the soil in late winter/early spring, will slow the warming of the soil by acting as an insulator, and will hold in moisture by preventing evaporation. Mulch, when applied at the time of peak soil temperatures in mid-summer, will maintain high soil temperatures further into the autumn (fall). The effect of mulch upon soil moisture content in mid-summer is complex however. Mulch prevents sunlight from reaching the soil surface, thus reducing evaporation. However, mulch can absorb much of the rainfall provided during light rainfall, which will later quickly evaporate when exposed to sunlight, thus preventing absorption into the soil, whilst heavy rainfall is able to saturate the mulch layer, and reach the soil below.

In order to maximise the benefits of mulch, whilst minimising its negative influences, it is often applied in late spring/early summer when soil temperatures have risen sufficiently, but soil moisture content is still relatively high. Furthermore, at this point in the growing season, plants should be well enough established to be able to cope with the increase in the numbers of slugs and snails owing to the habitat provided for them by the mulch.

Plastic mulch used in large-scale commercial production is laid down with a tractor-drawn or standalone layer of plastic mulch. This is usually part of a sophisticated mechanical process, where raised beds are formed, plastic is rolled out on top, and seedlings are transplanted through it. Drip irrigation is often required, with drip tape laid under the plastic, as plastic mulch is impermeable to water.

In home gardens and smaller farming operations, organic mulch is usually spread by hand around emerged plants. For materials like straw and hay, a shredder may be used to chop up the material. Organic mulches are usually piled quite high, six inches (152 mm) or more, and settle over the season.

In some areas of the United States, such as central Pennsylvania and northern California, mulch is often referred to as 'tanbark', even by manufacturers and distributors. In these areas, the word 'mulch' is used specifically to refer to very fine tanbark or peat moss.

Mulch made with wood can contain or feed termites, so care must be taken about not placing mulch too close to houses or building that can be damaged by those insects. Some mulch manufacturers recommend putting mulch several inches away from buildings.

Anaerobic (Sour) Mulch

Mulch should normally smell like freshly cut wood, but sometimes develops a toxicity that causes it to smell like vinegar, ammonia, sulphur or silage. This happens when material with ample nitrogen content is not rotated often enough and it forms pockets of increased decomposition. When this occurs, the process may become anaerobic and produce these phytotoxic materials in small quantities. Once exposed to the air, the process quickly reverts to an aerobic process, but these toxic materials may be present for a period of time. If the mulch is placed around plants before the toxicity has had a chance to dissipate, then the plants could very likely be damaged or killed depending on their hardiness. Plants that are predominantly low to the ground or freshly planted are the most susceptible, and the phytotoxicity may prevent germination of some seeds.

If sour mulch is applied and there is plant kill, the best thing to do is to water the mulch heavily. Water dissipates the chemicals faster and refreshes the plants. Removing the offending mulch may have little effect, because by the time plant kill is noticed, most of the toxicity is already dissipated. While testing after plant kill will not likely turn up anything, a simple pH check may reveal high acidity, in the range of 3.8 to 5.6 instead of the normal range of 6.0 to 7.2. Finally, placing a bit of the offending mulch around another plant to check for plant kill will verify if the toxicity has departed. If the new plant is also killed, then sour mulch is probably not the problem.

Groundcovers (Living Mulches)

Groundcovers are plants which grow close to the ground, under the main crop, to slow the development of weeds and provide other benefits of mulch. They are usually fast-growing plants that continue growing with the main crops. By contrast, cover crops are incorporated into the soil or killed with herbicides. However, living mulches also may need to be mechanically or chemically killed eventually to prevent competition with the main crop.

Some groundcovers can perform additional roles in the garden such as nitrogen fixation in the case of clovers, dynamic accumulation of nutrients from the subsoil in the case of creeping comfrey (*Symphytum ibericum*), and even food production in the case of *Rubus Tricolour*.

On-site Mulch Production

Owing to the great bulk of mulch which is often required on a site, it is often impractical and expensive to source and import sufficient mulch materials. An alternative to importing mulch materials is to grow them on site in a 'mulch garden'— an area of the site dedicated entirely to the production of mulch which is then transferred to the growing area. Mulch gardens should be sited as close as possible to the growing area so as to facilitate transfer of mulch materials.

Mulching (Composting) Over Unwanted Plants

Sufficient mulch over plants will destroy them, and may be more advantageous than using herbicide, cutting, mowing, pulling, raking or tilling. The higher the temperature that this 'mulch' is composted, the quicker the reduction of undesirable materials. 'Undesirable materials' may include living seed, plant 'trash', as well as pathogens such as from animal feces, urine (e.g. hantavirus), fleas, lice, ticks, etc.

In some ways this improves the soil by attracting and feeding earthworms, and adding humus. Earthworms 'till' the soil, and their feces are among the best fertilisers and soil conditioners. Urine may be toxic to plants if applied to growing areas undiluted.

SOIL REGENERATION

Throughout history humanity has constantly fought a war with the elements of wind and rain in trying to conserve the small amount of topsoil that is the living skin of the earth. In many countries like Australia we have largely lost that war and are also facing other related problems of land degradation in the form of salinity and tree decline. The challenge before us is to look beyond the problem and seek commercial ways to regenerate or to literally 'grow' increasing volumes of topsoil in order to achieve sustainable plant production. This is no small challenge and [yet] many great names in agriculture have spoken of this need. The wording they use often involves a three part saying. For example there is a need to re-generate the soil, to rebuild the soil, renew the soil or reconstruct the soil.

The concept of soil regeneration involves the creation of new topsoil and to achieve this two main goals have to be achieved simultaneously: to increase both soil fertility and the granular structure of the soil. Soil regeneration is about building or making topsoil. For example where one inch of top soil is now on your farm the aim is to have twice as much in under three to five years.

In order to regenerate or rebuild your topsoil it is essential to have an understanding of the main natural principals that allow a soil to literally regenerate itself. Nature constantly uses these principles in order to build topsoil. For example consider how nature rebuilds a biological worn out paddock with nature's own mineral gatherers which we call 'weeds.' The plants collect minerals from the subsoil and bring them to the surface in addition to their root systems aerating the soil.

It is also essential for you to interpret and assess your own soil's current state of regeneration, as by doing this you can identify its potential limitation with a view to determine what form of long and short term cost effective soil regenerative strategy can be taken. These soil regenerative strategies can be described as 'triggering mechanisms' that change the soil environment and so enable the further progression and development of a greater quality and volume of top soil within a given soil type. Triggering mechanisms may including soil aeration, appropriate fertiliser applications, i.e. lime if it is needed and grazing.

Soil regeneration greatly relies on soil forming plant life and active beneficial soil organisms. Many soils are biologically dead and part of our aim is to create the right environment so the 'life' in the soil can continue to recycle nutrients and thus increase the sustainability of the property. The end result of nutrient recycling is a farmer's most valuable asset— humus. The on going formation of humus in the form of polysaccharides is the natural outcome of a biologically healthy and productive soil.

In summary we will be focusing on exploring different management strategies in the form of triggering mechanisms that can be used to enhance your soil's own natural self perpetuating regenerative process. Learning to farm with nature (Fig. 24.2).

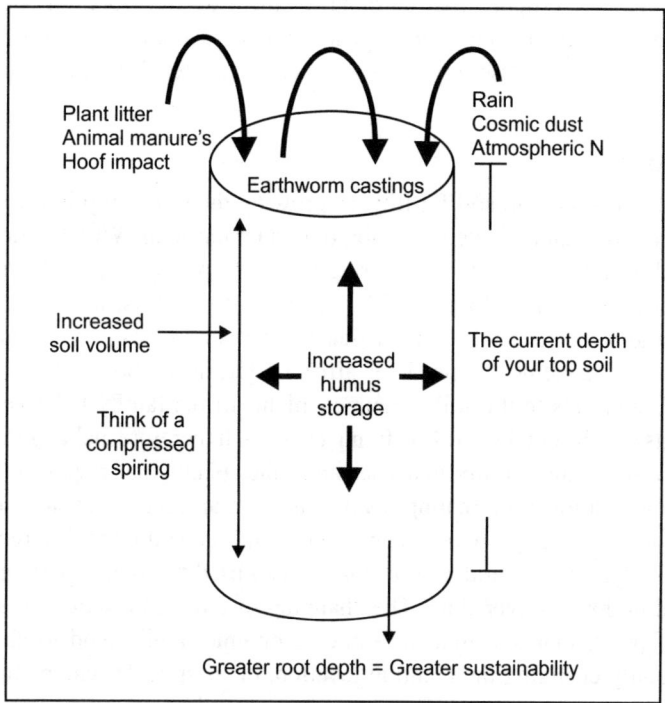

Fig. 24.2. Learning to farm with nature.

TERRACE

Terraces are used in farming to cultivate sloped land. Graduated terrace steps are commonly used to farm on hilly or mountainous terrain. Terraced fields decrease erosion and surface runoff, and are effective for growing crops requiring much water, such as rice. Natural terracing, the result of small-scale erosion, can occur where cattle are grazed for long periods on steep sloping pasture, such as at Glastonbury Tor.

Terraced paddy fields are used widely in rice farming in east, south, and southeast Asia, as well as other places. Drier-climate terrace farming is common throughout the Mediterranean Basin, e.g. in Cadaqués, Catalonia, where they were used for vineyards, olive trees, cork oak, etc. on Mallorca or in Cinque Terre, Italy.

In the Andes, farmers have used terraces known as andenes for over a thousand years to farm potatoes, maize, and other native crops. The Inca also used terraces for soil conservation, along with a system of canals and aqueducts to direct water through dry land and increase fertility.

Terracing is also used for gardening on sloping terrain. Terraced fields are common in islands with vigorous slopes. The Canary Islands present a complex system of terraces covering the landscape from the coastal irrigated plantations to the dry fields in the highlands. These terraces, which are named 'cadenas' (chains), are built with stone walls of skillful design, which include attached stairs and channels.

CROP ROTATION

Crop rotation is the practice of growing a series of dissimilar types of crops in the same area in sequential seasons for various benefits such as to avoid the build up of pathogens and pests that often occurs when one species is continuously cropped. A traditional element of crop rotation is the replenishment of nitrogen through the use of green manure in sequence with cereals and other crops. It is one component of polyculture. Crop rotation can also improve soil structure and fertility by alternating deep-rooted and shallow-rooted plants.

Method and Purpose

Crop rotation avoids a decrease in soil fertility, as growing the same crop in the same place for many years in a row disproportionately depletes the soil of certain nutrients. With rotation, crops that leaches the soil of one kind of nutrient is followed during the next growing season by a dissimilar crop that returns that nutrient to the soil or draws a different ratio of nutrients, for example, rice followed by cotton. By crop rotation farmers can keep their fields under continuous production, without the need to let them lie fallow, and reducing the need for artificial fertilisers, both of which can be expensive. Rotating crops adds nutrients to the soil. Legumes, plants of the family Fabaceae, for instance, have nodules on their roots which contain nitrogen-fixing bacteria. It therefore makes good sense agriculturally to alternate them with cereals (family Poaceae) and other plants that require nitrates. An extremely common modern crop rotation is alternating soyabeans and maize (corn). In subsistence farming, it also makes good nutritional sense to grow beans and grain at the same time in different fields.

Crop rotation is a type of cultural control that is also used to control pests and diseases that can become established in the soil over time. The changing of crops in a sequence tends to decrease the population level of pests. Plants within the same taxonomic family tend to have similar pests and pathogens. By regularly changing the planting location, the pest cycles can be broken or limited. For example, root-knot nematode is a serious problem for some plants in warm climates and sandy soils, where it slowly builds up to high levels in the soil, and can severely damage plant productivity by cutting off circulation from the plant roots. Growing a crop that is not a host for root-knot nematode for one season greatly reduces the level of the nematode in the soil, thus making it possible to grow a susceptible crop the following season without needing soil fumigation.

It is also difficult to control weeds similar to the crop which may contaminate the final produce. For instance, ergot in weed grasses is difficult to separate from harvested grain. A different crop allows the weeds to be eliminated, breaking the ergot cycle. This principle is of particular use in organic farming, where pest control may be achieved without synthetic pesticides.

A general effect of crop rotation is that there is a geographic mixing of crops, which can slow the spread of pests and diseases during the growing season. The different crops can also reduce the effects of adverse weather for the individual farmer and, by requiring planting and harvest at different times, allow more land to be farmed with the same amount of machinery and labour. The choice and sequence of rotation crops depends on the nature of the soil, the climate, and precipitation which together determine the type of plants that may be cultivated. Other important aspects of farming such as crop marketing and economic variables must also be considered when deciding crop rotations.

Effects on Soil Erosion

Crop rotation can greatly affect the amount of soil lost from erosion by water. In areas that are highly susceptible to erosion, farm management practices such as zero and reduced tillage can be supplemented with specific crop rotation methods to reduce raindrop impact, sediment detachment, sediment transport, surface runoff, and soil loss. Protection against soil loss is maximised with rotation methods that leave the greatest mass of crop stubble (plant residue left after harvest) on top of the soil. Stubble cover in contact with the soil minimises erosion from water by reducing overland flow velocity, stream power, and thus the ability of the water to detach and transport sediment. Soil Erosion and Cill prevent the disruption and detachment of soil aggregates that cause macrospores to block, infiltration to decline, and runoff to increase. This significantly improves the resilience of soils when subjected to periods of erosion and stress.

The effect of crop rotation on erosion control varies by climate. In regions under relatively consistent climate conditions, where annual rainfall and temperature levels are assumed, rigid crop rotations can produce sufficient plant growth and soil cover. In regions where climate conditions are less predictable, and unexpected periods of rain and drought may occur, a more flexible approach for soil cover by crop rotation is necessary. An opportunity cropping system promotes adequate soil cover under these erratic climate conditions. In an opportunity cropping system, crops are grown when soil water is adequate and there is a reliable sowing window. This form of cropping system is likely to produce better soil cover than a rigid crop rotation because crops are only sown under optimal conditions, whereas rigid systems are sown in the best conditions available.

Crop rotations also affect the timing and length of when a field is subject to fallow. This is very important because depending on a particular region's climate, a field could be the most vulnerable to erosion when it is under fallow. Efficient fallow management is an essential part of reducing erosion in a crop rotation system. Zero tillage is a fundamental management practice that promotes crop stubble retention under longer unplanned fallows when crops cannot be planted. Such management practices that succeed in retaining suitable soil cover in areas under fallow will ultimately reduce soil loss.

CONTOUR PLOWING

Contour plowing (or contour ploughing) or contour farming is the farming practice of plowing across a slope following its elevation contour lines. The rows formed slows water run-off during rainstorms to prevent soil erosion and allows the water time to settle into the soil. In contour plowing, the ruts made by the plow run perpendicular rather than parallel to slopes, generally resulting in furrows that curve around the land and are level. A similar practice is contour bunding where stones are placed around the contours of slopes (Fig. 24.3).

Fig. 24.3. Contour ploughing, schematic.

In the Gina River catchment contour ploughing is also practised to reduce soil erosion. However, it could only once be monitored due to the period of field investigation and thus was not considered in the assessment of soil erosion risk in the drainage basin. Besides these measures, it is also important to consider some socio-economic, institutional and political aspects of soil and water conservation.

FLOOD CONTROL

Flood control refers to all methods used to reduce or prevent the detrimental effects of flood waters.

Causes of Floods

Floods are caused by many factors: heavy rainfall, highly accelerated snowmelt, severe winds over water, unusual high tides, tsunamis or failure of dams, levees, retention ponds or other structures that retained the water. Flooding can be exacerbated by increased amounts of impervious surface or by other natural hazards such as wildfires, which reduce the supply of vegetation that can absorb rainfall. Periodic floods occur on many rivers, forming a surrounding region known as the flood plain.

During times of rain or snow, some of the water is retained in ponds or soil, some is absorbed by grass and vegetation, some evaporates, and the rest travels over the land as surface runoff. Floods occur when ponds, lakes, riverbeds, soil, and vegetation cannot absorb all the water. Water then runs off the land in quantities that cannot be carried within stream channels or retained in natural ponds, lakes, and man-made reservoirs. About 30 per cent of all precipitation becomes runoff and that amount might be increased by water from melting snow. River flooding is often caused by heavy rain, sometimes increased by melting snow. A flood that rises rapidly, with little or no advance warning, is called a flash flood. Flash floods usually result from intense rainfall over a relatively small area or if the area was already saturated from previous precipitation.

Severe winds over water

Even when rainfall is relatively light, the shorelines of lakes and bays can be flooded by severe winds — such as during hurricanes — that blow water into the shore areas.

Unusual high tides

Coastal areas are sometimes flooded by unusually high tides, such as spring tides, especially when compounded by high winds and storm surges.

Tsunamis

Tsunamis are high, large waves, typically caused by undersea earthquakes or massive explosions, such as the eruption of an undersea volcano.

Effects of Floods

Flooding has many impacts. It damages property and endangers the lives of humans and other species. Rapid water runoff causes soil erosion and concomitant sediment deposition elsewhere (such as further downstream or down a coast). The spawning grounds for fish and other wildlife habitats can become polluted or completely destroyed. Some prolonged high floods can delay traffic in areas which lack elevated roadways. Floods can interfere with drainage and economic use of lands, such as interfering with farming. Structural damage can occur in bridge abutments, bank lines, sewer lines, and other structures within floodways. Waterway navigation and hydroelectric power are often impaired. Financial losses due to floods are typically millions of dollars each year, with the worst floods in recent US history having cost billions dollars.

Control of Floods

Some methods of flood control have been practiced since ancient times. These methods include planting vegetation to retain extra water, terracing hillsides to slow flow downhill, and the construction of floodways (man-made channels to divert floodwater). Other techniques include the construction of levees, dikes, dams, reservoirs or retention ponds to hold extra water during times of flooding.

Methods of control

Dams

Many dams and their associated reservoirs are designed wholly or partially to aid in flood protection and control.

River defences

In many countries, rivers prone to floods are often carefully managed. Defences such as levees, bunds, reservoirs, and weirs are used to prevent rivers from overflowing their banks. When these defences fail, emergency measures such as sandbags or portable inflatable tubes are used. A weir, also known as a lowhead dam, is most often used to create millponds, but on the Humber River in Toronto, a weir was built near Raymore Drive to prevent a recurrence of the flood damage caused by Hurricane Hazel in 1954.

Coastal defences

Coastal Flooding has been addressed in Europe and the Americas with coastal defences, such as sea walls, beach nourishment, and barrier islands. Tide gates are used in conjunction with dykes and culverts. They can be placed at the mouth of streams or small rivers, where an estuary begins or where tributary streams or drainage ditches connect to sloughs. Tide gates close during incoming tides to prevent tidal waters from moving upland, and open during outgoing tides to allow waters to drain out via the culvert and into the estuary side of the dike. The opening and closing of the gates is driven by a difference in water level on either side of the gate.

Flood control by continent

Americas

Another elaborate system of floodway defenses can be found in the Canadian province of Manitoba. The Red River flows northward from the United States, passing through the city of Winnipeg (where it meets the Assiniboine River) and into Lake Winnipeg. As is the case with all north-flowing rivers in the temperate zone of the Northern Hemisphere, snowmelt in southern sections may cause river levels to rise before northern sections have had a chance to completely thaw. This can lead to devastating flooding, as occurred in Winnipeg during the spring of 1950. To protect the city from future floods, the Manitoba government undertook the construction of a massive system of diversions, dikes, and floodways (including the Red River Floodway and the Portage Diversion). The system kept Winnipeg safe during the 1997 flood which devastated many communities upriver from Winnipeg, including Grand Forks, North Dakota and Ste. Agathe, Manitoba.

In the US the New Orleans Metropolitan Area, 35 per cent of which sits below sea level, is protected by hundreds of miles of levees and flood gates. This system failed catastrophically, with numerous breaks, during Hurricane Katrina in the city proper and in eastern sections of the Metro Area, resulting in the inundation of approximately 50 per cent of the Metropolitan area, ranging from a few inches to twenty feet in coastal communities.

The Morganza Spillway provides a method of diverting water from the Mississippi river when a river flood threatens New Orleans, Baton Rouge and other major cities on the lower Mississippi. It is the largest of a system of spillways and floodways along the Mississippi. Completed in 1954, the spillway has been opened twice, in 1973 and in 2011.

In an act of successful flood prevention, the Federal Government of the United States offered to buy out flood-prone properties in the United States in order to prevent repeated disasters after the 1993 flood across the Midwest. Several communities accepted and the government, in partnership with the state, bought 25,000 properties which they converted into wetlands. These wetlands act as a sponge in storms and in 1995, when the floods returned, the government did not have to expend resources in those areas.

Asia

In China, flood diversion areas are rural areas that are deliberately flooded in emergencies in order to protect cities.

The consequences of deforestation and changing land use on the risk and severity are prone to discussion. In assessing the impacts of Himalayan deforestation on the Ganges-Brahmaputra Lowlands, it was found that forests would not have prevented or significantly reduced flooding in the case of an extreme weather event. However, more general or overview studies agree on the negative impacts deforestation has on flood safety — and the positive effects of wise land use and reforestation.

Europe

London is protected from flooding by a huge mechanical barrier across the River Thames, which is raised when the water level reaches a certain point.

Venice has a similar arrangement, although it is already unable to cope with very high tides. The defenses of both London and Venice will be rendered inadequate if sea levels continue to rise.

The largest and most elaborate flood defenses can be found in the Netherlands, where they are referred to as Delta Works with the Oosterschelde dam as its crowning achievement. These works were built in response to the North Sea flood of 1953, in the southwestern part of the Netherlands. The Dutch had already built one of the world's largest dams in the north of the country. The Afsluitdijk closing occurred in 1932.

Currently the Saint Petersburg Flood Prevention Facility Complex finished by 2008, in Russia, to protect Saint Petersburg from storm surges. It also has a main traffic function, as it completes a ring road around Saint Petersburg. Eleven dams extend for 25.4 km (15.8 mi) and stand 8 metres (26 ft) above water level.

Flood clean-up safety

Clean-up activities following floods often pose hazards to workers and volunteers involved in the effort. Potential dangers include electrical hazards, carbon monoxide exposure, musculoskeletal hazards, heat or cold stress, motor vehicle-related dangers, fire, drowning, and exposure to hazardous materials. Because flooded disaster sites are unstable, clean-up workers might encounter sharp jagged debris, biological hazards in the flood water, exposed electrical lines, blood or other body fluids, and animal and human remains. In planning for and reacting to flood disasters, managers provide workers with hard hats, goggles, heavy work gloves, life jackets, and watertight boots with steel toes and insoles.

Future

Europe is at the forefront of the flood control technology, with low-lying countries such as the Netherlands developing techniques that can serve as examples to other countries facing similar problems.

After hurricane Katrina, the US state of Louisiana sent politicians to the Netherlands to take a tour of the complex and highly developed flood control system in place in the Netherlands. With a BBC article quoting experts as saying 70 per cent more people will live in delta cities by 2050, the number of people impacted by a rise in sea level will greatly increase. The Netherlands has one of the best flood control systems in the world and new ways to deal with water are constantly being developed and tested, such as the underground storage of water, storing water in reservoirs in large parking garages or on playgrounds. Rotterdam started a project to construct a floating housing development of 120 acres (0.49 km^2) to deal with rising sea levels. Several approaches, from high-tech sensors detecting imminent levee failure to movable semi-circular structures closing an entire river, are being developed or used around the world. Regular maintenance of hydraulic structures, however, is another crucial part of flood control.

Benefits of Flooding

There are many disruptive effects of flooding on human settlements and economic activities. However, flooding can bring benefits, such as making soil more fertile and providing nutrients in which it is deficient. Periodic flooding was essential to the well-being of ancient communities along the Tigris-Euphrates Rivers, the Nile River, the Indus River, the Ganges and the Yellow River, among others. The viability for hydrologically based renewable sources of energy is higher in flood-prone regions.

SOIL CONSERVATION METHODS

Soil conservation is maintaining good soil health, by various practices. The aim of soil conservation methods is to prevent soil erosion, prevent soil's overuse and prevent soil contamination from chemicals. There are various measures that are used to maintain soil health, and prevent the above harms to soil. Here are the soil conservation methods which are practiced for soil management.

Soil Conservation Strategies

There are many ways to conserve soil, some are suited to those areas where farming is done, and some are according to soil needs. Here are the various soil conservation methods that are practiced.

Planting vegetation: This is one of the most effective and cost saving soil conservation methods. This measure is among soil conservation methods used by farmers. By planting trees, grass, plants, soil erosion can be greatly prevented. Plants help to stabilise the properties of soil and trees also act as a wind barrier and prevents soil from being blown away.

This is also among strategies used for soil conservation methods in urban areas, one can plant trees and plants in the landscape areas of the residential places. The best choices for vegetation are herbs, small trees, plants with wild flowers, and creepers which provide a ground cover.

Contour ploughing: Contour farming or ploughing is used by farmers, wherein they plough across a slope and follow the elevation contour lines. This methods prevents water run off, and thus prevents soil erosion by allowing water to slowly penetrate the soil.

Maintaining the soil pH: The measurement of soil's acidity or alkalinity is done by measuring the soil pH levels. Soil gets polluted due to the addition of basic or acidic pollutants which can be countered by maintaining the desirable pH of soil.

Soil organisms: Without the activities performed by soil organisms, the organic material required by plants will litter and won't be available for plant growth. Using beneficial soil organisms like earthworms, helps in aeration of soil and makes the macro-nutrients available for the plants. Thus, the soil becomes more fertile and porous.

Crop rotation practice: Crop rotation is the soil conservation method where a series of different crops are planted one after the other in the same soil area, and is used greatly in organic farming. This is done to prevent the accumulation of pathogens, which occur if the same plants are grown in the soil, and also depletion of nutrients.

Watering the soil: We water plants and trees, but it is equally important to water soil to maintain its health. Soil erosion occurs if the soil is blown away by wind. By watering and settling the soil, one can prevent soil erosion from the blowing away of soil by wind. One of the effective soil conservation methods in India is the drip irrigation system which provides water to the soil without the water running off.

Salinity management: Excessive collection of salts in the soil has harmful effects on the metabolism of plants. Salinity can lead to death of the vegetation and thus cause soil erosion, which is why salinity management is important.

Terracing: Terracing is among one of the best soil conservation methods, where cultivation is done on a terrace leveled section of land. In terracing, farming is done on a unique step like structure and the possibility of water running off is slowed down.

Bordering from indigenous crops: It is preferable to plant native plants, but when native plants are not planted then bordering the crops with indigenous crops is necessary. This helps to prevent soil erosion, and this measure is greatly opted in poor rural areas.

No-tilling farming method: The process of soil being ploughed for farming is called tilling, wherein the fertilisers get mixed and the rows for plantation are created. However, this method leads to death of beneficial soil organisms, loss of organic matter and compaction of soil. Due to these side effects, the no-tilling strategy is used to conserve soil health.

These were the 10 ways to conserve soil used across the world. Soil is a very important constituent, and is developed by a long process of weathering and disintegration of rocks which turn into sand or clay. The clay like fertile soil provides home to organisms like earthworms, beetles, ants which live in it. Soil provides anchorage to plants and trees. The plants and trees provide home to birds and animals. The crops growing on the soil provide us food and clothes. Thus, soil defines the quality of life around it, which is why it is important to use these soil conservation methods. Branches of environmental science like earth science are constantly trying to find new methods, for maintaining the ecological balance. In different parts of world people studying soil science, are coming up with different new beneficial soil conservation methods.

WATER CONSERVATION

Water conservation refers to reducing the usage of water and recycling of waste water for different purposes such as cleaning, manufacturing, and agricultural irrigation.

Water conservation can be defined as:

1. Any beneficial reduction in water loss, use or waste as well as the preservation of water quality.
2. A reduction in water use accomplished by implementation of water conservation or water efficiency measures.
3. Improved water management practices that reduce or enhance the beneficial use of water. A water conservation measure is an action, behavioural change, device, technology or improved

design or process implemented to reduce water loss, waste or use. Water efficiency is a tool of water conservation. This results in more efficient water use and thus reduces water demand. The value and cost-effectiveness of a water efficiency measure must be evaluated in relation to its effects on the use and cost of other natural resources (e.g. energy or chemicals).

Water Efficiency

The goals of water conservation efforts include as follows:
1. Sustainability: To ensure availability for future generations, the withdrawal of fresh water from an ecosystem should not exceed its natural replacement rate.
2. Energy conservation: Water pumping, delivery, and waste-water treatment facilities consume a significant amount of energy. In some regions of the world over 15 per cent of total electricity consumption is devoted to water management.
3. Habitat conservation: Minimising human water use helps to preserve fresh water habitats for local wildlife and migrating waterfowl, as well as reducing the need to build new dams and other water diversion infrastructure.

Social Solutions

Water conservation programs are typically initiated at the local level, by either municipal water utilities or regional governments. Common strategies include public outreach campaigns, tiered water rates (charging progressively higher prices as water use increases) or restrictions on outdoor water use such as lawn watering and car washing. Cities in dry climates often require or encourage the installation of xeriscaping or natural landscaping in new homes to reduce outdoor water usage.

One fundamental conservation goal is universal metering. The prevalence of residential water metering varies significantly worldwide. Recent studies have estimated that water supplies are metered in less than 30 per cent of UK households, and about 61 per cent of urban Canadian homes. Although individual water meters have often been considered impractical in homes with private wells or in multifamily buildings, the US Environmental protection agency estimates that metering alone can reduce consumption by 20 to 40 per cent. In addition to raising consumer awareness of their water use, metering is also an important way to identify and localise water leaks.

Some researchers have suggested that water conservation efforts should be primarily directed at farmers, in light of the fact that crop irrigation accounts for 70 per cent of the world's fresh water use. The agricultural sector of most countries is important both economically and politically, and water subsidies are common. Conservation advocates have urged removal of all subsidies to force farmers to grow more water-efficient crops and adopt less wasteful irrigation techniques.

Household Applications

Water-saving technology for the home includes:
1. Low-flow shower heads sometimes called energy-efficient shower heads as they also use less energy.
2. Low-flush toilets and composting toilets. These have a dramatic impact in the developed world, as conventional Western toilets use large volumes of water.
3. Dual flush toilets created by Caroma includes two buttons or handles to flush different levels of water. Dual flush toilets use up to 67 per cent less water than conventional toilets.
4. Saline water (sea water) or rain water can be used for flushing toilets.

5. Faucet aerators, which break water flow into fine droplets to maintain 'wetting effectiveness' while using less water. An additional benefit is that they reduce splashing while washing hands and dishes.
6. Waste-water reuse or recycling systems, allowing:
    (a) Reuse of graywater for flushing toilets or watering gardens.
    (b) Recycling of Waste-water through purification at a water treatment plant.
7. Rainwater harvesting.
8. High-efficiency clothes washers.
9. Weather-based irrigation controllers.
10. Garden hose nozzles that shut off water when it is not being used, instead of letting a hose run.
11. Using low flow taps in wash basins.
12. Automatic faucet is a water conservation faucet that eliminates water waste at the faucet. It automates the use of faucets without the use of hands.
13. A valve which reduces water, gas, time, money and CO_2 known as a Combisave.

Water can also be conserved by landscaping with native plants and by changing behaviour, such as shortening showers and not running the faucet while brushing teeth.

Commercial Applications

Many water-saving devices (such as low-flush toilets) that are useful in homes can also be useful for business water saving. Other water-saving technology for businesses includes:

1. Waterless urinals.
2. Waterless car washes.
3. Infrared or foot-operated faucets, which can save water by using short bursts of water for rinsing in a kitchen or bathroom.
4. Pressurised waterbrooms, which can be used instead of a hose to clean sidewalks.
5. X-ray film processor re-circulation systems.
6. Cooling tower conductivity controllers.
7. Utilisation of lake water and or sea water for cooling towers.
8. Water-saving steam sterilisers, for use in hospitals and health care facilities.

One of the method of water conservation is rain water harvesting. However, ultra-low flow sink faucets, particularly those whose flow rate is less than 75 GPM have been shown to have serious undesired consequences, including increased wash time, hands not completely cleaned, and some users choosing to forgo washing altogether to avoid the inconvenience.

Avoid contamination of water. Some tips on conserving water are to fill one sink with wash water and one sink with rinse water when you are washing dishes, upgrade to air-cooled appliances, adjust your sprinklers so only your lawn is watered, and not your house, street or sidewalk. Choose shrubs or groundcovers only, because turfs can absorb too much water, and when you walk on it, becomes squishy and moist.

Agricultural Applications

For crop irrigation, optimal water efficiency means minimising losses due to evaporation, runoff or subsurface drainage while maximising production. An evaporation pan in combination with specific crop correction factors can be used to determine how much water is needed to satisfy plant requirements. Flood irrigation, the oldest and most common type, is often very uneven in distribution, as parts of a field may receive excess water in order to deliver sufficient quantities to other parts. Overhead irrigation, using center-pivot or lateral-moving sprinklers, has the potential for a much more equal and controlled distribution pattern. Drip irrigation is the most expensive and least-used type, but offers the ability to deliver water to plant roots with minimal losses. However, drip irrigation is increasingly affordable, especially for the home gardener and in light of rising water rates. There are also cheap effective methods similar to drip irrigation such as the use of soaking hoses that can even be submerged in the growing medium to eliminate evaporation.

As changing irrigation systems can be a costly undertaking, conservation efforts often concentrate on maximising the efficiency of the existing system. This may include chiseling compacted soils, creating furrow dikes to prevent runoff, and using soil moisture and rainfall sensors to optimise irrigation schedules. Usually large gains in efficiency are possible though measurement and more effective management of the existing irrigation system. Infiltration basins, also called recharge pits, capture rainwater and recharge ground water supplies. Use of these management practices reduces soil erosion caused by stormwater runoff and improves water quality in nearby surface waters.

SECTION VII

Biodiversity and Ecological Perspective

Chapter 25

Biodiversity

INTRODUCTION

Biodiversity is the degree of variation of life forms within a given ecosystem, biome, or an entire planet. Biodiversity is a measure of the health of ecosystems. Greater biodiversity implies greater health. Biodiversity is in part a function of climate. In terrestrial habitats, tropical regions are typically rich whereas polar regions support fewer species. Rapid environmental changes typically cause extinctions. One estimate is that less than one per cent of the species that have existed on earth are extant.

Since life began on earth, five major mass extinctions and several minor events have led to large and sudden drops in biodiversity. The Phanerozoic eon (the last 540 million years) marked a rapid growth in biodiversity via the Cambrian explosion — a period during which nearly every phylum of multicellular organisms first appeared. The next 400 million years included repeated, massive biodiversity losses classified as mass extinction events. In the Carboniferous, rainforest collapse led to a great loss of plant and animal life. The Permian–Triassic extinction event, 251 million years ago, was the worst; vertebrate recovery took 30 million years. The most recent, the Cretaceous–Tertiary extinction event, occurred 65 million years ago, and has often attracted more attention than others because it resulted in the extinction of the dinosaurs.

The period since the emergence of humans has displayed an ongoing biodiversity reduction and an accompanying loss of genetic diversity. Named the Holocene extinction, the reduction is caused primarily by human impacts, particularly habitat destruction. Biodiversity's impact on human health is a major international issue.

'Biological diversity' or 'biodiversity' can have many interpretations. It is most commonly used to replace the more clearly defined and long established terms, species diversity and species richness. Biologists most often define biodiversity as the 'totality of genes, species, and ecosystems of a region'. An advantage of this definition is that it seems to describe most circumstances and presents a unified view of the traditional three levels at which biological variety has been identified:

1. Species diversity.
2. Ecosystem diversity.
3. Genetic diversity.

Measuring diversity at one level in a group of organisms may not precisely correspond to diversity at other levels. However, tetrapod (terrestrial vertebrates) taxonomic and ecological diversity shows a very close correlation.

DISTRIBUTION

Selection bias amongst researchers may contribute to biased empirical research for modern estimates of biodiversity. Biodiversity is not evenly distributed. Flora and fauna diversity depends on climate, altitude, soils and the presence of other species. Diversity consistently measures higher in the tropics and in other localised regions such as Cape Floristic Province and lower in polar regions generally. In 2008 many species were formally classified as rare or endangered or threatened; moreover, scientists have estimated that millions more species are at risk which have not been formally recognised. About 40 per cent of the 40,177 species assessed using the IUCN Red List criteria are now listed as threatened with extinction — a total of 16,119.

Even though terrestrial biodiversity declines from the equator to the poles, this characteristic is unverified in aquatic ecosystems, especially in marine ecosystems. In addition, several assessments reveal tremendous diversity in higher latitudes. Generally terrestrial biodiversity is up to 25 times greater than ocean biodiversity.

A biodiversity hotspot is a region with a high level of endemic species. Hotspots were first named in 1988 by Dr. Norman Myers. Many hotspots have large nearby human populations. Most hotspots are located in the tropics and most of them are forests.

Brazil's Atlantic Forest is considered one such hotspot, containing roughly 20,000 plant species, 1350 vertebrates, and millions of insects, about half of which occur nowhere else. The island of Madagascar, particularly the unique Madagascar dry deciduous forests and lowland rainforests, possess a high ratio of endemism. Since the island separated from mainland Africa 65 million years ago, many species and ecosystems have evolved independently. Indonesia's 17,000 islands cover 7,35,355 square miles (1,904,560 km^2) contain 10 per cent of the world's flowering plants, 12 per cent of mammals and 17 per cent of reptiles, amphibians and birds — along with nearly 240 million people. Many regions of high biodiversity and/or endemism arise from specialised habitats which require unusual adaptations, for example alpine environments in high mountains or Northern European peat bogs.

EVOLUTION

Biodiversity is the result of 3.5 billion years of evolution. The origin of life has not been definitely established by science, however some evidence suggests that life may already have been well-established only a few hundred million years after the formation of the earth. Until approximately 600 million years ago, all life consisted of archaea, bacteria, protozoans and similar single-celled organisms.

The fossil record suggests that the last few million years featured the greatest biodiversity in history. However, not all scientists support this view, since there is uncertainty as to how strongly the fossil record is biased by the greater availability and preservation of recent geologic sections. Some scientists believe that corrected for sampling artifacts, modern biodiversity may not be much different from biodiversity 300 million years ago, whereas others consider the fossil record reasonably reflective of the diversification of life. Estimates of the present global macroscopic species diversity vary from 2 million to 100 million, with a best estimate of somewhere near 13–14 million, the vast majority arthropods. Diversity appears to increase continually in the absence of natural selection.

Evolutionary Diversification

The existence of a 'global carrying capacity', limiting the amount of life that can live at once, is debated, as is the question of whether such a limit would also cap the number of species. While records of life in the sea shows a logistic pattern of growth, life on land (insects, plants and tetrapods) shows an exponential rise in diversity. As one author states, 'Tetrapods have not yet invaded 64 per cent of potentially habitable modes, and it could be that without human influence the ecological and taxonomic diversity of tetrapods would continue to increase in an exponential fashion until most or all of the available ecospace is filled'.

On the other hand, changes through the Phanerozoic correlate much better with the hyperbolic model (widely used in population biology, demography and macrosociology, as well as fossil biodiversity) than with exponential and logistic models. The latter models imply that changes in diversity are guided by a first-order positive feedback (more ancestors, more descendants) and/or a negative feedback arising from resource limitation. Hyperbolic model implies a second-order positive feedback. The hyperbolic pattern of the world population growth arises from a second-order positive feedback between the population size and the rate of technological growth. The hyperbolic character of biodiversity growth can be similarly accounted for by a feedback between diversity and community structure complexity. The similarity between the curves of biodiversity and human population probably comes from the fact that both are derived from the interference of the hyperbolic trend with cyclical and stochastic dynamics.

Most biologists agree however that the period since human emergence is part of a new mass extinction, named the Holocene extinction event, caused primarily by the impact humans are having on the environment. It has been argued that the present rate of extinction is sufficient to eliminate most species on the planet earth within 100 years.

New species are regularly discovered (on average between 5–10,000 new species each year, most of them insects) and many, though discovered, are not yet classified (estimates are that nearly 90 per cent of all arthropods are not yet classified). Most of the terrestrial diversity is found in tropical forests.

HUMAN BENEFITS

Biodiversity supports ecosystem services including air quality, climate (e.g. CO_2 sequestration), water purification, pollination, and prevention of erosion. Since the stone age, species loss has accelerated above the prior rate, driven by human activity. Estimates of species loss are at a rate 100–10,000 times as fast as is typical in the fossil record. Non-material benefits include spiritual and aesthetic values, knowledge systems and the value of education.

Agriculture

The reservoir of genetic traits present in wild varieties and traditionally grown landraces is extremely important in improving crop performance. Important crops, such as potato, banana and coffee, are often derived from only a few genetic strains. Improvements in crop species over the last 250 years have been largely due to incorporating genes from wild varieties and species into cultivars. Crop breeding for beneficial traits has helped to more than double crop production in the last 50 years as a result of the Green Revolution. A biodiverse environment preserves the genome from which such productive genes are drawn.

Crop diversity aids recovery when the dominant cultivar is attacked by a disease or predator:

1. The Irish potato blight of 1846 was a major factor in the deaths of one million people and the emigration of another million. It was the result of planting only two potato varieties, both vulnerable to the blight.
2. When rice grassy stunt virus struck rice fields from Indonesia to India in the 1970s, 6,273 varieties were tested for resistance. Only one was resistant, an Indian variety, and known to science only since 1966. This variety formed a hybrid with other varieties and is now widely grown.
3. Coffee rust attacked coffee plantations in Sri Lanka, Brazil, and Central America in 1970. A resistant variety was found in Ethiopia. Although the diseases are themselves a form of biodiversity.

Monoculture was a contributing factor to several agricultural disasters, including the European wine industry collapse in the late 19th century, and the US Southern Corn Leaf Blight epidemic of 1970.

Although about 80 per cent of humans' food supply comes from just 20 kinds of plants, humans use at least 40,000 species. Many people depend on these species for food, shelter, and clothing. Earth's surviving biodiversity provides resources for increasing the range of food and other products suitable for human use, although the present extinction rate shrinks that potential.

Human Health

Biodiversity's relevance to human health is becoming an international political issue, as scientific evidence builds on the global health implications of biodiversity loss. This issue is closely linked with the issue of climate change, as many of the anticipated health risks of climate change are associated with changes in biodiversity (e.g. changes in populations and distribution of disease vectors, scarcity of fresh water, impacts on agricultural biodiversity and food resources, etc.) Some of the health issues influenced by biodiversity include dietary health and nutrition security, infectious disease, medical science and medicinal resources, social and psychological health. Biodiversity is also known to have an important role in reducing disaster risk, and in post-disaster relief and recovery efforts.

Biodiversity provides critical support for drug discovery and the availability of medicinal resources. A significant proportion of drugs are derived, directly or indirectly, from biological sources: at least 50 per cent of the pharmaceutical compounds on the US market are derived from plants, animals, and micro-organisms, while about 80 per cent of the world population depends on medicines from nature (used in either modern or traditional medical practice) for primary healthcare. Only a tiny fraction of wild species has been investigated for medical potential. Biodiversity has been critical to advances throughout the field of bionics. Evidence from market analysis and biodiversity science indicates that the decline in output from the pharmaceutical sector since the mid-1980s can be attributed to a move away from natural product exploration (bioprospecting) in favour of genomics and synthetic chemistry; meanwhile, natural products have a long history of supporting significant economic and health innovation. Marine ecosystems are particularly important, although inappropriate bioprospecting can increase biodiversity loss, as well as violating the laws of the communities and states from which the resources are taken. Higher biodiversity also limits the spread of infectious diseases as many different species act as buffers to them.

Business and Industry

Many industrial materials derive directly from biological sources. These include building materials, fibres, dyes, rubber and oil. Biodiversity is also important to the security of resources such as water, timber, paper, fibre, and food. As a result, biodiversity loss is a significant risk factor in business development and a threat to long term economic sustainability.

Leisure, Cultural and Aesthetic Value

Biodiversity enriches leisure activities such as hiking, birdwatching or natural history study. Biodiversity inspires musicians, painters, sculptors, writers and other artists. Many cultures view themselves as an integral part of the natural world which requires them to respect other living organisms.

Popular activities such as gardening, fishkeeping and specimen collecting strongly depend on biodiversity. The number of species involved in such pursuits is in the tens of thousands, though the majority do not enter commerce. The relationships between the original natural areas of these often exotic animals and plants and commercial collectors, suppliers, breeders, propagators and those who promote their understanding and enjoyment are complex and poorly understood. The general public responds well to exposure to rare and unusual organisms, reflecting their inherent value.

Philosophically it could be argued that biodiversity has intrinsic aesthetic and spiritual value to mankind in and of itself. This idea can be used as a counterweight to the notion that tropical forests and other ecological realms are only worthy of conservation because of the services they provide.

Other Services

Biodiversity supports many ecosystem services that are often not readily visible. It plays a part in regulating the chemistry of our atmosphere and water supply. Biodiversity is directly involved in water purification, recycling nutrients and providing fertile soils. Experiments with controlled environments have shown that humans cannot easily build ecosystems to support human needs; for example insect pollination cannot be mimicked, and that activity alone represents tens of billions of dollars in ecosystem services per year to humankind. Ecosystem stability is also positively related to biodiversity, protecting against disruption by extreme weather or human exploitation.

According to the Global Taxonomy Initiative and the European Distributed Institute of Taxonomy, the total number of species for some phyla may be much higher than what was known in 2010:

1. 10–30 million insects; (of some 0.9 million we know today).
2. 5–10 million bacteria.
3. 1.5 million fungi; (of some 0.075 million we know today).
4. 1 million mites.
5. The number of microbial species is not reliably known, but the Global Ocean Sampling Expedition dramatically increased the estimates of genetic diversity by identifying an enormous number of new genes from near-surface plankton samples at various marine locations, initially over the 2004–2006 period. The findings may eventually cause a significant change in the way science defines species and other taxonomic categories.

Since the rate of extinction has increased, many extant species may become extinct before they are described.

SPECIES LOSS RATES

During the last century, decreases in biodiversity have been increasingly observed. In 2007, German Federal Environment Minister Sigmar Gabriel cited estimates that up to 30 per cent of all species will be extinct by 2050. Of these, about one eighth of known plant species are threatened with extinction. Estimates reach as high as 1,40,000 species per year (based on Species-area theory). This figure indicates unsustainable ecological practices, because few species emerge each year. Almost all scientists acknowledge that the rate of species loss is greater now than at any time in human history, with extinctions occurring at rates hundreds of times higher than background extinction rates.

THREATS

Habitat Destruction

Habitat destruction has played a key role in extinctions, especially related to tropical forest destruction. Factors contributing to habitat loss are: overpopulation, deforestation, pollution (air pollution, water pollution, soil contamination) and global warming or climate change.

Habitat size and numbers of species are systematically related. Physically larger species and those living at lower latitudes or in forests or oceans are more sensitive to reduction in habitat area. Conversion to 'trivial' standardised ecosystems (e.g. monoculture following deforestation) effectively destroys habitat for the more diverse species that preceded the conversion. In some countries lack of property rights or lax law/regulatory enforcement necessarily leads to biodiversity loss (degradation costs having to be supported by the community).

A 2007 study conducted by the National Science Foundation found that biodiversity and genetic diversity are codependent — that diversity among species requires diversity within a species, and vice versa. 'If any one type is removed from the system, the cycle can break down, and the community becomes dominated by a single species'. At present, the most threatened ecosystems are found in fresh water, according to the Millennium Ecosystem Assessment 2005, which was confirmed by the 'Freshwater Animal Diversity Assessment', organised by the biodiversity platform, and the French Institute of Research Development.

Co-extinctions are a form of habitat destruction. Co-extinction occurs when the extinction or decline in one accompanies the other, such as in plants and beetles.

Introduced and Invasive Species

Barriers such as large rivers, seas, oceans, mountains and deserts encourage diversity by enabling independent evolution on either side of the barrier. Invasive species occur when those barriers are blurred. Without barriers such species occupy new niches, substantially reducing diversity. Repeatedly humans have helped these species circumvent these barriers, introducing them for food and other purposes. This has occurred on a time scale much shorter than the eons that historically have been required for a species to extend its range.

Not all introduced species are invasive, nor all invasive species deliberately introduced. In cases such as the zebra mussel, invasion of US waterways was unintentional. In other cases, such as mongooses in Hawaii, the introduction is deliberate but ineffective (nocturnal rats were not vulnerable to the diurnal mongoose). In other cases, such as oil palms in Indonesia and Malaysia, the introduction produces substantial economic benefits, but the benefits are accompanied by costly unintended consequences.

Finally, an introduced species may unintentionally injure a species that depends on the species it replaces. In Belgium, Prunus spinosa from Eastern Europe leafs much sooner than its West European counterparts, disrupting the feeding habits of the Thecla betulae butterfly (which feeds on the leaves). Introducing new species often leaves endemic and other local species unable to compete with the exotic species and unable to survive. The exotic organisms may be predators, parasites or may simply outcompete indigenous species for nutrients, water and light.

At present, several countries have already imported so many exotic species, particularly agricultural and ornamental plants, that the own indigenous fauna/flora may be outnumbered.

Genetic pollution

Endemic species can be threatened with extinction through the process of genetic pollution, i.e. uncontrolled hybridisation, introgression and genetic swamping. Genetic pollution leads to homogenisation or replacement of local genomes as a result of either a numerical and/or fitness advantage of an introduced species. Hybridisation and introgression are side-effects of introduction and invasion. These phenomena can be especially detrimental to rare species that come into contact with more abundant ones. The abundant species can interbreed with the rare species, swamping its gene pool. This problem is not always apparent from morphological (outward appearance) observations alone. Some degree of gene flowis normal adaptation, and not all gene and genotype constellations can be preserved. However, hybridisation with or without introgression may, nevertheless, threaten a rare species existence.

Overexploitation

Overexploitation occurs when a resource is consumed at an unsustainable rate. This occurs on land in the form of overhunting, excessive logging, poor soil conservation in agriculture and the illegal wildlife trade. Joe Walston, director of the Wildlife Conservation Society's Asian programs, called the latter the 'single largest threat' to biodiversity in Asia. The international trade of endangered species is second in size only to drug trafficking.

About 25 per cent of world fisheries are now overfished to the point where their current biomass is less than the level that maximises their sustainable yield. The overkill hypothesis explains why earlier megafaunal extinctions occurred within a relatively short period of time. This can be connected with human migration.

Hybridisation, Genetic Pollution/Erosion and Food Security

In agriculture and animal husbandry, the green revolution popularised the use of conventional hybridisation to increase yield. Often hybridised breeds originated in developed countries and were further hybridised with local varieties in the developing world to create high yield strains resistant to local climate and diseases. Local governments and industry have been pushing hybridisation. Formerly huge gene pools of various wild and indigenous breeds have collapsed causing widespread genetic erosion and genetic pollution. This has resulted in loss of genetic diversity and biodiversity as a whole.

GM organisms have genetic material altered by genetic engineering procedures such as recombinant DNA technology. GM crops have become a common source for genetic pollution, not only of wild varieties but also of domesticated varieties derived from classical hybridisation.

Genetic erosion coupled with genetic pollution may be destroying unique genotypes, thereby creating a hidden crisis which could result in a severe threat to our food security. Diverse genetic material could cease to exist which would impact our ability to further hybridise food crops and livestock against more resistant diseases and climatic changes.

Climate Change

Global warming is also considered to be a major threat to global biodiversity. For example coral reefs—which are biodiversity hotspots will be lost in 20 to 40 years if global warming continues at the current trend. In 2007, an international collaborative study on four continents estimated that 10 per cent of species would become extinct by 2050 because of global warming. 'We need to limit climate change or we wind up with a lot of species in trouble, possibly extinct', said Dr. Lee Hannah, a co-author of the paper and chief climate change biologist at the Center for Applied Biodiversity Science at Conservation International.

Overpopulation

From 1950 to 2005, world population increased from 2.5 billion to 6.5 billion and is forecast to reach a plateau of more than 9 billion during the 21st century. Sir David King, former chief scientific adviser to the UK government, told a parliamentary inquiry: 'It is self-evident that the massive growth in the human population through the 20th century has had more impact on biodiversity than any other single factor'.

HOLOCENE EXTINCTION

Rates of decline in biodiversity in this sixth mass extinction match or exceed rates of loss in the five previous mass extinction events in the fossil record. Loss of biodiversity results in the loss of natural capital that supplies ecosystem goods and services. The economic value of 17 ecosystem services for earth's biosphere (calculated in 2007) has an estimated value of US$ 37 trillion (3.7×10^{13}) per year.

CONSERVATION

Conservation biology matured in the mid-20th century as ecologists, naturalists, and other scientists began to research and address issues pertaining to global biodiversity declines. The conservation ethic advocates management of natural resources for the purpose of sustaining biodiversity in species, ecosystems, the evolutionary process, and human culture and society.

Conservation biology is reforming around strategic plans to protect biodiversity. Preserving global biodiversity is a priority in strategic conservation plans that are designed to engage public policy and concerns affecting local, regional and global scales of communities, ecosystems, and cultures. Action plans identify ways of sustaining human well-being, employing natural capital, market capital, and ecosystem services.

PROTECTION AND RESTORATION TECHNIQUES

The most powerful technique is to preserve habitat. Exotic species removal allows less competitive species to recover their ecological niches. Exotic species that have become a pest can be identified taxonomically (e.g. with digital automated identification system (DAISY), using the barcode of life. Removal is practical only given large groups of individuals due to the econimic cost.

Once the preservation of the remaining native species in an area is assured. Missing species can be identified and reintroduced using databases such as the Encyclopedia of Life and the Global Biodiversity Information Facility.

Other techniques include:

1. Biodiversity banking places a monetary value on biodiversity. One example is the Australian Native Vegetation Management Framework.
2. Gene banks are collections of specimens and genetic material. Some banks intend to reintroduce banked species to the ecosystem (e.g. via tree nurseries).
3. Reducing and better targeting of pesticides allows more species to survive in agricultural and urbanised areas.
4. Location-specific approaches are less useful for protecting migratory species. One approach is to create wildlife corridors that correspond to the animals' movements. National and other boundaries can complicate corridor creation.

Resource Allocation

Focusing on limited areas of higher potential biodiversity promises greater immediate return on investment than spreading resources evenly or focusing on areas of little diversity but greater interest in biodiversity.

A second strategy focuses on areas that retain most of their original diversity, which typically require little or no restoration. These are typically non-urbanised, non-agricultural areas. Tropical areas often fit both criteria, given their natively high diversity and relative lack of development.

LEGAL STATUS

Biodiversity is taken into account in some political and judicial decisions:

1. The relationship between law and ecosystems is very ancient and has consequences for biodiversity. It is related to private and public property rights. It can define protection for threatened ecosystems, but also some rights and duties (for example, fishing and hunting rights).

2. Law regarding species is more recent. It defines species that must be protected because they may be threatened by extinction. The US Endangered Species Act is an example of an attempt to address the 'law and species' issue.

3. Laws regarding gene pools are only about a century old. Domestication and plant breeding methods are not new, but advances in genetic engineering has led to tighter laws covering distribution of genetically modified organisms, gene patents and process patents. Governments struggle to decide whether to focus on for example, genes, genomes or organisms and species.

Global agreements such as the (Convention on Biological Diversity), give 'sovereign national rights over biological resources' (not property). The agreements commit countries to 'conserve biodiversity', 'develop resources for sustainability' and 'share the benefits' resulting from their use. Biodiverse countries that allow bioprospecting or collection of natural products, expect a share of the benefits rather than allowing the individual or institution that discovers/exploits the resource to capture them privately. Bioprospecting can become a type of biopiracy when such principles are not respected.

Sovereignty principles can rely upon what is better known as Access and Benefit Sharing Agreements (ABAs). The Convention on Biodiversity implies informed consent between the source country and the collector, to establish which resource will be used and for what, and to settle on a fair agreement on benefit sharing.

Uniform approval for use of biodiversity as a legal standard has not been achieved, however. Bosselman argues that biodiversity should not be used as a legal standard, claiming that the remaining areas of scientific uncertainty cause unacceptable administrative waste and increase litigation without promoting preservation goals.

ANALYTICAL LIMITS

Taxonomic and Size Relationships

Less than one per cent of all species that have been described have been studied beyond simply noting their existence. The vast majority of earth's species are microbial. Contemporary biodiversity physics is 'firmly fixated on the visible [macroscopic] world'. For example, microbial life is metabolically and environmentally more diverse than multicellular life . 'On the tree of life, based on analyses of small-subunit ribosomal RNA, visible life consists of barely noticeable twigs. The inverse relationship of size and population recurs higher on the evolutionary ladder — to a first approximation, all multicellular species on earth are insects'. Insect extinction rates are high — supporting the Holocene extinction hypothesis.

Chapter 26

Threats to Global Biodiversity

INTRODUCTION

Biodiversity refers to the number and variety of species, of ecosystems, and of the genetic variation contained within species. Roughly 1.4 million species are known to science, but because many species are undescribed, an estimated 10–30 million species likely exists at present. Biodiversity is threatened by the sum of all human activities. It is useful to group threats into the categories of over-hunting, habitat destruction, invasion of non-native species, domino effects, pollution, and climate change. Habitat loss presents the single greatest threat to world biodiversity, and the magnitude of this threat can be approximated from species-area curves and rates of habitat loss. The spread of non-native species threatens many local species with extinction, and pushes the world's biota toward a more homogeneous and widely distributed subset of survivors. Climate change threatens to force species and ecosystems to migrate toward higher latitudes, with no guarantee of suitable habitat or access routes. These three factors thus are of special concern.

Three pie charts Figs 26.1 to 26.3 (number of species of all kinds, number of animal species, number of plant species) show the same information diagramatically.

These numbers reflect a huge amount of scientific study. In order for a species to be included on the above list, it was examined by a specialist, who carefully compared its features to those of all similar species. Then, once satisfied that the species in hand was different from all known forms, the specialist published a thorough description including drawings, counts of hairs along forelegs, length of measurable parts such as limbs and antennae, photographs, etc. The new species has then been described, and is added to our list of 'known' species.

Remarkably, our estimates of the number of unknown species greatly exceed our count of the number of known species. Most experts estimate the world's species diversity at 10 to 30 million, but that is very approximate. Only 1.4 million species are 'known to science'—meaning that they have been classified by a specialist. The estimates of 10 to 30 million species are based on expert opinion of how many species are yet to be formally identified. One study of insects in the forest canopy found 5 out of 6 to be new species. Even vertebrates are not completely known—it is estimated that nearly half of the freshwater fishes of South America are undescribed. New finds are made continuously in the tropics, and exploration of deep-sea hydrothermal vents recently led to the discovery not just of new species, but of new life forms at the family level (20 families or subfamilies). When you consider that virtually every species has its own parasite, and how many groups such as nematodes and bacteria have yet to be well-studied, it is apparent that the estimates of 10 to 30 million are not out of line.

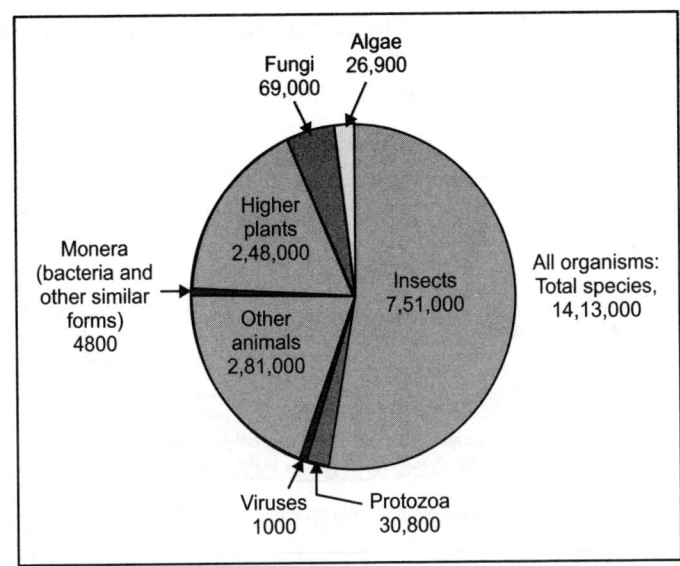

Fig. 26.1. Number of living species of all organisms currently known.

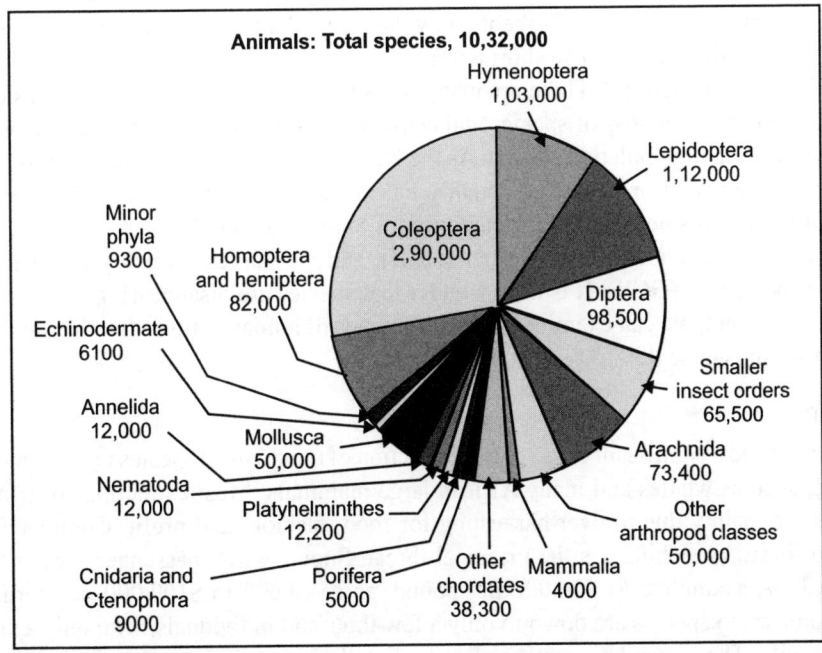

Γig. 26.2. Number of living animal species currently known.

THREATS TO BIODIVERSITY

Extinction is a natural event and, from a geological perspective, routine. We now know that most species that have ever lived have gone extinct. The average rate over the past 200 million years is 1–2 species per year, and 3–4 families per million years. The average duration of a species is 2–10 million years (based on last 200 million years). There have also been occasional episodes of mass extinction, when many taxa representing a wide array of lifeforms have gone extinct in the same blink of geological time.

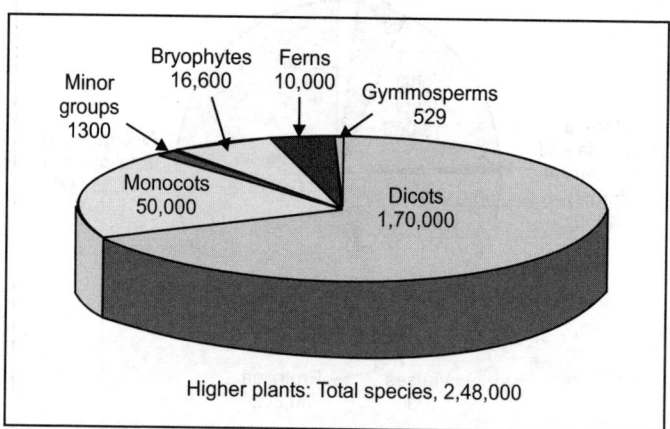

Fig. 26.3. Number of living species of higher plants currently known.

In the modern era, due to human actions, species and ecosystems are threatened with destruction to an extent rarely seen in earth history. Probably only during the handful of mass extinction events have so many species been threatened, in so short a time.

What are these human actions? There are many ways to conceive of these—let's consider two.

First, we can attribute the loss of species and ecosystems to the accelerating transformation of the earth by a growing human population (GCII). As the human population passes the six billion mark, we have transformed, degraded or destroyed roughly half of the word's forests (GCII). We appropriate roughly half of the world's net primary productivity for human use (GCII).

We appropriate most available fresh water (GCII), and we harvest virtually all of the available productivity of the oceans (GCII). It is little wonder that species are disappearing and ecosystems are being destroyed. Second, we can examine six specific types of human actions that threaten species and ecosystems - the 'sinister sextet'.

Over-Hunting

Over-hunting has been a significant cause of the extinction of hundreds of species and the endangerment of many more, such as whales and many African large mammals. Most extinctions over past several hundred years are mainly due to over-harvesting for food, fashion, and profit. Commercial hunting, both legal and illegal (poaching), is the principal threat. Snowy egret, passenger pigeon (Fig. 26.4), heath hen are USA examples. At $16,000 per pound, and $40,000 to $100,000 per horn, it is little wonder that some rhino species are down to only a few thousand individuals, with only a slim hope of survival in the wild. The pet and decorative plant trade falls within this commercial hunting category, and includes a mix of legal and illegal activities. The annual trade is estimated to be at least $5 billion, with perhaps 1/4 to 1/3 of it illegal.

Sport or recreational hunting causes no endangerment of species where it is well regulated, and may help to bring back a species from the edge of extinction. Many wildlife managers view sport hunting as the principal basis for protection of wildlife.

While over-hunting, particularly illegal poaching, remains a serious threat to certain species, for the future, it is less important than other factors mentioned next.

Fig. 26.4. Passenger pigeon.

Habitat Loss/Degradation/Fragmentation

Habitat loss/degradation/fragmentation is an important cause of known extinctions. As deforestation proceeds in tropical forests, this promises to become the cause of mass extinctions caused by human activity. All species have specific food and habitat needs. The more specific these needs and localised the habitat, the greater the vulnerability of species to loss of habitat to agricultural land, livestock, roads and cities. In the future, the only species that survive are likely to be those whose habitats are highly protected, or whose habitat corresponds to the degraded state associated with human activity (human commensals).

Habitat damage, especially the conversion of forested land to agriculture (and, often, subsequent abandonment as marginal land), has a long human history. It began in China about 4000 years ago, was largely completed in Europe by about 400 years ago, and swept across USA over the past 200 years or so. Viewed in this historical context, we are now mopping up the last forests of Pacific Northwest.

In the new world tropics, lowland, seasonal, deciduous forest began to disappear with Spanish and Portuguese colonisation of the New World. These were the forested regions most easily converted to agriculture, and with a more welcoming climate. The more forbidding, tropical humid forests came under attack mainly in 20th century, under the combined influences of population growth, inequitable land and income distribution, and development policies that targeted rain forests as the new frontier to colonise.

Tropical forests are so important because they harbour at least 50 per cent, and perhaps more, of world's biodiversity. Direct observations, reinforced by satellite data, documents that these forests are declining. The original extent of tropical rain forests was 15 million km². Now there remains about 7.5–8 million km², so half is gone. The current rate of loss is estimated at near 2 per cent annually (100,000 km² destroyed, another 100,000 km² degraded). While there is uncertainty regarding the rate of loss, and what it will be in future, the likelihood is that tropical forests will be reduced to 10–25 per cent of their original extent by late 21st century. Habitat fragmentation is a further aspect of habitat loss that often goes unrecognised. The forest, meadow, or other habitat that remains generally is in small, isolated bits rather than in large, intact units.

Each is a tiny island that can at best maintain a very small population. Environmental fluctuations, disease, and other chance factors make such small isolates highly vulnerable to extinction. Any species that requires a large home range, such as a grizzly bear, will not survive if the area is too small. Finally, we know that small land units are strongly affected by their surroundings, in terms of climate, dispersing species, etc. As a consequence, the ecology of a small isolate may differ from that of a similar ecosystem on a larger scale.

For the future, habitat loss, degradation, and fragmentation combined is the single most important factor in the projected extinction crisis.

Invasion of Non-native Species

Invasion of non-native species is an important and often-overlooked cause of extinctions. The African Great Lakes—Victoria, Malawi and Tanganyika—are famous for their great diversity of endemic species, termed 'species flocks', of cichlid fishes.

In Lake Victoria, a single, exotic species, the Nile Perch, has become established and may cause the extinction of most of the native species, by simply eating them all. It was a purposeful introduction for subsistence and sports fishing, and a great disaster. Of all documented extinctions since 1600, introduced species appear to have played a role in at least half. The clue is the disproportionate number of species lost from islands: some 93 per cent of 30 documented extinctions of species and subspecies of amphibians and reptiles, 93 per cent of 176 species and subspecies of land and freshwater birds, but only 27 per cent of 114 species and subspecies of mammals. Why are island species so vulnerable, and why is this evidence of the role of non-indigenous species?

Islands are laboratories for evolution (occur when the removal of one species (an extinction event) or the addition of one species (an invasion event) affects the entire biological system. Domino effects are especially likely when two or more species are highly interdependent, or when the affected species is a 'keystone' species, meaning that it has strong connections to many other species (GCI).

The seeds of the tree Calvaria major, now found exclusively on the island of Mauritius, must pass through the abrasive gut of a large animal in order to germinate. Their tough seed coats are protection against digestion, but also a kind of living coffin, for the seed cannot germinate unless abraded. None of the animals currently on Mauritius have that ability.

The dodo (a 25 kg pigeon), hunted to extinction in the late 17th century, probably was the key to recruitment in this species. Some seeds, abraded, roughened, and excreted by dodos, germinated and grew. Today, no seeds germinate, and only a few very old trees now survive. The blackfooted ferret was once very abundant in the western prairies. It preyed upon prairie dogs and used their burrows to nest in. Poisoning of prairie dogs has greatly reduced their abundance, and the blackfooted ferret is now the rarest mammal in North America.

Pollution

Pollution from chemical contaminants certainly poses a further threat to species and ecosystems. While not commonly a cause of extinction, it likely can be for species whose range is extremely small, and threatened by contamination. Several species of desert pupfish, occurring in small isolated pools in the US southwest, are examples.

Climate Change

A changing global climate threatens species and ecosystems. The distribution of species (biogeography) is largely determined by climate, as is the distribution of ecosystems and plant vegetation zones (biomes) [GCI]. Climate change may simply shift these distributions but, for a number of reasons, plants and animals may not be able to adjust.

The pace of climate change almost certainly will be more rapid than most plants are able to migrate. The presence of roads, cities, and other barriers associated with human presence may provide no opportunity for distributional shifts. Parks and nature reserves are fixed locations. The climate that characterises present-day Yellowstone Park will shift several hundred miles northward. The park itself is a fixed location. For these reasons, some species and ecosystems are likely to be eliminated by climate change. Agricultural production likely will show regional variation in gains and losses, depending upon crop and climate. As a consequence of these multiple forces, many scientists fear that by end of next century, perhaps 25 per cent of existing species will be lost.

HOW CAN WE ESTIMATE RATES OF SPECIES LOSS?

Estimates of current and future extinction rates are based on well-documented relationships between the number of species in a region and habitat area, and on reasonably well-known rates of habitat loss. We must also employ some ratio to approximate the total number of species (described and undescribed), from the number of described species (Fig. 26.5).

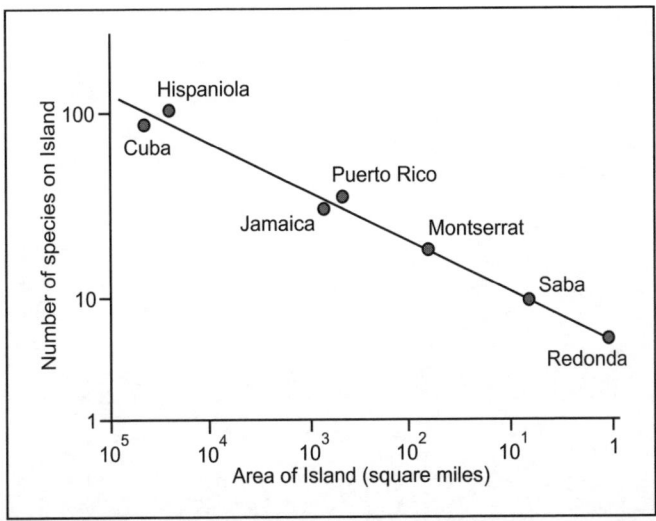

Fig. 26.5. The number of species living on islands increases or decreases with the area of the island. The diversity of reptiles and amphibians in the West Indies is depicted here. A reduction of 90 per cent in area from one island to the next results in a 50 per cent loss of species.

The relationship between species (S) and area (A) is described by the equation:

$$S = cA^z$$

where, z is the slope of the log-linear relationship, and c is a constant which described the height of the line. Based on censuses of species on islands, the number of species found on an island increases log-linearly with island area. Conversely, as island (or habitat area) is reduced, so is the number of species that will be found there. The slope (z) usually varies between 0.15 to 0.35. When combined with current rates of loss of tropical forest (this calculation uses 1.8 per cent per year), these values of the slope translate into species extinction rates of roughly 0.5 per cent annually. Extrapolated to the year 2020, roughly 20 per cent of remaining species will disappear. Simply using the most conservative values of the slope, and assuming the true biodiversity of tropical forests is roughly 10 million species, the projected rate of loss of species is 27,000 per year, and three during this hour.

SECTION VIII

Environmental Pollution and Its Control

Chapter 27

Conservation of Natural Resources

INTRODUCTION

Conservation is an ethic of resource use, allocation, and protection. Its primary focus is upon maintaining the health of the natural world: its, fisheries, habitats, and biological diversity. Secondary focus is on materials conservation and energy conservation, which are seen as important to protect the natural world. Those who follow the conservation ethic and, especially, those who advocate or work toward conservation goals are termed conservationists.

Thus, conservation of nature means nothing but protecting and preserving of natural resources for further generations. The first principle of conservation of nature is the development and use of natural resources which is existing now for the benefit of people.

To conserve habitat in terrestrial ecoregions and stop deforestation is a goal widely shared by many groups with a wide variety of motivations. To protect sea life from extinction due to overfishing is another commonly stated goal of conservation—ensuring that 'some will be available for our children' to continue a way of life.

The consumer conservation ethic is sometimes expressed by the four R's: ' Rethink, Reduce, Reuse, Recycle'. This social ethic primarily relates to local purchasing, moral purchasing, the sustained, and efficient use of renewable resources, the moderation of destructive use of finite resources, and the prevention of harm to common resources such as air and water quality, the natural functions of a living earth, and cultural values in a built environment.

The principal value underlying most expressions of the conservation ethic is that the natural world has intrinsic and intangible worth along with utilitarian value—a view carried forward by the scientific conservation movement and some of the older Romantic schools of ecology movement.

More utilitarian schools of conservation seek a proper valuation of local and global impacts of human activity upon nature in their effect upon human well being, now and to our posterity. How such values are assessed and exchanged among people determines the social, political, and personal restraints and imperatives by which conservation is practiced. This is a view common in the modern environmental movement. These movements have diverged but they have deep and common roots in the conservation movement. The term 'conservation' itself may cover the concepts such as cultural diversity, genetic diversity and the concept of movements environmental conservation, seedbank (preservation of seeds). These are often summarised as the priority to respect diversity, especially by Greens.

Much recent movement in conservation can be considered a resistance to commercialism and globalisation. Slow food is a consequence of rejecting these as moral priorities, and embracing a slower and more locally focused lifestyle. Distinct trends exist regarding conservation development. While many countries' efforts to preserve species and their habitats have been government-led, those in the North Western Europe tended to arise out of the middle-class and aristocratic interest in natural history, expressed at the level of the individual and the national, regional or local learned society. Thus countries like Britain, the Netherlands, Germany, etc. had what we would today term NGOs—in the shape of the RSPB, National Trust and County Naturalists' Trusts Natuurmonumenten, Provincial Conservation Trusts for each Dutch province, Vogelbescherming, etc. a long time before there were National Parks and National Nature Reserves. This in part reflects the absence of wilderness areas in heavily cultivated Europe, as well as a long-standing interest in laissez-faire government in some countries, like the UK, leaving it as no coincidence that John Muir, the British-born founder of the National Park movement (and hence of government-sponsored conservation) did his sterling work in the USA, where he was the motor force behind the establishment of such NPs as Yosemite and Yellowstone. Nowadays, officially more than 10 per cent of the world is legally protected in some way or the other, and in practice private fundraising is insufficient to pay for the effective management of so much land with protective status.

Protected areas in developing countries, where probably as many as 70–80 per cent of the species of the world live, still enjoy very little effective management and protection. Although some countries such as Mexico have non-profit civil organisations and land owners dedicated to protect vast private property.

CONSERVATION BIOLOGY

Conservation biology is the scientific study of the nature and status of earth's biodiversity with the aim of protecting species, their habitats, and ecosystems from excessive rates of extinction. It is an interdisciplinary subject drawing on sciences, economics, and the practice of natural resource management.

The rapid decline of established biological systems around the world means that conservation biology is often referred to as a 'discipline with a deadline'. Conservation biology is tied closely to ecology in researching the dispersal, migration, demographics, effective population size, inbreeding depression, and minimum population viability of rare or endangered species. Conservation biology is concerned with phenomena that affect the maintenance, loss, and restoration of biodiversity and the science of sustaining evolutionary processes that engender genetic, population, species, and ecosystem diversity. The concern stems from estimates suggesting that up to 50 per cent of all species on the planet will disappear within the next 50 years, which has contributed to poverty, starvation, and will reset the course of evolution on this planet.

Conservation biologists research and educate on the trends and process of biodiversity loss, species extinctions, and the negative affect this is having on our capabilities to sustain the well-being of human society. Conservation biologists work in the field and office, in government, universities, non-profit organisations and industry. They are funded to research, monitor, and catalog every angle of the earth and its relation to society. The topics are diverse, because this is an interdisciplinary network with professional alliances in the biological as well as social sciences. Those dedicated to the cause and profession advocate for a global response to the current biodiversity crisis based on morals, ethics, and scientific reason. Organisations and citizens are responding to the biodiversity crisis through conservation action plans that direct research, monitoring, and education programs that engage concerns at local through global scales.

Context and Trends

Conservation biologists study trends and process from the paleontological past to the ecological present as they gain an understanding of the context related to species extinction. It is generally accepted that there have been five major global mass extinctions that register in earth's history. These include: the Ordovician (440 mya), Devonian (370 mya), Permian–Triassic (245 mya), Triassic–Jurassic (200 mya), and Cretaceous (65 mya) extinction spasms. Within the last 10,000 years, human influence over the earth's ecosystems has been so extensive that scientists have difficulty estimating the number of species lost; that is to say the rates of deforestation, reef destruction, wetland draining and other human acts are proceeding much faster than human assessment of species.

Sixth extinction

Conservation biologists are dealing with and have published evidence from all corners of the planet indicating that humanity may be living the sixth and greatest planetary extinction event. It has been suggested that we are living in an era of unprecedented numbers of species extinctions, also known as the Holocene extinction event. The global extinction rate may be approximately 1,00,000 times higher than the natural background extinction rate. It is estimated that two-thirds of all mammal genera and one-half of all mammal species weighing at least 44 kilograms (97 lb) have gone extinct in the last 50,000 years. It is speculated that this sixth extinction period is unique because it would be the first major extinction to be caused by another biotic agent over the course of the earth's 4 billion year history.

Status of oceans and reefs

Global assessments of coral reefs of the world continue to report drastic and rapid rates of decline. By 2000, 27 per cent of the world's coral reef ecosystems had effectively collapsed. The largest period of decline occurred in a dramatic 'bleaching' event in 1998, where approximately 16 per cent of all the coral reefs in the world disappeared in less than a year. Coral bleaching is caused by a mixture of environmental stresses, including increases in ocean temperatures and acidity, causing both the release of symbiotic algae and death of corals. Decline and extinction risk in coral reef biodiversity has risen dramatically in the past ten years. The loss of coral reefs, which are predicted to go extinct in the next century, will have huge economic impacts, threatens the balance of global biodiversity, and endangers food security for hundreds of millions of people. Conservation biology plays an important role in international agreements covering the world's oceans (and other issues pertaining to biodiversity).

The oceans are threatened by acidification due to an increase in CO_2 levels. This is a most serious threat to societies relying heavily upon oceanic natural resources. A concern is that the majority of all marine species will not be able to evolve or acclimate in response to the changes in the ocean chemistry.

The prospects of averting mass extinction seems unlikely when '90 per cent of all of the large (average approximately $\geq$50 kg), open ocean tuna, billfishes, and sharks in the ocean' are reportedly gone. Given the scientific review of current trends, the ocean is predicted to have few surviving multicellular organisms with only microbes left to dominate marine ecosystems.

Insects and other groups

There are serious concerns also being hailed from taxonomic groups that do not receive the same degree of social attention or attract funds as the vertebrates do, including fungi, lichen, plant and insect communities where the vast majority of biodiversity is represented. Insect conservation, in particular, is of pivotal importance for conservation biology. The value of insects in the biosphere is enormous because they outnumber all other living groups in measure of species richness. The greatest bulk of biomass on land is found in plants, which is sustained by insect relations. This great ecological value of insects is countered by a society that oftentimes reacts negatively toward these aesthetically 'unpleasant' creatures. One area of concern in the insect world that has caught the public eye is the mysterious case of missing honey bees (*Apis mellifera*). Honey bees provide an indispensable ecological services through their acts of pollination supporting a huge variety of agriculture crops. The sudden disappearance of bees leaving empty hives or colony collapse disorder (CCD) is not uncommon.

Conservation biology of parasites

A large proportion of parasite species are threatened by extinction. A few of them are being eradicated as pests of humans or domestic animals, however, most of them are harmless. Threats include the decline or fragmentation of host populations, or the extinction of host species.

Threats to biodiversity

Many of the threats to biodiversity, including disease and climate change, are reaching inside borders of protected areas, leaving them 'not-so protected' (e.g. Yellowstone National Park). Climate change, for example, is often cited as a serious threat in this regard, because there is a feedback loop between species extinction and the release of carbon dioxide into the atmosphere. Ecosystems store and cycle large amounts of carbon to regulate global conditions. The effects of global warming adds a catastrophic threat toward a mass extinction of global biological diversity. The extinction threat is estimated to range from 15 to 37 per cent of all species by 2050, or 50 per cent of all species over the next 50 years.

Some of the most significant and insidious threats to biodiversity and ecosystem processes include climate change, mass agriculture, deforestation, overgrazing, slash-and-burn agriculture, urban development, wildlife trade, light pollution and pesticide use. Habitat fragmentation poses one of the more difficult challenges, because the global network of protected areas only covers 11.5 per cent of the earth's surface. A significant consequence of fragmentation and lack of linked protected areas is the reduction of animal migration on a global scale. Considering that billions of tonnes of biomass are responsible for nutrient cycling across the earth, the reduction of migration is a serious matter for conservation biology. These figures do not imply, however, that human activities must necessarily cause irreparable harm to the biosphere. With conservation management and planning for biodiversity at all levels, from genes to ecosystems, there are examples where humans mutually coexist in a sustainable way with nature. However, it may be too late for human intervention to reverse the current mass extinction.

Concepts and Foundations

Measuring extinction rates

Extinction rates are measured in a variety of ways. Conservation biologists measure and apply statistical measures of fossil records, rates of habitat loss, and a multitude of other variables such as loss of biodiversity as a function of the rate of habitat loss and site occupancy to obtain such estimates. The Theory of Island Biogeography is possibly the most significant contribution toward the scientific understanding of both the process and how to measure the rate of species extinction. The current background extinction rate is estimated to be one species every few years. The measure of ongoing species loss is made more complex by the fact that most of the earth's species have not been described or evaluated.

Systematic conservation planning

Systematic conservation planning is an effective way to seek and identify efficient and effective types of reserve design to capture or sustain the highest priority biodiversity values and to work with communities in support of local ecosystems.

Approaches

Conservation may be classified as either *in situ* conservation, which is protecting an endangered species in its natural habitat, or *ex situ* conservation, which occurs outside the natural habitat. *In situ* conservation involves protecting or cleaning up the habitat itself which may include a great deal of environmental preservation, or by defending the species from predators. *Ex situ* conservation may be used on some or all of the population, when *in situ* conservation is too difficult, or impossible.

Also, non-interference may be used, which is termed a preservationist method. Preservationists advocate for giving areas of nature and species a protected existence that halts interference from the humans. In this regard, conservationists differ from preservationists in the social dimension, as conservation biology engages society and seeks equitable solutions for both society and ecosystems.

Ethics and values

Conservation biologists are interdisciplinary researchers that practice ethics in the biological and social sciences. Chan states that conservationists must advocate for biodiversity and can do so in a scientifically ethical manner by not promoting simultaneous advocacy against other competing values. A conservationist researches biodiversity and reasons through a Resource Conservation Ethic.

Some conservation biologists argue that nature has an intrinsic value that is independent of anthropocentric usefulness or utilitarianism. Intrinsic value advocates that a gene, or species, be valued because they have a utility for the ecosystems they sustain.

Conservation priorities

While most in the community of conservation science 'stress the importance' of sustaining biodiversity, there is debate on how to prioritise genes, species, or ecosystems, which are all components of biodiversity. While the predominant approach to date has been to focus efforts on endangered species by conserving biodiversity hotspots, some scientists and conservation organisations, such as the Nature Conservancy, argue that it is more cost effective, logical, and socially relevant to invest in biodiversity coldspots. The costs of discovering, naming, and mapping out the distribution every species, they argue, is an ill advised conservation venture. They reason it is better to understand the significance of the ecological roles of species.

Biodiversity hotspots and coldspots are a way of recognising that the spatial concentration of genes, species, and ecosystems is not uniformly distributed on the earth's surface. For example, '44 per cent of all species of vascular plants and 35 per cent of all species in four vertebrate groups are confined to 25 hotspots comprising only 1.4 per cent of the land surface of the earth'.

Those arguing in favour of setting priorities for coldspots point out that there are other measures to consider beyond biodiversity. They point out that emphasising hotspots downplays the importance of the social and ecological connections to vast areas of the earth's ecosystems where biomass, not biodiversity, reigns supreme. It is estimated that 36 per cent of the earth's surface, encompassing 38.9 per cent of the worlds vertebrates, lacks the endemic species to qualify as biodiversity hotspot. Moreover, measures show that maximising protections for biodiversity does not capture ecosystem services any better than targeting randomly chosen regions. Population level biodiversity (i.e. coldspots) are disappearing at a rate that is ten times that at the species level. The level of importance in addressing biomass versus endemism as a concern for conservation biology is highlighted in literature measuring the level of threat to global ecosystem carbon stocks that do not necessarily reside in areas of endemism. A hotspot priority approach would not invest so heavily in places such as steppes, the Serengeti, the Arctic or taiga. These areas contribute a great abundance of population (not species) level biodiversity and ecosystem services, including cultural value and planetary nutrient cycling.

ECOLOGY

Ecology is the scientific study of the relation of living organisms to each other and their surroundings. Ecology includes the study of plant and animal populations, plant and animal communities and ecosystems. Ecologists study a range of living phenomena from the role of bacteria in nutrient recycling to the effects of tropical rain forest on the earth's atmosphere.

Ecology is a subdiscipline of biology, which is the study of life, branching out from the natural sciences in the late 19th century. Ecology is not synonymous with environment, environmentalism, natural history or environmental science. Ecology is closely related to the biological disciplines of physiology, evolution, genetics and behaviour.

Ecology seeks to explain: (i) life processes and adaptations, (ii) distribution and abundance of organisms, (iii) the movement of materials and energy through living communities, (iv) the successional development of ecosystems, and (v) the abundance and distribution of biodiversity in context of the environment.

There are many practical applications of ecology in conservation biology, wetland management, natural resource management (agriculture, forestry, fisheries), city planning (urban ecology), community health, economics, basic and applied science and it provides a conceptual framework for understanding and researching human social interaction (human ecology).

Levels of Organisation and Study

Scale and complexity

The processes that influence ecological phenomena vary through space and time. It can take thousands of years for ecological processes to mature; the life-span of a tree, for example, can encompass different successional stages. The ecological process is extended even further through time as trees die, decay and provide habitat as nurse logs or coarse woody debris. The area of an ecosystem can vary greatly from tiny to vast. A single tree is of little consequence to the classification of a forest ecosystem, but it is far more significant to smaller organisms. Several generations of an aphid population can exist over the life-span of a single leaf. Each of those aphids, in turn, support diverse bacterial communities. Tree growth is related to local site variables, such as soil type, moisture content, slope of the land, and forest canopy closure. More complex global factors, such as climate, must be considered for the classification and understanding of processes leading to larger patterns spanning across a forested landscape.

Global patterns of biological diversity are complex. This biocomplexity stems from the interplay among ecological processes that operate and influence patterns that grade into each other, such as transitional areas or ecotones that stretch across different scales. Ecologists have identified emergent and self-organising phenomena that operate at different environmental scales of influence, ranging from molecular to planetary, and these require different sets of scientific explanation.

To structure the study of ecology into a manageable framework of understanding, the biological world is conceptually organised as a nested hierarchy of organisation, ranging in scale from genes, to cells, to tissues, to organs, to organisms, to species and up to the level of the biosphere. Together these hierarchical scales of life form a panarchy. Ecosystems are primarily researched at three key levels of organisation—organisms, populations, and communities. Ecologists study ecosystems by sampling a certain number of individuals that are representative of a population. Ecosystems consist of communities interacting with each other and the environment. In ecology, communities are created by the interaction of the populations of different species in an area.

Biodiversity (an abbreviation of biological diversity) describes the diversity of life from genes to ecosystems and spans every level of biological organisation. There are many ways to index, measure, and represent biodiversity. Biodiversity includes species diversity, ecosystem diversity, genetic diversity and the complex processes operating at and among these respective levels. Biodiversity plays an important role in ecological health as much as it does for human health. Preventing or prioritising species extinctions is one way to preserve biodiversity, but populations, the genetic diversity within them and ecological processes, such as migration, are being threatened on global scales and disappearing rapidly as well. Conservation priorities and management techniques require different approaches and considerations to address the full ecological scope of biodiversity. Populations and species migration, for example, are more sensitive indicators of ecosystem services that sustain and contribute natural capital toward the well-being of humanity. An understanding of biodiversity has practical application for ecosystem-based conservation planners as they make ecologically responsible decisions in management recommendations to consultant firms, governments and industry.

Ecological niche and habitat

'The niche is the set of biotic and abiotic conditions in which a species is able to persist and maintain stable population sizes'. The ecological niche is a central concept in the ecology of organisms and is subdivided into the fundamental and the realised niche. The fundamental niche is the set of environmental conditions under which a species is able to persist. The realised niche is the set of environmental plus ecological conditions under which a species persists.

The habitat of a species is a related but distinct concept that describes the environment over which a species is known to occur and the type of community that is formed as a result. More specifically, 'habitats can be defined as regions in environmental space that are composed of multiple dimensions, each representing a biotic or abiotic environmental variable; that is, any component or characteristic of the environment related directly (e.g. forage biomass and quality) or indirectly (e.g. elevation) to the use of a location by the animal'. For example, the habitat might refer to an aquatic or terrestrial environment that can be further categorised as montane or alpine ecosystems.

Biogeographical patterns and range distributions are explained or predicted through knowledge and understanding of a species traits and niche requirements. Species have functional traits that are uniquely adapted to the ecological niche. A trait is a measurable property of an organism that influences its performance. Traits of each species are suited are uniquely adapted to their ecological niche. This means that resident species are at an advantage and able to competitively exclude other similarly adapted species from having an overlapping geographic range. This is called the competitive exclusion principle.

Organisms are subject to environmental pressures, but they are also modifiers of their habitats. The regulatory feedback between organisms and their environment can modify conditions from local (e.g. a pond) to global scales (e.g. Gaia), over time and even after death, such as decaying logs or silica skeleton deposits from marine organisms. The process and concept of ecosystem engineering has also been called niche construction. Ecosystem engineers are defined as: 'organisms that directly or indirectly modulate the availability of resources to other species, by causing physical state changes in biotic or abiotic materials. In so doing they modify, maintain and create habitats'.

The ecosystem engineering concept has stimulated a new appreciation for the degree of influence that organisms have on the ecosystem and evolutionary process. The terms niche construction are more often used in reference to the under appreciated feedback mechanism of natural selection imparting forces on the abiotic niche. An example of natural selection through ecosystem engineering occurs in the nests of social insects, including ants, bees, wasps, and termites. There is an emergent homeostasis in the structure of the nest that regulates, maintains and defends the physiology of the entire colony. Termite mounds, for example, maintain a constant internal temperature through the design of air-conditioning chimneys. The structure of the nests themselves are subject to the forces of natural selection. Moreover, the nest can survive over successive generations, which means that ancestors inherit both genetic material and a legacy niche that was constructed before their time.

Population ecology

The population is the unit of analysis in population ecology. A population consists of individuals of the same species that live, interact and migrate through the same niche and habitat. A primary law of population ecology is the Malthusian growth model. This law states that: 'a population will grow (or decline) exponentially as long as the environment experienced by all individuals in the population remains constant'.

Metapopulation ecology

Populations are also studied and modelled according to the metapopulation concept. The metapopulation concept was introduced in 1969: 'as a population of populations which go extinct locally and recolonise'. Metapopulation ecology is another statistical approach that is often used in conservation research. Metapopulation research simplifies the landscape into patches of varying levels of quality.

Metapopulation models have been used to explain life-history evolution, such as the ecological stability of amphibian metamorphosis in small vernal ponds. Alternative ecological strategies have evolved. For example, some salamanders forgo metamorphosis and sexually mature as aquatic neotenes. The seasonal duration of wetlands and the migratory range of the species determines which ponds are connected and if they form a metapopulation. The duration of the life history stages of amphibians relative to the duration of the vernal pool before it dries up regulates the ecological development of metapopulations connecting aquatic patches to terrestrial patches.

In metapopulation terminology there are emigrants (individuals that leave a patch), immigrants (individuals that move into a patch) and sites are classed either as sources or sinks. A site is a generic term that refers to places where ecologists sample populations, such as ponds or defined sampling areas in a forest. Source patches are productive sites that generate a seasonal supply of juveniles that migrate to other patch locations. Sink patches are unproductive sites that only receive migrants and will go extinct unless rescued by an adjacent source patch or environmental conditions become more favourable. Metapopulation models examine patch dynamics over time to answer questions about spatial and demographic ecology. The ecology of metapopulations is a dynamic process of extinction and colonisation. Small patches of lower quality (i.e. sinks) are maintained or rescued by a seasonal influx of new immigrants. A dynamic metapopulation structure evolves from year to year, where some patches are sinks in dry years and become sources when conditions are more favourable. Ecologists use a mixture of computer models and field studies to explain metapopulation structure.

Community ecology

Community ecology is a subdiscipline of ecology which studies the distribution, abundance, demography, and interactions between coexisting populations. An example of a study in community ecology might measure primary production in a wetland in relation to decomposition and consumption rates. This requires an understanding of the community connections between plants (i.e. primary producers) and the decomposers (e.g. fungi and bacteria), or the analysis of predator-prey dynamics affecting amphibian biomass. Food webs and trophic levels are two widely employed conceptual models used to explain the linkages among species.

Food webs: A food web is the archetypal ecological network. They are a type of concept map that illustrate pathways of energy flows in an ecological community, usually starting with solar energy being used by plants during photosynthesis. As plants grow, they accumulate carbohydrates and are eaten by grazing herbivores. Step by step lines or relations are drawn until a web of life is illustrated.

There are different ecological dimensions that can be mapped to create more complicated food webs, including: species composition (type of species), richness (number of species), biomass (the dry weight of plants and animals), productivity (rates of conversion of energy and nutrients into growth), and stability (food webs over time). A food web diagram illustrating species composition shows how change in a single species can directly and indirectly influence many others. Microcosm studies are used to simplify food web research into semi-isolated units such as small springs, decaying logs, and laboratory experiments using organisms that reproduce quickly, such as daphnia feeding on algae grown under controlled environments in jars of water.

Principles gleaned from food web microcosm studies are used to extrapolate smaller dynamic concepts to larger systems. Food webs are limited because they are generally restricted to a specific habitat, such as a cave or a pond. Many of these species migrate into other habitats to distribute their effects on a larger scale. In other words, food webs are incomplete, but are nonetheless a valuable tool in understanding community ecosystems.

Ecosystem ecology

The concept of the ecosystem describe habitats within biomes that form an integrated whole and a dynamically responsive system having both physical and biological complexes. Within an ecosystem there are inseparable ties that link organisms to the physical and biological components of their environment to which they are adapted. Ecosystems are complex adaptive systems where the interaction of life processes form self-organising patterns across different scales of time and space. This section introduces key areas of ecosystem ecology that are used to inquire, understand and explain observed patterns of biodiversity and ecosystem function across different scales of organisation.

Biome

Ecological units of organisation are defined through reference to any magnitude of space and time on the planet. Communities of organisms, for example, are somewhat arbitrarily defined, but the processes of life integrate at different levels and organise into more complex wholes. Biomes, for example, are a larger unit of organisation that categorise regions of the earth's ecosystems mainly according to the structure and composition of vegetation. Different researchers have applied different methods to define continental boundaries of biomes dominated by different functional types of vegetative communities that are limited in distribution by climate, precipitation, weather and other environmental variables. Examples of biome names include: tropical rainforest, temperate broadleaf and mixed forests, temperate deciduous forest, taiga, tundra, hot desert, and polar desert. Other researchers have recently started to categorise other types of biomes, such as the human and oceanic microbiomes. To a microbe, the human body is a habitat and a landscape.

The microbiome has been largely discovered through advances in molecular genetics that have revealed a hidden richness of microbial diversity on the planet. The oceanic microbiome plays a significant role in the ecological biogeochemistry of the planet's oceans.

Biosphere

Ecological theory has been used to explain self-emergent regulatory phenomena at the planetary scale. The largest scale of ecological organisation is the biosphere: the total sum of ecosystems on the planet. Ecological relations regulate the flux of energy, nutrients, and climate all the way up to the planetary scale. For example, the dynamic history of the planetary CO_2 and O_2 composition of the atmosphere has been largely determined by the biogenic flux of gases coming from respiration and photosynthesis, with levels fluctuating over time and in relation to the ecology and evolution of plants and animals. When sub-component parts are organised into a whole there are oftentimes emergent properties that describe the nature of the system. This the Gaia hypothesis, and is an example of holism applied in ecological theory. The ecology of the planet acts as a single regulatory or holistic unit called Gaia. The Gaia hypothesis states that there is an emergent feedback loop generated by the metabolism of living organisms that maintains the temperature of the earth and atmospheric conditions within a narrow self-regulating range of tolerance.

Relation to Evolution

Ecology and evolution are considered sister disciplines of the life sciences. Natural selection, life history, development, adaptation, populations, and inheritance are examples of concepts that thread equally into ecological and evolutionary theory. Morphological, behavioural and/or genetic traits, for example, can be mapped onto evolutionary trees to study the historical development of a species in relation to their functions and roles in different ecological circumstances. In this framework, the analytical tools of ecologists and evolutionists overlap as they organise, classify and investigate life through common systematic principals, such as phylogenetics or the Linnaean system of taxonomy. There is no sharp boundary separating ecology from evolution and they differ more in their areas of applied focus. Both disciplines discover and explain emergent and unique properties and processes operating across different spatial or temporal scales of organisation. While the boundary between ecology and evolution is not always clear, it is understood that ecologists study the abiotic and biotic factors that influence the evolutionary process.

Ecosystem Services and the Biodiversity Crisis

The ecosystems of planet earth are coupled to human environments. Ecosystems regulate the global geophysical cycles of energy, climate, soil nutrients, and water that in turn support and grow natural capital (including the environmental, physiological, cognitive, cultural, and spiritual dimensions of life). Ultimately, every manufactured product in human environments comes from natural systems. Ecosystems are considered common-pool resources because ecosystems do not exclude beneficiaries and they can be depleted or degraded. For example, green space within communities provides common-pool health services. Research shows that people who are more engaged with regular access to natural areas have lower rates of diabetes, heart disease and psychological disorders. These ecological health services are regularly depleted through urban development projects that do not factor in the common-pool value of ecosystems.

The ecological commons delivers a diverse supply of community services that sustains the well-being of human society. The Millennium Ecosystem Assessment, an international UN initiative involving more than 1360 experts worldwide, identifies four main ecosystem service types having 30 sub-categories stemming from natural capital. The ecological commons includes provisioning (e.g. food, raw materials, medicine, water supplies), regulating (e.g. climate, water, soil retention, flood retention), cultural (e.g. science and education, artistic, spiritual), and supporting (e.g. soil formation, nutrient cycling, water cycling) services. Ecological economics is an economic science that uses many of the same terms and methods that are used in accounting.

Natural capital is the stock of materials or information stored in biodiversity that generates services that can enhance the welfare of communities. Population losses are the more sensitive indicator of natural capital than are species extinction in the accounting of ecosystem services. The prospect for recovery in the economic crisis of nature is grim. Populations, such as local ponds and patches of forest are being cleared away and lost at rates that exceed species extinctions.

The current wave of threats, including massive extinction rates and concurrent loss of natural capital to the detriment of human society, is happening rapidly. This is called a biodiversity crisis, because 50 per cent of the worlds species are predicted to go extinct within the next 50 years. The world's fisheries are facing dire challenges as the threat of global collapse appears imminent, with serious ramifications for the well-being of humanity.

HABITAT CONSERVATION

Habitat conservation is a land management practice that seeks to conserve, protect and restore, habitat areas for wild plants and animals, especially conservation reliant species, and prevent their extinction, fragmentation or reduction in range. It is a priority of many groups that cannot be easily characterised in terms of any one ideology.

Most of the species extinctions from 1000 AD to 2000 AD are due to human activities, in particular destruction of plant and animal habitats. Raised rates of extinction are being driven by human consumption of organic resources, especially related to tropical forest destruction. While most of the species that are becoming extinct are not food species, their biomass is converted into human food when their habitat is transformed into pasture, cropland, and orchards. It is estimated that more than a third of the earth's biomass is tied up in only the few species that represent humans, livestock and crops. Because an ecosystem decreases in stability as its species are made extinct, these studies warn that the global ecosystem is destined for collapse if it is further reduced in complexity. Factors contributing to loss of biodiversity are: overpopulation, deforestation, pollution (air pollution, water pollution, soil contamination) and global warming or climate change, driven by human activity. These factors, while all stemming from overpopulation, produce a cumulative impact upon biodiversity.

Conservation movement: Some of the conservation movement's goals are to protect habitats and promote continued recreational opportunities for people such as hiking, birdwatching, fishing and hunting.

Ecology movement: The global ecology movement is based upon environmental protection, and is one of several new social movements that emerged at the end of the 1960s. As a values-driven social movement, it should be distinguished from the pre-existing science of ecology. Aspects of the ecology movement view wild species as possessing natural life-rights to exist based upon the importance of maintaining and preserving biodiversity. Another argument for the preservation of species is based upon species competition: species tend to compete most intensely with their own kind, so therefore any cessation of competition between humans must be presaged by cessation of competition between humans and other species.

WATER CONSERVATION

Water conservation refers to reducing the usage of water and recycling of waste water for different purposes such as cleaning, manufacturing, and agricultural irrigation.

The goals of water conservation efforts include as follows:

1. Sustainability: To ensure availability for future generations, the withdrawal of fresh water from an ecosystem should not exceed its natural replacement rate.
2. Energy conservation: Water pumping, delivery, and waste-water treatment facilities consume a significant amount of energy. In some regions of the world (for example, California) over 15 per cent of total electricity consumption is devoted to water management.
3. Habitat conservation: Minimising human water use helps to preserve fresh water habitats for local wildlife and migrating waterfowl, as well as reducing the need to build new dams and other water diversion infrastructure.

Social Solutions

Water conservation programs are typically initiated at the local level, by either municipal water utilities or regional governments. Common strategies include public outreach campaigns, tiered water rates (charging progressively higher prices as water use increases), or restrictions on outdoor water use such as lawn watering and car washing. Cities in dry climates often require or encourage the installation of xeriscaping or natural landscaping in new homes to reduce outdoor water usage.

Some researchers have suggested that water conservation efforts should be primarily directed at farmers, in light of the fact that crop irrigation accounts for 70 per cent of the world's fresh water use. The agricultural sector of most countries is important both economically and politically, and water subsidies are common. Conservation advocates have urged removal of all subsidies to force farmers to grow more water-efficient crops and adopt less wasteful irrigation techniques.

Agricultural applications

For crop irrigation, optimal water efficiency means minimising losses due to evaporation, runoff or subsurface drainage. An evaporation pan can be used to determine how much water is required to irrigate the land. Flood irrigation, the oldest and most common type, is often very uneven in distribution, as parts of a field may receive excess water in order to deliver sufficient quantities to other parts. Overhead irrigation, using center-pivot or lateral-moving sprinklers, gives a much more equal and controlled distribution pattern. Drip irrigation is the most expensive and least-used type, but offers the best results in delivering water to plant roots with minimal losses.

As changing irrigation systems can be a costly undertaking, conservation efforts often concentrate on maximising the efficiency of the existing system. This may include chiselling compacted soils, creating furrow dikes to prevent runoff, and using soil moisture and rainfall sensors to optimise irrigation schedules. Infiltration basins, also called recharge pits, capture rainwater and recharge ground water supplies. Use of these management practices reduces soil erosion caused by stormwater runoff and improves water quality in nearby surface waters.

WILDLIFE

Wildlife includes all non-domesticated plants, animals and other organisms. Domesticating wild plant and animal species for human benefit has occurred many times all over the planet, and has a major impact on the environment, both positive and negative.

Wildlife can be found in all ecosystems. Deserts, rain forests, plains, and other areas including the most developed urban sites, all have distinct forms of wildlife. While the term in popular culture usually refers to animals that are untouched by human factors, most scientists agree that wildlife around the world is impacted by human activities.

Humans have historically tended to separate civilisation from wildlife in a number of ways including the legal, social, and moral sense. This has been a reason for debate throughout recorded history. Religions have often declared certain animals to be sacred, and in modern times concern for the natural environment has provoked activists to protest the exploitation of wildlife for human benefit or entertainment.

Food, Pets, Traditional Medicines

Anthropologists believe that the Stone Age peoples and hunter-gatherers relied on wildlife, both plant and animal, for their food. In fact, some species may have been hunted to extinction by early human hunters. Today, hunting, fishing, or gathering wildlife is still a significant food source in some parts of the world. In other areas, hunting and non-commercial fishing are mainly seen as a sport or recreation, with the edible meat as mostly a side benefit. Meat sourced from wildlife that is not traditionally regarded as game is known as bush meat. The increasing demand for wildlife as a source of traditional food in East Asia is decimating populations of sharks, primates, pangolins and other animals, which they believe have aphrodisiac properties.

Religion

Many wildlife species have spiritual significance in different cultures around the world, and they and their products may be used as sacred objects in religious rituals. For example, eagles, hawks and their feathers have great cultural and spiritual value to Native Americans as religious objects.

Media: Wildlife has long been a common subject for educational television shows. There are many magazines which cover wildlife including National Wildlife Magazine, Birds and Blooms, Birding (magazine), and Ranger Rick (for children).

Tourism: Fuelled by media coverage and inclusion of conservation education in early school curriculum, Wildlife tourism and Ecotourism has fast become a popular industry generating substantial income for developing nations with rich wildlife specially, Africa and India. This ever growing and ever becoming more popular form of tourism is providing the much needed incentive for poor nations to conserve their rich wildlife heritage and its habitat.

Destruction

This subsection focuses on anthropogenic forms of wildlife destruction. Destruction of wildlife does not always lead to an extinction of the species in question, however, the dramatic loss of entire species across earth dominates any review of wildlife destruction as extinction is the level of damage to a wild population from which there is no return. The four most general reasons that lead to destruction of wildlife include overkill, habitat destruction and fragmentation, impact of introduced species and chains of extinction.

Overkill

Overkill occurs whenever hunting occurs at rates greater than the reproductive capacity of the population is being exploited. The effects of this are often noticed much more dramatically in slow growing populations such as many larger species of fish. Initially when a portion of a wild population is hunted, an increased availability of resources (food, etc.) is experienced increasing growth and reproduction as density dependent inhibition is lowered. Hunting, fishing and so on, has lowered the competition between members of a population. However, if this hunting continues at rate greater than the rate at which new members of the population can reach breeding age and produce more young, the population will begin to decrease in numbers. Populations are confined to islands, whether literal islands or just areas of habitat that are effectively an 'island' for the species concerned have also been observed to be at greater risk of dramatic population declines following unsustainable hunting.

Habitat destruction and fragmentation

The habitat of any given species is considered its preferred area or territory. Many processes associated human habitation of an area cause loss of this area and the decrease the carrying capacity of the land for that species. In many cases these changes in land use cause a patchy break-up of the wild landscape. Agricultural land frequently displays this type of extremely fragmented, or relictual, habitat. Farms sprawl across the landscape with patches of uncleared woodland or forest dotted in-between occasional paddocks. Examples of habitat destruction include grazing of bushland by farmed animals, changes to natural fire regimes, forest clearing for timber production and wetland draining for city expansion.

Impact of introduced species

Mice, cats, rabbits, dandelions and poison ivy are all examples of species that have become invasive threats to wild species in various parts of the world. Frequently species that are uncommon in their home range become out-of-control invasions in distant but similar climates. The reasons for this have not always been clear and Charles Darwin felt it was unlikely that exotic species would ever be able to grow abundantly in a place in which they had not evolved. The reality is that the vast majority of species exposed to a new habitat do not reproduce successfully. Occasionally, however, some populations do take hold and after a period of acclimation can increase in numbers significantly, having destructive effects on many elements of the native environment of which they have become part.

Chains of extinction

This final group is one of secondary effects. All wild populations of living things have many complex intertwining links with other living things around them. Large herbivorous animals such as the hippopotamus have populations of insectivorous birds that feed off the many parasitic insects that grow on the hippo. Should the hippo die out, so too will these groups of birds, leading to further destruction as other species dependent on the birds are affected. Also referred to as a Domino effect, this series of chain reactions is by far the most destructive process that can occur in any ecological community. Another example is the black drongos and the cattle egrets found in India. These birds feed on insects on the back of cattle, which helps to keep them disease-free. If we destroy the nesting habitats of these birds, it will result a decrease in the cattle population because of the spread of insect-borne diseases.

FORESTRY IN INDIA

Forestry is a major government enterprise in India which faces the challenges of dwindling forest cover area due to overpopulation, farming and environmental factors.

Strategy to Increase Cover

India's long-term strategy for forestry development reflects three major objectives: to reduce soil erosion and flooding; to supply the growing needs of the domestic wood products industries; and to supply the needs of the rural population for fuelwood, fodder, small timber, and miscellaneous forest produce. To achieve these objectives, the National Commission on Agriculture in 1976 recommended the reorganisation of state forestry departments and advocated the concept of social forestry. The Commission itself worked on the first two objectives, emphasising traditional forestry and wildlife activities; in pursuit of the third objective, the Commission recommended the establishment of a new kind of unit to develop community forests. Following the leads of Gujarat and Uttar Pradesh, a number of other states also established community-based forestry agencies that emphasised programs on farm forestry, timber management, extension forestry, reforestation of degraded forests, and use of forests for recreational purposes.

Such socially responsible forestry was encouraged by state community forestry agencies. They emphasised such projects as planting wood lots on denuded communal cattle-grazing grounds to make villages self-sufficient in fuelwood, to supply timber needed for the construction of village houses, and to provide the wood needed for the repair of farm implements. Both individual farmers and tribal communities were also encouraged to grow trees for profit. For example, in Gujarat, one of the more aggressive states in developing programs of socioeconomic importance, the forestry department distributed 200 million tree seedlings in 1983. The fast-growing eucalyptus is the main species being planted nationwide, followed by pine and poplar.

Conservation

The role of forests in the national economy and in ecology was further emphasised in the 1988 National Forest Policy, which focused on ensuring environmental stability, restoring the ecological balance, and preserving the remaining forests. Other objectives of the policy were meeting the need for fuelwood, fodder, and small timber for rural and tribal people while recognising the need to actively involve local people in the management of forest resources. Also in 1988, the Forest Conservation Act of 1980 was amended to facilitate stricter conservation measures. A new target was to increase the forest cover to 33 per cent of India's land area from the then-official estimate of 23 per cent. In June 1990, the central government adopted resolutions that combined forest science with social forestry, that is, taking the sociocultural traditions of the local people into. The cumulative area afforested during the 1951–91 period was nearly 1,79,000 square kilometres. However, despite large-scale tree planting programs, forestry is one arena in which India has actually regressed since independence. Annual fellings at about four times the growth rate are a major cause. Widespread pilfering by villagers for firewood and fodder also represents a major decrement. In addition, the forested area has been shrinking as a result of land cleared for farming, inundations for irrigation and hydroelectric power projects, and construction of new urban areas, industrial plants, roads, power lines, and schools.

Timber Mafia

Protected forest areas in several parts of India, such as Jammu and Kashmir, Himachal Pradesh, Karnataka and Jharkhand, are vulnerable to illegal logging by timber mafias that have coopted or intimidated forestry officials, local politicians, businesses and citizenry. Clear-cutting is sometimes covered-up by conniving officials who report fictitious forest fires.

Forest Rights

In 2006 forestry in India underwent a major change with the passage of the Forest Rights Act, a new legislation that seeks to reverse the 'historical injustice' to forest dwelling communities that resulted from the failure to record their rights over forest land and resources. It also sought to bring in new forms of community conservation.

WETLAND CONSERVATION

Wetland conservation supports the functions of wetlands to provide habitat for numerous amounts of species, by their conservation. Wetlands help with aquifer replenishment, purification of water, flood control, and nutrient cycling. Increased amounts of wetlands with help create a healthier environment.

Wetlands have many definitions. Most wetland definitions include the area being covered in water for part of the year or having water in the root zone. These areas are suitable for the life of species that are adapted to wet soils or water. Water cover or saturation is the key factor that aids in soil development. With the development of the soil comes the type of plant and animal species suited for the soil. Wetlands cannot have a specific definition because of the changes that are made throughout the wetland systems due to climate, topography, hydrology, water chemistry, and vegetation as well as human interaction. Generally wetlands include areas such as bogs, swamps, and marshes. Wetlands are found in every continent except Antarctica. Wetlands are categorised into two categories; Coastal, and Inland.

Coastal wetlands mix freshwater with salt water making uneven levels of water salinity. The uneven salinity has most of these areas without vegetation. In some of the coastal wetlands which have tropical climates, salt loving trees and shrubs can be found.

Inland wetlands include flood plains along rivers and streams as well as playas, basins, potholes, etc. These areas are surrounded by dry land. Inland wetlands like marshes and wet meadows are covered by herbaceous plants, shrubs, and trees. There are different types of inland wetland based on their geography. Many inland wetlands are dry for part of the year.

Wetland Functions

Wetlands have a very broad list of uses and services. Wetlands play a very important role in storing and cleaning water. Wetlands normally lie in the path of water runoff and they intercept that run off before it can get to the rivers and streams. They use physical, chemical, and biological processes to remove pollutants. The filtering provided by the wetlands saves millions upon millions of dollars in water filtering every year. It has been estimated that between 70 and 90 per cent of the nitrogen in run off is removed by wetlands. Wetlands can also remove 45–80 per cent of the phosphorus. Wetlands serve as major storage areas. Wetlands are very important in recharging the ground water as well as streams and rivers.

Flood control is another major service provided by wetlands. Their flood control importance increases because of location as well as surrounding landforms. If they are close to frequently flooded streams or rivers and have great enough size to them they can take on the water and release it back to the ground water or eventually back to that live body of water.

Another important service that wetlands provide is shore stabilisation. These wetlands reduce the amount of shoreline erosion. This is an important type of erosion control because it still allows there to be habitat as well as having good erosion control.

Beneficial Species

Wetlands are home to hundreds of species of plants and animals. There is a wide variety of animals that benefit from the wetlands. The wetlands also become important because many animals use the wetlands as a water source. The bird population is one of the most beneficial species. Over 80 per cent of birds in the United States are dependant on wetlands. Ducks and geese use the wetlands for their migration as well as reproduction. These birds will use the wetlands to migrate north and south as well as a place to stay after the migration. Around 96 per cent of the commercially important fish in North America need wetlands to survive.

Destruction of Wetlands

Wetland destruction is a major issue. A majority of the loss of wetlands can be attributed to agriculture. With practices such as draining and diking more and more land has been put into production agriculture land. Another large destroyer of wetlands is urban development. The disruption of the flow of water is also a destroyer of wetlands. By building dikes and levees the normal water that would fill or replenish the wetland gets run straight back into the rivers and streams. This is also creates a problem with pollution. The wetlands play an important role in the cleaning and purifying water. When the watersheds and wetlands are destroyed or bypassed the water flows straight into the streams and rivers unpurified and then the pollutants that were in the runoff gets put into the rivers and streams.

Wetland destruction can also produce economic losses. Logging and mining are two things that create income for a short period of time but eventually they will run out and the wetland is still destroyed. When a wetland is destroyed it limits the amount of tourism that leads to a loss of income. Also it costs a lot of money to fill or drain the wetland and make it usable for something else and after that is done it cost even more to restore the wetland and make it good usable habitat.

Climate Issues

With the change in climate and the rising sea levels the coastal wetlands are in danger. Basins are very important forms of wetlands for waterfowl reproduction. The waterfowl use these basins to breed. With the climate getting warmer the basins are starting to disappear because of things like plant translocation as well as evaporation. The basins are precipitation fed making them vulnerable to dry spells. The continued warming of the planet is expected to radically change the wetland ecosystems. Thawing permafrost is expected to make some wetlands dry up while creating some new wetlands. The change in climate is going to leave some areas short of precipitation that is going to hurt wetlands in those areas. Wetlands store a large amount of carbon. With the increased temperature and increased wetland destruction there will be a lot of carbon released into the atmosphere. In fact wetlands, which only account for 6 per cent of the earth's terrestrial area, contain as much carbon as the earth's atmospheric store.

Conservation Solutions

The first area to conserve is the grassland. The grassland is important because of the breeding ground it creates for wetland animals. The next habitat to conserve is the flooded forest. These forests are ones that fill with water once a year due to overflowing river banks. These areas are good vegetative wetlands that provide a good winter habitat for wetland species like ducks.

The last area to consider when working with conserving the wetlands is the watersheds. These are the areas surrounding the wetlands that help to filter the water that is flowing into the wetland. When they are disturbed it drastically changes the output of plant material that hurts the species that live in that wetland. Private land owners are a very important part of wetland conservation because they hold 75 per cent of Americas remaining wetlands. Most of these wetlands are found on farmers and ranchers land. Conservation organisations like Ducks Unlimited and North American Waterfowl Management Plan work together and with other organisations to acquire land.

Organisations like Ducks Unlimited understand that some of the best wetlands are in danger of being developed. To conserve that land they use conservation easements. These easements are agreements with landowners that protects wetland habitat from development forever.

Ducks Unlimited also has management agreements with landowners that will give them financial incentives for conserving habitat. With this type of agreement the landowner receives benefits that should ensure the prolonged habitat conservation that will benefit hundreds of species.

The conservation reserve program was also a part of wetland conservation. The conservation reserve program or CRP is working towards a goal of restoring 250,000 acres (1000 km^2) of wetlands. They are also working to restore thousands of acres of playas with their plans as well. These playas are low lying areas that have been formed by wind and waves. These playas are most common in the great plains areas and they are very important when dealing with conserving wetlands because these plays average 17 acres (69,000 m^2) apiece that creates a large area of habitat area. The playas are very beneficial to the migratory birds. They use the playas for migration and over 90 per cent of migratory birds overwinter in these playas. The sand hill crane is another population that over 90 per cent of them rely on these for overwintering. The playa restoration is also helping to recharge the Ogallala aquifer. This aquifer is one of the largest in the world. It serves Nebraska, Oklahoma, Texas, Kansas, and Colorado. CRP has restored over 2 million acres (8100 km^2) of wetlands. The CRP wetland restoration is responsible for 2.2 million new ducks added to America. The wetland reserve program is another program that works with land owners to conserver and restore their wetlands. The wetland reserve program is working to set up long-term preservation of wetlands.

Wetlands are an important ecosystem. They are important for hundreds of species of wildlife as well as humans. To benefit wildlife wetlands provide a home, a place to stay during the migration, as well as a water source to outside wildlife. They are rich in vegetation that creates habitat for a broad range of wildlife.

SUSTAINABLE AGRICULTURE

Sustainable agriculture is the practice of farming using principles of ecology, the study of relationships between organisms and their environment. It has been defined as an integrated system of plant and animal production practices having a site-specific application that will, over the long term:

1. Satisfy human food and fibre needs.
2. Make the most efficient use of non-renewable resources and on-farm resources and integrate, where appropriate, natural biological cycles and controls.
3. Sustain the economic viability of farm operations.
4. Enhance the quality of life for farmers and society as a whole.

Sustainable agriculture in the United States was addressed by the 1990 farm bill. More recently, as consumer and retail demand for sustainable products has risen, organisations such as Food Alliance and Protected harvest have started to provide measurement standards and certification programs for what constitutes a sustainably grown crop.

Farming and Natural Resources

The physical aspects of sustainability are partly understood. Practices that can cause long-term damage to soil include excessive tillage (leading to erosion) and irrigation without adequate drainage (leading to salinisation). Long-term experiments have provided some of the best data on how various practices affect soil properties essential to sustainability. The most important factors for an individual site are sun, air, soil and water. Of the four, water and soil quality and quantity are most amenable to human intervention through time and labour.

Although air and sunlight are available everywhere on earth, crops also depend on soil nutrients and the availability of water. When farmers grow and harvest crops, they remove some of these nutrients from the soil. Without replenishment, land suffers from nutrient depletion and becomes either unusable or suffers from reduced yields. Sustainable agriculture depends on replenishing the soil while minimising the use of non-renewable resources, such as natural gas (used in converting atmospheric nitrogen into synthetic fertiliser), or mineral ores (e.g. phosphate). Possible sources of nitrogen that would, in principle, be available indefinitely, include:

1. Recycling crop waste and livestock or treated human manure.
2. Growing legume crops and forages such as peanuts or alfalfa that form symbioses with nitrogen-fixing bacteria called rhizobia.
3. Industrial production of nitrogen by the Haber process uses hydrogen, which is currently derived from natural gas, [but this hydrogen could instead be made by electrolysis of water using electricity (perhaps from solar cells or windmills)].
4. Genetically engineering (non-legume) crops to form nitrogen-fixing symbioses or fix nitrogen without microbial symbionts.

The last option was proposed in the 1970s, but is only recently becoming feasible. Sustainable options for replacing other nutrient inputs (phosphorus, potassium, etc.) are more limited.

More realistic, and often overlooked, options include long-term crop rotations, returning to natural cycles that annually flood cultivated lands (returning lost nutrients indefinitely) such as the Flooding of the Nile, the long-term use of biochar, and use of crop and livestock landraces that are adapted to less than ideal conditions such as pests, drought, or lack of nutrients.

Water

In some areas, sufficient rainfall is available for crop growth, but many other areas require irrigation. For irrigation systems to be sustainable they require proper management (to avoid salinisation) and must not use more water from their source than is naturally replenished, otherwise the water source becomes, in effect, a non-renewable resource. Improvements in water well drilling technology and submersible pumps combined with the development of drip irrigation and low pressure pivots have made it possible to regularly achieve high crop yields where reliance on rainfall alone previously made this level of success unpredictable.

Several steps should be taken to develop drought-resistant farming systems even in 'normal' years, including both policy and management actions: (i) improving water conservation and storage measures, (ii) providing incentives for selection of drought-tolerant crop species, (iii) using reduced-volume irrigation systems, (iv) managing crops to reduce water loss, and (v) not planting at all.

Soil

Soil erosion is fast becoming the one of the worlds greatest problems. It is estimated that 'more than a thousand million tons of southern Africa's soil are eroded every year. Experts predict that crop yields will be halved within thirty to fifty years if erosion continues at present rates'. Soil erosion is not unique to Africa but is occurring worldwide. The phenomenon is being called peak soil as present large scale factory farming techniques are jeopardising humanities ability to grow food in the present and in the future. Without efforts to improve soil management practices, the availability of arable soil will become increasingly problematic. Some soil management techniques:

1. Growing wind breaks to hold the soil.
2. Incorporating organic matter back into fields.
3. Stop using chemical fertilisers (which contain salt).

Economics

Socioeconomic aspects of sustainability are also partly understood. Regarding less concentrated farming, the best known analysis is Netting's study on smallholder systems through history. The Oxford Sustainable Group defines sustainability in this context in a much broader form, considering effect on all stakeholders in a 360 degree approach. Given the finite supply of natural resources at any specific cost and location, agriculture that is inefficient or damaging to needed resources may eventually exhaust the available resources or the ability to afford and acquire them. It may also generate negative externality, such as pollution as well as financial and production costs.

The way that crops are sold must be accounted for in the sustainability equation. Food sold locally does not require additional energy for transportation (including consumers). Food sold at a remote location, whether at a farmers' market or the supermarket, incurs a different set of energy cost for materials, labour, and transport.

Methods

What grows where and how it is grown are a matter of choice. Two of the many possible practices of sustainable agriculture are crop rotation and soil amendment, both designed to ensure that crops being cultivated can obtain the necessary nutrients for healthy growth. Soil amendments would include using locally available compost from community recycling centers. These community recyclying centers help produce the compost needed by the local organic farms.

Many scientists, farmers, and businesses have debated how to make agriculture sustainable. Using community recycling from yard and kitchen waste utilises a local area's commonly available resources. These resources in past were thrown away into large waste disposal sites, are now used to produce low cost organic compost for organic farming. Other practices includes growing a diverse number of perennial crops in a single field, each of which would grow in separate season so as not to compete with each other for natural resources. This system would result in increased resistance to diseases and decreased effects of erosion and loss of nutrients in soil. Nitrogen fixation from legumes, for example, used in conjunction with plants that rely on nitrate from soil for growth, helps to allow the land to be reused annually. Legumes will grow for a season and replenish the soil with ammonium and nitrate, and the next season other plants can be seeded and grown in the field in preparation for harvest.

Monoculture, a method of growing only one crop at a time in a given field, is a very widespread practice, but there are questions about its sustainability, especially if the same crop is grown every year. Today it is realised to get around this problem local cities and farms can work together to produce the needed compost for the farmers around them. This combined with growing a mixture of crops (polyculture) sometimes reduces disease or pest problems but polyculture has rarely, if ever, been compared to the more widespread practice of growing different crops in successive years (crop rotation) with the same overall crop diversity. Cropping systems that include a variety of crops (polyculture and/or rotation) may also replenish nitrogen (if legumes are included) and may also use resources such as sunlight, water, or nutrients more efficiently.

Replacing a natural ecosystem with a few specifically chosen plant varieties reduces the genetic diversity found in wildlife and makes the organisms susceptible to widespread disease. In practice, there is no single approach to sustainable agriculture, as the precise goals and methods must be adapted to each individual case. There may be some techniques of farming that are inherently in conflict with the concept of sustainability, but there is widespread misunderstanding on impacts of some practices. Today the growth of local farmers markets offer small farms the ability to sell the products that they have grown back to the cities that they got the recycled compost from. By using local recycling this will help move us away from, the slash-and-burn techniques that are the characteristic feature of shifting cultivators are often cited as inherently destructive, yet slash-and-burn cultivation has been practiced in the Amazon for at least 6000 years; serious deforestation did not begin until the 1970s, largely as the result of Brazilian government programs and policies. To note that it may not have been slash-and-burn so much as slash-and-char, which with the addition of organic matter produces terra preta, one of the richest soils on earth and the only one that regenerates itself.

There are also many ways to practice sustainable animal husbandry. Some of the key tools to grazing management include fencing off the grazing area into smaller areas called paddocks, lowering stock density, and moving the stock between paddocks frequently.

Soil treatment

Soil steaming can be used as an ecological alternative to chemicals for soil sterilisation. Different methods are available to induce steam into the soil in order to kill pests and increase soil health. Community and farm composting of kitchen, yard, and farm organic waste can provide most if not all the required needs of local farms.

Off-farm Impacts

A farm that is able to 'produce perpetually', yet has negative effects on environmental quality elsewhere is not sustainable agriculture. An example of a case in which a global view may be warranted is over-application of synthetic fertiliser or animal manures, which can improve productivity of a farm but can pollute nearby rivers and coastal waters (eutrophication). The other extreme can also be undesirable, as the problem of low crop yields due to exhaustion of nutrients in the soil has been related to rainforest destruction, as in the case of slash and burn farming for livestock feed.

Increased production may come from creating new farmland, which may ameliorate carbon dioxide emissions if done through reclamation of desert as in Palestine, or may worsen emissions if done through slash and burn farming, as in Brazil. Additionally, Genetically modified organism crops show promise for radically increasing crop yields, although many people and governments are apprehensive of this new farming method.

Some advocates of sustainable agriculture favour as the only system which can be sustained over the long-term. However, organic production methods, especially in transition, yield less than their conventional counterparts and raise the same problems of sustaining populations globally. While evidence suggests that organic farms handle periods of drought better.

Urban Planning

There has been considerable debate about which form of human residential habitat may be a better social form for sustainable agriculture. Many environmentalists advocate urban developments with high population density as a way of preserving agricultural land and maximising energy efficiency. However, others have theorised that sustainable ecocities, or ecovillages which combine habitation and farming with close proximity between producers and consumers, may provide greater sustainability.

The use of available city space (e.g. rooftop gardens, community gardens, garden sharing, and other forms of urban agriculture) for cooperative food production is another way to achieve greater sustainability. One of the latest ideas in achieving sustainable agricultural involves shifting the production of food plants from major factory farming operations to large, urban, technical facilities called vertical farms. The advantages of vertical farming include year-round production, isolation from pests and diseases, controllable resource recycling, and on-site production that reduces transportation costs. While a vertical farm has yet to become a reality, the idea is gaining momentum among those who believe that current sustainable farming methods will be insufficient to provide for a growing global population. For vertical farming to become a reality, billions of dollars in tax credits and subsidies will need to be made available to the operation.

SUSTAINABILITY

Sustainability is the capacity to endure. In ecology, the word describes how biological systems remain diverse and productive over time. Long-lived and healthy wetlands and forests are examples of sustainable biological systems. For humans, sustainability is the potential for long-term maintenance of well being, which has environmental, economic, and social dimensions.

Healthy ecosystems and environments provide vital goods and services to humans and other organisms. There are two major ways of reducing negative human impact and enhancing ecosystem services. The first is environmental management; this approach is based largely on information gained from earth science, environmental science, and conservation biology. The second approach is management of human consumption of resources, which is based largely on information gained from economics.

Sustainability interfaces with economics through the social and ecological consequences of economic activity. Sustainability economics involves ecological economics where social, cultural, health-related and monetary/financial aspects are integrated. Moving towards sustainability is also a social challenge that entails international and national law, urban planning and transport, local and individual lifestyles and ethical consumerism. Ways of living more sustainably can take many forms from reorganising living conditions (e.g. ecovillages, eco-municipalities and sustainable cities), reappraising economic sectors (permaculture, green building, sustainable agriculture), or work practices (sustainable architecture), using science to develop new technologies (green technologies, renewable energy), to adjustments in individual lifestyles that conserve natural resources.

The philosophical and analytic framework of sustainability draws on and connects with many different disciplines and fields; in recent years an area that has come to be called sustainability science has emerged. Sustainability science is not yet an autonomous field or discipline of its own, and has tended to be problem-driven and oriented towards guiding decision-making.

Scale and context: Sustainability is studied and managed over many scales (levels or frames of reference) of time and space and in many contexts of environmental, social and economic organisation. The focus ranges from the total carrying capacity (sustainability) of planet Earth to the sustainability of economic sectors, ecosystems, countries, municipalities, neighbourhoods, home gardens, individual lives, individual goods and services, occupations, lifestyles, behaviour patterns and so on. In short, it can entail the full compass of biological and human activity or any part of it.

Consumption—Population, Technology, Resources

The overall driver of human impact on earth systems is the destruction of biophysical resources, and especially, the earth's ecosystems. The total environmental impact of a community or of humankind as a whole depends both on population and impact per person, which in turn depends in complex ways on what resources are being used, whether or not those resources are renewable, and the scale of the human activity relative to the carrying capacity of the ecosystems involved. Careful resource management can be applied at many scales, from economic sectors like agriculture, manufacturing and industry, to work organisations, the consumption patterns of households and individuals and to the resource demands of individual goods and services.

Measurement

Sustainability measurement is a term that denotes the measurements used as the quantitative basis for the informed management of sustainability. The metrics used for the measurement of sustainability (involving the sustainability of environmental, social and economic domains, both individually and in various combinations) are still evolving: they include indicators, benchmarks, audits, indexes and accounting, as well as assessment, appraisal and other reporting systems. They are applied over a wide range of spatial and temporal scales.

Some of the best known and most widely used sustainability measures include corporate sustainability reporting, Triple Bottom Line accounting, and estimates of the quality of sustainability governance for individual countries using the Environmental Sustainability Index and Environmental Performance Index.

Population

It is the combination of population increase in the developing world and unsustainable consumption levels in the developed world that poses a stark challenge to sustainability.

Environmental Dimension

Healthy ecosystems provide vital goods and services to humans and other organisms. There are two major ways of reducing negative human impact and enhancing ecosystem services and the first of these is environmental management. This direct approach is based largely on information gained from earth science, environmental science and conservation biology. However, this is management at the end of a long series of indirect causal factors that are initiated by human consumption, so a second approach is through demand management of human resource use.

Management of human consumption of resources is an indirect approach based largely on information gained from economics. Herman Daly has suggested three broad criteria for ecological sustainability: renewable resources should provide a sustainable yield (the rate of harvest should not exceed the rate of regeneration); for non-renewable resources there should be equivalent development of renewable substitutes; waste generation should not exceed the assimilative capacity of the environment.

Environmental management

At the global scale and in the broadest sense environmental management involves the oceans, freshwater systems, land and atmosphere, but following the sustainability principle of scale it can be equally applied to any ecosystem from a tropical rainforest to a home garden.

Atmosphere: Management of the global atmosphere now involves assessment of all aspects of the carbon cycle to identify opportunities to address human-induced climate change and this has become a major focus of scientific research because of the potential catastrophic effects on biodiversity and human communities.

Other human impacts on the atmosphere include the air pollution in cities, the pollutants including toxic chemicals like nitrogen oxides, sulphur oxides, volatile organic compounds and particulate matter that produce photochemical smog and acid rain, and the chlorofluorocarbons that degrade the ozone layer. Anthropogenic particulates such as sulphate aerosols in the atmosphere reduce the direct irradiance and reflectance (albedo) of the earth's surface. Known as global dimming, the decrease is estimated to have been about 4 per cent between 1960 and 1990 although the trend has subsequently reversed. Global dimming may have disturbed the global water cycle by reducing evaporation and rainfall in some areas. It also creates a cooling effect and this may have partially masked the effect of greenhouse gases on global warming.

Freshwater and oceans: Water covers 71 per cent of the earth's surface. Of this, 97.5 per cent is the salty water of the oceans and only 2.5 per cent freshwater, most of which is locked up in the Antarctic ice sheet. The remaining freshwater is found in glaciers, lakes, rivers, wetlands, the soil, aquifers and atmosphere. Due to the water cycle, fresh water supply is continually replenished by precipitation, however there is still a limited amount necessitating management of this resource. Awareness of the global importance of preserving water for ecosystem services has only recently emerged as, during the 20th century, more than half the world's wetlands have been lost along with their valuable environmental services. Increasing urbanisation pollutes clean water supplies and much of the world still does not have access to clean, safe water. Greater emphasis is now being placed on the improved management of blue (harvestable) and green (soil water available for plant use) water, and this applies at all scales of water management.

Ocean circulation patterns have a strong influence on climate and weather and, in turn, the food supply of both humans and other organisms. Scientists have warned of the possibility, under the influence of climate change, of a sudden alteration in circulation patterns of ocean currents that could drastically alter the climate in some regions of the globe. Ten per cent of the world's population—about 600 million people—live in low-lying areas vulnerable to sea level rise.

Land use: Loss of biodiversity stems largely from the habitat loss and fragmentation produced by the human appropriation of land for development, forestry and agriculture as natural capital is progressively converted to man-made capital. Land use change is fundamental to the operations of the biosphere because alterations in the relative proportions of land dedicated to urbanisation, agriculture, forest, woodland, grassland and pasture have a marked effect on the global water, carbon and nitrogen biogeochemical cycles and this can impact negatively on both natural and human systems. At the local human scale, major sustainability benefits accrue from sustainable parks and gardens and green cities.

Food is essential to life and feeding more than six billion human bodies takes a heavy toll on the Earth's resources. This begins with the appropriation of about 38 per cent of the earth's land surface and about 20 per cent of its net primary productivity. Added to this are the resource-hungry activities of industrial agribusiness—everything from the crop need for irrigation water, synthetic fertilisers and pesticides to the resource costs of food packaging, transport (now a major part of global trade) and retail. Environmental problems associated with industrial agriculture and agribusiness are now being addressed through such movements as sustainable agriculture, organic farming and more sustainable business practices.

Management of human consumption

The underlying driver of direct human impacts on the environment is human consumption. This impact is reduced by not only consuming less but by also making the full cycle of production, use and disposal more sustainable. Consumption of goods and services can be analysed and managed at all scales through the chain of consumption, starting with the effects of individual lifestyle choices and spending patterns, through to the resource demands of specific goods and services, the impacts of economic sectors, through national economies to the global economy. Analysis of consumption patterns relates resource use to the environmental, social and economic impacts at the scale or context under investigation. The ideas of embodied resource use (the total resources needed to produce a product or service), resource intensity, and resource productivity are important tools for understanding the impacts of consumption. Key resource categories relating to human needs are food, energy, materials and water.

Energy. The sun's energy, stored by plants (primary producers) during photosynthesis, passes through the food chain to other organisms to ultimately power all living processes. Since the industrial revolution the concentrated energy of the Sun stored in fossilised plants as fossil fuels has been a major driver of technology which, in turn, has been the source of both economic and political power. In 2007 climate scientists of the IPCC concluded that there was at least a 90 per cent probability that atmospheric increase in CO_2 was human-induced, mostly as a result of fossil fuel emissions but, to a lesser extent from changes in land use. Stabilising the world's climate will require high-income countries to reduce their emissions by 60–90 per cent over 2006 levels by 2050 which should hold CO_2 levels at 450–650 ppm from current levels of about 380 ppm. Above this level, temperatures could rise by more than 2°C to produce 'catastrophic' climate change. Reduction of current CO_2 levels must be achieved against a background of global population increase and developing countries aspiring to energy-intensive high consumption Western lifestyles. Reducing greenhouse emissions, referred to as decarbonisation, is being tackled at all scales, ranging from tracking the passage of carbon through the carbon cycle to the commercialisation of renewable energy, developing less carbon-hungry technology and transport systems and attempts by individuals to lead carbon neutral lifestyles by monitoring the fossil fuel use embodied in all the goods and services they use.

Water. Water security and food security are inextricably linked. In the decade 1951–60 human water withdrawals were four times greater than the previous decade. This rapid increase resulted from scientific and technological developments impacting through the economy—especially the increase in irrigated land, growth in industrial and power sectors, and intensive dam construction on all continents. This altered the water cycle of rivers and lakes, affected their water quality and had a significant impact on the global water cycle. Currently towards 35 per cent of human water use is unsustainable, drawing on diminishing aquifers and reducing the flows of major rivers: this percentage is likely to increase if climate change impacts become more severe, populations increase, aquifers become progressively depleted and supplies become polluted and unsanitary.

Water efficiency is being improved on a global scale by increased demand management, improved infrastructure, improved water productivity of agriculture, minimising the water intensity (embodied water) of goods and services, addressing shortages in the non-industrialised world, concentrating food production in areas of high productivity, and planning for climate change. At the local level, people are becoming more self-sufficient by harvesting rainwater and reducing use of mains water.

Food. The American Public Health Association (APHA) defines a 'sustainable food system' as 'one that provides healthy food to meet current food needs while maintaining healthy ecosystems that can also provide food for generations to come with minimal negative impact to the environment. A sustainable food system also encourages local production and distribution infrastructures and makes nutritious food available, accessible, and affordable to all. Further, it is humane and just, protecting farmers and other workers, consumers, and communities'. Concerns about the environmental impacts of agribusiness and the stark contrast between the obesity problems of the western world and the poverty and food insecurity of the developing world have generated a strong movement towards healthy, sustainable eating as a major component of overall ethical consumerism. The environmental effects of different dietary patterns depend on many factors, including the proportion of animal and plant foods consumed and the method of food production.

At the global level the environmental impact of agribusiness is being addressed through sustainable agriculture and organic farming. At the local level there are various movements working towards local food production, more productive use of urban wastelands and domestic gardens including permaculture, urban horticulture, local food, slow food, sustainable gardening, and organic gardening.

Materials, toxic substances, waste: As global population and affluence has increased, so has the use of various materials increased in volume, diversity and distance transported. Included here are raw materials, minerals, synthetic chemicals (including hazardous substances), manufactured products, food, living organisms and waste.

Sustainable use of materials has targeted the idea of dematerialisation, converting the linear path of materials (extraction, use, disposal in landfill) to a circular material flow that reuses materials as much as possible, much like the cycling and reuse of waste in nature. This approach is supported by product stewardship and the increasing use of material flow analysis at all levels, especially individual countries and the global economy. Synthetic chemical production has escalated following the stimulus it received during the second World War. Chemical production includes everything from herbicides, pesticides and fertilisers to domestic chemicals and hazardous substances.

Apart from the build-up of greenhouse gas emissions in the atmosphere, chemicals of particular concern include: heavy metals, nuclear waste, chlorofluorocarbons, persistent organic pollutants and all harmful chemicals capable of bioaccumulation. Although most synthetic chemicals are harmless there needs to be rigorous testing of new chemicals, in all countries, for adverse environmental and health effects. International legislation has been established to deal with the global distribution and management of dangerous goods.

Every economic activity produces material that can be classified as waste. To reduce waste industry, business and government are now mimicking nature by turning the waste produced by industrial metabolism into resource. Dematerialisation is being encouraged through the ideas of industrial ecology, ecodesign and ecolabelling. In addition to the well-established 'reduce, reuse and recycle' shoppers are using their purchasing power for ethical consumerism.

Economic Dimension

On one account, sustainability 'concerns the specification of a set of actions to be taken by present persons that will not diminish the prospects of future persons to enjoy levels of consumption, wealth, utility, or welfare comparable to those enjoyed by present persons'. Sustainability interfaces with economics through the social and ecological consequences of economic activity. Sustainability economics represents: 'a broad interpretation of ecological economics where environmental and ecological variables and issues are basic but part of a multidimensional perspective. Social, cultural, health-related and monetary/financial aspects have to be integrated into the analysis'.

Decoupling environmental degradation and economic growth

Historically there has been a close correlation between economic growth and environmental degradation: as communities grow, so the environment declines. This trend is clearly demonstrated on graphs of human population numbers, economic growth, and environmental indicators. Unsustainable economic growth has been starkly compared to the malignant growth of a cancer because it eats away at the earth's ecosystem services which are its life-support system. There is concern that, unless resource use is checked, modern global civilisation will follow the path of ancient civilisations that collapsed through over exploitation of their resource base. While conventional economics is concerned largely with economic growth and the efficient allocation of resources, ecological economics has the explicit goal of sustainable scale (rather than continual growth), fair distribution and efficient allocation, in that order.

Nature as an economic externality

The economic importance of nature is indicated by the use of the expression ecosystem services to highlight the market relevance of an increasingly scarce natural world that can no longer be regarded as both unlimited and free. In general, as a commodity or service becomes more scarce the price increases and this acts as a restraint that encourages frugality, technical innovation and alternative products. However, this only applies when the product or service falls within the market system. As ecosystem services are generally treated as economic externalities they are unpriced and therefore overused and degraded, a situation sometimes referred to as the Tragedy of the Commons.

Economic opportunity

Treating the environment as an externality may generate short-term profit at the expense of sustainability. Sustainable business practices, on the other hand, integrate ecological concerns with social and economic ones (i.e. the triple bottom line). Growth that depletes ecosystem services is sometimes termed 'uneconomic growth' as it leads to a decline in quality of life. Minimising such growth can provide opportunities for local businesses. For example, industrial waste can be treated as an 'economic resource in the wrong place'. The benefits of waste reduction include savings from disposal costs, fewer environmental penalties, and reduced liability insurance. This may lead to increased market share due to an improved public image. Energy efficiency can also increase profits by reducing costs.

Social Dimension

Sustainability issues are generally expressed in scientific and environmental terms, but implementing change is a social challenge that entails, among other things, international and national law, urban planning and transport, local and individual lifestyles and ethical consumerism. 'The relationship between human rights and human development, corporate power and environmental justice, global poverty and citizen action, suggest that responsible global citizenship is an inescapable element of what may at first glance seem to be simply matters of personal consumer and moral choice'.

Peace, security, social justice

Social disruptions like war, crime and corruption divert resources from areas of greatest human need, damage the capacity of societies to plan for the future, and generally threaten human well-being and the environment. Broad-based strategies for more sustainable social systems include: improved education and the political empowerment of women, especially in developing countries; greater regard for social justice, notably equity between rich and poor both within and between countries; and intergenerational equity. Depletion of natural resources including fresh water increases the likelihood of 'resource wars'. This aspect of sustainability has been referred to as environmental security and creates a clear need for global environmental agreements to manage resources such as aquifers and rivers which span political boundaries, and to protect global systems including oceans and the atmosphere.

Human relationship to nature

According to Murray Bookchin, the idea that humans must dominate nature is common in hierarchical societies. Bookchin contends that capitalism and market relationships, if unchecked, have the capacity to reduce the planet to a mere resource to be exploited. Nature is thus treated as a commodity: 'The plundering of the human spirit by the market place is paralleled by the plundering of the earth by capital'. Still more basically, Bookchin argued that most of the activities that consume energy and destroy the environment are senseless because they contribute little to quality of life and well being. The function of work is to legitimise, even create, hierarchy. For this reason understanding the transformation of organic into hierarchical societies is crucial to finding a way forward.

Social ecology, founded by Bookchin, is based on the conviction that nearly all of humanity's present ecological problems originate in, indeed are mere symptoms of, dysfunctional social arrangements. Whereas most authors proceed as if our ecological problems can be fixed by implementing recommendations which stem from physical, biological, economic, etc. studies, Bookchin's claim is that these problems can only be resolved by understanding the underlying social processes and intervening in those processes by applying the concepts and methods of the social sciences. Deep ecology establishes principles for the well-being of all life on earth and the richness and diversity of life forms. This is only compatible with a substantial decrease of the human population and the end of human interference with the nonhuman world. To achieve this, deep ecologists advocate policies for basic economic, technological, and ideological structures that will improve the quality of life rather than the standard of living. Those who subscribe to these principles are obliged to make the necessary change happen.

Human settlements

One approach to sustainable living, exemplified by small-scale urban transition towns and rural ecovillages, seeks to create self-reliant communities based on principles of simple living, which maximise self-sufficiency particularly in food production. These principles, on a broader scale, underpin the concept of a bioregional economy. Other approaches, loosely based around new urbanism, are successfully reducing environmental impacts by altering the built environment to create and preserve sustainable cities which support sustainable transport. Residents in compact urban neighbourhoods drive fewer miles, and have significantly lower environmental impacts across a range of measures, compared with those living in sprawling suburbs. Ultimately, the degree of human progress towards sustainability will depend on large scale social movements which influence both community choices and the built environment. Eco-municipalities may be one such movement. Eco-municipalities take a systems approach, based on sustainability principles. The eco-municipality movement is participatory, involving community members in a bottom-up approach.

Chapter 28

Air Pollution and Its Control

INTRODUCTION

Air pollution is the presence of substances in air in sufficient concentration and for sufficient time, so as to be, or threaten to be injurious to human, plant or animal life, or to property, or which reasonably interferes with the comfortable enjoyment of life and property. Air pollutants arise from both man made and natural processes. Pollutants are also defined as primary pollutants resulting from combustion of fuels and industrial operations and secondary pollutants, those which are produced due to reaction of primary pollutants in the atmosphere. The ambient air quality may be defined by the concentration of a set of pollutants which may be present in the ambient air we breath in. These pollutants may be called criteria pollutants. Emission standards express the allowable concentrations of a contaminant at the point of discharge before any mixing with the surrounding air. Table 28.1 lists names of some common air pollutants, their sources and classification.

Table 28.1. Common pollutants and their sources.

Pollutants	Sources
Suspended particulate matter, SPM[a]	Automobile, power plants, boilers, industries requiring crushing and grinding such as quarry, cement.
Chlorine	Chlor-alkali plants.
Fluoride	Fertiliser, aluminium refining
Sulphur dioxide	Power plants, boilers, sulphuric acid manufacture, ore refining, petroleum refining.
Lead	Ore refining, battery manufacturing, automobiles.
Oxides of nitrogen,[a] NO_1 NO_2 (NO_x)	Automobiles, power plants, nitric acid manufacture, also a secondary pollutant
Peroxyacetyl nitrate, PAN	Secondary pollutant
Formaldehyde	Secondary pollutant
Ozone[a]	Secondary pollutant
Carbon monoxide[a]	Automobiles
Hydrogen sulphide	Pulp and paper, petroleum refining.
Hydrocarbons	Automobiles, petroleum refining
Ammonia	Fertiliser plant

[a] Criteria pollutants.

Air pollution is the introduction of chemicals, particulate matter, or biological materials that cause harm or discomfort to humans or other living organisms, or damages the natural environment into the atmosphere. The atmosphere is a complex dynamic natural gaseous system that is essential to support life on planet earth. Stratospheric ozone depletion due to air pollution has long been recognised as a threat to human health as well as to the earth's ecosystems. An air pollutant is known as a substance in the air that can cause harm to humans and the environment. Pollutants can be in the form of solid particles, liquid droplets, or gases. In addition, they may be natural or man-made.

CLASSIFICATION OF POLLUTANTS

Pollutants can be classified as either primary or secondary. Usually, primary pollutants are substances directly emitted from a process, such as ash from a volcanic eruption, the carbon monoxide gas from a motor vehicle exhaust or sulphur dioxide released from factories. Secondary pollutants are not emitted directly. Rather, they form in the air when primary pollutants react or interact. An important example of a secondary pollutant is ground level ozone—one of the many secondary pollutants that make up photochemical smog. Note that some pollutants may be both primary and secondary: that is, they are both emitted directly and formed from other primary pollutants.

Primary Pollutants

Major primary pollutants produced by human activity include:

1. Sulphur oxides (SO_x)—especially sulphur dioxide, a chemical compound with the formula SO_2. SO_2 is produced by volcanoes and in various industrial processes. Since coal and petroleum often contain sulphur compounds, their combustion generates sulphur dioxide. Further oxidation of SO_2, usually in the presence of a catalyst such as NO_2, forms H_2SO_4, and thus acid rain. This is one of the causes for concern over the environmental impact of the use of these fuels as power sources.

2. Nitrogen oxides (NO_x)—especially nitrogen dioxide are emitted from high temperature combustion. Can be seen as the brown haze dome above or plume downwind of cities. Nitrogen dioxide is the chemical compound with the formula NO_2. It is one of the several nitrogen oxides. This reddish-brown toxic gas has a characteristic sharp, biting odour. NO_2 is one of the most prominent air pollutants.

3. Carbon monoxide—is a colourless, odourless, non-irritating but very poisonous gas. It is a product by incomplete combustion of fuel such as natural gas, coal or wood. Vehicular exhaust is a major source of carbon monoxide.

4. Carbon dioxide (CO_2)—a greenhouse gas emitted from combustion but is also a gas vital to living organisms. It is a natural gas in the atmosphere.

5. Volatile organic compounds—VOCs are an important outdoor air pollutant. In this field they are often divided into the separate categories of methane (CH_4) and non-methane volatile organic compounds (MNVOCs). Methane is an extremely efficient greenhouse gas which contributes to enhanced global warming. Other hydrocarbon VOCs are also significant greenhouse gases via their role in creating ozone and in prolonging the life of methane in the atmosphere, although the effect varies depending on local air quality. Within the NMVOCs, the aromatic compounds benzene, toluene and xylene are suspected carcinogens and may lead to leukemia through prolonged exposure. 1,3-butadiene is another dangerous compound which is often associated with industrial uses.

6. Particulate matter— Particulates, alternatively referred to as particulate matter (PM) or fine particles, are tiny particles of solid or liquid suspended in a gas. In contrast, aerosol refers to particles and the gas together. Sources of particulate matter can be man-made or natural. Some particulates occur naturally, originating from volcanoes, dust storms, forest and grassland fires, living vegetation, and sea spray. Human activities, such as the burning of fossil fuels in vehicles, power plants and various industrial processes also generate significant amounts of aerosols. Averaged over the globe, anthropogenic aerosols—those made by human activities—currently account for about 10 per cent of the total amount of aerosols in our atmosphere. Increased levels of fine particles in the air are linked to health hazards such as heart disease, altered lung function and lung cancer.

7. Persistent free radicals connected to airborne fine particles could cause cardiopulmonary disease.

8. Toxic metals, such as lead, cadmium and copper.

9. Chlorofluorocarbons (CFCs)—harmful to the ozone layer emitted from products currently banned from use.

10. Ammonia (NH_3)—emitted from agricultural processes. Ammonia is a compound with the formula NH_3. It is normally encountered as a gas with a characteristic pungent odour. Ammonia contributes significantly to the nutritional needs of terrestrial organisms by serving as a precursor to foodstuffs and fertilizers. Ammonia, either directly or indirectly, is also a building block for the synthesis of many pharmaceuticals. Although in wide use, ammonia is both caustic and hazardous.

11. Odours—such as from garbage, sewage, and industrial processes.

12. Radioactive pollutants—produced by nuclear explosions, war explosives, and natural processes such as the radioactive decay of radon.

Secondary Pollutants

Secondary pollutants include:

1. Particulate matter formed from gaseous primary pollutants and compounds in photochemical smog. Smog is a kind of air pollution; the word 'smog' is a portmanteau of smoke and fog. Classic smog results from large amounts of coal burning in an area caused by a mixture of smoke and sulphur dioxide. Modern smog does not usually come from coal but from vehicular and industrial emissions that are acted on in the atmosphere by sunlight to form secondary pollutants that also combine with the primary emissions to form photochemical smog.

2. Ground level ozone (O_3) formed from NO_x and VOCs. Ozone (O_3) is a key constituent of the troposphere (it is also an important constituent of certain regions of the stratosphere commonly known as the Ozone layer). Photochemical and chemical reactions involving it drive many of the chemical processes that occur in the atmosphere by day and by night. At abnormally high concentrations brought about by human activities (largely the combustion of fossil fuel), it is a pollutant, and a constituent of smog.

3. Peroxyacetyl nitrate (PAN)—similarly formed from NO_x and VOCs.

SOURCES OF PARTICULATES

Natural processes inject 800–2000 million tons of particulate matter each year into the atmosphere. These processes include volcanic eruptions, blowing of dust and soil by the wind, spraying of salt and various other solid particles by the seas and oceans, etc.

Human activities, however, emit 450 million tons of particulates every year. For example, particulates in the form of dust and asbestoes are formed during construction, fly-ash rich in particulates is given out from power plants, smelters, mining process and smoke from incomplete combustion processes.

According to an estimate: (i) stationary combustion sources (coal, wood, fuel, oil, natural gas), (ii) industrial process, and (iii) miscellaneous sources (coal refuse burning, agricultural burning, forest fires, structural fires etc.) contribute equally (nearly one-third) of the total particulate emission (450 mn tons) by human activities.

Effects of Particulate Pollutants

Some of the effects of particulate pollutants are:

Effects on plants

1. Plants are adversely affected by gaseous pollutants and deposition of particulates on soil. This deposition of toxic metals on the soil makes it unsuitable for plant growth.
2. Several particulate pollutants fall on the soil by acid rain which tend to lower its pH, making it more acidic and infertile.
3. Particulates such as dust, fog, soot, deposited on plant leaves block the stomata of plants, thus inhibiting the rate of transpiration of minerals from the soil.
4. Deposited particulates restrict the absorption of CO_2, thereby reducing the rate of photosynthesis, retarding plant growth and crop production.
5. Dust mixed with mist or light rain forms a thick crust on the upper leaf surfaces, which shields the bright sunlight necessary for carbon assimilation.
6. Some plants are very sensitive to the traces of toxic metals, where particulates inhibit the action of their enzyme system.
7. Arsenic is a cumulative, potent, protoplasmic poison which inhibits SH-group in enzymes. This is present in almost all types of soils in minute quantities and affects plant growth.

Effects on humans

The effects of particulate pollutants are largely dependent on the particle size. Airborne particles, i.e. dust, soot, fumes and mists, are potentially dangerous for human health.

1. Particulate pollutants have a bearing on the penetration of particles beyond the respiratory passages into the lungs. Nasal passage prevents coarser particulates bigger than 5 microns from entering into the respiratory system. Particles with a size of about 1 micron enter into lungs easily and rapidly. Actually, aerosols less than 1 micron may reach the alveoli of lungs and damage lung tissues.
2. Soluble aerosols will be absorbed into the blood from the alveoli while insoluble aerosols are carried to the lymphatic stream and get deposited in pulmonary lymphatic depot points or in the lymph glands, where they create toxicity on the respiratory system.
3. Workers exposed to pollutant asbestos mostly develop cancer called mesothelioma, which occurs in the tissue lining the abdomen. They also have a greatly elevated risk of lung cancer.
4. Insoluble particulates which can not be phagocytised by white blood corpuscles (WBCs) pass through the alveolar walls into lymph channels. They also accumulate in various specific organs and their increased concentration exerts actions on lungs.
5. Lead, the most serious pollutant released from automobile exhaust, is reported to have detrimental effect on children's brains.

6. Lead interferes with the development and maturation of red blood cells.
7. Workers exposed to lead excrete porphyrins, the precursors of haemoglobin, in their urine.
8. It has been reported that smokers can more easily develop symptoms of asthma, which is also due to excess concentration of lead, than nonsmokers.
9. Silcosis, a chronic disease of lungs, is caused by inhalation of dust containing free silica, SiO_2.
10. Acid particulates and aldehydes cause eye, nose and throat irritation.
11. Formaldehyde is extremely toxic to human health. Acrolein causes irritation to mucus membranes and also poses bronchioconstriction.
12. Lead and asbestos act as cumulative poison and are dangerous to children, causing brain damage and cancer. Small fibres of asbestos irritate lung tissues causing asbestosis, a condition characterised by lung fibrosis.
13. Black lung disease is common among coal miners, while white lung disease occurs frequently among textile workers.
14. Carbon, in the form of soot, deposits in the nose, throat and respiratory tract.
15. Particulates such as silica, asbestos and different forms of carbon are capable of exerting a noxious or fibrotic local action in the interstitial areas of the lungs and in the lymphatic tissues.
16. Beryllium compounds like $BeSO_4$ and $BeCl_2$ cause acute inflammation of the lungs.
17. Small exposure of cadmium causes cardiovascular diseases and hypertension. It also interferes with copper and zinc metabolism.
18. Traces of mercury cause nerve damage and death, while arsenic creates acute and chronic cancer.
19. Fine particulates of less than 2 μ size are the worst causes of lung damage while the larger particles (3 μ) are trapped in the nose and throat. These particles create various breathing troubles due to nasal tract blockage and irritation of lung capillaries.
20. Arsenic is absorbed through the lungs and skin, and causes diarrhoea, peripheral neuritis, conjunctivitis, hyperketosis, lung and skin cancer. Chronic exposure to arsenic leads to the so-called 'black foot' disease.

Effects on materials

1. Particulates affect a variety of materials in various ways. They damage buildings, paints, furniture, etc. Painted surfaces are very susceptible to damage in wet conditions. Particulate fumes and mists react directly with any painted surfaces and cause cracks in them.
2. Particulates accelerate corrosion of metals, mainly in urban and industrial areas.
3. Particles, including fumes, dust, soot, mists and aerosols, can cause severe damage to soil, buildings, sculpture and monuments.
4. As a result of the 1984 Bhopal disaster, soil within 160 km, radius was coated with thick dust due to methyl isocyanate leakage and soil fertility was lost for at least the next ten years.
5. Particulates cause extensive damage when they themselves are corrosive and carry toxic harmful chemicals along with them.
6. Particulate pollutants are responsible for polluted cloud formation, rain and snow in which the particles act as the nuclei on which water droplets condense.
7. Particulates accumulate on soil surfaces causing erosion. The particles are generally sticky, tarry and acidic, so they adhere to surfaces and act as acid reservoirs.
8. Precipitation effect over localities or suburbs is due to particulate pollutants. They also influence the formation of rain, dew and snow.

Effects on solar radiation

1. Particulates reduce visibility by absorption and scattering of solar radiation. Decreased visibility creates chronic and annoying problems.
2. They cause illumination problems by reducing sunlight by nearly one-third.
3. In winter, increased heating requires more power which releases more particulate particles in the atmosphere.
4. They also upset the delicate heat balance of the earth's atmosphere. Increased content of CO_2 (10 per cent) causes global warming but the particulates (i.e. aerosols) have the ability to reject more solar radiations thus compensating the climatic effect of increased carbon dioxide concentration.

CONTROL OF PARTICULATES

There are three broad approaches to the control of particulates: (i) dilution in the atmosphere, (ii) control at source, and (iii) control by using pollution control equipment.

Dilution in the Atmosphere

Dilution of particulates and gases can be accomplished by the use of tall stacks. Pollutants released from taller stacks disperse easily and hence low ground level concentrations are observed. Tall stacks penetrate the inversion layer and disperse the contaminants easily so that the ground level concentrations are less harmful. However, dilution is only a short-term control measure and has highly undesirable long-range effects. In India, industries have to ensure a minimum stack height of 30 metres. Theoretically, the minimum height of stack, H, required for effective dispersion of particulates is given by $H = 74 \, Q^{0.27}$ where, Q is particulate emission rate in tons per hour. This often demands stack heights in excess of 400 m, especially in cement industries and thermal power plants, and therefore may prove more uneconomical when compared to particulate control by treatment.

Control at Source

The most effective means of dealing with the problem of air pollution is to prevent emission at the source itself. In the case of industrial pollutants, this can often be achieved by investigating various approaches at the early stage of process design and development, and selecting those methods which do not contribute to air pollution, (or) have the minimum air pollution potential.

Raw material changes

Some raw materials are primarily responsible for causing air pollution. Use of pure grade of raw material is often beneficial and may reduce the formation of undesirable impurities and by-products, and may even eliminate troublesome effluents.

Process changes

Changing the processing methods is still another important method of controlling emissions at their source. For example, petroleum/chemical industries have undergone radical changes in processing methods which emphasise continuous automatic operations, often computer controlled and completely enclosed systems that minimise the release of materials into the atmosphere.

Particulate Control by Using Equipment

The most effective methods of particulate control are reduction at the source by the application of control equipment and process control. If air pollution problems are properly considered in an industry prior to its design, real economy can be effected. But unfortunately in most cases, air pollution control is an afterthought and the ways and means to control it are derived and designed hurriedly at the eleventh hour.

To remove particulate matter from gas streams, various types of control equipment are available. But to select the required equipment, certain basic data must be available. The required data are:

1. Quantity of gas to be treated and its variation with time.
2. Nature and concentration of the particulate matter to be removed.
3. Temperature and pressure of the gas stream.
4. Nature of the gas phase (for solubility and corrosive effects).
5. Desired quality of the treated effluent, i.e. efficiency of removal of particulates required.

Filters

Filters are now widely used to collect extremely fine particulates. Solid dispersoids can be removed from the carrier gas by filtration of the effluents through porous cloth or fibre. Next to cyclonic separators, filters constitute the most effective devices of controlling particulate pollutants.

Scrubbers

Scrubbers may be classified as wet washers, cyclonic scrubbers, gravity spray towers, impingement scrubbers, disintegrator scrubbers and spray de-dusters.

Electrostatic Precipitators (ESP)

The electrostatic precipitator is one of the most widely used devices for controlling particulate emissions from industrial installations manufacturing household appliances to power plants, cement and paper mills and oil refineries. In most cases, the particulates to be collected are by-products of combustion. In others, they are dust fibres or other small particles such as acid mists from process industries. The electrostatic precipitators are particulate collection devices that utilise electrical energy directly to assist in the removal of particulate matter.

GASEOUS POLLUTANTS

Among the gaseous pollutants are carbon monoxide, sulphur dioxide and nitrogen dioxide, etc.

Carbon Monoxide

Carbon monoxide is another such gas which, although was present in the atmosphere earlier, is now considered to be a major pollutant. An excess of the same has a harmful effect on our system. There are many reasons why carbon monoxide can be released into the atmosphere as a result of human activities. This is also produced due to any fuel burning appliance and appliances such as gas water heaters, fireplaces, woodstoves, gas stoves, gas dryers, yard equipments as well as automobiles, which add to the increased proportion of this gas into the atmosphere.

Sulphur Dioxide

Sulphur dioxide is yet another harmful pollutant that causes air pollution. Sulphur dioxide is emitted largely to the excessive burning of fossil fuels, petroleum refineries, chemical and coal burning power plants, etc. Nitrogen dioxide when combined with sulphur dioxide can even cause a harmful reaction in the atmosphere that can cause acid rain.

Nitrogen Dioxide

Nitrogen dioxide is one more gas that is emitted into the atmosphere as a result of various human activities. An excess of nitrogen dioxide mainly happens due to most power plants seen in major cities, the burning of fuels due to various motor vehicles and other such sources, whether industrial or commercial that cause the increase in the levels of nitrogen dioxide.

Control of Gaseous Pollutants

Emission of gaseous pollutants can be controlled by the following methods: (i) absorption, (ii) adsorption, and (iii) combustion.

Absorption

Absorption involves the transfer of pollutants from the gas phase to the liquid phase across the interface in response to a concentration gradient, with the concentration decreasing in the direction of mass transfer. Absorption of a gaseous contaminant by a liquid occurs because the latter is not saturated with the contaminant at the conditions existing in the absorber. The difference between the concentration of the contaminant in the liquid if it was saturated and the actual concentration provides the driving force for absorption. Hence, the more soluble a contaminant is in the liquid phase, the greater is the overall efficiency that can be attained. In the absorption process, effluent gases are passed through absorbers (scrubbers), which contain liquid absorbents, that remove one or more pollutants from the gas stream. Absorbents are being used to remove SO_2, H_2S, SO_3, F and oxides of nitrogen. The absorbents may be either reactive or non-reactive with the pollutants removed by them. The reactive type absorbents may be either regenerative or non-generative. Various equipment which use the principle of absorptions are: (i) spray towers, (ii) packed towers, (iii) plate towers, (iv) bubble plate towers, and (v) ventury scrubbers.

Adsorption

An alternative to absorption by liquids is the adsorption of air pollutants on solids. The commonly used adsorbers include activated carbon, molecular sieves such as dehydrated zeolites, silica gel, activated alumina, lithium, chloride, bauxite, etc. Adsorption is a surface phenomenon by which gas or liquid molecules are captured by and adhere to the surface of the solid adsorbent. The attractive forces holding the molecules on the surface may be either physical (physical adsorption) or chemical (chemisorption) in nature. The steps necessary for the effective removal of gaseous pollutants by adsorbents are: (i) intimate contact between gaseous pollutants and the solid adsorbent, (ii) separation of unadsorbed gases from the adsorbent and adsorbate, and for final disposal, and (iii) separation (or desorption) of the adsorbed gaseous pollutant from the solid adsorbent by regeneration or replacement of the adsorbent.

Combustion or incineration

This is used when the pollutants in the gas stream are oxidisable to an inert gas. Pollutants like hydrocarbons, and carbon monoxide can be easily burned, oxidised and removed from the combustion equipment. This is achieved by: (i) direct flame incineration, and (ii) catalytic incineration.

Sulphur Compounds (SO$_x$) as Pollutants

Oxides of sulphur, i.e. sulphur dioxide (SO_2) and sulphur trioxide (SO_3), represented as SO_x, hydrogen sulphide (H_2S), carbonyl sulphide (COS), carbon disulphide (CS_2), dimethyl sulphide [$(CH_3)_2S$] and sulphates (SO_4) are the most serious air pollutants.

Oxides of Carbon as Pollutants

CO, a significant contaminant is produced in the atmosphere by natural processes, i.e. forest fires, natural gas emission, marsh gas production and volcanic actions contribute to form CO. However, human activities, mainly automobile exhausts, contribute significantly (about 80 per cent) to CO emission. Its concentration varies depending upon the density of vehicular traffic. CO is removed by converting it into CO_2 which then metabolises. Thus, the fixation of CO by plants is of immense importance as green plants are a major global sink of CO. This CO absorption by plants increases linearly with the increase of CO concentration. Therefore, in cities where CO concentration is higher, the rate of CO absorption by plants may be greater by a factor of 10 to 100. Thus, plants play a vital role in the global CO sink.

Effects of CO pollutants

1. On plants: CO has some detrimental effects on plants, when they are exposed to CO for long duration. For example, it inhibits the nitrogen fixation ability of bacteria when they are exposed to CO levels of 2000 ppm for 33–38 hours.
2. On humans: All gaseous pollutants cause severe damage to the respiratory system. But the adverse effects of CO in the human body are unique. On inhalation, it passes through the lungs into the blood stream.

Effects of carbon dioxide (CO_2)

Carbon dioxide is nontoxic, therefore it is not harmful to human health, unlike CO. CO_2 is utilised by green plants to prepare starch during photosynthesis. But today its increasing concentration in the atmosphere (10 per cent) has long-term effects. It produces adverse physiological effects only at very high levels. The increased amount of CO_2 in air is mainly responsible for global warming. CO_2 molecules absorb heat energy and tend to prevent the long wave infrared heat radiation from earth from escaping into space and it deflects these radiations back to earth.

CONTROL/REMOVAL OF SO_x POLLUTANTS

There are six procedures for controlling SO_x emissions. They are either in-plant control measures or effluent treatment methods.
1. Natural dispersion by dilution
2. Using alternate fuels
3. Removal of sulphur from fuels (desulphurisation)
4. Process modifications
5. Control of SO_x in the combustion process
6. Treatment of flue gas emissions.

Natural Dispersion by Dilution

The control method is based on natural dispersion at high elevation so that the ground-level concentrations are acceptable at all times. In India, minimum stack heights of 30 m are recommended.

It is also a common practice to stop discharging the effluents into the atmosphere during adverse meteorological conditions. However, this technique is not possible in large-scale power plants, etc.

Using Alternate Fuels

A switch to natural gas from the conventional high-sulphur fuels like coal and petroleum to lessen SO_x emissions is an available alternative. Liquefied natural gas, LNG, also is an effective alternative. However, for utility use, its cost will be much higher than that of other alternatives. Low sulphur coal is another alternative, but obtaining low sulphur coal from the ground is neither quick nor cheap.

Removal of Sulphur from Fuels

The process of removing sulphur prior to combustion is theoretically attractive but practically ineffective. Coal consists of sulphur in both organic and inorganic forms. The inorganic form of sulphur is iron disulphide (FeS_2) mostly available in the forms of pyrites and marcasites. Apparently, washing seems to be an effective process with more than 30 per cent of sulphur being removed. But this results in a loss of combustible material and may increase the requirement of coal and thus the cost. Organic sulphur is present in the form of crystine, thiols, sulphides and some other cyclic compounds, which can only be removed by chemical processing.

OXIDES OF NITROGEN AS POLLUTANTS

Of the six or seven oxides of nitrogen known, only three—nitrous oxide (N_2O), nitric oxide (NO), and nitrogen oxide (NO_2)—are formed in any appreciable quantities in the atmosphere. Often NO and NO_2 are analysed together in air and are referred to as NO_x.

The various reasons for the increase of oxides of nitrogen in the ambient atmosphere may be the increased emission of NO_x from industries and automobiles, and various other sources. Well over 90 per cent of all man-made nitrogen oxides that enter our atmosphere are produced by the combustion of various fuels.

The real danger posed by NO_x at the concentration found in metropolitan areas lies in its role in photochemical reactions leading to smog formation. These atmospheric reactions lead to the formation of chemical compounds that have a direct adverse effect on human beings and plants. In some situations, NO_x may be present in a high enough concentration, yet not react to form smog because other necessary conditions for the reaction are absent. However, nearly every major city in India experiences the effects induced by the presence of NO_x.

Sources of NO_x Pollution

Natural stratospheric NO_x are also produced by the action of cosmic rays in the upper atmosphere. Man-made sources of NO_x varies depending upon global areas. NO_x are 10 to 100 times greater in urban atmosphere as compared to rural areas. Major man-made activities include combustion of coal, oil, natural gas and gasoline, etc. which produce up to 50 ppm of nitrogen.

Effects of nitrogen oxides (NO_x)

Mostly the oxides of nitrogen are not so dangerous, but the role they play in the formation of photochemical oxidants, etc. constitute the most harmful effect. These NO_x affect plants, human health, and the atmosphere.

NO_x and acid rain

Oxides of nitrogen (NO, NO_2 and N_2O) through photochemical chain reactions, produce irritating gases which are toxic and corrosive. These gases produce nitric acid which is washed down and contributes to acid rain. Acidity of rain water increases when the amount of oxides of nitrogen and sulphur increases in the atmosphere to produce more nitric and sulphuric acid. Acid rain reduces fertility of the soil. It is also affecting the Taj Mahal in Agra. NO_2 is a part of the photolytic NO_2 cycle.

Control of NO$_x$ Pollution

NO$_x$ emissions from flue gases can be reduced by: (i) dilution in the atmosphere by increasing stack height or by discharging effluents into the atmosphere only during favourable meteorological conditions, (ii) modification of operating and design conditions, and (iii) treatment of flue gases. The method of control of NO$_x$ by dilution is similar to that of SO$_x$, particulates and other pollutants.

NO$_x$ control by treatment

The modification of design and combustion techniques, in general, is to prevent or to suppress the emission of nitrogen oxides in the combustion chambers. However, if the measures are not enough or too costly, stack gas treatment is necessary. Stack gas treatment is an efficient method for the control of NO$_x$. The methods are: (i) absorption by liquids, (ii) adsorption by solids, (iii) catalytic reduction — selective and nonselective, and (iv) electron beam irradiation.

PHOTOCHEMICAL SMOG

Photochemical smog is initiated by the photochemical dissociation of NO$_2$ and the consequent secondary reactions involving unsaturated hydrocarbons, other organic compounds and free radicals, leading to the formation of organic peroxides and ozone. This phenomenon takes place during sunny days with low winds and low level inversion. Photochemical smog and the consequent formation of aerosols reduce visibility, cause irritation to eyes and damage plants and rubber goods.

The oxidation of SO$_2$ can also take place by interaction with the free radical HO$\cdot$ present in photochemical smog

$$SO_2 + HO\cdot \rightarrow HOSO_2\cdot$$

$$HOSO_2\cdot + O_2 \rightarrow HOSO_2\,O_2\cdot$$

$$HOSO_2\,O_2\cdot \rightarrow HOSO_2\,O\cdot + NO_2\cdot$$
$$\text{(sulphate)}$$

Chemical oxidation of SO$_2$ may also take place in water droplets, present in aerosols. This reaction is accelerated in the presence of NH$_3$ and catalysts, e.g. oxides of Mn, Fe, Cu, Ni.

Solid particles, such as soot, bring about catalytic oxidation of SO$_2$ by providing a heterogeneous phase for contact. Soot is formed during combustion of solid and liquid fuels in domestic and industrial operations, and automobile emissions. Sulphur dioxide is a pollutant responsible for smog formation, acid rains and corrosion of metals and alloys.

Oxidation of Organic Compounds

Organic compounds such as hydrocarbons, aldehydes and ketones absorb solar radiation and undergo various photochemical and chemical reactions involving free radicals. Some of these reactions are catalysed by particulate matter such as soot and metal oxides. Some of the resultant intermediates and final products contribute to photochemical smog formation.

GREENHOUSE EFFECT

The earth is heated by sunlight and some of the heat that is absorbed by the earth is radiated back into space. However, some of the gases in the lower atmosphere, acting like glass in a greenhouse, allow solar radiations (in the range 300 to 2,500 nm, i.e., near UV, visible and near infrared region, while filtering the dangerous u.v. radiations, i.e. < 300 nm) but do not allow the earth to re-radiate the heat into space. In other words, these gases in the atmosphere are transparent to the sunlight coming in, but they strongly absorb infrared radiation, which the earth sends back as heat. A part of the heat so trapped in these atmospheric gases is re-emitted to the earth's surface. The net result is the heating of the earth's surface by this phenomenon, called the 'greenhouse effect'. The gases that are responsible for this greenhouse effect are CO_2, water vapour, CH_4 and man-made chlorofluorocarbons (CFCs). Water vapour strongly absorbs infrared radiations in the range 4000 to 8000 nm and CO_2 in the range 12,000 to 16,300 nm. The radiations in the range 8000 to 12,000 nm escape unabsorbed and this is known as the region of atmospheric window.

Carbon dioxide is released by volcanoes, oceans, decaying plants as well as human activities, such as deforestation and combustion of fossil fuels. Automobile exhausts account for 30 per cent of CO_2 emissions in developed countries. Methane is released from coal mines, decomposition of organic matter in swamps, rice paddy cultivation, guts of termites in forest debris and stomachs of ruminants.

Chlorofluorocarbons (CFCs) are used as coolants in refrigerators, propellants in aerosol sprays, plastic foam materials like 'thermocoles' or 'styrofoam' and in automobile air-conditioners.

In fact, the 'greenhouse gases' (particularly CO_2 and water vapour) are responsible for keeping our planet warm and thus sustaining life on the earth. If the greenhouse gases were very less or totally absent then the average temperature on the earth would have been at sub-zero levels. But, however, if the concentration of greenhouse gases increases, they may trap too much of heat, which may threaten the very existence of life on earth. For instance, the CO_2 present in the atmosphere of the planet *Venus*, is about 60,000 times more than that on earth. Hence, the average temperature of *Venus* is about 425°C, making the existence of life impossible there.

Oceans and bio-mass are the major sinks for atmospheric CO_2. Oceans convert CO_2 into soluble bicarbonates. The photosynthetic activity in the green plants increases with the increase in CO_2 level in the atmosphere. Forests are the places where lot of photosynthetic activity occurs. They also act as vast reservoirs of fixed but readily oxidisable carbon in the form of vegetation, wood and humus. Hence, forests maintain a balance in the atmospheric CO_2 level, and deforestation upsets this balance and increases the atmospheric CO_2 level.

It is estimated that the atmospheric CO_2 content has increased by 25 per cent during the last two centuries. This is mostly attributed to the industrial revolution and is one of the reasons for the slight increase in the global temperature (about 0.5°C). Since the concentrations of greenhouse gases have been continuously increasing because of deforestation, industrialisation, increased burning of fossil fuels, mining, exhausts from increasing number of automobiles and other anthropogenic activities, there is an increasing concern about the possible 'global warming'. Some scientists fear that if proper precautions are not taken, the concentration of greenhouse gases in the atmosphere may double within the next 50–100 years. If this happens, the average global temperature may increase by 4°–5°C. This will increase the evaporation of surface waters, which may influence climatic changes depending upon the pattern of cloud formation. For instance, low-level dense clouds may exert cooling effect whereas high-level thin cloud formation may exert heating effect due to increased greenhouse effect.

The projections from computer modelling regarding the climatic changes that could be triggered off due to 'global warming' reveal alarming scenarios. Even a 1.5°C rise in surface temperature can adversely affect food production in the world. Thus, the wheat growing zones in the northern latitude may be shifted from the USSR and Canada to the polar regions, i.e. from fertile soils to poor soils near the North Pole. The biological productivity of the ocean would also decrease due to warming of the earth's surface layer, which in turn, may reduce the transport of nutrients from deeper layers to the surface by vertical circulation. Computer modelling also indicates the following effects due to 'global warming': melting of the polar ice caps; dry areas becoming drier; humid areas like the Amazon suffering more intense tropical storms; drastic drop in food production, particularly in lands within 35 degrees north and south of the Equator; increased breeding of pests and diseases due to more humid conditions; shorter, wetter and warmer winters and longer, hotter and drier summers, particularly in mid-continental areas. Global warming may also trigger increased thermal expansion of oceans and melting of glaciers, which may result in an increase in the sea-level by 20 cm to 1.5 metres by the latter part of the 21st century. Thus, cities like Mumbai, Miami, London, Venice, Bangkok and Leningrad may become extremely vulnerable. Defences against the rising sea-levels and expanding oceans are very difficult and expensive, which many nations cannot afford. Further, a global temperature rise, is likely to cause more floods, hurricanes, and tornadoes.

There are differences of opinion among experts regarding the dynamics and effects of 'global warming' due to the complexity of natural phenomena that might be operating simultaneously. More accurate future climatic projections will be possible with better super-computer models, based on greater understanding of the complex natural climatic forces involved. But until that time, the possible devastating effect due to 'global warming' by the 'greenhouse effect' cannot be underestimated. Some of the steps suggested to minimise the 'greenhouse effect' include reduction in the use of fossil fuels, encouraging the use of alternative sources of energy (e.g. solar, geothermal, wind, biogas, etc.), conservation of forests, extensive afforestation, encouraging community forestry, reduction in the use of automobiles, research in the development of more efficient automobile engines, ban on CFCs and nuclear explosions, development of environmentally compatible technologies with the help of intensive inter-disciplinary research, effective check on the growth of population and imparting of non-formal and formal environmental education.

FORMATION AND DEPLETION OF OZONE IN THE STRATOSPHERE

Ozone is an important chemical species present in the stratosphere. At an altitude of about 30 km, its concentration is about 10 ppm. The ozone layer present in the stratosphere acts as a protective shield for life on earth. It strongly absorbs ultraviolet radiations from the sun in the region 220–330 nm and thereby protects life on earth from severe radiation damage, such as DNA mutation and skin cancer. Thus only a small fraction of UV radiation reaches the lower atmosphere and the earth's surface.

Ozone is formed in the stratosphere by photochemical reaction:

$$O_2 + h\nu\,(242\ nm) \rightarrow O + O$$

$$O + O_2 + M\ (third\ body,\ such\ as\ N_2\ or\ O_2) \rightarrow O_3 + M$$

The third body absorbs the excess energy liberated by the above reaction and thereby the ozone molecule is stabilised. Thus, ozone is constantly formed in the stratosphere. However, it is also destroyed by chlorine, released due to volcanic activity and also by reaction with: (i) nitric oxide; (ii) atomic oxygen, and (iii) reactive hydroxyl radical, which are also present in the atmosphere. In the atmosphere, nitrogen oxide (NO) comes from chemical and photochemical reactions, supersonic jets, nuclear explosions, etc; Cl_2 comes from CFC's and volcanoes; and OH comes from biomass burning and from natural water systems by the following reactions:

1. $O_3 + NO \rightarrow NO_2 + O_2$
2. $O_3 + O \rightarrow O_2 + O_2$

3. $O_3 + HO\cdot \rightarrow HO_2 + HOO\cdot$
4. $HOO\cdot + O \rightarrow HO\cdot + O_2$

Ozone, in the stratosphere, is also destroyed by man-made chlorofluorocarbons (CFCs), which are used as coolants in refrigerators, air-conditioners, propellants in aerosol sprays and in plastic foams, such as 'thermocole' or 'styrofoam'. The CFC molecules, escaping into the atmosphere, decompose to release chlorine in the ozone layer (by photo-dissociation) and each atom of chlorine, thus liberated is capable of attacking several ozone molecules.

$$Cl + O_3 \rightarrow ClO + O_2$$

This reaction is followed by:

$$ClO + O \rightarrow Cl + O_2$$

which regenerates Cl atoms, so that a long chain process is involved, which conserves Cl atoms. The environmental hazards of CFCs were recognised as early as 1970. In fact, temporary thinning in the stratospheric ozone layer, leading to the formation of 'Ozone hole' was actually detected over the Antarctica during September to November 1985. Reported increase in cases of skin cancer in South Australia are also attributed to UV radiations reaching the earth, due to depletion of the ozone layer.

The detection of the 'Ozone hole' over Antarctica in 1985 attracted the attention of global scientific community. The US immediately banned the use of CFCs in spray cans. Further, in the year 1987, 24 nations signed the Montreal Protocol, which aimed at 35 per cent reduction in the global production of the CFCs by the year 1999. Simultaneously, efforts to produce chlorine-free substitutes have also started. In fact, synthesis of a product called HFC-134a has already been reported as an effective substitute for CFC. The use of hydrofluorocarbons (HFCs), hydrochlorofluorocarbons (HCFCs), and methyl cyclohexane (MCH) as substitutes for CFCs is envisaged for several applications.

Almost all the sulphur present in liquid and gaseous fuels and about 80 per cent of sulphur present in solid fuels appears as SO_2 in the flue gases. Depending on the sulphur content of the fuel burnt and the conditions of combustion (e.g. percentage of excess air used), the concentration of SO_x in flue gases varies from 0.05 to 0.4 per cent. However, in metallurgical operations such as smelting of sulphide ores, the SO_2 concentration in stack gases may be 5 to 10 per cent.

SO_2 is oxidised to SO_3 in atmospheric air by photolytic and catalytic processes involving ozone, NO_x and hydrocarbons, giving rise to the formation of photochemical smog. Oxidation of SO_2 can take place in presence of catalysts such as NO_x, metal oxides, soot and dust. Under normal humid conditions, SO_3 reacts with water vapour to produce droplets of H_2SO_4 aerosol which gives rise to the so-called 'acid rain'. The sulphuric acid and sulphate aerosols present in urban air are smaller than 2μ and hence can easily reach the pulmonary region of lungs, causing serious respiratory problems, particularly in older people.

$$SO_2 + O_3 \rightarrow SO_3 + O_2$$

$$SO_3 + H_2O \rightarrow H_2SO_4 \rightarrow (H_2SO_4)_n$$

$$SO_2 + 1/2\, O_2 + H_2O \xrightarrow[\text{metal oxide, soot, etc.}]{\text{Catalyst such as}} H_2SO_4 \xrightarrow{\text{Aerosol}} (H_2SO_4)_n$$

Control of SO_x emissions from the anthropogenic activities is contemplated along the following lines:

1. Removing SO_x from flue gases before letting them out into the atmosphere: Chemical scrubbers such as (i) Lime stone or (ii) Citric acid are suggested to absorb SO_2 from the flue gases.
   (a) $2CaCO_3 + 2SO_2 + O_2 \rightarrow 2CaSO_4 + CO_2$
   (b) $SO_2 + H_2O \rightarrow HSO_3^- + H^+$
   (c) $HSO_3^- + H_2\, cit^- \rightarrow (HSO_3 . H_2\, cit)^{-2}$

2. Removing sulphur from the fuels used for combustion: Pyritic sulphur in coal can be removed by grinding and washing in coal washeries. However, organically bound sulphur cannot be easily removed from coals. Research is in progress to synthesise special type of micro-organisms using bio-technology, which are capable of converting organically bound sulphur into soluble form.
3. Utilising low-sulphur fuels.
4. Generation of power by alternative energy sources and discouraging fossil-fuel based thermal power-plants.

ACID RAIN

Rain has always been valued by mankind, because good crops and abundant water supplies are possible only due to timely and plentiful rainfall. Summer rains refresh people. Spring rains recharge the aquifers and cleanse the groundwater. Autumn rains and winter snow help clean the air. Rain, in general, brings with it a sense of hope, vitality and a promise for the future.

Over the last few decades, simple rainfall has taken on a threatening complexity in some parts of the world. In these locales, rain must pass through an atmosphere polluted with oxides of sulphur (SO_x) and of nitrogen (NO_x). The falling rain and snow often react with these oxide pollutants to produce often a mixture of sulphuric acid, nitric acid and water. This is known as *acid precipitation* or *acid rain*.

Rain tends to be naturally acidic with a pH of 5.6 to 5.7 due to the reaction of atmospheric CO_2 with water to produce carbonic acid. This small amount of acidity is sufficient to dissolve minerals the earth's crust and make them available to plant and animal life, but it is not acidic enough to inflict any major damage. Other atmospheric substances from volcanic eruptions, forest fires and other similar natural phenomena also contribute to the acidity in rain. Thus, even with the enormous amounts of acids created by nature annually, normal rainfall is able to assimilate them to the point where they cause little, if any, known damage. But, it is the contributions of SO_x, NO_x, etc., from anthropogenic activities that disturb this acid balance and convert natural and mildly acidic rain into precipitation with far-reaching environmental consequences.

Acid rain represents one of the major consequences of air pollution, because of large SO_x and NO_x emissions from big industrial areas. The longer the SO_x and NO_x remain in the atmosphere, the greater are the chances of their oxidation to H_2SO_4 and HNO_3 due to photochemical and catalytic chemical reactions. Acid rains may cause extensive damage to materials and terrestrial ecosystems, such as water, fish, vegetation, stone, steel, paint, soil and mankind.

The only practical approach to counter the problem of acid rain is to reduce SO_x and NO_x emissions. The following three general options are considered for this purpose:

1. *Energy conservation* resulting in reduced fuel consumption and hence slower emissions of SO_x and NO_x. Conservation via more efficient fuel use and through improved thermal insulation is also being studied.
2. *Desulphurisation and denitrification of fuels* of stack gases and increased use of fuels naturally low in sulphur content or use of technologies that reduce SO_x and NO_x emissions. Desulphurisation and use of low NO_x-producing technologies are the only viable control options today and will perhaps continue to be so for some more time.
3. *Substitutions for fossil fuels* by other alternative energy forms may offer future solutions to this problem.

Reduction of SO_x emissions can be accomplished by: (i) removing the sulphur content before the fuel is burnt with the help of techniques such as coal cleaning, coal gasification and desulphurisation of liquid fuels; (ii) **removing the sulphur content during combustion, as in fluidised-bed combustion; and** (iii) removal of sulphur emissions after combustion, as in stack or flue gas desulphurisation systems or scrubbers. The future of SO_x control from traditional fuel sources lies in the perfection of these techniques.

Reduction of NO_x emissions from stationary combustion sources can be achieved by modification of furnace and burner design, and/or modification of operating conditions. The combustion modification techniques available now include using two-stage combustion, precisely controlling air, injecting water during combustion, recirculating flue gases, and/or by altering design of firing chambers. Reductions in NO_x emissions from mobile combustion sources may be achieved by lowering the combustion temperatures in the engine and catalytic removal of NO_x from exhaust gases using devices such as a three-way system that simultaneously reduces carbon monoxide, hydrocarbons and NO_x.

CONTROL DEVICES FOR AIR POLLUTION

The following items are commonly used as pollution control devices by industry or transportation devices. They can either destroy contaminants or remove them from an exhaust stream before it is emitted into the atmosphere.

1. Particulate control:
   (a) Mechanical collectors (dust cyclones, multicyclones).
   (b) Electrostatic precipitators: An electrostatic precipitator (ESP), or electrostatic air cleaner is a particulate collection device that removes particles from a flowing gas (such as air) using the force of an induced electrostatic charge. Electrostatic precipitators are highly efficient filtration devices that minimally impede the flow of gases through the device, and can easily remove fine particulate matter such as dust and smoke from the air stream.
   (c) Baghouses designed to handle heavy dust loads, a dust collector consists of a blower, dust filter, a filter-cleaning system, and a dust receptacle or dust removal system (distinguished from air cleaners which utilise disposable filters to remove the dust).
   (d) Particulate scrubbers/Wet scrubber is a form of pollution control technology. The term describes a variety of devices that use pollutants from a furnace flue gas or from other gas streams. In a wet scrubber, the polluted gas stream is brought into contact with the scrubbing liquid, by spraying it with the liquid, by forcing it through a pool of liquid or by some other contact method, so as to remove the pollutants.

2. Scrubbers:
   (a) Baffle spray scrubber.
   (b) Cyclonic spray scrubber.
   (c) Ejector venturi scrubber.
   (d) Mechanically aided scrubber.
   (e) Spray tower.
   (f) Wet scrubber.

3. NO_x control:
   (a) Low NO_x burners.
   (b) Selective catalytic reduction (SCR).
   (c) Selective non-catalytic reduction (SNCR).
   (d) NO_x scrubbers.
   (e) Exhaust gas recirculation.
   (f) Catalytic converter (also for VOC control).

4. VOC abatement:
    (a) Adsorption systems, such as activated carbon.
    (b) Flares.
    (c) Thermal oxidisers.
    (d) Catalytic oxidisers.
    (e) Biofilters.
    (f) Absorption (scrubbing).
    (g) Cryogenic condensers.
    (h) Vapour recovery systems.
5. Acid gas/SO_2 control:
    (a) Wet scrubbers.
    (b) Dry scrubbers.
    (c) Flue gas desulphurisation.
6. Mercury control:
    (a) Sorbent injection technology.
    (b) Electro-catalytic oxidation (ECO).
    (c) K-fuel.

5. Write ....
PA alkanaline ....

Chapter 29

Water Pollution and Its Control

INTRODUCTION

Water pollution is the contamination of water bodies (e.g. lakes, rivers, oceans and groundwater). Water pollution affects plants and organisms living in these bodies of water; and, in almost all cases the effect is damaging not only to individual species and populations, but also to the natural biological communities. Water pollution occurs when pollutants are discharged directly or indirectly into water bodies without adequate treatment to remove harmful compounds. Water pollution is a major problem in the global context.

WATER POLLUTION CATEGORIES

Surface water and groundwater have often been studied and managed as separate resources, although they are interrelated. Sources of surface water pollution are generally grouped into two categories based on their origin.

Point Source Pollution

Point source pollution refers to contaminants that enter a waterway through a discrete conveyance, such as a pipe or ditch. Examples of sources in this category include discharges from a sewage treatment plant, a factory, or a city storm drain.

Non-point Source Pollution

Non-point source (NPS) pollution refers to diffuse contamination that does not originate from a single discrete source. NPS pollution is often the cumulative effect of small amounts of contaminants gathered from a large area. The leaching out of nitrogen compounds from agricultural land which has been fertilised is a typical example. Nutrient runoff in stormwater from 'sheet flow' over an agricultural field or a forest are also cited as examples of NPS pollution.

Contaminated storm water washed off of parking lots, roads and highways, called urban runoff, is sometimes included under the category of NPS pollution. However, this runoff is typically channelled into storm drain systems and discharged through pipes to local surface waters, and is a point source. However where such water is not channelled and drains directly to ground it is a non-point source.

GROUNDWATER POLLUTION

Interactions between groundwater and surface water are complex. Consequently, groundwater pollution, sometimes referred to as groundwater contamination, is not as easily classified as surface water pollution.

By its very nature, groundwater aquifers are susceptible to contamination from sources that may not directly affect surface water bodies, and the distinction of point vs. non-point source may be irrelevant. A spill or ongoing releases of chemical or radionuclide contaminants into soil (located away from a surface water body) may not create point source or non-point source pollution, but can contaminate the aquifer below, defined as a toxin plume. The movement of the plume, a plume front, can be part of a Hydrological transport model or Groundwater model. Analysis of groundwater contamination may focus on the soil characteristics and site geology, hydrogeology, hydrology, and the nature of the contaminants.

CAUSES OF WATER POLLUTION

The specific contaminants leading to pollution in water include a wide spectrum of chemicals, pathogens, and physical or sensory changes such as elevated temperature and discolouration. While many of the chemicals and substances that are regulated may be naturally occurring (calcium, sodium, iron, manganese, etc.) the concentration is often the key in determining what is a natural component of water, and what is a contaminant. Oxygen-depleting substances may be natural materials, such as plant matter (e.g. leaves and grass) as well as man-made chemicals. Other natural and anthropogenic substances may cause turbidity (cloudiness) which blocks light and disrupts plant growth, and clogs the gills of some fish species. Many of the chemical substances are toxic. Pathogens can produce waterborne diseases in either human or animal hosts. Alteration of water's physical chemistry includes acidity (change in pH), electrical conductivity, temperature, and eutrophication. Eutrophication is an increase in the concentration of chemical nutrients in an ecosystem to an extent that increases in the primary productivity of the ecosystem. Depending on the degree of eutrophication, subsequent negative environmental effects such as anoxia (oxygen depletion) and severe reductions in water quality may occur, affecting fish and other animal populations.

Pathogens

Coliform bacteria are a commonly used bacterial indicator of water pollution, although not an actual cause of disease. Other micro-organisms sometimes found in surface waters which have caused human health problems include:

1. *Burkholderia pseudomallei.*
2. *Cryptosporidium parvum.*
3. *Giardia lamblia.*
4. *Salmonella.*
5. *Novovirus* and other viruses.
6. Parasitic worms (helminths).

High levels of pathogens may result from inadequately treated sewage discharges. This can be caused by a sewage plant designed with less than secondary treatment (more typical in less-developed countries). In developed countries, older cities with ageing infrastructure may have leaky sewage collection systems (pipes, pumps, valves), which can cause sanitary sewer overflows. Some cities also have combined sewers, which may discharge untreated sewage during rain storms.

Pathogen discharges may also be caused by poorly managed livestock operations.

Chemical and Other Contaminants

Contaminants may include organic and inorganic substances.

Organic water pollutants

Organic water pollutants include:

1. Detergents.
2. Disinfection by-products found in chemically disinfected drinking water, such as chloroform.
3. Food processing waste, which can include oxygen-demanding substances, fats and grease.
4. Insecticides and herbicides, a huge range of organohalides and other chemical compounds.
5. Petroleum hydrocarbons, including fuels (gasoline, diesel fuel, jet fuels, and fuel oil) and lubricants (motor oil), and fuel combustion by-products, from stormwater runoff.
6. Tree and bush debris from logging operations.
7. Volatile organic compounds (VOCs), such as industrial solvents, from improper storage. Chlorinated solvents, which are dense nonaqueous phase liquids (DNAPLs), may fall to the bottom of reservoirs, since they do not mix well with water and are denser.
8. Various chemical compounds found in personal hygiene and cosmetic products.

Inorganic water pollutants

Inorganic water pollutants include:

1. Acidity caused by industrial discharges (especially sulphur dioxide from power plants).
2. Ammonia from food processing waste.
3. Chemical waste as industrial by-products.
4. Fertilisers containing nutrients—nitrates and phosphates—which are found in stormwater runoff from agriculture, as well as commercial and residential use.
5. Heavy metals from motor vehicles (via urban stormwater runoff) and acid mine drainage.
6. Silt (sediment) in runoff from construction sites, logging, slash and burn practices or land clearing sites.

Macroscopic pollution—large visible items polluting the water—may be termed 'floatables' in an urban stormwater context, or marine debris when found on the open seas, and can include such items as:

1. Trash (e.g. paper, plastic, or food waste) discarded by people on the ground, and that are washed by rainfall into storm drains and eventually discharged into surface waters.
2. Nurdles, small ubiquitous waterborne plastic pellets.
3. Shipwrecks, large derelict ships.

Thermal pollution

Thermal pollution is the rise or fall in the temperature of a natural body of water caused by human influence. A common cause of thermal pollution is the use of water as a coolant by power plants and industrial manufacturers. Elevated water temperatures decreases oxygen levels (which can kill fish) and affects ecosystem composition, such as invasion by new thermophilic species. Urban runoff may also elevate temperature in surface waters.

Thermal pollution can also be caused by the release of very cold water from the base of reservoirs into warmer rivers.

WASTE-WATER TREATMENT

The principal objective of waste-water treatment is generally to allow human and industrial effluents to be disposed of without danger to human health or unacceptable damage to the natural environment.

Irrigation with waste-water is both disposal and utilisation and indeed is an effective form of waste-water disposal (as in slow-rate land treatment). However, some degree of treatment must normally be provided to raw municipal waste-water before it can be used for agricultural or landscape irrigation or for aquaculture. The quality of treated effluent used in agriculture has a great influence on the operation and performance of the waste-water-soil-plant or aquaculture system. In the case of irrigation, the required quality of effluent will depend on the crop or crops to be irrigated, the soil conditions and the system of effluent distribution adopted. Through crop restriction and selection of irrigation systems which minimise health risk, the degree of pre-application waste-water treatment can be reduced. A similar approach is not feasible in aquaculture systems and more reliance will have to be placed on control through waste-water treatment.

The most appropriate waste-water treatment to be applied before effluent use in agriculture is that which will produce an effluent meeting the recommended microbiological and chemical quality guidelines both at low cost and with minimal operational and maintenance requirements. Adopting as low a level of treatment as possible is especially desirable in developing countries, not only from the point of view of cost but also in acknowledgement of the difficulty of operating complex systems reliably. In many locations it will be better to design the reuse system to accept a low-grade of effluent rather than to rely on advanced treatment processes producing a reclaimed effluent which continuously meets a stringent quality standard. Nevertheless, there are locations where a higher-grade effluent will be necessary and it is essential that information on the performance of a wide range of waste-water treatment technology should be available. The design of waste-water treatment plants is usually based on the need to reduce organic and suspended solids loads to limit pollution of the environment. Pathogen removal has very rarely been considered an objective but, for reuse of effluents in agriculture, this must now be of primary concern and processes should be selected and designed accordingly. Treatment to remove waste-water constituents that may be toxic or harmful to crops, aquatic plants (macrophytes) and fish is technically possible but is not normally economically feasible. Unfortunately, few performance data on waste-water treatment plants in developing countries are available and even then they do not normally include effluent quality parameters of importance in agricultural use.

The short-term variations in waste-water flows observed at municipal waste-water treatment plants follow a diurnal pattern. Flow is typically low during the early morning hours, when water consumption is lowest and when the base flow consists of infiltration-inflow and small quantities of sanitary waste-water. A first peak of flow generally occurs in the late morning, when waste-water from the peak morning water use reaches the treatment plant, and a second peak flow usually occurs in the evening. The relative magnitude of the peaks and the times at which they occur vary from country to country and with the size of the community and the length of the sewers. Small communities with small sewer systems have a much higher ratio of peak flow to average flow than do large communities. Although the magnitude of peaks is attenuated as waste-water passes through a treatment plant, the daily variations in flow from a municipal treatment plant make it impracticable, in most cases, to irrigate with effluent directly from the treatment plant. Some form of flow equalisation or short-term storage of treated effluent is necessary to provide a relatively constant supply of reclaimed water for efficient irrigation, although additional benefits result from storage.

Conventional Waste-water Treatment Processes

Conventional waste-water treatment consists of a combination of physical, chemical, and biological processes and operations to remove solids, organic matter and, sometimes, nutrients from waste-water.

General terms used to describe different degrees of treatment, in order of increasing treatment level, are preliminary, primary, secondary, and tertiary and/or advanced waste-water treatment. In some countries, disinfection to remove pathogens sometimes follows the last treatment step. A generalised waste-water treatment diagram is shown in Fig. 29.1.

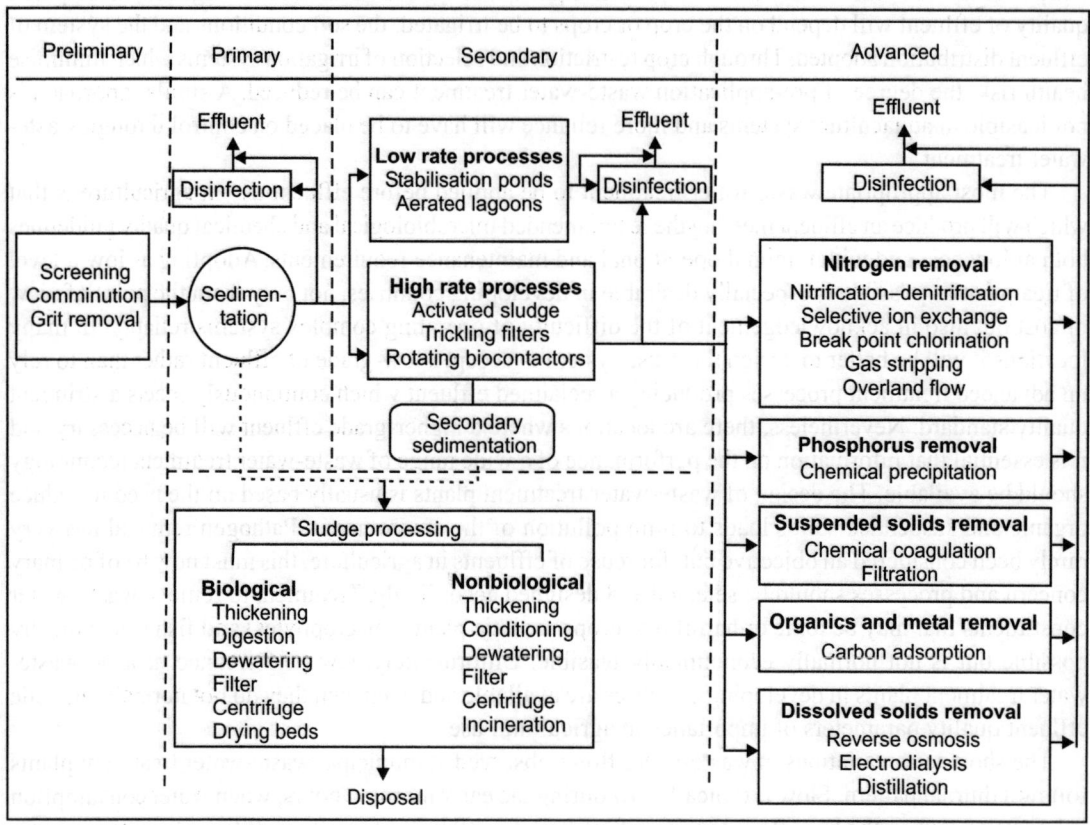

Fig. 29.1. Generalised flow diagram for municipal waste-water treatment.

Preliminary treatment

The objective of preliminary treatment is the removal of coarse solids and other large materials often found in raw waste-water. Removal of these materials is necessary to enhance the operation and maintenance of subsequent treatment units. Preliminary treatment operations typically include coarse screening, grit removal and, in some cases, comminution of large objects. In grit chambers, the velocity of the water through the chamber is maintained sufficiently high, or air is used, so as to prevent the settling of most organic solids. Grit removal is not included as a preliminary treatment step in most small waste-water treatment plants. Comminutors are sometimes adopted to supplement coarse screening and serve to reduce the size of large particles so that they will be removed in the form of a sludge in subsequent treatment processes. Flow measurement devices, often standing-wave flumes, are always included at the preliminary treatment stage.

Primary treatment

The objective of primary treatment is the removal of settleable organic and inorganic solids by sedimentation, and the removal of materials that will float (scum) by skimming. Approximately 25 to 50 per cent of the incoming biochemical oxygen demand (BOD_5), 50 to 70 per cent of the total suspended solids (SS), and 65 per cent of the oil and grease are removed during primary treatment. Some organic nitrogen, organic phosphorus, and heavy metals associated with solids are also removed during primary sedimentation but colloidal and dissolved constituents are not affected. The effluent from primary sedimentation units is referred to as primary effluent.

In many industrialised countries, primary treatment is the minimum level of preapplication treatment required for waste-water irrigation. It may be considered sufficient treatment if the waste-water is used to irrigate crops that are not consumed by humans or to irrigate orchards, vineyards, and some processed food crops. However, to prevent potential nuisance conditions in storage or flow-equalising reservoirs, some form of secondary treatment is normally required in these countries, even in the case of non-food crop irrigation. It may be possible to use at least a portion of primary effluent for irrigation if off-line storage is provided.

Secondary treatment

The objective of secondary treatment is the further treatment of the effluent from primary treatment to remove the residual organics and suspended solids. In most cases, secondary treatment follows primary treatment and involves the removal of biodegradable dissolved and colloidal organic matter using aerobic biological treatment processes. Aerobic biological treatment is performed in the presence of oxygen by aerobic micro-organisms (principally bacteria) that metabolise the organic matter in the waste-water, thereby producing more micro-organisms and inorganic end-products (principally CO_2, NH_3, and H_2O). Several aerobic biological processes are used for secondary treatment differing primarily in the manner in which oxygen is supplied to the micro-organisms and in the rate at which organisms metabolise the organic matter.

High-rate biological processes are characterised by relatively small reactor volumes and high concentrations of micro-organisms compared with low rate processes. Consequently, the growth rate of new organisms is much greater in high-rate systems because of the well controlled environment. The micro-organisms must be separated from the treated waste-water by sedimentation to produce clarified secondary effluent. The sedimentation tanks used in secondary treatment, often referred to as secondary clarifiers, operate in the same basic manner as the primary clarifiers described previously. The biological solids removed during secondary sedimentation, called secondary or biological sludge, are normally combined with primary sludge for sludge processing.

Common high-rate processes include the activated sludge processes, trickling filters or biofilters, oxidation ditches, and rotating biological contactors (RBC). A combination of two of these processes in series (e.g. biofilter followed by activated sludge) is sometimes used to treat municipal waste-water containing a high concentration of organic material from industrial sources.

1. Activated sludge: In the activated sludge process, the dispersed-growth reactor is an aeration tank or basin containing a suspension of the waste-water and micro-organisms, the mixed liquor. The contents of the aeration tank are mixed vigorously by aeration devices which also supply oxygen to the biological suspension. Aeration devices commonly used include submerged diffusers that release compressed air and mechanical surface aerators that introduce air by agitating the liquid surface. Hydraulic retention time in the aeration tanks usually ranges from

3 to 8 hours but can be higher with high BOD_5 waste-waters. Following the aeration step, the micro-organisms are separated from the liquid by sedimentation and the clarified liquid is secondary effluent. A portion of the biological sludge is recycled to the aeration basin to maintain a high mixed-liquor suspended solids (MLSS) level. The remainder is removed from the process and sent to sludge processing to maintain a relatively constant concentration of micro-organisms in the system. Several variations of the basic activated sludge process, such as extended aeration and oxidation ditches, are in common use, but the principles are similar.

2. Trickling filters: A trickling filter or biofilter consists of a basin or tower filled with support media such as stones, plastic shapes or wooden slats. Waste-water is applied intermittently or sometimes continuously, over the media. Micro-organisms become attached to the media and form a biological layer or fixed film. Organic matter in the waste-water diffuses into the film, where it is metabolised. Oxygen is normally supplied to the film by the natural flow of air either up or down through the media, depending on the relative temperatures of the waste-water and ambient air. Forced air can also be supplied by blowers but this is rarely necessary. The thickness of the biofilm increases as new organisms grow. Periodically, portions of the film slough off the media. The sloughed material is separated from the liquid in a secondary clarifier and discharged to sludge processing. Clarified liquid from the secondary clarifier is the secondary effluent and a portion is often recycled to the biofilter to improve hydraulic distribution of the waste-water over the filter.

3. Rotating biological contactors: Rotating biological contactors (RBCs) are fixed-film reactors similar to biofilters in that organisms are attached to support media. In the case of the RBC, the support media are slowly rotating discs that are partially submerged in flowing waste-water in the reactor. Oxygen is supplied to the attached biofilm from the air when the film is out of the water and from the liquid when submerged, since oxygen is transferred to the waste-water by surface turbulence created by the discs' rotation. Sloughed pieces of biofilm are removed in the same manner described for biofilters.

High-rate biological treatment processes, in combination with primary sedimentation, typically remove 85 per cent of the BOD_5 and SS originally present in the raw waste-water and some of the heavy metals. Activated sludge generally produces an effluent of slightly higher quality, in terms of these constituents, than biofilters or RBCs. When coupled with a disinfection step, these processes can provide substantial but not complete removal of bacteria and virus. However, they remove very little phosphorus, nitrogen, non-biodegradable organics or dissolved minerals.

Tertiary and/or advanced treatment

Tertiary and/or advanced waste-water treatment is employed when specific waste-water constituents which cannot be removed by secondary treatment must be removed. Individual treatment processes are necessary to remove nitrogen, phosphorus, additional suspended solids, refractory organics, heavy metals and dissolved solids. Because advanced treatment usually follows high-rate secondary treatment, it is sometimes referred to as tertiary treatment. However, advanced treatment processes are sometimes combined with primary or secondary treatment (e.g. chemical addition to primary clarifiers or aeration basins to remove phosphorus) or used in place of secondary treatment (e.g. overland flow treatment of primary effluent).

An adaptation of the activated sludge process is often used to remove nitrogen and phosphorus and an example of this approach is the 23 Ml/d treatment plant commissioned in 1982 in British Columbia,

Canada. Effluent from primary clarifiers flows to the biological reactor, which is physically divided into five zones by baffles and weirs. In sequence these zones are: (i) anaerobic fermentation zone (characterised by very low dissolved oxygen levels and the absence of nitrates), (ii) anoxic zone (low dissolved oxygen levels but nitrates present), (iii) aerobic zone (aerated), (iv) secondary anoxic zone, and (v) final aeration zone. The function of the first zone is to condition the group of bacteria responsible for phosphorus removal by stressing them under low oxidation-reduction conditions, which results in a release of phosphorus equilibrium in the cells of the bacteria. On subsequent exposure to an adequate supply of oxygen and phosphorus in the aerated zones, these cells rapidly accumulate phosphorus considerably in excess of their normal metabolic requirements. Phosphorus is removed from the system with the waste activated sludge.

Most of the nitrogen in the influent is in the ammonia form, and this passes through the first two zones virtually unaltered. In the third aerobic zone, the sludge age is such that almost complete nitrification takes place, and the ammonia nitrogen is converted to nitrites and then to nitrates. The nitrate-rich mixed liquor is then recycled from the aerobic zone back to the first anoxic zone. Here denitrification occurs, where the recycled nitrates, in the absence of dissolved oxygen, are reduced by facultative bacteria to nitrogen gas, using the influent organic carbon compounds as hydrogen donors. The nitrogen gas merely escapes to atmosphere. In the second anoxic zone, those nitrates which were not recycled are reduced by the endogenous respiration of bacteria. In the final reaeration zone, dissolved oxygen levels are again raised to prevent further denitrification, which would impair settling in the secondary clarifiers to which the mixed liquor then flows.

Disinfection

Disinfection normally involves the injection of a chlorine solution at the head end of a chlorine contact basin. The chlorine dosage depends upon the strength of the waste-water and other factors, but dosages of 5 to 15 mg/l are common. Ozone and ultra violet (UV) irradiation can also be used for disinfection but these methods of disinfection are not in common use. Chlorine contact basins are usually rectangular channels, with baffles to prevent short-circuiting, designed to provide a contact time of about 30 minutes. However, to meet advanced waste-water treatment requirements, a chlorine contact time of as long as 120 minutes is sometimes required for specific irrigation uses of reclaimed waste-water. The bactericidal effects of chlorine and other disinfectants are dependent upon pH, contact time, organic content, and effluent temperature.

Effluent storage

Although not considered a step in the treatment process, a storage facility is, in most cases, a critical link between the waste-water treatment plant and the irrigation system. Storage is needed for the following reasons:

1. To equalise daily variations in flow from the treatment plant and to store excess when average waste-water flow exceeds irrigation demands; includes winter storage.
2. To meet peak irrigation demands in excess of the average waste-water flow.
3. To minimise the effects of disruptions in the operations of the treatment plant and irrigation system. Storage is used to provide insurance against the possibility of unsuitable reclaimed waste-water entering the irrigation system and to provide additional time to resolve temporary water quality problems.

BIOLOGICAL WASTE-WATER TREATMENT

The idea behind all biological methods of waste-water treatment is to introduce contact with bacteria (cells), which feed on the organic materials in the waste-water, thereby reducing its BOD content. In other words, the purpose of biological treatment is BOD reduction. Typically, waste-water enters the treatment plant with a BOD higher than 200 mg/l, but primary settling has already reduced it to about 150 mg/l by the time it enters the biological component of the system. It needs to exit with a BOD content no higher than about 20–30 mg/l, so that after dilution in the nearby receiving water body (river, lake), the BOD is less than 2–3 mg/l. Thus, the biological treatment needs to accomplish a 6-fold decrease in BOD.

Principle: Simple bacteria (cells) eat the organic material present in the waste-water. Through their metabolism, the organic material is transformed into cellular mass, which is no longer in solution but can be precipitated at the bottom of a settling tank or retained as slime on solid surfaces or vegetation in the system. The water exiting the system is then much clearer than it entered it.

A key factor is the operation of any biological system is an adequate supply of oxygen. Indeed, cells need not only organic material as food but also oxygen to breathe, just like humans. Without an adequate supply of oxygen, the biological degradation of the waste is slowed down, thereby requiring a longer residency time of the water in the system. For a given flowrate of water to be treated, this translates into a system with a larger volume and thus taking more space.

Advantages: Like all biological systems, operation takes place at ambient temperature. There is no need to heat or cool the water, which saves on energy consumption. Because waste-water treatment operations take much space, they are located outdoor, and this implies that the system must be able to operate at seasonally varying temperatures. Cells come in a mix of many types, and accommodation to a temperature change is simply accomplished by self adaptation of the cell population.

Similarly, a change in composition of the organic material (due to people's changing activities) leads to a spontaneous change in cell population, with the types best suited to digest the new material growing in larger numbers than other cell types.

BIOLOGICAL METHODS OF WASTE-WATER TREATMENT

Biological treatment — the use of bacteria and other micro-organisms to remove contaminants by assimilating them — has long been a mainstay of waste-water treatment in the chemical process industries (CPI). Because they are effective and widely used, many biological-treatment options are available today. They are, however, not all created equal, and the decision to install a biological-treatment system requires ample thought. When considering biological waste-water treatment for a particular application, it is important to understand the sources of the waste-water generated, typical waste-water composition, discharge requirements, events and practices within a facility that can affect the quantity and quality of the waste-water, and pretreatment ramifications. Consideration of these factors will allow you to maximise the benefits your plant gains from effective biological treatment. Those benefits can include:

1. Low capital and operating costs compared to those of chemical-oxidation processes.
2. True destruction of organics, versus mere phase separation, such as with air stripping or carbon adsorption.
3. Oxidation of a wide variety of organic compounds.
4. Removal of reduced inorganic compounds, such as sulphides and ammonia, and total nitrogen removal possible through denitrification.
5. Operational flexibility to handle a wide range of flows and waste-water characteristics.
6. Reduction of aquatic toxicity.

All biological-treatment processes take advantage of bacteria's remarkable ability to use diverse waste-water constituents to provide the energy for microbial metabolism and the building blocks for cell synthesis. This metabolic activity can remove contaminants that are as varied as the raw materials, by-products and products generated by the CPI.

Selection Criteria

Biological-treatment technologies vary greatly in their strengths and weaknesses. The following are application criteria, which are normally relevant in evaluating various biological-treatment options for the chemical process industries (CPI):

1. Bioassay/toxicity control: The ability to control and minimise the impact of toxic constituents in waste-water on indicating organisms when the treated water is released.
2. BOD removal efficiency: The ability to remove biodegradable, organic compounds.
3. COD removal efficiency: The ability to remove chemically oxidisable substances that may or may not be biodegradable.
4. O&M costs: The cost to operate and maintain the treatment method.
5. Sludge production: The amount of residual biological solids generated by the biological-treatment process.
6. Sludge disposal costs: The cost to collect, dewater and dispose of residual sludge from the treatment method, either on-site or off-site.
7. Performance in winter and summer: The degree in which high or low ambient temperatures will affect biological treatment.
8. Performance on high- and low-temperature water: The degree in which high and low waste-water temperature will affect biological treatment.
9. Operator attention: The relative amount of time required to operate the biological treatment system.
10. Upset recovery: The amount of time it takes for a treatment method to recover from upset conditions. Upset conditions are defined as abnormal variations in the flow or characteristics of the waste-water, which can detrimentally affect a biological treatment system.
11. Expandability: The ease of expanding the treatment capacity to accommodate either an overall plant expansion or an increase in loading.
12. Nitrification efficiency: The relative ease of converting ammonia contained in waste-water to nitrates.
13. VOC containment: The relative ease with which the biological-treatment equipment can be enclosed to contain and collect VOC emissions.
14. VOC stripping potential: The relative ease with which the biological-treatment system will strip volatile organic compounds from the waste-water.
15. Ease of installation: The total amount of time and labour required to install the treatment method.
16. Energy efficiency: The amount of energy used by a treatment method.
17. Ease of secondary containment: The ability and ease with which the treatment system can be provided with secondary containment in case of overflow, spills or leaks.
18. Space requirements: The area required by the treatment method.

Knowing the composition of the water to be handled is essential for planning a treatment process. In petroleum refineries, for example, excessive amounts of spent caustic can quickly overwhelm a waste-water treatment system due to the normally high chemical oxygen demand (COD) of the spent caustic. Another issue can be a significant increase in ammonia and sulphide loads that result from upsets in the operation of sour-water strippers. These loads can, in turn, upset a biological-treatment system if it is not designed to handle ammonia and sulphide.

In addition to **understanding the source and composition of the waste-water, one must also recognise** when pretreatment steps are needed to provide adequate protection for a biological treatment system. In most petroleum and petrochemical facilities, for example, raw waste-water normally contains free oil, which can have serious, detrimental effects. Oil can coat and kill bacteria, causing the micro-organisms to float out of the system, and can interfere with oxygen-transfer efficiency. Another source for concern at refineries is a potential upset in desalter operations that can lead to significant oil/water emulsions in the waste-water and thereby negatively impact the biological-treatment system. To prevent these types of problems in petroleum-industry systems, process steps prior to biological treatment are normally included. Pretreatment for this industry typically includes the use of oil/water separators, an equalisation tank to moderate spikes in waste-water composition, and off-spec waste-water storage. Figure 29.2 shows a typical waste-water treatment system for a petroleum facility. Even properly pretreated waste-water can still contain a wide variety of compounds which may or may not be biodegradable. There can also be significant concentrations of sulphides, ammonia, amines, mercaptans and other compounds that require modifications to the treatment process in order to meet discharge objectives. Vendors can be helpful in setting up pilot-plant or bench-scale tests to assist in determining if biotreatment is a viable option for a particular waste-water composition. The types of compounds present, the concentration of each and the ultimate discharge requirements are key to selecting a proper biological-treatment system. This is true whether the waste-water is being discharged directly to the environment or to a publicly owned treatment works (POTW) or if it is to be reused within the facility.

Typical water treatment system

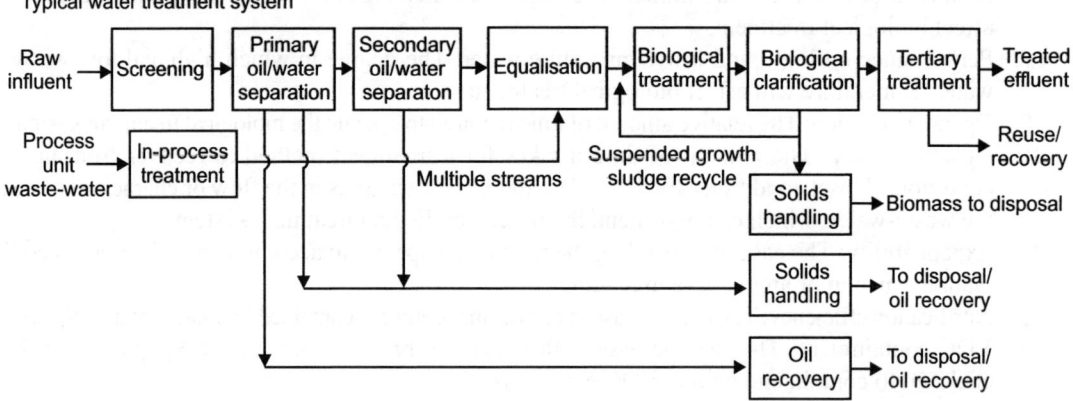

Fig. 29.2. In order to protect the micro-organisms, some biological-treatment processes include pretreatment steps such as screening, oil/water separation and equalisation to moderate fluctuations in waste-water composition. This figure shows a schematic of a waste-water treatment system which is typical for hydrocarbon related industries.

Once the factors discussed above have been resolved, selection from the many available options can begin. Biological treatment methods vary widely, ranging from fixed-film technologies like rotating and submerged biological contactors to technologies like sequencing batch reactors and continuous flow activated-sludge systems. When evaluating the options, one needs to consider their effectiveness in the presence of constraints such as toxicity, COD, biochemical oxygen demand (BOD), and levels of nitrogen and sulphur compounds. Perhaps less obvious, but equally important, is to answer questions such as these:

1. How will the treatment method operate in cold or warm climates?

2. Can the system treat low- and high-temperature waste-waters?
3. How much sludge will the treatment method produce?
4. Can the system recover from up-sets?
5. How much will the system cost to operate and maintain, and does it require extensive operator attention?

All of these factors affect a company's bottom line and the quality of its end product. When evaluating treatment methods, it is essential to examine all of these criteria before making a technology selection.

Evaluation Guidelines

Table 29.1 is an application guide to help determine which biological-treatment technologies are most applicable for typical waste-water applications in the CPI. This table will assist in identifying the top two or three biological treatment technologies to investigate first. Subsequent detailed applications engineering is still critical to the process, but the guide is meant to prioritise potential technologies for further analysis. The parameters listed in Table 29.1 and explained in the box on selection criteria are those typically encountered when evaluating the addition or upgrade of a waste-water biological-treatment system in the CPI.

Influent conditions used in the product/process evaluation are 600 mg/l COD and 250 mg/l BOD, amounts most commonly found in CPI waste-waters after appropriate pretreatment. For applications in which the influent conditions fall substantially outside of these criteria, the results in the table may not apply. In such cases, other biological processes such as anaerobic treatment and a fixed-film, fluidised-bed of activated carbon, should also be considered.

Biological Treatment Options

There are three basic categories of biological treatment: aerobic, anaerobic and anoxic. Aerobic biological treatment, which may follow some form of pretreatment such as oil removal, involves contacting waste-water with microbes and oxygen in a reactor to optimise the growth and efficiency of the biomass. The micro-organisms act to catalyse the oxidation of biodegradable organics and other contaminants such as ammonia, generating innocuous by-products such as carbon dioxide, water, and excess biomass (sludge).

Anaerobic (without oxygen) and anoxic (oxygen deficient) treatments are similar to aerobic treatment, but use micro-organisms that do not require the addition of oxygen. These micro-organisms use the compounds other than oxygen to catalyse the oxidation of biodegradable organics and other contaminants, resulting in innocuous by-products. The three individual types of biological-treatment technologies— aerobic, anaerobic or anoxic—can be run in combination or in sequence to offer greater levels of treatment. Regardless of the type of system selected, one of the keys to effective biological treatment is to develop and maintain an acclimated, healthy biomass, sufficient in quantity to handle maximum flows and the organic loads to be treated.

Maintaining the required population of 'workers' in a bioreactor is accomplished in one of two general ways:

1. Fixed film processes—micro-organisms are held on a surface, the fixed film, which may be mobile or stationary with waste-water flowing past the surface/media. These processes are designed to actively contact the biofilm with the waste-water and with oxygen, when needed.
2. Suspended growth processes—biomass is freely suspended in the waste-water and is mixed and can be aerated by a variety of devices that transfer oxygen to the bioreactor contents.

It is also possible to combine both methods in a single reactor for more effective treatment.

Table 29.1. Evaluation of biological treatment and aeration technologies for the CPI.

Evaluation parameter	Trickling filter	Rotating biological contactor	Submerged biological contactor	Disc aeration	Surface aeration	Fine bubble aeration	Coarse bubble aeration	Jet aeration	Sequencing batch reactor*	Membrane bio-reactor*	PACT*
Effective bioassay/toxicity control				✓	✓			✓		✓	✓
Effective BOD removal efficiency				✓	✓		✓	✓	✓	✓	✓
Effective COD removal efficiency				✓	✓		✓	✓	✓	✓	
Low O&M costs		✓	✓							✓	
Low sludge production	✓	✓	✓							✓	
Low sludge disposal costs	✓	✓	✓							✓	
Good operability: winter					✓					✓	
Good operability: summer	✓	✓	✓		✓		✓		✓	✓	
Good performance: high water temperature										✓	
Good performance: low water temperature								✓	✓	✓	
Minimal operator attention	✓	✓	✓							✓	
Quick upset recovery		✓	✓							✓	
Easy expandability		✓	✓							✓	
Efficient nitrification		✓								✓	✓
Easy to enclose for VOC containment			✓							✓	

(Contd...)

Evaluation parameter	Trickling filter	Rotating biological contactor	Submerged biological contactor	Disc aeration	Surface aeration	Fine bubble aeration	Coarse bubble aeration	Jet aeration	Sequencing batch reactor *	Membrane bio-reactor*	PACT*
Low VOC stripping potential	✓					✓			✓	✓	✓
Easy installation		✓	✓		✓						
Energy efficient	✓	✓	✓			✓					
Ease to secondary containment		✓	✓			✓	✓	✓	✓	✓	✓
Minimal space requirements											

Comparisons are made based on waste-water with a COD of 600 mg/l and BOD of 250 mg/l.
*All listed technologies are products, except for sequencing batch reactors (SBRs), membrane bioreactors (MBRs) and powdered activated-carbon treatment (PACT), which are processes that can incorporate the remaining listed products. SBRs can incorporate fine-bubble, coarse-bubble and jet aeration. MBRs can incorporate all listed products except for trickling filters. PACT can potentially incorporate all listed products except for trickling filters, RBCs and SBCs.

Fixed-film Options

Biotowers (trickling filters)

Biotowers, or trickling filters as they are often called, consist of a layer of media in a tank. Waste-water flowing into the biotower may have gone through an earlier treatment step to remove oil and coarse or settleable solids. Rotary distributor arms or fixed nozzles are used to spray the pretreated waste-water over the surface of the media.

The water then trickles downward through the bed. Air circulates upward through the media as treated water is removed by an underdrain system. As the waste-water trickles downward through the bed, a biological slime of microbes develops on the surface of the media. Continuous flow provides the needed contact between the microbes and the organics. As the slime layer gets thicker, it occasionally sloughs off of the media surface, requiring settling to remove the sloughed biosolids.

While biotowers generally are less efficient at removal of BOD and COD than other technologies, they do generate very little sludge and have a very low potential for stripping volatile organic compounds (VOC). Low VOC stripping potential can be an advantage for environmental reasons.

Rotating biological contactors

Rotating biological contactors (RBCs) consist of vertically arranged, plastic media on a horizontal, rotating shaft. The biomass-coated media are alternately exposed to waste-water and atmospheric oxygen as the shaft slowly rotates at 1–1.5 rpm, with about 40 per cent of the media submerged. High surface area allows a large, stable biomass population to develop, with excess growth continuously and automatically shed and removed in a downstream clarifier.

RBC systems have been installed in many petroleum facilities because of their ability to quickly recover from upset conditions. The RBC system is easily expandable should the need arise, and RBCs are also very easy to enclose should VOC containment become necessary.

Submerged biological contactors

Submerged biological contactors (SBCs), big brothers of the RBC, operate at nearly 90 per cent submergence with coarse-bubble diffused aeration providing a means of both aeration and motive force for rotation. Because of greater submergence, the load on the shaft is significantly less than that of an RBC. The SBC also provides nearly three times the surface area of a conventional RBC per foot of shaft length. With its compact design, the SBC is very easy to cover for VOC and odour containment. Unlike the RBC, the SBC system is driven completely by air, making it one of lowest maintenance and lowest operation-intensive, biological-treatment systems available. Like the RBC, the SBC is modular and can easily be expanded.

Suspended-growth Options

Diffused aeration

Diffused aerators add air to waste-water, increasing dissolved oxygen content and supplying micro-organisms with oxygen necessary for aerobic biological treatment. Fine-bubble diffused-aeration systems are available in various types including ceramic and membranes, and are highly efficient. More reliable, but less efficient, coarse-bubble aeration systems are also available, and are normally manufactured of corrosion-resistant, stainless-steel components. Both systems are compatible with new installations and replacement of existing gas-aeration equipment. Fine-bubble aerators offer very low VOC stripping potential, and both fine and coarse diffusers provide good BOD and COD removal efficiency.

Jet aeration

The jet-aeration system is designed to provide required aeration as well as maintain suspension of biological solids, with the flexibility to either aerate or mix independently without the need for additional equipment. Air flowrates to the system can be varied. When aeration requirements decrease and air is completely shut off, pumps provide the required mixing action to enhance process control and save energy. The subsurface discharge leads to smooth and quiet operation, with no misting, splashing or spray from the basin. This also translates to low VOC release to the atmosphere. Since jet aeration requires no moving parts in the basin, the system offers long life with no in-basin routine maintenance required.

Surface aeration

For efficient surface aeration, high- and low-speed floating aerators provide pumping action that transfers oxygen by breaking up the waste-water into a spray of droplets. The large surface area of the spray allows oxygen to enter the waste-water from the atmosphere. At the same time, the oxygen-enriched water is dispersed and mixed, resulting in effective oxygen delivery. High- and low-speed surface aerators offer excellent oxygen transfer and low operating costs. They are able to handle environmental extremes such as high temperatures.

Another alternative for surface aeration is the use of horizontally mounted aeration discs or rotors. These disc or rotor aerators can be used in oxidation ditches known as looped, 'race track' reactor configurations. They provide stable operation with resulting high-quality effluent. The aerators are above water for easy maintenance and are energy efficient. Other multichannel processes use a concentric arrangement of looped reactors, which is particularly energy efficient and designed to achieve total nitrogen removal through simultaneous nitrification/denitrification. Disc and rotor surface aerators offer good BOD and COD removal efficiencies, and are very easy to replace if necessary.

Reactors in a vertical-loop configuration are also available for surface aeration. They are essentially oxidation ditches flipped on their sides. Upper and lower compartments separated by a horizontal baffle run the length of the tank. Surface-mounted discs or rotors provide mixing and deliver oxygen. Typically, two or more basins make up the system. The first basin operates as an aerated anoxic reactor and the second basin is operated under aerobic conditions. These types of reactors also have high BOD/COD removal efficiency.

Biological Treatment Processes

All of the previously noted aeration technologies, both fixed film and suspended growth, can be considered treatment products. However, there are some technologies that are actually processes because they can incorporate a number of different aeration technologies in their design. These processes include sequencing batch reactors, membrane bioreactors and powdered activated-carbon treatment (PACT) systems.

Sequencing batch reactors

A variation of the conventional activated-sludge system (in such systems, a clarifier is used to settle and recycle biomass back to an aeration basin) is the sequencing batch reactor (SBR). The SBR is a fill-and-draw, non-steady-state, activated-sludge process in which one or more reactor basins are filled with waste-water during a discrete time period and then operated in batch mode. In a single reactor basin, the

SBR accomplishes equalisation, aeration and clarification in a timed sequence. Depending upon desired treatment objectives, the SBR can be operated in aerobic, anoxic or anaerobic conditions to encourage the growth of desirable micro-organisms. Aeration in this system is typically achieved with jet aeration, fine-bubble diffused aeration or coarse-bubble diffused aeration.

One of the advantages of SBRs is good operability in the winter, making them well suited for installations in colder climates. SBRs also take up little space because all of the treatment steps take place in a single reactor basin. Additionally, since the process is controlled by microprocessors, the plant operator is given tremendous flexibility to modify the treatment scheme to match changes in influent flow and loading characteristics.

PACT systems

When conventional biological treatment alone does not meet desired treatment requirements, powdered activated carbon can be added to enhance treatment efficiency. Activated carbon can be used in most suspended-growth, biological-treatment systems. The addition of carbon allows both physical adsorption and biological assimilation to occur simultaneously. PACT systems can be operated either aerobically or anaerobically.

Using powdered activated carbon in conjunction with traditional biological treatment provides excellent effluent bioassay results, provides for toxicity control within the bioreactor, and promotes higher nitrification efficiency than that of a conventional activated-sludge system. PACT systems also provide a buffering effect to shock or upset conditions, allowing the treatment system to recover quickly or even continue treatment with little or no detrimental effects. The use of activated carbon also decreases VOC emissions and improves COD removal efficiency.

Membrane bioreactors

In addition to the traditional types of biological treatment, speciality products have also been introduced to perform more than just one treatment step. Membrane bioreactor (MBR) systems are unique processes, which combine anoxic- and aerobic-biological treatment with an integrated membrane system that can be used with most suspended-growth, biological waste-water-treatment systems.

In the MBR, waste-water is screened before entering the biological treatment tank. Aeration within the aerobic-reactor zone provides oxygen for biological respiration and maintains solids in suspension. To retain active biomass in the process, the MBR relies on submerged membranes rather than clarifiers, eliminating sludge-settling issues. This allows the biological process to operate at longer than normal sludge ages (typically 20–100 days for a MBR) and to increase mixed-liquor, suspended-solids (MLSS) concentrations (typically 8000–15,000 mg/l in a MBR) for more effective removal of pollutants.

High MLSS concentrations and long-solids retention time promote numerous process benefits including stable operation, complete nitrification and reduced biosolids production. High MLSS concentrations also reduce biological-volume requirements and the associated space needed to only 20–30 per cent of conventional biological processes.

CONTROL OF WATER POLLUTION

Domestic Sewage

Domestic sewage is 99.9 per cent pure water, the other 0.1 per cent are pollutants. While found in low concentrations, these pollutants pose risk on a large scale. In urban areas, domestic sewage is typically treated by centralised sewage treatment plants. In the US, most of these plants are operated by local government agencies, frequently referred to as publicly owned treatment works (POTW). Municipal treatment plants are designed to control conventional pollutants: BOD and suspended solids. Well-designed and operated systems (i.e. secondary treatment or better) can remove 90 per cent or more of these pollutants. Some plants have additional subsystems to treat nutrients and pathogens. Most municipal plants are not designed to treat toxic pollutants found in industrial waste-water.

Cities with sanitary sewer overflows or combined sewer overflows employ one or more engineering approaches to reduce discharges of untreated sewage, including:

1. Utilising a green infrastructure approach to improve stormwater management capacity throughout the system, and reduce the hydraulic overloading of the treatment plant.
2. Repair and replacement of leaking and malfunctioning equipment.
3. Increasing overall hydraulic capacity of the sewage collection system (often a very expensive option).

A household or business not served by a municipal treatment plant may have an individual septic tank, which treats the waste-water on site and discharges into the soil. Alternatively, domestic waste-water may be sent to a nearby privately owned treatment system (e.g. in a rural community).

Industrial Waste-water

Some industrial facilities generate ordinary domestic sewage that can be treated by municipal facilities. Industries that generate waste-water with high concentrations of conventional pollutants (e.g. oil and grease), toxic pollutants (e.g. heavy metals, volatile organic compounds) or other nonconventional pollutants such as ammonia, need specialised treatment systems. Some of these facilities can install a pre-treatment system to remove the toxic components, and then send the partially treated waste-water to the municipal system. Industries generating large volumes of waste-water typically operate their own complete on-site treatment systems. Some industries have been successful at redesigning their manufacturing processes to reduce or eliminate pollutants, through a process called pollution prevention.

Heated water generated by power plants or manufacturing plants may be controlled with:

1. Cooling ponds, man-made bodies of water designed for cooling by evaporation, convection, and radiation.
2. Cooling towers, which transfer waste heat to the atmosphere through evaporation and/or heat transfer.
3. Cogeneration, a process where waste heat is recycled for domestic and/or industrial heating purposes.

Agricultural Waste-water

Non-point source controls

Sediment (loose soil) washed off fields is the largest source of agricultural pollution in the United States. Farmers may utilise erosion controls to reduce runoff flows and retain soil on their fields. Common techniques include contour plowing, crop mulching, crop rotation, planting perennial crops and installing riparian buffers.

Nutrients (nitrogen and phosphorus) are typically applied to farmland as commercial fertiliser; animal manure; or spraying of municipal or industrial waste-water (effluent) or sludge. Nutrients may also enter runoff from crop residues, irrigation water, wildlife, and atmospheric deposition. Farmers can develop and implement nutrient management plans to reduce excess application of nutrients.

To minimise pesticide impacts, farmers may use Integrated Pest Management (IPM) techniques (which can include biological pest control) to maintain control over pests, reduce reliance on chemical pesticides, and protect water quality.

Point source waste-water treatment

Farms with large livestock and poultry operations, such as factory farms, are called concentrated animal feeding operations or confined animal feeding operations in the US and are being subject to increasing government regulation. Animal slurries are usually treated by containment in lagoons before disposal by spray or trickle application to grassland. Constructed wetlands are sometimes used to facilitate treatment of animal wastes, as are anaerobic lagoons. Some animal slurries are treated by mixing with straw and composted at high temperature to produce a bacteriologically sterile and friable manure for soil improvement.

Construction Site Stormwater

Sediment from construction sites is managed by installation of:
1. Erosion controls, such as mulching and hydroseeding.
2. Sediment controls, such as sediment basins and silt fences.

Discharge of toxic chemicals such as motor fuels and concrete washout is prevented by use of:
1. Spill prevention and control plans.
2. Specially designed containers (e.g. for concrete washout) and structures such as overflow controls and diversion berms.

MEASUREMENT OF WATER POLLUTION

Water pollution may be analysed through several broad categories of methods: physical, chemical and biological. Most involve collection of samples, followed by specialised analytical tests. Some methods may be conducted *in situ*, without sampling, such as temperature. Government agencies and research organisations have published standardised, validated analytical test methods to facilitate the comparability of results from disparate testing events.

Sampling

Sampling of water for physical or chemical testing can be done by several methods, depending on the accuracy needed and the characteristics of the contaminant. Many contamination events are sharply restricted in time, most commonly in association with rain events. For this reason 'grab' samples are often inadequate for fully quantifying contaminant levels. Scientists gathering this type of data often employ auto-sampler devices that pump increments of water at either time or discharge intervals.

Sampling for biological testing involves collection of plants and/or animals from the surface water body. Depending on the type of assessment, the organisms may be identified for biosurveys (population counts) and returned to the water body, or they may be dissected for bioassays to determine toxicity.

Physical Testing

Common physical tests of water include temperature, solids concentration like total suspended solids (TSS) and turbidity.

Chemical Testing

Water samples may be examined using the principles of analytical chemistry. Many published test methods are available for both organic and inorganic compounds. Frequently used methods include pH, biochemical oxygen demand (BOD), chemical oxygen demand (COD), nutrients (nitrate and phosphorus compounds), metals (including copper, zinc, cadmium, lead and mercury), oil and grease, total petroleum hydrocarbons (TPH), and pesticides.

Biological Testing

Biological testing involves the use of plant, animal, and/or microbial indicators to monitor the health of an aquatic ecosystem.

Chemical Testing

Water samples may be examined using the principles of analytical chemistry. Many professed test methods are available for inorganic and microbiological compounds. Frequently used methods include pH, conductivity, oxygen (dissolved oxygen, Biochemical oxygen demand (COD)), nutrients (nitrate and phosphorus compounds), metals (including copper, zinc, cadmium, lead, mercury), and physical properties (temperature, pH, turbidity and pesticides).

Biological Testing

Biological testing involves the use of plant, animal, and/or microbial indicators to monitor the health of an aquatic ecosystem.

Chapter 30

Thermal and Radioactive Pollution

INTRODUCTION

Thermal pollution is the degradation of water quality by any process that changes ambient water temperature. A common cause of thermal pollution is the use of water as a coolant by power plants and industrial manufacturers. When water used as a coolant is returned to the natural environment at a higher temperature, the change in temperature: (i) decreases oxygen supply, and (ii) affects ecosystem composition. Urban runoff—stormwater discharged to surface waters from roads and parking lots—can also be a source of elevated water temperatures.

When a power plant first opens or shuts down for repair or other causes, fish and other organisms adapted to particular temperature range can be killed by the abrupt rise in water temperature known as 'thermal shock'.

MAJOR SOURCES OF THERMAL POLLUTION

The major sources of thermal pollution are electric power plants and industrial factories. In most electric power plants, heat is produced when coal, oil or natural gas is burned or nuclear fuels undergo fission to release huge amounts of energy. This heat turns water to steam, which in turn spins turbines to produce electricity. After doing its work, the spent steam must be cooled and condensed back into water. To condense the steam, cool water is brought into the plant and circulated next to the hot steam. In this process, the water used for cooling warms 5° to 10°C (9° to 18°F), after which it may be dumped back into the lake, river, or ocean from which it came. Similarly, factories contribute to thermal pollution when they dump water used to cool their machinery.

The second type of thermal pollution is much more widespread. Streams and small lakes are naturally kept cool by trees and other tall plants that block sunlight. People often remove this shading vegetation in order to harvest the wood in the trees, to make room for crops, or to construct buildings, roads, and other structures. Left unshaded, the water warms by as much as 10°C (18°F). In a similar manner, grazing sheep and cattle can strip streamsides of low vegetation, including young trees. Even the removal of vegetation far away from a stream or lake can contribute to thermal pollution by speeding up the erosion of soil into the water, making it muddy. Muddy water absorbs more energy from the sun than clear water does, resulting in further heating. Finally, water running off of artificial surfaces, such as streets, parking lots, and roofs, is warmer than water running off vegetated land and, thus, contributes to thermal pollution.

IMPACTS OF THERMAL POLLUTION

All plant and animal species that live in water are adapted to temperatures within a certain range. When water in an area warms more than they can tolerate, species that cannot move, such as rooted plants and shellfish, will die. Species that can move, such as fish, will leave the area in search of cooler conditions, and they will die if they cannot find them. Typically, other species, often less desirable, will move into the area to fill the vacancy.

In general, cold waters are better habitat for plants and animals than warm ones because cold waters contain more dissolved oxygen. Many freshwater fish species that are valued for sport and food, especially trout and salmon, do poorly in warm water. Some organisms do thrive in warm water, often with undesirable effects. Algae and other plants grow more rapidly in warm water than in cold, but they also die more rapidly; the bacteria that decompose their dead tissue use up oxygen, further reducing the amount available for animals. The dead and decaying algae make the water look, taste, and smell unpleasant.

ECOLOGICAL EFFECTS—WARM AND COLD WATER

Ecological Effects—Warm Water

Elevated temperature typically decreases the level of dissolved oxygen (DO) in water. The decrease in levels of DO can harm aquatic animals such as fish, amphibians and copepods. Thermal pollution may also increase the metabolic rate of aquatic animals, as enzyme activity, resulting in these organisms consuming more food in a shorter time than if their environment were not changed. An increased metabolic rate may result in fewer resources; the more adapted organisms moving in may have an advantage over organisms that are not used to the warmer temperature. As a result one has the problem of compromising food chains of the old and new environments. Biodiversity can be decreased as a result.

It is known that temperature changes of even one to two degrees Celsius can cause significant changes in organism metabolism and other adverse cellular biology effects. Principal adverse changes can include rendering cell walls less permeable to necessary osmosis, coagulation of cell proteins, and alteration of enzyme metabolism. These cellular level effects can adversely affect mortality and reproduction.

Primary producers are affected by warm water because higher water temperature increases plant growth rates, resulting in a shorter lifespan and species overpopulation. This can cause an algae bloom which reduces oxygen levels. A large increase in temperature can lead to the denaturing of life-supporting enzymes by breaking down hydrogen- and disulphide-bonds within the quaternary structure of the enzymes. Decreased enzyme activity in aquatic organisms can cause problems such as the inability to break down lipids, which leads to malnutrition.

In limited cases, warm water has little deleterious effect and may even lead to improved function of the receiving aquatic ecosystem. This phenomenon is seen especially in seasonal waters and is known as thermal enrichment. An extreme case is derived from the aggregational habits of the manatee, which often uses power plant discharge sites during winter. Projections suggest that manatee populations would decline upon the removal of these discharges.

Ecological Effects—Cold Water

Releases of unnaturally cold water from reservoirs can dramatically change the fish and macroinvertebrate fauna of rivers, and reduce river productivity. In Australia, where many rivers have warmer temperature regimes, native fish species have been eliminated, and macroinvertebrate fauna have been drastically altered.

Effects of Thermal Pollution

Heat and hot water result from many industrial processes. They are in particular by-products of the activity of the power stations and nuclear. The water rejected into the marine mediums has harmful effects, primarily on the marine animal-life. The pollution thermal has effects difficult to define with precision. One observes significant interferences of the reduction in the salinity of water and variations of this one in time and space. Several actions of such a pollution were however determined. It was in particular established that the hotter the temperature of rejected water is and more harmful is its effect on the benthic organisations. On the algae, the action of a thermal pollution is very variable. One noted opposite effects is strong proliferations or on the contrary a significant mortality. The metabolism of the algae, the phytoplanktons, the benthic macrophagesis very faded by heated water. One will thus be satisfied to point out some facts of observation which seem to have characters of general data:

1. One observes a numerical imbalance between the existing species in consequence of the differences in tolerances to the heating of water, from where modification with in the community of which the structure and balance are faded.
2. The distribution of the settlement is established in three zones: a died zone, a zone with reduced specific diversity and finally a zone where one observes a re-establishment in a gradual way of the flora of the biotope considered.

CONTROL OF THERMAL POLLUTION

Industrial Waste-water

In the United States, thermal pollution from industrial sources is generated mostly by power plants, petroleum refineries, pulp and paper mills, chemical plants, steel mills and smelters. Heated water from these sources may be controlled with:

1. Cooling ponds, man-made bodies of water designed for cooling by evaporation, convection, and radiation.
2. Cooling towers, which transfer waste heat to the atmosphere through evaporation and/or heat transfer
3. Cogeneration, a process where waste heat is recycled for domestic and/or industrial heating purposes.

Some facilities use once-through cooling (OTC) systems which do not reduce temperature as effectively as the above systems. For example, the Potrero Generating Station in San Francisco, which uses OTC, discharges water to San Francisco Bay approximately 10°C (20°F) above the ambient bay temperature.

Urban Runoff

During warm weather, urban runoff can have significant thermal impacts on small streams, as stormwater passes over hot parking lots, roads and sidewalks. Stormwater management facilities that absorb runoff or direct it into groundwater, such as bioretention systems and infiltration basins, can reduce these thermal effects. Retention basins tend to be less effective at reducing temperature, as the water may be heated by the sun before being discharged to a receiving stream.

RADIOACTIVE POLLUTION

The menace of radioactive pollution spreading into the environment has increased extensively as a result of the discovery of artificial radioactivity, particularly due to the development of the atom bomb, hydrogen bomb and of techniques of harnessing nuclear energy. Actually, this dangerous pollution enters into the environment in waste streams and stack gases from operations of power processing plants. From neutron bombardment of atomic fuel, heavy radio-nuclides are produced which are extremely toxic. Once these radio-elements find access into the environment, they enter the ecocycling processes and ultimately into the food chain and metabolic pathways.

Radioactive pollution poses a serious threat to the environment and future generation. Radioactive wastes from nuclear plants, reactors, etc. are, however, of a special kind in the sense that they do not smell bad or pollute the atmosphere like smoke, but are extremely lethal to living beings even in minute quantities. These wastes persist in the environment for a long-time. Non-radiation pollutants and short half-life radio nuclides are being constantly released into the air and it is expected that they will disperse or degrade in a short span of time. Consequently, they are not regarded as a future pollution threat until their concentration exceeds the limit. They cannot be disposed of into the environment like other industrial wastes. The nuclear establishments produce various types of solid and liquid radioactive wastes which contain different amounts of radioactivity and special care has to be taken for their safe disposal.

Radioactive wastes require royal disposal methods. During the phenomenon of radioactivity, some naturally unstable elements tend to become stable by emitting alpha (α), beta (β) and gamma (γ) rays. These rays can ionise the air and disarry the life activities in a cell when they pass through them.

The radioactive pollution is defined as the physical pollution of air, water and the other radioactive materials. The ability of certain materials to emit the proton, gamma rays and electrons by their nuclei is known as the radioactivity. The protons are known as the alpha particle and the electrons are also known as the beta particle. Those materials are known as the radioactive elements. The environmental radiations can be from different sources and can be natural or manmade. The natural radiations are also known as the background radiations. In this the cosmic rays are involved and reach the surface of earth from space. It includes the radioactive elements like radium, uranium, thorium, radon, potassium and carbon. These occur in the rock, soil and water. The man made radiations include the mining and refining of plutonium and thorium. This production and explosion of nuclear weapons include the nuclear fuels, power plants and radioactive isotopes. The first atom bomb was exploded in the Japan in the year 1945. It affected the Hiroshima and Nagasaki cities. It adversely affected the flora, fauna and humans of that area. In spite of these destructions the nuclear race is still going on between different nations. The nuclear arms are tested with the production of nuclear weapons. The radioactive elements are produced in the environment and affect other materials also. It includes the strontium, radium and iodine. The gases and particles are produced by the radioactive materials. They are carried by the wind and the rain brings down the radioactive particles to the ground which is referred as nuclear fallout. The soil transfers these radioactive substances to the plants and ultimately they reach the human body and cause many side effects. The iodine may affect the white blood cells, bone marrow, spleen, lymph, skin cancer, sterility, eye and damage to the lung. The strontium has the ability to aggregate in the bones and form a bone cancer and leads to tissue degeneration. The radioactive materials are passed through the land to water and cause an adverse effect on the aquatic animals. They reach to human through the food chain. The nuclear power generates a lot of energy which is used to run turbines and produces electricity. The fuel and the coolant produce a large amount of pollution in the environment. The atomic reactors are also rich in the radioactive materials. There biggest problem is in their disposal and if they are not properly disposed they can harm the living organisms. If they escape they can cause a hell lot of destruction. The gases escape as a vapour and cause pollution on the land and water. The use of radioactive isotopes is multipurpose. They are of a great scientific value and they may be present in the waste water.

From these water resources they reach to the human body via food chain. The people who work in power plants have more chances of the exposure to harmful radiations. The human beings also receive the radiation and radiotherapy from the X-rays.

Effects of Radioactive Pollution

There are many effects of the radioactive pollution which is broadly classified as short and long range. The first effects were noted in the early 20th century. The people who were working in the uranium mines suffer from skin burn and cancer. These occur due to the radiations from the radioactive material. The different organisms show different sensitivity to the radiations. There are certain conditions in which the oak trees can survive but the pine trees are not able to do so. The plants which are present at the high altitudes have a multiple set of chromosomes which is referred as a polyploidy. It helps in the protection from radiations. The southern part of our country has a large number of radiations which are harmful and are background in nature and occur in the coastal areas. The cells which divide rapidly are also damaged easily. It includes the skin cells, intestinal cells, bone marrow and gonads. The cells which do not divide rapidly are also not damaged easily. It includes the bone, muscle and nervous cells. The short range effects are known as the immediate effects and occur within the few days. It includes the loss of hair, nails, subcutaneous bleeding, and change in the number of cells and metabolism, change in the proportion of cells. The long range effects are known as the delayed effects and do not occur within the few days. It takes few months to some years to occur. They cause genetic changes, mutations, decrease the life span and form tumors. The human race possesses mutations. The radiation affects all the organisms. In some animals the radioactive materials aggregate and are transferred via food chain. It includes the zinc, iron and strontium.

Biological Effects of Radioactivity

The amount of injury caused by a radioactive isotope depends on its physical half-life, and on how quickly it is absorbed and then excreted by an organism. Most studies of the harmful effects of radiation have been performed on single-celled organisms. Obviously, the situation is more complex in humans and other multicellular organisms, because a single cell damaged by radiation may indirectly affect other cells in the individual. The most sensitive regions of the human body appear to be those which have many actively dividing cells, such as the skin, gonads, intestine, and tissues that grow blood cells (spleen, bone marrow, lymph organs). Radioactivity is toxic because it forms ions when it reacts with biological molecules. These ions can form free radicals, which damage proteins, membranes, and nucleic acids. Radioactivity can damage DNA (deoxyribonucleic acid) by destroying individual bases (particularly thymine), by breaking single strands, by breaking double strands, by cross-linking different DNA strands, and by cross-linking DNA and proteins. Damage to DNA can lead to cancers, birth defects, and even death. However, cells have biochemical repair systems which can reverse some of the damaging biological effects of low-level exposures to radioactivity. This allows the body to better tolerate radiation that is delivered at a low dose rate, such as over a longer period of time. In fact, all humans are exposed to radiation in extremely small doses throughout their life. The biological effects of such small doses over such a long-time are almost impossible to measure, and are essentially unknown at present. There is, however, a theoretical possibility that the small amount of radioactivity released into the environment by normally operating nuclear power plants, and by previous atmospheric testing of nuclear weapons, has slightly increased the incidence of certain cancers in human populations. However, scientists have not been able to conclusively show that such an effect has actually occurred.

Currently, there is disagreement among scientists about whether there is a threshold dose for radiation damage to organisms. In other words, is there a dose of radiation below which there are no harmful biological effects? Some scientists maintain that there is no such threshold, and that radiation at any dose carries a finite risk of causing some biological damage. Furthermore, the damage caused by very low doses of radiation may be cumulative, or additive to the damage caused by other harmful agents to which humans are exposed. Other scientists maintain that there is a threshold dose for radiation damage. They believe that biological repair systems, which are presumably present in all cells, can fix the biological damage caused by extremely low doses of radiation. Thus, these scientists claim that the extremely low doses of radiation to which humans are commonly exposed are not harmful.

One of the most informative studies of the harmful effects of radiation is a long-term investigation of the survivors of the 1945 atomic blasts at Hiroshima and Nagasaki by James Neel and his colleagues. The survivors of these explosions had abnormally high rates of cancer, leukemia, and other diseases. However, there seemed to be no detectable effect on the occurrence of genetic defects in children of the survivors. The radiation dose needed to cause heritable defects in humans is higher than biologists originally expected. Radioactive pollution is an important environmental problem. It could become much worse if extreme vigilance is not utilised in the handling and use of radioactive materials, and in the design and operation of nuclear power plants.

Control of Radioactive Pollution

The radioactive pollution can be controlled by number of ways. It includes the stoppage of leakage from the radioactive materials including the nuclear reactors, industries and laboratories. The disposal of radioactive material must be safe and secure. They must be stored in the safe places and must be changed into harmless form. The wastes with a very low radiation must be put into the sewage. The nuclear power plants must follow all the safe instructions. The protective garments must be worn by the workers who work in the nuclear plants. The natural radiation must be at the permissible limits and they must not cross it. The radioactive wastes generated during the various nuclear fuel cycle operations are highly variable in nature, composition, volume, radioactivity levels and half-lives of radio-nuclides contained. They are classified into solid, liquid and gaseous wastes. Solid and liquid wastes are nominally labelled as low, medium and high-level wastes, depending upon the surface dose and or radio-nuclide content. An elaborate segregation and collection system is necessary before choosing the treatment and disposal methods. The basic scheme for management of radioactive wastes is given in Fig. 30.1. The widely accepted principles for management of radioactive wastes are: (i) dilute and disperse, (ii) delay and decay, and (iii) concentrate and contain.

The first principle is generally applied for gaseous and low-level liquid effluents. It is ascertained that the dilution by air or water is to a level sufficiently low so that the resulting dose will be below the acceptable limits. These practices should conform to regulatory standards. 'Delay and decay' applies to those waste which contain only short life nuclides. 'Concentrate and contain' is generally applied for medium and high-level wastes, both solid and liquid. The wastes are processed, treated and or conditioned before containment.

A variety of techniques have been developed for treatment, storage and disposal of radioactive wastes. The gaseous wastes generated are generally released into the atmosphere through stacks. On the basis of the meteorological data applicable to the site, the height of releases are optimised so that the resulting dose from atmospheric releases are maintained well within the stipulated limits.

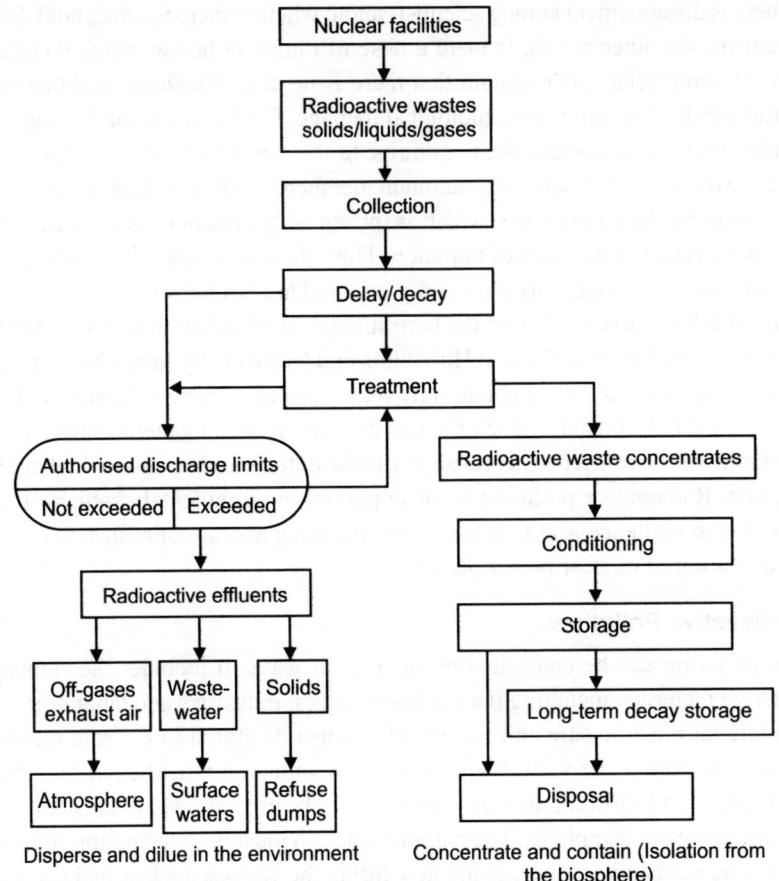

Fig. 30.1 Basic scheme for the management of radioactive wastes.

Liquid effluents, especially low and medium categories, are treated by chemical coagulation, ion exchange or evaporation to remove a bulk the of radionuclides. The treated liquid effluents that conform to authorised limits are only allowed to be discharged. The residual solids are conditioned to immobilise the radio-nuclides by cementation, bituminisation or by polymerisation. The high-level liquid effluents generated in reprocessing operations, which contain over 99.9 per cent of the non-gaseous fission products and traces of urecovered actinides generated in power reactors, are converted into a non-leachable glass form. The solid wastes include a wide variety of materials including contaminated materials, filters, resins, chemical sludges, incineration ash, etc. The combustible items are incinerated in suitably designed incinerators with air cleaning facilities and the ash is conditioned in cement matrix for proper storage/disposal.

For low and medium-level solid wastes, underground disposal is the most viable option. Both reinforced concrete trenches as well as steel-lined concrete tile holes are used depending upon the activity of the solid matrix. In the case of vitrified high-level wastes, after an engineered storage for a period of 25 years, disposal in suitable deep underground geological formation will ensure isolation from groundwater sources. Thus, the main objective of waste management in nuclear operations is to contain (isolate from biosphere) as is much radioactive materials as is practicable so as to minimise releases into the environment.

Chapter 31

Wastes from the Anthrosphere

INTRODUCTION

Human activities produce large quantities of wastes. Some of these are rejected into the atmosphere, where they may reside for sometime as pollutants or are transformed by atmospheric chemical processes to other pollutants. Some of the potentially harmful by-products of manufacturing and other human activities are dumped into water and become water pollutants. Other potential pollutants end up on land. A major concern, therefore, with improperly controlled human activities is the production and distribution of hazardous substances and hazardous wastes—wastes from the anthrosphere. These materials and their potential environmental impact are discussed in this chapter.

Hazardous substances and hazardous wastes: A hazardous substance is a material that may pose a danger to living organisms, materials, structures, or the environment by explosion or fire hazards, corrosion, toxicity to organisms, or other detrimental effects. A simple definition of a hazardous waste is that it is a hazardous substance that has been discarded, abandoned, neglected, released or designated as a waste material, or one that may interact with other substances to be hazardous. In a simple sense a hazardous waste is a material that has been left somewhere that it may cause harm if encountered.

CLASSIFICATION OF HAZARDOUS SUBSTANCES AND WASTES

Many specific chemicals in widespread use are hazardous because of their chemical reactivities, fire hazards, toxicities, and other properties. There are numerous kinds of hazardous substances, usually consisting of mixtures of specific chemicals. These include the following:

1. Explosives: Such as dynamite, or ammunition.
2. Compressed gases: Such as hydrogen and sulphur dioxide.
3. Flammable liquids: Such as gasoline and aluminium alkyls.
4. Flammable solids: Such as magnesium metal, sodium hydride, and calcium carbide, that burn readily, are water-reactive or are spontaneously combustible.
5. Oxidising materials: Such as lithium peroxide, that supply oxygen for the combustion of normally nonflammable materials.
6. Corrosive materials: Including oleum, sulphuric acid, and caustic soda, which may wound exposed flesh or cause disintegration of metal containers.
7. Poisonous materials: Such as hydrocyanic acid or aniline.
8. Etiologic agents: Including causative agents of anthrax, botulism, or tetanus.
9. Radioactive materials: Including plutonium, cobalt-60, and uranium hexafluoride.

Potentially hazardous waste substances are shown in Fig. 31.1.

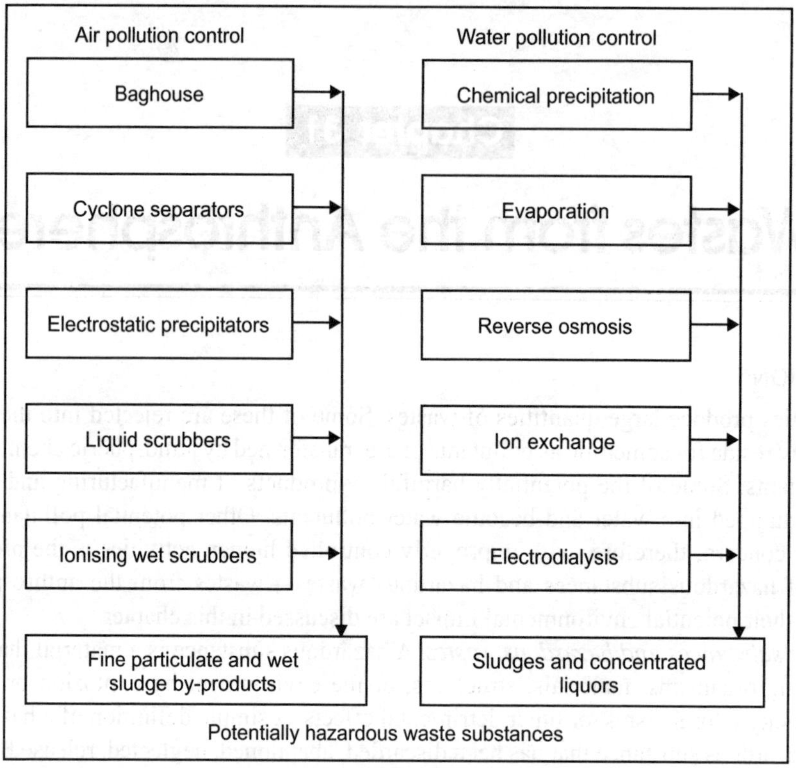

Fig. 31.1. Potential contributions of air and water pollution control measures to hazardous wastes production.

Wastes Defined by Characteristics

EPA defines hazardous substances in terms of characteristics:

1. Ignitability: Characteristic of substances that are liquids whose vapours are likely to ignite in the presence of ignition sources, nonliquids that may catch fire from friction or contact with water and that burn vigorously or persistently, ignitable compressed gases, and oxidisers.
2. Corrosivity: Characteristic of substances that exhibit extremes of acidity or basicity or a tendency to corrode steel.
3. Reactivity: Characteristic of substances that have a tendency to undergo violent chemical, change (an explosive substance is an obvious example).
4. Toxicity: Defined in terms of a standard extraction procedure followed by chemical analysis for specific substances.

Listed Wastes

In addition to classification by characteristics, EPA designates more than 450 listed wastes that are specific substances or classes of substances known to be hazardous. Each such substance is assigned an EPA hazardous waste number in the format of a letter followed by 3 numerals, where a different letter is assigned to substances from each of the four following lists:

1. F-type wastes from nonspecific sources: For example, quenching waste-water treatment sludges from metal heat treating operations where cyanides are used in the process (F012).
2. K-type wastes from specific sources: For example, heavy ends from the distillation of ethylene dichloride in ethylene dichloride production (K019).
3. P-type acute hazardous wastes: These are mostly specific chemical species such as fluorine (P056) or 3-chloropropane nitrile (P027).
4. U-type generally hazardous wastes: These are predominantly specific compounds such as calcium chromate (U032) or phthalic anhydride (U190).

FLAMMABLE AND COMBUSTIBLE SUBSTANCES

In a broad sense a flammable substance is something that will burn readily, whereas a combustible substance requires relatively more persuasion to burn. Before trying to sort out these definitions it is necessary to define several other terms. Most chemicals that are likely to burn accidentally are liquids. Liquids form vapours, which are usually more dense than air and thus tend to settle. The tendency of a liquid to ignite is measured by a test in which the liquid is heated and periodically exposed to a flame until the mixture of vapour and air ignites at the liquid's surface. The temperature at which this occurs is called the flash point. With these definitions in mind it is possible to divide ignitable materials into four major classes. A flammable solid is one that can ignite from friction or from heat remaining from its manufacture, or which may cause a serious hazard if ignited. Explosive materials are not included in this classification. A flammable liquid is one having a flash point below 37.8°C (100°F). A combustible liquid has a flash point in excess of 37.8°C, but below 93.3°C. Gases are substances that exist entirely in the gaseous phase at 0°C and 1 atm pressure. A flammable compressed gas meets specified criteria for lower flammability limit, flammability range, and flame projection.

In considering the ignition of vapours, two important concepts are those of flammability limit and flammability range. Values of the vapour/air ratio below which ignition cannot occur because of insufficient fuel define the lower flammability limit. Similarly, values of the vapour/air ratio above which ignition cannot occur because of insufficient air define the upper flammability limit. The difference between upper and lower flammability limits at a specified temperature is the flammability range. Table 31.1 gives some examples of these values for common liquid chemicals.

Table 31.1. Flammabilities of some common organic liquids.

		Volume per cent in air	
Liquid	Flash point (°C)[a]	LFL[b]	UFL[b]
Diethyl ether	−43	1.9	36
Pentane	−40	1.5	7.8
Acetone	−20	2.6	13
Toluene	4	1.27	7.1
Methanol	12	6.0	37
Gasoline (2,2,4-trimethylpentane)	–	1.4	7.6
Naphthalene	157	0.9	5.9

[a] Closed-cup flash point test.
[b] LFL, lower flammability limit; UFL, upper flammability limit at 25°C.

Combustion of Finely Divided Particles

Finely divided particles of combustible materials are somewhat analogous to vapours in respect to flammability. One such example is a spray or mist of hydrocarbon liquid in which oxygen has the opportunity for intimate contact with the liquid particles; the liquid may ignite at a temperature below its flash point. Dust explosions can occur with a large variety of solids that have been ground to a finely divided state. Many metal dusts, particularly those of magnesium and its alloys, zirconium, titanium, and aluminium, can burn explosively in air. In the case of aluminium, for example, the highly exothermic (heat-releasing) reaction is the following:

$$4Al\,(powder) + 3O_2\,(from\ air) \rightarrow 2Al_2O_3 \qquad\qquad ...\,(31.1)$$

Coal dust and grain dusts have caused many fatal fires and explosions in coal mines and grain elevators, respectively. Dusts of polymers such as cellulose acetate, polyethylene, and polystyrene can also be explosive.

Oxidisers

Combustible substances are reducing agents that react with oxidisers (oxidising agents or oxidants) to produce heat. Diatomic oxygen, O_2, from air is the most common oxidiser. Many oxidisers are chemical compounds that contain oxygen in their formulas. The halogens and many of their compounds are oxidiser. An example of a reaction of an oxidiser is that of concentrated HNO_3 with copper metal, which gives toxic NO_2 gas as a product:

$$4HNO_3 + Cu \rightarrow Cu(NO_3)_2 + 2H_2O + 2NO_2 \qquad\qquad ...\,(31.2)$$

The toxic effects of some oxidisers are due to their ability to oxidise biomolecules.

Whether or not a substance acts as an oxidiser depends upon the reducing strength of the material that it contacts. For example, carbon dioxide is a common fire extinguishing material that can be sprayed onto a burning substance to keep air away. However, aluminum is such a strong reducing agent that carbon dioxide in contact with hot, burning aluminum reacts as an oxidising agent to give off toxic combustible carbon monoxide gas:

$$2Al + 3CO_2 \rightarrow Al_2O_3 + 3CO \qquad\qquad ...\,(31.3)$$

Oxidisers can contribute strongly to fire hazards because fuels may burn explosively in contact with an oxidiser.

Spontaneous Ignition

Pyrophoric substances catch fire spontaneously in air without an ignition source. These include several elements—white phosphorus, the alkali metals (group 1A), and powdered forms of magnesium, calcium, cobalt, manganese, iron, zirconium, and aluminium. Also included are some organometallic compounds, such as lithium ethyl (LiC_2H_4); some metal carbonyl compounds, such as iron pentacarbonyl, $Fe(CO)_5$; and metal and metalloid hydrides including lithium hydride, LiH; pentaborane, B_5H_9; and arsine, AsH_3. Moisture in air is often a factor in spontaneous ignition. For example, lithium hydride undergoes the following reaction with water from moist air:

$$LiH + H_2O \rightarrow LiOH + H_2 + heat \qquad\qquad ...\,(31.4)$$

The heat generated from this reaction can be sufficient to ignite the hydride so that it burns in air:

$$2LiH + O_2 \rightarrow Li_2O + H_2O \qquad\qquad ...\,(31.5)$$

Many mixtures of oxidisers and oxidisable chemicals catch fire spontaneously and are called hypergolic mixtures. Nitric acid and phenol form such a mixture.

.ic Products of Combustion

Some of the greater dangers of fires are from toxic products and by-products of combustion. The most obvious of these is toxic carbon monoxide, CO. Toxic SO_2, P_4O_{10}, and HCl are formed by the combustion of sulphur, phosphorus, and organochloride compounds, respectively. A large number of noxious organic compounds such as aldehydes are generated as by-products of combustion. In addition to forming carbon monoxide, combustion under oxygen-deficient conditions produces polycyclic aryl hydrocarbons consisting of fused ring structures. Some of these compounds, such as benzo(a)pyrene, below, are precarcinogens that are acted upon by enzymes in the body to yield cancer-producing metabolites.

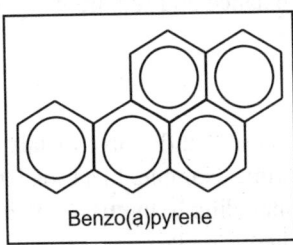

Benzo(a)pyrene

REACTIVE SUBSTANCES

Reactive substances are those that tend to undergo rapid or violent reactions under certain conditions. Such substances include those that react violently or form potentially explosive mixtures with water. An example is sodium metal, which reacts strongly with water as follows:

$$2Na + 2H_2O \rightarrow 2NaOH + H_2 + heat \qquad \text{... (31.6)}$$

This reaction usually generates enough heat to ignite the sodium. Explosives constitute another class of reactive substances. For regulatory purposes substances are also classified as reactive that react with water, acid, or base to produce toxic fumes, particularly those of hydrogen sulphide or hydrogen cyanide.

Many reactions require energy of activation from heat to get them started, the rates of most reactions increase sharply with increasing temperature, and most chemical reactions give off heat. Therefore, once a reaction is started in a reactive mixture, the rate may increase exponentially with time, leading to an uncontrollable event. Other factors that may affect reaction rate include physical form of reactants (for example, a finely divided metal powder that reacts explosively with oxygen, whereas a single mass of metal barely reacts), rate and degree of mixing of reactants, degree of dilution with nonreactive media (solvent), presence of a catalyst, and pressure.

Chemical Structure and Reactivity

Some chemical structures are associated with high reactivity. High reactivity in some organic compounds results from unsaturated bonds in the carbon skeleton, particularly where multiple bonds are adjacent (allenes, C=C=C) or separated by only one carbon–carbon single bond (dienes, C=C–C=C). Some organic structures involving oxygen are very reactive. Examples are oxiranes, such as ethylene oxide:

Ethylene oxide

Many different classes of inorganic compounds are reactive. These include some of the halog
compounds of nitrogen (shock-sensitive nitrogen triiodide, NI_3, is an outstanding example), compounds
with metal-nitrogen bonds, halogen oxides (ClO_2), and compounds with oxyanions of the halogens. An
example of the last group of compounds is ammonium perchlorate, NH_4ClO_4.

Explosives such as nitroglycerin or TNT that are single compounds containing both oxidising and
reducing functions in the same molecule are called redox compounds. Some redox compounds have
more oxygen than is needed for a complete reaction and are said to have a positive balance of oxygen,
some have exactly the stoichiometric quantity of oxygen required (zero balance, maximum energy
release), and others have a negative balance and require oxygen from outside sources to completely
oxidise all components.

CORROSIVE SUBSTANCES

Corrosive substances are regarded as those that dissolve metals or cause oxidised material to form on
the surface of metals-rusted iron is a prime example. In a broader sense corrosives are defined as those
that cause deterioration of materials, including living tissue, that they contact. Most corrosives belong
to at least one of the four following chemical classes: (i) strong acids, (ii) strong bases, (iii) oxidants,
(iv) dehydrating agents.

TOXIC SUBSTANCES

Toxicity is of the utmost concern in dealing with hazardous substances. This includes both long-term
chronic effects from continual or periodic exposures to low levels of toxicants and acute effects from a
single large exposure.

Toxicity Characteristic Leaching Procedure

For regulatory and remediation purposes a standard test is needed to measure the likelihood of toxic
substances getting into the environment and causing harm to organisms. The test required by the
US EPA is the toxicity characteristic leaching procedure (TCLP) designed to determine the mobility of
both organic and inorganic contaminants present in liquid, solid, and multiphasic wastes. For analysis
of toxic species a solution is leached from the waste or filtered from it and is designated as the TCLP
extract. After the TCLP extract is separated from the solids, it is analysed for a number of specified
volatile organic compounds, semivolatile organic compounds, and metals to determine if the waste
exceeds specified levels of these contaminants.

PHYSICAL FORMS AND SEGREGATION OF WASTES

Three major categories of wastes based upon their physical forms are organic materials, aqueous wastes,
and sludges. These forms largely determine the course of action taken in treating and disposing of the
wastes. The level of segregation, a concept illustrated in Fig. 31.2, is very important in treating, storing,
and disposing of different kinds of wastes. It is relatively easy to deal with wastes that are not mixed
with other kinds of wastes, that is, those that are highly segregated. For example, spent hydrocarbon
solvents can be used as fuel in boilers. However, if these solvents are mixed with spent organochloride
solvents, the production of contaminant hydrogen chloride during combustion may prevent fuel use and
require disposal in special hazardous waste incinerators. Further mixing with inorganic sludges adds
mineral matter and water. These impurities complicate the treatment processes required by producing
mineral ash in incineration or lowering the heating value of the material incinerated because of the

Toxence of water. Among the most difficult types of wastes to handle and treat are those with the least segregation, of which a 'worst-case scenario' would be 'dilute sludge consisting of mixed organic and inorganic wastes', as shown in Fig. 31.2.

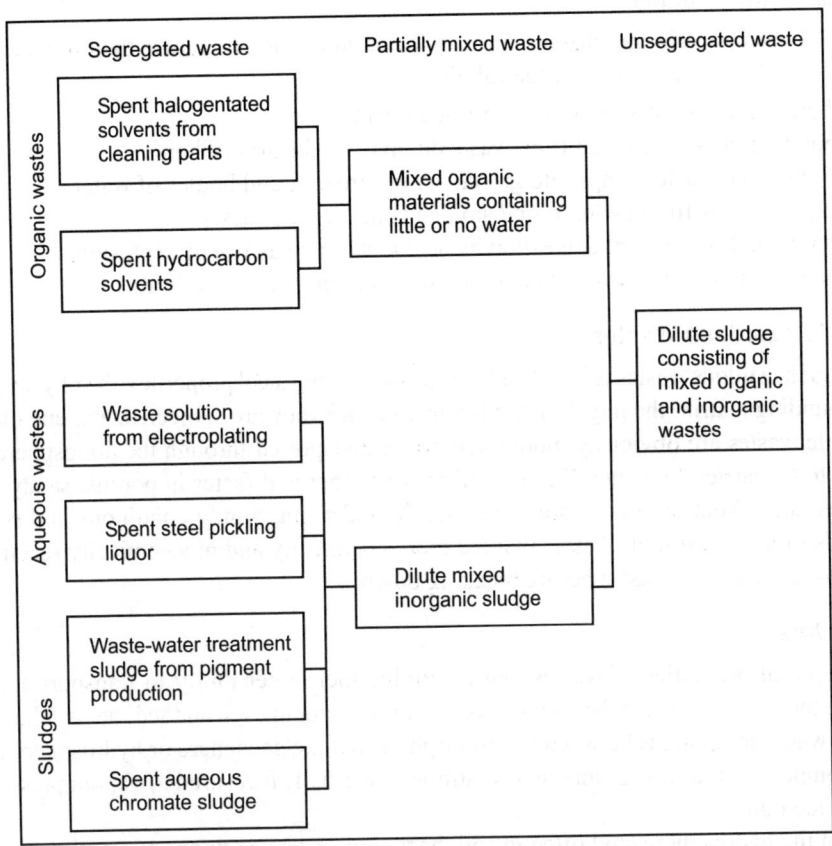

Fig. 31.2. Illustration of waste segregation.

Concentration of wastes is an important factor in their management. A waste that has been concentrated or preferably never diluted is generally much easier and more economical to handle than one that is dispersed in a large quantity of water or soil. Dealing with hazardous wastes is greatly facilitated when the original quantities of wastes are minimised and the wastes remain separated and concentrated insofar as possible.

HAZARDOUS WASTES IN THE ENVIRONMENT

Having outlined the nature and sources of hazardous substances and hazardous wastes earlier in this chapter, it is now possible to discuss their environmental behaviour according to the following factors:

1. Origin.
2. Transport.
3. Reactions.
4. Effects.
5. Ultimate fate.

In addition, consideration must be given to the distribution of hazardous wastes among the geosphere, hydrosphere, atmosphere, and biosphere.

Origin of Hazardous Wastes

For purposes of discussion in this chapter, origin of hazardous wastes refers to their points of entry into the environment. These may consist of the following:

1. Deliberate addition to soil, water or air by humans.
2. Evaporation or wind erosion from waste dumps into the atmosphere.
3. Leaching from waste dumps into groundwater, streams, and bodies of water.
4. Leakage, such as from underground storage tanks or pipelines.
5. Evolution and subsequent deposition by accidents: such as fire or explosion.
6. Release from improperly operated waste treatment or storage facilities.

Transport of Hazardous Wastes

The transport of hazardous wastes is largely a function of their physical properties, the physical properties of their surrounding matrix, the physical conditions to which they are subjected, and chemical factors. Highly volatile wastes are obviously more likely to be transported through the atmosphere and more soluble ones to be carried by water. Wastes will move farther and faster in porous, sandy formations than in denser soils. Volatile wastes are more mobile under hot, windy conditions and soluble ones during periods of heavy rainfall. Wastes that are more chemically and biochemically reactive will not move so far as less reactive wastes before breaking down.

Physical factors

The major physical properties of wastes that determine their amenability to transport are volatility, solubility, and the degree to which they are sorbed to solids, including soil and sediments. The distribution of hazardous waste compounds between the atmosphere and the geosphere or hydrosphere is largely a function of compound volatility. Compound volatilities are usually measured by vapour pressures, which vary over a wide range.

Usually, in the hydrosphere, and often in soil, hazardous waste compounds are dissolved in water; therefore, the tendency of water to hold the compound is a factor in its mobility. For example, although ethyl alcohol has a higher evaporation rate and lower boiling temperature than toluene, vapour of the latter compound is more readily evolved from soil because of its limited solubility in water compared to ethanol, which is totally miscible with water.

Chemical factors

As an illustration of chemical factors involved in transport of wastes, consider largely cationic inorganic species. Inorganic species can be divided into three groups based upon their retention by clay minerals. Elements that tend to be highly retained by clay include cadmium, mercury, lead, and zinc. Potassium, magnesium, iron, silicon, and NH_4^+ are moderately retained by clay, whereas sodium, chloride, calcium, manganese, and boron are poorly retained. The apparent retention of the last three elements is probably biased in that they are leached from clay, so that negative retention (elution) is often observed. It should be noted, however, that the retention of iron and manganese is a strong function of oxidation state in that the reduced forms of Mn and Fe are somewhat mobile, whereas the oxidised forms of $Fe_2O_3 \cdot xH_4O$ and MnO_2 are very insoluble and stay on soil as solids.

Reactions of Hazardous Wastes

Many environmental chemical and environmental biochemical processes operate on hazardous wastes in water, air, and soil. One of the important results of this is that, in addition to primary pollutants added directly to the environment, there are secondary pollutants as well, which may be even more harmful than their precursor species. An example of the formation of a very damaging secondary pollutant from a relatively less dangerous primary pollutant is the oxidation of primary pollutant sulphur dioxide,

$$2SO_2 + O_2 + 2H_2O \rightarrow 2H_2SO_4 \qquad \qquad \text{... (31.7)}$$
(several steps and intermediates)

to yield secondary pollutant sulphuric acid, the prime ingredient of acid rain.

Effects of Hazardous Wastes

The effects of hazardous wastes in the environment may be divided among effects on organisms, effects on materials, and effects on the environment. These are addressed briefly here and in greater detail in later sections.

The ultimate concern with wastes has to do with their toxic effects to animals, plants, and microbes. Virtually all hazardous waste substances are poisonous to a degree, some extremely so. The toxicity of a waste is a function of many factors, including the chemical nature of the waste, the matrix in which it is contained, circumstances of exposure, the species exposed, manner of exposure, degree of exposure, and time of exposure.

Many hazardous wastes are corrosive to materials, usually because of extremes of pH or because of dissolved salt content. Oxidant wastes can cause combustible substances to burn uncontrollably. Highly reactive wastes can explode, causing damage to materials and structures. Contamination by wastes, such as by toxic pesticides in grain, can result in substances becoming unfit for use.

In addition to their toxic effects in the biosphere, hazardous wastes can damage air, water, and soil. Wastes that get into air can cause deterioration of air quality, either directly, or by the formation of secondary pollutants. Hazardous waste compounds dissolved in, suspended in, or as surface films on, the surface of water can render it unfit for use and for sustenance of aquatic organisms. Soil exposed to hazardous wastes can be severely damaged by alteration of its physical and chemical properties and ability to support plants. For example, soil exposed to concentrated brines from petroleum production may become unable to support plant growth so that the soil becomes extremely susceptible to erosion.

Fates of Hazardous Wastes

The fates of hazardous waste substances are addressed in more detail in subsequent sections. As with all environmental pollutants, such substances eventually reach a state of physical and chemical stability, although that may take many centuries to occur. In some cases, the fate of a hazardous waste material is a simple function of its physical properties and surroundings.

The fate of a hazardous waste substance in water is a function of the substance's solubility, density, biodegradability, and chemical reactivity. Dense, water-immiscible liquids may simply sink to the bottoms of bodies of water or aquifers and accumulate there as 'blobs' of liquid. This has happened, for example, with hundreds of tons of PCB wastes that have accumulated in sediments in the Hudson River in New York State. Biodegradable substances are broken down by bacteria, a process for which the availability of oxygen is an important variable. Substances that readily undergo bioaccumulation are taken up by organisms, exchangeable cationic materials become bound to sediments, and organophilic materials may be sorbed by organic matter in sediments.

The fates of hazardous waste substances in the atmosphere are often determined by photochemical reactions. Ultimately, such substances may be converted to nonvolatile, insoluble matter and precipitate from the atmosphere onto soil or plants.

HAZARDOUS WASTES IN THE GEOSPHERE

The sources, transport, interactions, and fates of contaminant hazardous wastes in the geosphere involve a complex scheme, some aspects of which are illustrated in Fig. 31.3. The primary environmental concern regarding hazardous wastes in the geosphere is the possible contamination of groundwater aquifers by waste leachates and leakage from wastes. As the figure shows, there are a number of possible contamination sources. The most obvious one is leachate from landfills containing hazardous wastes. In some cases, liquid hazardous materials are placed in lagoons, which can leak into aquifers. Leaking sewers can also result in contamination, as can the discharge from septic tanks. Hazardous wastes spread on land can result in aquifer contamination by leachate. Hazardous chemicals are sometimes deliberately disposed of underground in waste disposal wells. This means of disposal can result in interchange of contaminated water between surface water and groundwater at discharge and recharge points.

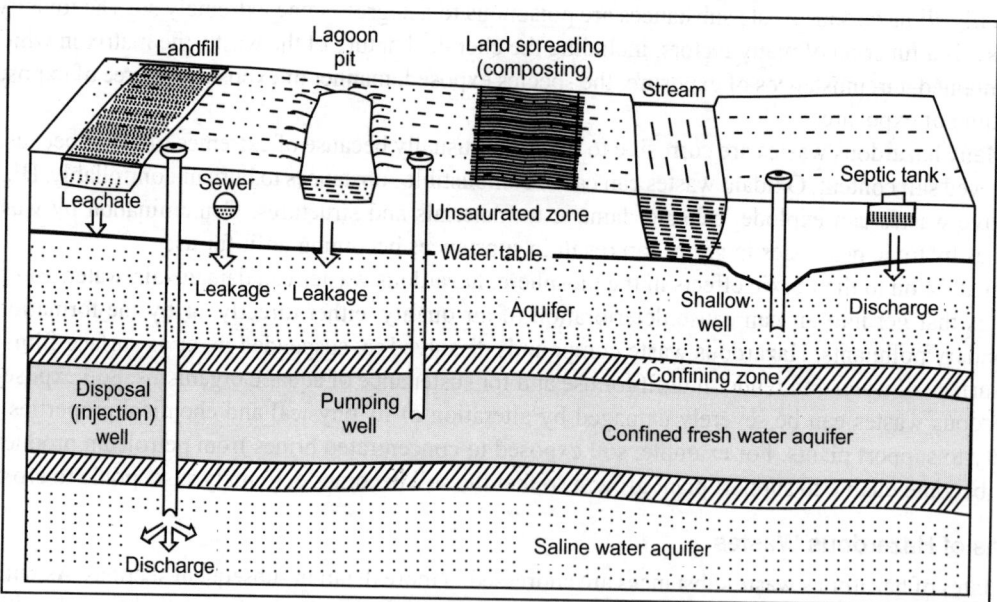

Fig. 31.3. Sources, disposal, and movement of hazardous wastes in the geosphere.

The transport of contaminants in the geosphere depends largely upon the hydrologic factors governing the movement of water underground and the interactions of hazardous waste constituents with geological strata, particularly unconsolidated earth materials. As shown in Figure 31.4, groundwater contaminated with hazardous wastes tends to flow as a relatively undiluted plug or plume along with the groundwater in an aquifer. The groundwater flow rate depends upon the water gradient and aquifer characteristics, such as permeability and cross-section area. The rate of flow is generally relatively slow; 1 meter per day would be considered fast. Contaminated groundwater can result in contamination of a surface water source. This can occur at a discharge area where the groundwater flows into a lake or stream.

As discussed in the preceding section, hazardous waste dissolved in groundwater can be attenuated by soil or rock by means of various sorption mechanisms.

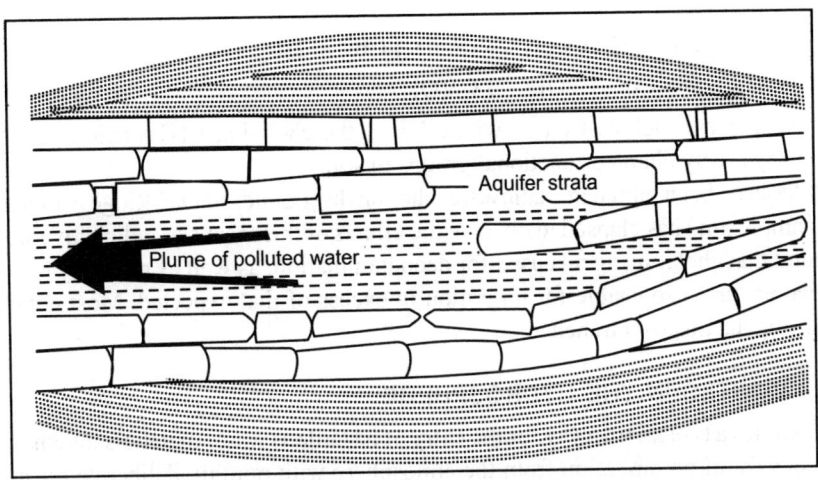

Fig. 31.4. Plug-flow of hazardous wastes in groundwater.

Mathematically, the distribution of a solute between groundwater or leachate water and soil is expressed by a distribution coefficient, K_d,

$$K_d = \frac{C_S}{C_W} \qquad \qquad ...\,(31.8)$$

where, C_S is the concentration of the species in the solid phase and C_W is its concentration in water. This equation assumes that the relative degree of sorption is independent of C_W.

The degree of attenuation depends upon the surface properties of the solid, particularly its surface area. The chemical nature of the attenuating solid is also important because attenuation is a function of the organic matter (humus) content, presence of hydrous metal oxides, and the content and types of clays present. The chemical characteristics of the leachate also affect attenuation greatly. For example, attenuation of metals is very poor in acidic leachate because precipitation reactions, such as,

$$M^{2+} + 2OH^- \rightarrow M(OH)_2(s) \qquad \qquad ...\,(31.9)$$

are reversed in acid:

$$M(OH)_2(s) + 2H^+ \rightarrow M^{2+} + 2H_2O \qquad \qquad ...\,(31.10)$$

Organic solvents in leachates tend to prevent attenuation of organic hazardous waste constituents.

The degree of attenuation of a pollutant by soil depends upon the water content of the soil. There is an unsaturated zone of soil in which attenuation is more highly favoured. Normally, soil has a greater surface area at liquid-solid interfaces in this zone so that absorption and ion-exchange processes are favoured. Aerobic degradation is possible in the unsaturated zone, enabling more rapid and complete degradation of biodegradable hazardous wastes.

Codisposal of chelating agents with heavy metals can have a strong effect upon the mobility of metal ions in soil. This effect was observed resulting from codisposal of intermediate-level nuclear wastes with chelating agents (generally organic species that bind strongly with metal ions) during the period 1951–1965 at Oak Ridge National Laboratory. The presence of chelating agents resulted from the use of salts of chelating ethylenediaminetetraacetic acid (EDTA) in decontaminating facilities exposed to nuclear wastes. Whereas metal cations are readily held by ion exchange processes and precipitation on soil,

$$2Soil\{^-H^+ + Co^{2+} \rightarrow (Soil\}^-)_2Co^{2+} + 2H^+ \qquad \qquad ... (31.11)$$
$$Co^{2+} + 2OH^- \rightarrow Co(OH)_2(s) \qquad \qquad ... (31.12)$$

chelated anionic species, such as CoY^{2-} (where Y^{4-} is the chelating EDTA anion), are not strongly retained by the negatively charged functional groups in soil.

Radionuclides have been buried in shallow trenches on the grounds of Oak Ridge National Laboratory since 1944, so ample time has elapsed to observe the effects of this means of radioactive waste disposal. It has been found that chelating agents used for decontamination, as well as naturally occurring humic substance chelators, are responsible for migration in excess of that expected. Most notably, ^{60}Co has been found outside the disposal trenches.

HAZARDOUS WASTES IN THE HYDROSPHERE

Figure 31.5 illustrates a typical pathway for the entry of hazardous waste materials into the hydrosphere. Other sources consist of precipitation from the atmosphere with rainfall, deliberate release to streams and bodies of water, runoff from soil, and mobilisation from sediments. Once in an aquatic system, hazardous waste species are subject to a number of chemical and biochemical processes, including acid-base, oxidation-reduction, precipitation-dissolution, and hydrolysis reactions, as well as biodegradation.

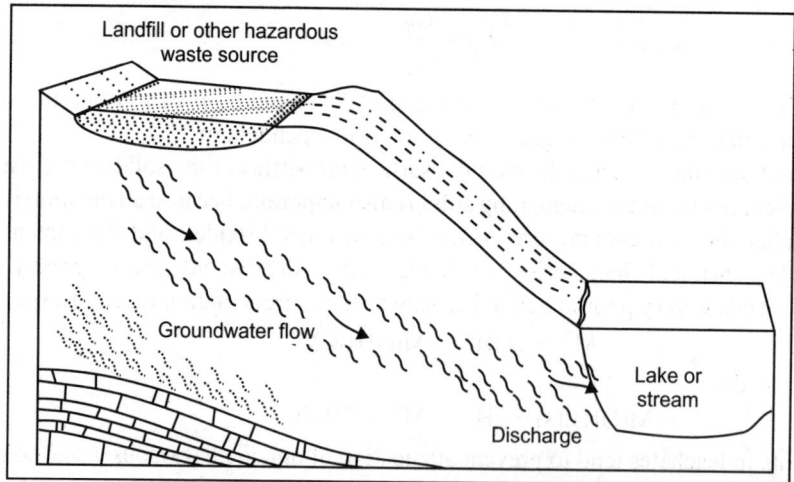

Fig. 31.5. Discharge of groundwater contaminated from hazardous waste landfill into a body of water.

The presence of organic matter in water has a tendency to increase the solubility of hazardous organic substances. Typically, the solubility of hexachlorobenzene is 1.8 micrograms per litre in pure water at 25°C, whereas it is 2.3 μg/L in creek water containing organic solutes and 4–4.5 μg/L in landfill leachate.

In considering the processes that hazardous wastes undergo in water, it is important to recall the nature of aquatic systems and the unique properties of water. Water in the environment is far from pure. Just as the atmosphere is a constantly changing mass of bodies of moving air with different temperatures, pressures, and humidities, bodies of water are highly dynamic systems. Rivers, impoundments, and groundwater aquifers are subject to the input and loss of a variety of materials from both natural and anthropogenic sources. These materials may be gases, liquids or solids. They interact chemically with each other and with living organisms—particularly bacteria—in the water. They are subject to dispersion and transport by stream flow, convection currents, and other physical phenomena. Hazardous substances or their by-products in water may undergo bioaccumulation through food chains involving aquatic organisms.

Several physical, chemical, and biochemical processes are particularly important in determining the transformations and ultimate fates of hazardous chemical species in the hydrosphere. These include hydrolysis reactions, through which a molecule is cleaved with the addition of H_2O; precipitation reactions, generally accompanied by aggregation of colloidal particles suspended in water; oxidation-reduction reactions, generally mediated by micro-organisms; sorption (adsorption and/or adsorption) of hazardous solutes by sediments and by suspended mineral and organic matter; biochemical processes, often involving hydrolysis and oxidation-reduction reactions; photolysis reactions; and miscellaneous chemical phenomena.

The hydrolysis of hazardous waste acetic anhydride is illustrated by the following reaction:

$$\underset{\underset{H}{\overset{H}{\vert}}}{H-C}\overset{\overset{O}{\Vert}}{-C}-O-\overset{\overset{O}{\Vert}}{C}-\underset{\underset{H}{\overset{H}{\vert}}}{C}-H \; + 2HOH \; \longrightarrow \; 2H-\underset{\underset{H}{\overset{H}{\vert}}}{C}-\overset{\overset{O}{\Vert}}{C}-OH \qquad \text{... (31.13)}$$

The rates at which compounds hydrolyse in water vary widely. Acetic anhydride hydrolyses very rapidly. In fact, the great affinity of this compound for water (including water in skin) is one of the reasons that it is hazardous. Once in the aquatic environment, though, acetic anhydride is converted very rapidly to essentially harmless acetic acid. Many ethers, esters, and other compounds formed originally by the joining together of two or more molecules with the loss of water hydrolyse very slowly, although the rate may be greatly increased by the action of enzymes in micro-organisms (biochemical processes).

The formation of precipitates in the form of sludges is one of the most common means of isolating hazardous components from an unsegregated waste. Although solid inorganic ionic compounds are often discussed in terms of very simple formulas, such as $PbCO_3$ for lead carbonate, much more complicated species (for example, $2PbCO_3 \cdot Pb(OH)_2$) generally result when precipitates are formed in the aquatic environment. For example, a hazardous heavy metal ion in the hydrosphere may be precipitated as a relatively complicated compound, coprecipitated as a minor constituent of some other compound, or sorbed by the surface of another solid.

The major anions present in natural waters and waste-waters are OH^-, CO_3^{2-}, and SO_4^{2-}. Since these anions are all capable of forming precipitates with cationic impurities, such pollutants tend to precipitate as hydroxides, carbonates, and sulphates. Sorption processes are particularly common methods for the removal of low level hazardous materials from water. Many heavy metals are sorbed by or coprecipitated with hydrated iron (III) oxide ($Fe_2P_3 \cdot xH_2O$) or manganese (IV) oxide ($MnO_2 \cdot xH_2O$). Oxidation-reduction reactions are very important means of transformation of hazardous wastes in water.

Under many circumstances, biochemical processes largely determine the fates of hazardous chemical species in the hydrosphere. The most important such processes are those mediated by micro-organisms. In particular, the oxidation of biodegradable hazardous organic wastes in water generally occurs by means of micro-organism-mediated biochemical reactions. Bacteria produce organic acids and chelating agents, such as citrate, which have the effect of solubilising hazardous heavy metal ions. Some mobile methylated forms, such as compounds of methylated arsenic and mercury, are produced by bacterial action. Photolysis reactions are those initiated by the absorption of light. The effect of photolytic processes on the destruction of hazardous wastes in the hydrosphere is minimal, although some photochemical reactions of hazardous waste compounds can occur when the compounds are present as surface films on water exposed to sunlight. Groundwater is the part of the hydrosphere most vulnerable to damage from hazardous wastes. Although surface water supplies are subject to contamination, groundwater can become almost irreversibly contaminated by the improper land disposal of hazardous chemicals.

HAZARDOUS WASTES IN THE ATMOSPHERE

Some chemicals found in hazardous waste sites may enter the atmosphere by evaporation or even as windblown particles. Three major areas of interest in respect to hazardous waste compounds in the atmosphere are their pollution potential, atmospheric fate, and residence time. These strongly interrelated factors are discussed in this section.

Air Pollution Potential of Hazardous Waste Compounds

The pollution potential of hazardous wastes in the atmosphere depends upon whether they are primary pollutants that have a direct effect or secondary pollutants that are converted to harmful substances by atmospheric chemical processes. Hazardous waste sites do not usually evolve sufficient quantities of air pollutants to give significant amounts of secondary pollutants, so primary air pollutants are the greater concern. Examples of primary air pollutants include toxic organic vapours (vinyl chloride), corrosive acid gases (HCl), and toxic inorganic gases, such as H_2S released by the accidental mixing of waste acid and waste metal sulphides:

$$2HCl + FeS \rightarrow FeCl_2 + H_2S(g) \qquad \text{... (31.14)}$$

Primary air pollutants are most dangerous in the immediate vicinity of a site, usually to workers involved in disposal or cleanup or people living adjacent to the site. Quantities are rarely sufficient to pose any kind of regional air pollution hazard.

The two major kinds of secondary air pollutants from hazardous wastes are those that are oxidised in the atmosphere to corrosive substances and organic materials that undergo photochemical oxidation. Plausible examples of the former are sulphur dioxide released from the action of waste strong acids on sulphites and subsequently oxidised in the atmosphere to corrosive sulphuric acid,

$$SO_2 + \tfrac{1}{2}O_2 + H_2O \rightarrow H_2SO_4(\text{aerosol}) \qquad \text{... (31.15)}$$

and nitrogen dioxide (itself a toxic primary air pollutant) produced by the reaction of waste nitric acid with reducing agents such as metals and oxidised to corrosive nitric acid or converted to corrosive nitrate salts:

$$4HNO_3 + Cu \rightarrow Cu(NO_3)_2 + 2NO_2(g) + 2H_2O \qquad \text{... (31.16)}$$

$$2NO_2(g) + \tfrac{1}{2}O_2 + H_2O \rightarrow 2HNO_3(\text{aerosol}) \qquad \text{... (31.17)}$$

$$HNO_3(\text{aerosol}) + NH_3(g) \rightarrow NH_4NO_3(\text{aerosol}) \qquad \text{... (31.18)}$$

Organic species that produce secondary air pollutants are those that form photochemical smog. The more reactive of these are unsaturated compounds that react with atomic oxygen or hydroxyl radical in air,

$$R-CH=CH_2 + HO^\bullet \rightarrow RCH_2CH_2O^\bullet \qquad \text{... (31.19)}$$

to yield reactive radicals that participate in chain reactions to eventually yield ozone, organic oxidants, noxious aldehydes, and other products characteristic of photochemical smog.

Fate and Residence Times of Hazardous Waste Compounds in the Atmosphere

An obvious means by which hazardous waste species may be removed from the atmosphere is by dissolution in water in the form of cloud or rain droplets. Inorganic acid, base, and salt compounds, such as H_2SO_4, HNO_3, and NH_4NO_3 mentioned above, are readily removed from the atmosphere by dissolution. For vapours of compounds that are not highly soluble in water, solubility information combined with information about rainfall amounts and mixing in the atmosphere can be used to estimate the atmospheric half-life, $\tau_{1/2}$, of the species. Solubility rates may be used to estimate half-lives for substances that are more miscible in water. For poorly water-soluble compounds, such calculations tend to drastically underestimate lifetimes, which indicates that other removal mechanisms must predominate.

The lifetimes of vapourised hazardous waste species removed from the atmosphere through adsorption by aerosol particles are limited to that of the sorbing aerosol particles (typically about 7 days) plus the time spent in the vapour phase before adsorption. This mechanism appears to be viable only for highly nonvolatile constituents such as benzo(a)pyrene.

Sorptive removal by soil, water, or plants on the earth's surface, called dry deposition, is another means for physical removal of hazardous substances from the atmosphere. Predictions of dry deposition rate vary greatly with type of compound, type of surface, and weather conditions. For highly volatile organic compounds, such as low-molecular-mass organohalide compounds predicted rates of dry deposition give atmospheric lifetimes many-fold higher than those actually observed, so, for such compounds, dry deposition is probably not a common removal mechanism.

Predicted rates of physical removal of a number of volatile organic compounds that are not very soluble in water are far too slow to account for the loss of such compounds from the atmosphere, so chemical processes must predominate. The most important of these processes is reaction with hydroxyl radical, $HO^{\bullet}$, in the troposphere. Ozone can react with compounds having a double bond. Other oxidant species that might react with hazardous waste compounds in the troposphere and stratosphere are atomic oxygen (O), peroxyl radicals ($HOO^{\bullet}$), alkylperoxyl radicals ($ROO^{\bullet}$), and NO_3.

Despite the fact that its concentration in the troposphere is relatively low, $HO^{\bullet}$ is so reactive that it tends to initiate most of the reactions leading to the chemical removal of most refractory organic compounds from the atmosphere. When hydroxyl radical reacts with organic compounds in the atmosphere, new reactive free radicals are formed that undergo further reactions, leading to nonvolatile and/or water-soluble species, which are scavenged from the atmosphere by physical means. These scavengeable species tend to be aldehydes, ketones or acids. Halogenated organic compounds may lose halogen atoms in the form of halo-oxy radicals and undergo further reactions to form scavengeable species. In general, reactions with species other than $HO^{\bullet}$ or O_3 are not considered significant in the removal of hazardous organic waste compounds from the troposphere.

Photolytic transformations involve direct cleavage (photodissociation) of compounds by reactions with light and ultraviolet radiation:

$$R\text{-}X + hv \rightarrow R^{\bullet} + X^{\bullet} \qquad \qquad ...(31.20)$$

The extent of these reactions varies greatly with light intensity, quantum yields (chemical reactions per quantum of radiation energy absorbed), and other factors.

In order for photolysis to be an important process for its removal from the atmosphere, a molecule must have a chromophore (light-absorbing group) that absorbs light in a wavelength region of significant intensity in the impinging light spectrum. This requirement limits the importance of photolysis as a removal mechanism to only a few classes of compounds, including conjugated alkenes, carbonyl compounds, some halides, and some nitrogen compounds, particularly nitro compounds. However, these do include a number of the more important hazardous waste compounds.

HAZARDOUS WASTES IN THE BIOSPHERE

One of the most crucial aspects of fate and toxic effects of environmental chemicals is their accumulation by organisms from their surroundings, including bioaccumulation and biomagnification phenomena. Biodegradation of wastes is their conversion by biological processes to simple inorganic molecules and, to a certain extent, to biological materials (Fig. 31.6). The complete bioconversion of a substance to inorganic species such as CO_2, NH_3, and phosphate is called mineralisation.

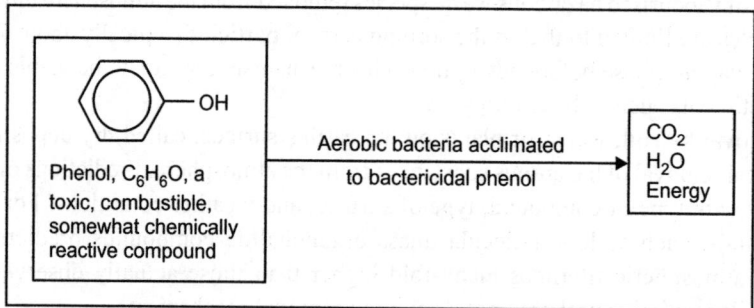

Fig. 31.6. Illustration of biological action on a hazardous waste constituent. This example shows mineralisation in which the organic compound is degraded completely to simple inorganic species.

Detoxification refers to the biological conversion of a toxic substance to a less toxic species, which may still be relatively complex, or biological conversion to an even more complex material. An example of detoxification is illustrated below for the enzymatic conversion of paraoxon (a highly toxic organophosphate insecticide) to p-nitrophenol, which has only about 1/200 the toxicity of the parent compound:

$$... (31.21)$$

Usually, the products of biodegradation are simpler molecular forms that tend to occur in nature. The definition of biodegradation is illustrated by an example in Fig. 31.6. Biodegradation is usually carried out by the action of micro-organisms, particularly bacteria and fungi.

Biodegradation Processes

The biotransformations of environmental chemicals, including pesticides and industrial chemicals, in vertebrates (birds, mammals, fish, reptiles) can be of the utmost importance in determining their fates and effects. Biotransformation is what happens to any substance that is metabolised and thereby altered by biochemical processes in an organism. Metabolism is divided into the two general categories of catabolism, which is the breaking down of more complex molecules, and anabolism, which is the building up of life molecules from simpler materials. The substances subjected to biotransformation may be naturally occurring or anthropogenic (made by human activities). They may consist of xenobiotic molecules that are foreign to living systems.

An important biochemical process that occurs in the biodegradation of many synthetic and hazardous waste materials is cometabolism. Cometabolism does not serve a useful purpose to an organism in terms of providing energy or raw material to build biomass, but occurs concurrently with normal metabolic processes. An example of cometabolism of hazardous wastes is provided by the white rot fungus, *Phanerochaete chrysosporium*. This organism, which has been investigated for its hazardous waste treatment potential, degrades a number of kinds of organochlorine compounds—including DDT, PCBs, and chlorodioxins—under the appropriate conditions. The enzyme system responsible for this degradation is one that the fungus uses to break down lignin in plant material under normal conditions.

Enzymes in Waste Degradation

Enzyme systems hold the key to biodegradation of hazardous wastes. For most biological treatment processes currently in use, enzymes are present in living organisms in contact with the wastes. However, in some cases it is possible to use cell-free extracts of enzymes removed from bacterial or fungal cells to treat hazardous wastes. For this application the enzymes may be present in solution or more commonly, immobilised in biochemical reactors. Biodegradation of municipal waste-water and solid wastes in landfills occurs by design.

Biodegradation of any kind of waste that can be metabolised takes place whenever the wastes are subjected to conditions conducive to biological processes. The most common type of biodegradation is that of organic compounds in the presence of air, that is, aerobic processes. However, in the absence of air, anaerobic biodegradation may also take place. Furthermore, inorganic species are subject to both aerobic and anaerobic biological processes.

Although biological treatment of wastes is normally regarded as degradation to simple inorganic species such as carbon dioxide, water, sulphates, and phosphates, the possibility must always be considered of forming more complex or more hazardous chemical species. An example of the latter is the production of volatile, soluble, toxic methylated forms of arsenic and mercury from inorganic species of these elements by bacteria under anaerobic conditions.

For the most part, anthropogenic compounds resist biodegradation much more strongly than do naturally occurring compounds. This is generally due to the absence of enzymes that can bring about an initial attack on the compound. A number of physical and chemical characteristics of a compound are involved in its amenability to biodegradation. Such characteristics include hydrophobicity, solubility, volatility, and affinity for lipids. Some organic structural groups impart particular resistance to biodegradation. These include branched carbon chains, ether linkages, chlorine, amines, methoxy groups, sulphonates, and nitro groups.

Several groups of micro-organisms are capable of partial or complete degradation of hazardous organic compounds. Among the aerobic bacteria, those of the *Pseudomonas* family are the most widespread and most adaptable to the degradation of synthetic compounds. Anaerobic bacteria catabolise biomass through hydrolytic processes, breaking down proteins, lipids, and saccharides. They are also known to reduce nitro compounds to amines, degrade nitrosamines, promote reductive dechlorination, reduce epoxide groups to alkenes, and break down aryl structures. Actinomycetes are micro-organisms that are morphologically similar to both bacteria and fungi. They are involved in the degradation of a variety of organic compounds, including degradation-resistant alkanes, and lignocellulose. Other compounds attacked include pyridines, phenols, nonchlorinated aromatics, and chlorinated aromatics. Fungi are particularly noted for their ability to attack long-chain and complex hydrocarbons and are more successful than bacteria in the initial attack on PCB compounds. Phototrophic micro-organisms, which include algae, photosynthetic bacteria, and cyanobacteria (blue-green algae) tend to concentrate organophilic compounds in their lipid stores and induce photochemical degradation of the stored compounds. For example, *Oscillatoria* can initiate the biodegradation of naphthalene by the attachment of –OH groups.

Practically all classes of synthetic organic compounds can be at least partially degraded by various micro-organisms. These classes include nonhalogenated alkanes, halogenated alkanes (trichloroethane, dichloromethane), nonhalogenated aryl compounds (benzene, naphthalene, benzo(a)pyrene), halogenated aromatic compounds (hexachlorobenzene, pentachlorophenol), phenols (phenol, cresols), polychlorinated biphenyls, phthalate esters, and pesticides (chlordane, parathion).

Glossary

Abiocen	:	Non-living (abiotic) component of the environment.
Abiotic environment	:	Non-biological surroundings of an organism, such as temperature, light intensity and rainfall.
Abyssopelagic	:	Organisms at water depths more than 3000 m.
Aeolin	:	A sediment deposit carried by wind.
Aflatoxin	:	Toxin made by grain-inhabiting fungi.
Alien species	:	This term refers to species that spread beyond their native range, not necessarily harmful or species introduced to a new range that establish themselves and spread; similar terms include exotic species, foreign species, introduced species, non indigenous species, and non native species.
Alleles	:	Alternative forms of a gene found at the same locus.
Allogenic succession	:	Changes in the species that make up a community as a result of changes in the external environment.
Alpine	:	Part of mountain above tree line, but below permanent snow.
Amensalism	:	Symbiosis in which one organism is inhibited by the other but the latter remains unaffected.
Anemophilous	:	Wind pollinated.
Antagonistic	:	Co-evolution in which the relationship harms one of the species.
Anthropocentric	:	Reasons that focus on the advantages to humans of conservation.
Aphytic zone	:	The area of lake floor, that due to its depth factor is devoid of plants.
Apomixis	:	Asexual reproduction in plants in which cells develop into embryos without undergoing meiosis and fertilisation.
Aquatic nuisance species	:	It is a nonindigenous species that threatens the diversity or abundance of native species or the ecological stability of infested waters, or commercial, agricultural, aquacultural or recreational activities dependent on such waters.
Arid zone	:	A zone of very low rainfall with most of the deserts.
Artiodactyla	:	Mammalian herbivores with hooves and an even number of toes, such as antelopes and deer.
Asexual reproduction	:	Production of new individuals without the formation of haploid cells by meiosis.
Assimilation	:	Uptake by cells of the products of absorption which are then synthesised into macro-molecules.
Autecology	:	Study of the ecological relationships of a single species.

Autogenic succession	:	Change in the species that make up a community as a result of effects of the organisms themselves.
Autotroph	:	Organism, such as a green plant, that can synthesise its organic compounds from inorganic ones.
Batesian mimicry	:	Mimicry in which a palatable mimic imitates a harmful or distasteful species to avoid predation.
Bathyal	:	Sea bed and sediments deposited between edge of continental shelf and the start of abyssal zone at a water depth of 2000 m.
Bathypelagic	:	Organisms living at water depths between 1000 m and 3000 m.
Behavioural ecology	:	Science which studies how the behaviour of organisms is related to their ecology.
Benthic	:	Living on the bottom of the ocean.
Biodiversity	:	A measure of the number of species and range of life forms found in an area.
Biological control or biocontrol	:	In general, the control of the numbers of one organism as a result of natural predation by another or others. Specifically, the human use of natural predators for the control of pests or weeds. Also applied to the introduction of large numbers of sterilised males of the pest species, whose matings result in the laying of infertile eggs.
Biological diversity or biodiversity	:	Used to describe species richness, ecosystem complexity, and genetic variation.
Biological index	:	Use of the presence or absence of certain organisms to indicate something about the environment, such as the level of pollution.
Biological invasion or bioinvasion	:	A broad term that refers to both human-assisted introductions and natural range expansions.
Bioregion	:	A biological subdivision of the earth's surface delineated by the flora and fauna of the region.
Biota	:	The plants and animals of a specific region or period, or the total aggregation of organisms in the biosphere.
Biotic (ecological) pyramid	:	Graphic representations of the trophic structure and function at successive trophic levels of an ecosystem. This may be shown in terms of number, biomass or energy content.
Boreal forest	:	Biome, also known as taiga, found at high latitudes and dominated by evergreen conifers.
Bourgeois	:	Strategy for an animal of fighting in a conflict only when it is in its own territory.
Brown earth	:	Fertile brown soil type with a characteristic soil profile.
Buzz pollination	:	Pollination technique in which the vibrations of a buzzing bee shake dry pollen from anthers on to the bee.
Calcifuge	:	Plant associated with acidic soil.
Casual species	:	This term is becoming less common in usage. A non native species that does not form self-replacing populations. Similar terms include introduced species, non indigenous species, and non native species.
Chaparral	:	Vegetation dominated by shrubs, with small, broad, hard, ever green leaves, found in mediterranean areas.
Chemical control	:	Control method that employs herbicides to control exotic plants.
Chernozem	:	Fertile, deep, black soil with a characteristic soil profile.
Cleistogamy	:	Fertilisation within unopened flowers.

Climatic climax	:	Community at the end of a succession, whose composition is determined by the climate.
Cohort life table	:	Life table produced using data from a single cohort.
Colonial growth	:	Growth shown in organisms such as corals where many genetically identical multicellular subunits remain attached to one another.
Community	:	Interacting collection of species found in a common environment or habitat.
Competition	:	Utilisation of a resource in short supply by two or more organisms with the result that at least one of the organisms grows or reproduces less.
Competitive exclusion	:	Absence of one species due to presence of another species utilising the same resources.
Cryptogenic species	:	Species that are neither clearly native nor exotic.
Cultivar	:	A variety of a plant produced and maintained by horticultural techniques and not normally found in wild populations.
Deep ecology	:	The belief that we need to examine the ways in which humans have distanced themselves from nature and recover the experience of being in communion with nature.
Deforestation	:	Destruction of forest cover and the undergrowth.
Deleterious allele	:	Allele the possession of which (usually in the homozygous state) harms an individual.
Density dependent	:	Describes an environmental factor whose influence on population size increases as the population grows in size.
Density independent	:	Describes an environmental factor whose influence on population size is unrelated to the number of individuals in the population.
Desertification	:	The spread of deserts into semi-arid lands.
Disturbance	:	An event or change in the environment that alters the composition and successional status of a biological community and may deflect succession onto a new trajectory, such as a forest fire or hurricane, glaciation, agriculture, and urbanisation.
Dominance hierarchy	:	Linear order of individuals in a group such that individuals higher in the order have preferential access to a specified resource.
Ecocline	:	Continuous variation in plant phenotype and associated genotype, along an environmental gradient.
Ecological isolation	:	Coexistence of two or more species by virtue of differences in their ecology.
Ecospecies	:	One or more ecotype in a single coenospecies.
Ecosystem	:	A discrete unit, or community of organisms and their physical environment (living and non-living parts), that interact to form a stable system.
Ecotoxicology	:	Study of the passage of toxic substances and their transformations in different trophic levels in a food chain.
Endemic	:	A species or taxonomic group that is restricted to a particular geographic areas because of such factors as isolation or response to soil or climatic conditions; this species is said to be endemic to the place and would be native.
Entomophilous	:	Insect pollinated.
Erosion (soil)	:	Removal of top soil by any external agent.

Eusocial	:	Species in which there is co-operation in looking after the young, some individuals are permanently sterile and there is an overlap of at least two generations contributing to colony labour.
Euthenics	:	Science of improving the human race by improving environment.
Eutrophication	:	Release of large amounts of organic matter or phosphate and nitrate into water resulting in a lowering of dissolved oxygen concentration.
Exponential growth	:	Increase in population size by a constant factor each unit of time.
Fauna	:	The animal life of a region or geological period.
Feminism	:	A movement whose principal aims are to transform human relationships and society through the exposure and analysis of inequalities, principally those resulting from the exploitation of women by men.
Flora	:	Plant or bacterial life forms of a region or geological period.
Folivore	:	Herbivore that feeds on the leaves of shrubs or trees.
Forbs	:	Herbaceous plants, excluding grasses, sedges and other grasslike plants.
Foreign species	:	A species introduced to a new area or country. Similar terms include alien species, exotic species, introduced species, non indigenous species, and non native species.
Frugivore	:	Herbivore that feeds on fruit.
Fundamental niche	:	Total niche an organism could occupy in the absence of competition and predation by other species.
Games theory	:	Theoretical approach which investigates the strategies that individuals should adopt in situations when the consequence of an individual's behaviour depends on what other individuals are doing.
Gene	:	Length of DNA that codes for a polypeptide.
Glacial till	:	Debris left by retreating glacier.
Greenhouse effect	:	Trapping of heat within the Earth's atmosphere.
Habit	:	Appearance (external outlook) of an organism.
Habitat	:	The place, including physical and biotic conditions, where a plant or an animal usually occurs.
Haplodiploid	:	Species in which males have one set of chromosomes and females two.
Hekistotherm	:	A plant which grows under very low temperatures (generally in alpine areas).
Herbicide	:	Pesticide that specifically targets vegetation.
Herbivore	:	Organism, such as a camel, that feeds on vegetation.
Heterotroph	:	Organism, such as an animal, that needs to take in organic compounds from its environment.
Holozoon	:	Organism, such as a wolf or rat, which feeds on relatively large pieces of dead organic matter.
Homoiothermic animals:		Animals which can regulate and maintain their body temperature at a higher level than the environment by altering their metabolic rate. Also called endotherms.
Homozygous	:	Gene which carries two identical alleles in a diploid individual.
Humification	:	Production of humus during the decay of biological materials.
Hygroscopic water hypervolume	:	Water held by surface forces of soil particles.
Inbreeding depression	:	Decrease in viability or fertility of an individual as a result of a high degree of homozygosity of its genes due to its parents being closely related.

Indeterminate growth	:	Growth, as occurs in most vascular plants, where the organism continues to increase in size until resources run out.
Indicator (species)	:	Species with a narrow range of tolerance for some environmental factors; their presence in a particular place indicates the existence of that particular condition of the habitat.
Infanticide	:	Killing of offspring by other individuals in the same species.
Injurious species	:	An introduced species that causes economic or environmental harm to humans. Similar terms include aquatic nuisance species, noxious weed, and invasive species.
Intentional introduction	:	A species that is brought to a new area, country, or bioregion for a specific purpose, such as for a garden or lawn; a crop species; a landscaping species; a species that provides food; a groundcover species; for soil stabilisation or hydrological control; for aesthetics or familiarity of the species; or other purposeful reasons.
Interglacial	:	Warmer interval between ice ages.
Introduced species	:	This term, along with the terms introduced species and non-indigenous species, is one of the most commonly used terms to describe a plant or animal species that is not originally from the area in which it occurs. This terms means those species that have been transported by human activities, either intentionally or unintentionally, into a region in which they did not occur in historical time and are now reproducing in the wild. Similar terms include alien species, exotic species, foreign species, non indigenous species, and non native species.
Invasibility	:	The ease with which a habitat is invaded.
Invasion	:	The expansion of a species into an area not previously occupied by it.
Inverted pyramid of numbers	:	Situation in which more individuals at one trophic level exist than at the trophic level beneath.
IPM	:	Integrated Pest Management. IPM focuses on long-term prevention or suppression of pests. The integrated approach to weed management incorporates the best suited cultural, biological and chemical controls that have minimum impact on the environment and on people.
Keystone species	:	A species that plays a pivotal role in the maintenance of an ecosystem and on which many other species depend.
Land reclamation	:	Treatment of barren land usually by filling with refuse so that land could be made productive.
Late successional	:	Mature communities that develop at the end of a succession.
Leaching	:	Loss of minerals as water drains through a soil.
Leghaemoglobin	:	Haemoglobin-like protein in leguminous root nodules which functions in nitrogen fixation.
Life table	:	Presentation of numerical data collected during a population study to show age-specific survival and reproduction.
Lincoln index	:	Estimate of population size calculated by the mark, release and recapture of organisms.
Littoral	:	Occurring at the border of the land and sea.
Macrofauna	:	Large soil invertebrates, such as earthworms.
Manual control	:	Removal that involves the use of tools such as shovels, axes, rakes, grubbing hoes, and hand clippers to expose, cut, and remove flowers, fruits, stems, leaves, and/or roots from target plants.

Marsupial	:	Mammal that gives birth to young in a very immature state and usually possesses a pouch over her teats.
Mechanical control	:	Removal that involves the use of motorised equipment such as mowers, 'weed-whackers', and tractor-mounted plows, disks, and sweepers. Burning is also categorised here.
Megatherm	:	A tropical plant, needing continuous higher temperatures.
Mesosphere	:	Part of atmosphere extending from the ionosphere to exosphere (400 to 1000 km above earth surface). Also part of atmosphere between the stratosphere and thermosphere (40–80 km above the earth surface.)
Metapopulation	:	A network of population with occasional movements of individuals between them.
Migration	:	Movement of animals carried out regularly often between breeding place and winter feeding grounds.
Monotreme	:	Mammal that lays eggs and lacks mammary glands.
Mullerian mimicry	:	Mimicry in which both the model and the mimic are unpalatable or potentially harmful to a prospective predator.
Mycorrhiza	:	Association between a non-pathogenic or weakly pathogenic fungus and living plant root cells.
Native range	:	The ecosystem that a species inhabits.
Native weed (invasive native)	:	A species that is native to an area or bioregion that has increased in number dramatically. In cases of disturbance or change to a landscape, a ruderal species can increase in cover and compete with other native plants, threatening members the diversity of a community. In other cases, landscape level changes can cause the increase of the population of a species, such as white-tailed deer in the northeastern part of the United State, which are at the highest levels historically and cause damage to humans, crops, and structures, suffer high disease levels, and pose threats to humans through interactions on roads.
Naturalised species	:	A species that was originally introduced from a different country, a different bioregion, or a different geographical area, but now behaves like a native species in that it maintains itself without further human intervention and now grows and reproduces in native communities.
Niche	:	Complete account of how an organism uses its environment.
Niche opportunity	:	Defines conditions that promote invasions in terms of resources, natural enemies, the physical environment, interactions between these factors, and the manner in which they vary in time and space.
Nitrifying bacteria	:	Bacteria that oxidise ammonium ions to nitrite and/or nitrate ions.
Nonindigenous species	:	Any species or other viable biological material that enters an ecosystem beyond its historic range, including any such organism transferred from one country into another.
Non native species	:	This term, along with the terms introduced species and non-indigenous species, is one of the most commonly used terms to describe a plant or animal species that is not originally from the area in which it occurs. Similar terms also include alien species, exotic species, and foreign species.
Obligate anaerobe	:	Organism that cannot utilise oxygen, indeed which may be poisoned by it.
Oestrus	:	Condition of a female mammal when she is prepared to mate. Human are not said to show oestrus because women are sexually receptive throughout their reproductive cycle.

Optimal foraging	:	Way in which an organism searches for food so as to maximise energy and nutrient intake.
Osmoregulation	:	Control of the water content of body of an organism.
Ovoviviparous	:	Producing eggs which hatch within the mother so that the young are born having already left their eggs.
Pairwise co-evolution	:	Tight co-evolution between two species.
Parasite load	:	Heavy infestation of parasites.
Parthenogenesis	:	Asexual reproduction in animals when sex cells start to develop into an embryo without undergoing meiosis and fertilisation.
Pedogenesis	:	Process of soil development.
Perissodactyla	:	Mammalian herbivores with hooves and an odd number of toes, such as horses and tapirs.
Pest	:	An animal that competes with humans by consuming or damaging food, fibre, or other materials intended for human consumption or use, such as an insect pest on a cropfield.
Pesticide	:	A chemical or biological agent intended to prevent, destroy, repel, or mitigate plant or animal life and any substance intended for use as a plant regulator, defoliant, or desiccant, including insecticides, fungicides, rodenticides, herbicides, nematocides, and biocides.
Phenology	:	Study of periodical changes in plants in relation to seasons of a year.
Pheromone	:	Airborne chemical produced by organisms and which alters the behaviour of other individuals in the same species.
Photosynthesis	:	Manufacture of organic compounds by organisms using the energy from sunlight.
Phytoplankton	:	Small photosynthetic organisms in lakes, oceans or seas which drift almost passively.
Pneumatophore	:	Stick-like structure that projects from the mud in mangrove forests and obtains oxygen for the mangrove trees.
Polymorphic	:	Existing in several forms, e.g. a polymorphic gene has several alleles.
Population	:	A group of potentially inter-breeding individuals of the same species found in the same place at the same time.
Population cycle	:	Regular pattern of change in population size over several years found in some organisms such as lemmings.
Population dynamics	:	Study of changes in population densities in an area.
Population pyramid	:	Representation of demographic data in which the ages of the individuals in a population are plotted against their abundance.
Predation	:	One organism is eaten by another.
Pride	:	Group of lions containing adult females.
Primary succession	:	Succession that begins on bare uncolonised ground or in newly formed lakes.
Profundal zone	:	The zone of a lake lying below the compensation depth.
Prohibited weed	:	A specific legal term applied to a plant or plant part that may not be brought into a state.
Psammophyte	:	A plant growing in sand.
Pyramid of biomass	:	Diagrammatic representation of the weight or mass of organisms at the different trophic levels in a community.

Pyramid of numbers	:	Diagrammatic representation of the numbers of organisms at the different trophic levels in a community.
Quaternary consumer	:	Organism that eats tertiary consumers.
Reciprocal altruism	:	Helping behaviour in which individuals subsequently help each other back.
Red Queen hypothesis	:	Notion that organisms are constantly having to evolve to keep up with other organisms.
Rendzina	:	Shallow, stony, fertile soils formed over limestone or chalk.
Reproductive efficiency	:	Percentage of the energy assimilated that an organism invests in reproduction.
Resource partitioning	:	Division of a resource such as food among two or more species so that each species has access to a different part of the resource.
Restricted weed	:	A specific legal term applied to a plant or plant part that may only be brought into a state in limited quantities.
Rhizome	:	Underground root-like stem of a vascular plant.
Ruderal species	:	A plant associated with human dwellings, construction, or agriculture, that usually colonises disturbed or waste ground. Ruderals are often weeds which have high demands for nutrients and are intolerant of competition.
Ruminant	:	Mammal, such as a sheep or deer, that possesses a rumen.
Sanitary land fill	:	Dumping domestic refuse on land.
Saprotroph	:	Organism, such as an earthworm, that feeds off dead organic matter which is either absorbed in solution or ingested as very small pieces.
Scavenger	:	Organism which feeds off quite large pieces of dead organic matter which it has not killed itself.
Secondary succession	:	Succession that begins in areas that have previously supported vegetation and retain at least some of the effects of this previous colonisation.
Seed bank	:	Seeds that become incorporated into the soil.
Self-fertilisation	:	Sexual reproduction in which both gametes come from the same individual.
Semelparous	:	Species in which individuals reproduce on only one occasion during their lives.
Sexual reproduction	:	Production of offspring by the fusion of two haploid cells, typically from two different individuals.
Shade intolerance	:	Inability of a plant to grow well in the shade.
Sociobiology	:	Study of the reasons for social behaviour.
Soil erosion	:	The loss of soil from an area, often as an indirect result of a human activity such as the removal of trees or the introduction of domestic herbivores.
Species	:	A group of organisms formally recognised as distinct from other groups; the taxon rank in the hierarchy of biological classification below genus; the basic unit of biological classification, defined by the reproductive isolation of the group from all other groups of organisms.
Statutory protection	:	Protection afforded, for example to a species or area, by law.
Stochastic	:	Discontinuous.
Suboptimal success	:	Structure or behaviour that could be improved on.
Succulent	:	Plant which minimises the loss of water through the possession of a thick cuticle, a very low surface area to volume ratio and sunken stomata.
Super-stimulus	:	Signal that is larger than the one an animal is used to and gives rise to an excessive behavioural response.
Swamp	:	Wetland ecosystem dominated by trees.

Taiga	:	Boreal forest dominated by evergreen coniferous trees.
Tertiary consumer	:	Organism that eats only secondary consumers.
Thermosphere	:	Part of upper atmosphere in which temperature increases with height.
Topography	:	Physical features of the environment such as altitude, slope and aspect.
Unintentional introduction	:	An introduction of nonindigenous species that occurs as the result of activities other than the purposeful or intentional introduction of the species involved, such as the transport of nonindigenous species in ballast or in water used to transport fish, mollusks or crustaceans for aquaculture or other purposes.
Vector	:	The physical means or agent by which a species is transported, such as ballast water, ships' hulls, boats, hiking boats, cars, vehicles, packing material, or soil in nursery stock.
Vegetation	:	The collective and continuous growth of plants in space, i.e. totality of plant growth or sum total of plant population covering a region.
Vegetative reproduction	:	Asexual production of new plants without the involvement of seeds.
Weathering	:	Breakdown (mechanical or chemical) of parent rock matter by any agent into smaller particles.
Xerosere	:	Series of communities which develop on dry land as a result of primary succession.
Zonation	:	Clear orientation of communities along a steep environmental gradient, such as occurs at the sea shore.
Zooplankton	:	Microscopic animals feeding on the phytoplanktons in aquatic ecosystem.
Zygomorphic	:	Bilaterally symmetrical.

References

Allen, W. and Lloyd, T., *Basic Concepts of Ecology*, Butterworths, London.

Andrew, D., *Ecology and Ecosystems*, Marcel Dekker Inc., New York.

Baker, J.M., *Population Ecology*, Applied Science Publishers, London.

Benn, R.L., *Acidic Precipitation, Charles C. Thomas*, Springfield, New York.

Blackman, G.W., *Study of Plant Community*, Academic Press, London.

Brill, G., *Grassland Ecosystems*, Cambridge University Press, Cambridge.

Burmel, I., *Global Ecology*, Progress Publishers, Moscow.

Cavallaro, S., Cavallaro, A. and Galli, G., *Ecological Imperialism*, Pergamon Press, New York.

Connell, D.W. and Miller, G.J., *Ecological Applications*, John Wiley & Sons, New York.

Coolingwood, R.T., *Biological Aspects of Air Pollution*, John Wiley & Sons, New York.

Curtis, C. and Hawkes, H.A., *Ecological Aspects of Air Pollution*, Academic Press, London.

Dix, N.M., *Environmental Pollution*, John Wiley & Sons, New York.

Dugan, P.P., *Animal Ecology*, Plenum Publishing Corporation, London.

Fruh, E., Gloyna, E.F. and Eckenfelder, W.W., *Global Environment*, University of Texas Press, Austin.

Gascon, Tyler, E.G. and Wayne, A.P., *Trends in Ecology and Evolution*, McGraw-Hill, New York.

Golterman, E.D., *Atmosphere and Pollution*, Gordon and Breach, Science Publishers, New York.

Hardin, C.M., *Concepts of Ecology*, Castle Housing Publications Ltd., London.

Jim, A. and Evison L., *Air Pollution Technology*, John Wiley & Sons, New York.

Kenn, R.L., *Population Ecology, Distribution and Surveillance*, Harwood Academic Publishers, New York.

Ledbetter, K. and Fletcher, W.W., *Advances in Ecological Research*, Blackie, Glasgow and London.

Lenihan, F.G., *Introduction to Atmosphere*, National Academy of Sciences, Washington DC.

Mark, M., *Introduction to Environmental Science and Technology*, John Wiley & Sons, New York.

McCaull, J. and Crossland, J., *Air Pollution*, Harcourt Brace Jovanovich, New York.

Michael, C., *Ecological Methods of Field and Laboratory Investigations*, Tata McGraw-Hill, New Delhi.

Neil, J., *Atmospheric Dispersion of Pollution*, Applied Science Publishers, London.

Palmer, E.P., *Fundamentals of Ecology*, Saunders, Philadelphia.

Smith, T.E., *Ecological Methods*, Chapman and Hall, London.

Stumm, C., *Ecology and Sustainable Development*, Applied Science Publishers, London.

Tayler, H.Y., *Conservation Biology*, Van Nostrand Reinhold, New York.

Vinogradov, N.L., *Food Webs*, Addison-Wesley Publishing Company, Philippines.

Watts, B.G., *Ecology and Tropical Biology*, Chilton Book Company, Radnor, Pennsylvania.

Index